Incompressible computational fluid dynamics is an emerging and important discipline, with numerous applications in industry and science. This book contains articles on vortex methods, finite elements, and spectral methods among others. Contributions from leading experts in the various subfields, covering both theoretical and practical developments, portray the wide ranging nature of the subject. The book provides an entree into the current research in the field. It can also serve as a source book for researchers and others who require up-to-date information on methods and techniques.

Incompressible Computational Fluid Dynamics
Trends and Advances

Incompressible Computational Fluid Dynamics
Trends and Advances

Edited by

MAX D. GUNZBURGER
Virginia Polytechnic Institute and State University

and

ROY A. NICOLAIDES
Carnegie Mellon University

CAMBRIDGE
UNIVERSITY PRESS

CAMBRIDGE UNIVERSITY PRESS
Cambridge, New York, Melbourne, Madrid, Cape Town, Singapore, São Paulo, Delhi

Cambridge University Press
The Edinburgh Building, Cambridge CB2 8RU, UK

Published in the United States of America by Cambridge University Press, New York

www.cambridge.org
Information on this title: www.cambridge.org/9780521404075

First published 1993
This digitally printed version 2008

A catalogue record for this publication is available from the British Library

ISBN 978-0-521-40407-5 hardback
ISBN 978-0-521-09622-5 paperback

Contents

Preface

Numerical methods for incompressible fluid dynamics have developed to the point at which a survey of the field is both timely and appropriate. A major stimulus to the field has been the large number of applications in which incompressible flows play a crucial role, and this has spurred the interest of numerous computational engineers and mathematicians. The articles which follow provide a reasonably broad view of algorithmic and theoretical aspects of incompressible flow calculations.

It should be noted at the outset that it can be dangerous to define an algorithm for simulating incompressible flows by setting, for example, the density to be constant in a successful compressible flow algorithm. The nature of the pressure as a Lagrange multiplier rather than as a thermodynamic variable as well as the infinite speed of propagation of disturbances and other factors peculiar to incompressible flows, make algorithmic development and implementation in this context a unique undertaking (see Appendix 7A).

Perhaps the first major advance in the application of large scale digital computation to incompressible flows occured in the late 1950s with the introduction of staggered mesh techniques, exemplified, for example, by the Marker-and-Cell (MAC) scheme. The use of staggered meshes in the context of the primitive variable formulation was found to provide a stable discretization of the incompressibility constraint. Shortly thereafter, it was realized that the use of staggered meshes could be avoided by employing the streamfunction-vorticity formulation in which the incompressibility constraint does not explicitly appear. Numerous finite difference algorithms were proposed and used based on this formulation of the Navier-Stokes equations.

Despite the sucess of the stream function-vorticity approach, many difficulties associated with practical computations remained unresolved, including complex geometries, boundary condition treatments and flows in three dimensions. The success of finite element methods in addressing these problems in the structural mechanics context naturally led to interest in applying them to incompressible flows. Thus, starting in the early 1970s and continuing to the present day, there has been an evergrowing understanding and use of finite element methods for incompressible flow problems.

A different methodology, whose modern development started during the 1960s, involves the use of point vortex and related singular functions to approximate solutions of the Navier-Stokes equations at relatively small viscosities. These methods have the apparent advantage of being grid-free, at least for certain types of problems involving simple boundaries. Vortex methods are currently enjoying a period of intense development.

More recently, the potentially high accuracy possible with spectral methods has generated considerable interest in their application to incompressible flows. The need to account for complicated boundaries naturally leads to the spectral element method which in turn is closely related to the p-version of the finite element method.

Another recent development has been covolume algorithms employing Voronoi-Delaunay dual tesselations of general domains. The discrete equations of the covolume technique can be chosen so that when specialized to rectangular geometries, they coincide with those of classical staggered mesh methods. In this way a generalization of the MAC scheme to triangular and tetrahedral domains is obtained.

The chapters in the book are arranged alphabetically. The chapter by Engelman provides an overview of some of the many real world applications in which the numerical simulation of incompressible flows plays a significant role.

A relatively large proportion of the papers is devoted to specific algorithms, or to components of algorithms for incompressible flows. Thus, Puckett reviews the state-of-the-art in vortex methods while Karniadakis, Orszag, Rønquist, and Patera do likewise for spectral methods. The latter also give a brief account of lattice gas methods. Nicolaides describes the use of Voronoi-Delaunay tesselations as a basis for covolume discretizations of the equations of incompressible flow. Temam considers methods based on notions stemming from dynamical systems theory.

There are four papers addressing basic algorithmic issues in finite element methods for incompressible flows. Dean and Glowinski give a review of some of the large body of theoretical results concerning the finite element discretization and solution of incompressible flows. Thatcher focuses on algorithms and elements that have proved to be useful for three-dimensional flows; Verfürth reviews some recent developments in adative mesh-refinement techniques based on a posteriori error estimation; Franca, Hughes, and Stenberg give a thorough account of "stabilized" finite element methods for which some difficulties associated with discretizing the incompressibility constraint are circumvented.

Cardot, Mohamadi, and Pironneau discuss how to efficiently incorporate classical turbulence models into existing laminar flow codes. Gunzburger discusses the application of finite element analyses and methodologies to optimal design and control problems for incompressible flows.

There are three chapters devoted to implementation issues in the context of engineering applications. Löhner discusses practical aspects of the design of incompressible flow solvers; Hafez and Soliman discuss issues arising in both the discretization and solution phases of an unstaggered grid method; a similar treatment of an algorithm using pressure dissipation is given by Habashi, Peeters, Robichaud, and Nguyen.

There are some aspects of incompressible computational fluid dynamics that are not considered, or are only briefly considered in the book. A non-exhaustive list of these includes the application of wavelets, cellular automata, turbulence modeling and details about the direct simulation of turbulent flows. We apologize to anyone offended by these omissions; in some cases we tried to obtain contributions but were thwarted by unwilling or tardy authors; in other cases we felt that the subject matter has not reached a sufficient stage of development to merit inclusion in this volume. We also note incompressible computational fluid dynamics is still a developing field so that necessarily, due to the time it takes to put together such a book, some very recent developments could not be included.

We, the editors, wish to take this opportunity to thank all the authors who contributed to this book; we feel that we have gathered here a group of papers by outstanding authorities and which are representative of the current state of computational methods for incompressible flows. We also wish to thank our editor at Cambridge University Press, Lauren Cowles, for her patience, understanding, and encouragement over the duration of the project.

Acknowledgement

We would like to express our sincere thanks to the United States Air Force Office of Scientific Research (M. D. G. and R. A. N.) and the Office of Naval Research (M. D. G.) for supporting our research in incompressible CFD over the past several years.

M. D. Gunzburger, Blacksburg

R. A. Nicolaides, Pittsburgh

Contributing Authors

B. Cardot
STCAN
Paris, France

Edward J. Dean
Department of Mathematics, University of Houston
Houston, TX 77204, USA

Roland Glowinski
Department of Mathematics, University of Houston
Houston, TX 77204, USA

Max D. Gunzburger
Department of Mathematics, Virginia Polytechnic Institute and State University
Blacksburg, VA 24061, USA

Michael S. Engelman
Fluid Dynamics International
Evanston, IL 60201, USA

Leopoldo P. Franca
Laboratório Nacional de Computação Científica
22290 Rio de Janeiro, Brazil

W. G. Habashi
Concordia University
Montreal, Quebec, Canada

M. Hafez
Department of Mechanical and Aerospace Engineering
University of California at Davis
Davis, CA 95616, USA

L. Steven Hou
Department of Mathematics and Statistics, York University
Toronto, Ontario M3J 1P3, Canada

Thomas J. R. Hughes
Division of Applied Mechanics, Stanford University
Stanford, CA 94305, USA

George Em Karniadakis
Department of Mechanical and Aerospace Engineering, Princeton University
Princeton, NJ 08544, USA

Rainald Löhner
School of Engineering and Applied Science, The George Washington University
Washington, DC 20052, USA

B. Mohammadi
INRIA
78153 Le Chesnay, France

V-N. Nguyen
Computational Methods Group, Pratt & Whitney
Montreal, Canada

Roy A. Nicolaides
Department of Mathematics, Carnegie Mellon University
Pittsburgh, PA 15213, USA

Steven A. Orszag
Department of Mechanical and Aerospace Engineering, Princeton University
Princeton, NJ 08544, USA

Anthony T. Patera
Department of Mechanical Engineering, Massachusetts Institute of Technology
Cambridge, MA 02139, USA

M. F. Peeters
Computational Methods Group, Pratt & Whitney
Montreal, Canada

Oliver Pironneau
INRIA
78153 Le Chesnay, France

Elbridge Gerry Puckett
Department of Mathematics, University of California at Davis
Davis, CA 95616, USA

M. P. Robichaud
Computational Methods Group, Pratt & Whitney
Montreal, Canada

Einar M. Rønquist
Nektonics, Inc.
Cambridge, MA, USA

M. Soliman
Department of Mechanical and Aerospace Engineering
University of California at Davis
Davis, CA 95616, USA

Rolf Stenberg
Faculty of Mechanical Engineering, Helsinki University of Technology
02150 Espoo, Finland

Thomas P. Svobodny
Department of Mathematics and Statistics, Wright State University
Dayton, OH 45435, USA

Roger Temam
Department of Mathematics, Indiana University
Bloomington, IN 47405, USA

R. W. Thatcher
Department of Mathematics, UMIST
Manchester M60 1QD, UK

R. Verfürth
Institut für Angewandte Mathematik, Universität Zürich
Zürich CH-8001, Switzerland

1 A Few Tools For Turbulence Models In Navier-Stokes Equations

B. Cardot, B. Mohammadi, and O. Pironneau

Abstract

This article is for those who have already a computer program for incompressible viscous transient flows and want to put a turbulence model into it. We discuss some of the implementation problems that can be encountered when the Finite Element Method is used on classical turbulence models except Reynolds stress tensor models. Particular attention is given to boundary conditions and to the stability of algorithms.

1.1 Introduction

Many scientists or engineers turn to turbulence modeling after having written a Navier-Stokes solver for laminar flows.

For them turbulence modeling is an external module into the computer program. Generally, the main ingredients to built a good Navier-Stokes solver are known; this includes tools like mixed approximations for the velocity u and pressure p to avoid checker board oscillations and also upwinding to damp high Reynolds number oscillations; however the problems that one may meet while implementing a turbulence model are not so well known because these models have not been studied much theoretically.

Judging from the literature [3][11][12][15][19][22] the most commonly used turbulence models seem to be

- algebraic eddy viscosity models (zero equation models)
- $k - \varepsilon$ models (two equations models)
- Reynolds stress models

All three start from a decomposition of u and p into a mean part and a fluctuating part u'. However oscillations are understood either as time oscillations or space oscillations or even variations due to changes in initial conditions. In any case, the decomposition $u+u'$ is applied to the Navier-Stokes equations. After averaging (and several handwaving steps) some closed set of equations are obtained for the mean flow variables (still denoted u and p here). We do not intend to discuss the validity of the models or how well they compares with experiments; we want to discuss the discretization of the equations and the

1

stability properties if any. So we shall take them one by one and make a few comments along the way. Note however that this paper is by no means a review as the literature is way too rich on this subject.

In Section 1.2 we shall start with the easiest, the algebraic models of Smagorinsky [21] and Baldwin-Lowmax [2]. Wall laws will also be discussed in this section.

Then in Section 1.3, we shall discuss the $k - \varepsilon$ model and its variations. Emphasis will be on the positivity of the variables.

In Section 1.4 some numerical results are presented including a $k - \varepsilon$ simulation of the flow behing a cylinder at moderate Reynolds number.

We shall not discuss the Reynolds stress models because they are still controversial and because it appears that they may be easier to implement in a compressible flow solver directly. This is because they make the Navier-Stokes equations hyperbolic even in the case of incompressible fluids. Also the complexity of the equations of the $R_{ij} - \varepsilon$ model makes the mathematical analysis quite difficult and messy.

For laminar flow the Navier-Stokes equations are

$$\partial_t u + u\nabla u + \nabla p - \mu\Delta u = 0 \tag{1.1.1}$$

$$\nabla \cdot u = 0 \text{ in } \Omega \times]0, T[\tag{1.1.2}$$

$$u = u_\Gamma \text{ on } \Gamma \times]0, T[(\Gamma = \partial\Omega) \tag{1.1.3}$$

$$u|_{t=0} = u^o \text{ in } \Omega \tag{1.1.4}$$

In our numerical tests these have been discretized by piecewise biquadratic quadrilateral elements for u and piecewise linear discontinuous triangular elements for the pressure p (each quadrilateral is divided into two triangles) (see [9][18] or [23] for example).

Furthermore upwinding was implemented by using the characteristic Galerkin method [5], [8], [17].

Then (1.1.1)–(1.1.4) is approximated by

$$\frac{1}{\delta t}(u_h^{m+1}, v_h) - (p_h^{m+1}, \nabla \cdot v_h) + \nu(\nabla u_h^{m+1} : \nabla v_h) = \\ \frac{1}{\delta t}(u_h^m o X_h^m, v_h), \quad \forall v_h \in V_{oh} \tag{1.1.5}$$

$$(\nabla \cdot u_h^{m+1}, q_h) = 0, \quad \forall q_h \in Q_h \tag{1.1.6}$$

Here δt is the time step size, (f, g) stands for $\int_\Omega f(x)g(x)dx$, V_{oh} is the space of piecewise biquadratic velocities on the quadrangulation of Ω which are zero on Γ; Q_h is the space of piecewise linear discontinuous pressure on the triangulation of Ω. Finally, $X_h^m(x)$ is an approximation at t^n of the solution of

$$\frac{dX}{d\tau} = u_h^m(X_h, \tau); \quad X_t(t^{m+1}) = x; \quad X_h^m(x) \simeq X(t^m) \tag{1.1.7}$$

Note that $X_h^m(x) \simeq x - u_h^m(x)\delta t$ and that $(u_h^{m+1}(x) - u_h^m o X_h^m)/\delta t$ is an approximation of $\partial_t u_h + u_h \nabla u_h$.

Alternatively we could have used a Galerkin least square upwinding or a Newton method without upwinding with an implicit in time discretization of $\partial_t u$. Other popular elements for spatial discretizations include the $P^1 - iso - P^2/P^1$ element or the mini-element [1] on triangles. Much of what will be said applies also if these alternative choices are made.

One advantage of (1.1.5)–(1.1.6) is that it yields a symmetric linear system at each time step; the price to pay is the computation of a complicated integral: $\int_\Omega u_h^n(X_h^n(x))v_h(x)dx$. The linear system has the form

$$\begin{pmatrix} A & B \\ B^T & 0 \end{pmatrix} \begin{pmatrix} U \\ P \end{pmatrix} = \begin{pmatrix} W \\ 0 \end{pmatrix} \qquad A = \begin{pmatrix} D & 0 \\ 0 & D \end{pmatrix} \qquad B = \begin{pmatrix} B^1 \\ B^2 \end{pmatrix} \qquad (1.1.8)$$

where v is the vector of degrees of freedom of u_h and P for p_h. The matrix A is also a bloc matrix where

$$D_{ij} = \frac{1}{\delta_t}(w^i, w^j) + \nu(\nabla w^i, \nabla w^j) \qquad (1.1.9)$$

where w^i is the hat finite element scalar basis function at vertex i. Thus the two components of u_h are coupled through the pressure only by $B_{ij}^k = -(q^i, \partial w^j/\partial x_k)$.

1.2   Zero Equation Models

1.2.1   Eddy viscosity

Most zero equation models consist of the Navier-Stokes equations

$$\partial_t u + u\nabla u + \nabla p - \nabla \cdot [\mu_T(\nabla u + \nabla u^T)] = 0 \qquad (1.2.1)$$

$$\nabla \cdot u = 0 \qquad (1.2.2)$$

with a non constant viscosity (eddy viscosity) μ_T which is a given function of position x and velocity μ.

In his subgrid scaled model Smagorinsky [21] (see also [7][15][20]) suggested

$$\mu_T = \nu + ch^2|\nabla u + \nabla u^T| \qquad (1.2.3)$$

where ν is the molecular (reduced) viscosity c is a numerical constant ($c = 0.01$) and $h(x)$ is the average mesh size around x. In order to have a reasonably smooth function $h(x)$, one method is to define it at the vertices of the quadrangulation as the average of the edges which are issued from the vertex.

The Baldwin-Lomax [2] model is well suited to turbulent boundary layers. There ν_T is also an algebraic function of $\nabla \times u$ and the distance $y(x)$ between x and Γ.

$$\nu_T = l^2 |\nabla \times u| \text{ if } y \leq y_c, \quad = \frac{0.0269\delta_i}{(1 + 5.5(\frac{y}{\delta}))^6} \text{ otherwise} \qquad (1.2.4)$$

If $\kappa = 0.41$ is the Von-Karman constant, $l = \kappa y(1 - e^{-\frac{y}{A}})$ with $A = 26(\nu / \frac{\partial u}{\partial n})^{1/2}$. The boundary layer thickness is δ. Finally $\delta_i = \int_0^\delta (1 - \frac{u}{u(\delta)})dy$ and y_c is such that both functions in (1.2.4) match.

To discretize (1.2.1), (1.2.2) it is usually sufficient to take ν_T at time m in the equations which define u^{m+1}. Thus the analogue of (1.1.5), (1.1.6) is

$$\begin{aligned}
&\frac{1}{\delta t}(u_h^{m+1}, v_h) - (p_h^{m+1}\nabla \cdot v_h) \\
&+ \tfrac{1}{2}(\nu_T^m[\nabla u_h^{m+1} + (\nabla u_h^{m+1})^T] : [\nabla v_h + \nabla v_h^T]) = \frac{1}{\delta t}(u_h^m o X_h^m, v_h)
\end{aligned} \qquad (1.2.5)$$

$$(\nabla \cdot u_h^{m+1}, q_h) = 0. \qquad (1.2.6)$$

There are two additional difficulties here

- the matrix of the linear system now depends upon m through μ_T^m
- the components of u_h^{m+1} are coupled through the viscous terms also.

These difficulties may be removed by considering the scheme

$$\begin{aligned}
&\frac{1}{\delta t}(u_h^{m+1}, v_h) - (p_h^{m+1}\nabla \cdot v_h) + (\bar{\nu}_T \nabla u_h^{m+1} : \nabla v_h) = \\
&\frac{1}{\delta t}(u_h^m o X_h^m, v_h) + (\bar{\nu}_T \nabla u_h^{m+1} : \nabla v_h) - \tfrac{1}{2}(\mu_h^m([\nabla u_h^m + \nabla u_h^{mT}] : [\nabla v_h + \nabla v_h^T])
\end{aligned} \qquad (1.2.7)$$

where $\bar{\nu}_T$ is close to μ_h^m and recomputed every say 5 or 6th time step.

Convergence however is not guaranteed, unlike (1.2.3), (1.2.5) which is more stable [13] than (1.1.5), (1.1.6).

1.2.2   Wall laws

Equations (1.2.1), (1.2.2) may develop boundary layers near the walls. An attempt can be made to remove them from the computational domain by replacing the no slip condition (1.1.3) by slip conditions of the type

$$u \cdot n = 0 \qquad (1.2.8)$$

$$\nu_T \frac{\partial u}{\partial n} \cdot \tau + \alpha u \cdot \tau = \beta \qquad (1.2.9)$$

where n is the outward normal to Γ, τ is the tangent, and α, β may be non linear function of $u \cdot \tau$ and even ∇u. Parès [16] has shown that such boundary conditions lead to a well posed problem for the Navier-Stokes equations provided some conditions on the growth of α and β are satisfied.

Condition (1.2.8) must be enforced at vertices but n is defined on the edges of Γ; some sort of average is needed to define n at the vertices.

Another way is to notice that when $u \in H^1(\Omega)^2$

$$(u, \nabla q) = 0, \quad \forall q \in H^1(\Omega) \Rightarrow \nabla \cdot u = 0 \text{ in } \Omega \text{ and } u \cdot n|_\Gamma = 0 \qquad (1.2.10)$$

because

$$(u, \nabla q) = -(\nabla \cdot u, q) + \int_\Gamma u \cdot nq. \qquad (1.2.11)$$

So consider the implementation of (1.2.8)–(1.2.9) in weak form in (1.2.7) in which the normal n has completly disappeared:

$$\frac{1}{\delta t}(u_h^{m+1}, v_h) - (\nabla p_h^{m+1}, v_h) + (\bar{\nu}_T \nabla u_h^{m+1}, \nabla v_h) + \int_\Gamma \alpha^m u_h^{m+1} v_h = f(v_h) + \int_\Gamma \beta^m v_h \cdot \tau \qquad (1.2.12)$$

for all v_h continuous piecewise bilinear

$$(u_h^{m+1}, \nabla q_h) = 0 \quad \forall q_h \text{ piecewise linear.} \qquad (1.2.13)$$

In (1.2.13), $f(v_h)$ denotes the right hand side of (1.2.7). Numerical test with this formulation can be found in Parès [16] and in [4].

1.3 The $k - \varepsilon$ Model

Let k the turbulent kinetic energy and ε the turbulent rate of dissipated energy, so if u' denotes the time oscillations of u:

$$k = \frac{1}{2} < |u'|^2 > \qquad (1.3.1)$$

$$\varepsilon = \frac{\nu}{2} < |\nabla u' + \nabla u'^T|^2 >; \qquad (1.3.2)$$

In the $k - \varepsilon$ model it is assumed that the small time oscillations of u, p are equivalent to an eddy viscosity:

$$\mu_T = c_\mu \frac{k^2}{\varepsilon} \qquad (1.3.3)$$

and that, away from the walls $k - \varepsilon$ are governed by:

$$k_{,t} + u\nabla k - \frac{c_\mu}{2} \frac{k^2}{\varepsilon} |\nabla u + \nabla u^T|^2 - \nabla \cdot (c_\mu \frac{k^2}{\varepsilon} \nabla k) + \varepsilon = 0 \qquad (1.3.4)$$

$$\varepsilon_{,t} + u\nabla\varepsilon - \frac{c_1}{2} k |\nabla u + \nabla u^T|^2 - \nabla \cdot (c_\varepsilon \frac{k^2}{\varepsilon} \nabla\varepsilon) + c_2 \frac{\varepsilon^2}{k} = 0 \qquad (1.3.5)$$

with $c_\mu = 0.09$, $c_1 = 0.1296$, $c_2 = 1.92$, $c_\varepsilon = 0.07$.

Natural boundary conditions could be

$$k, \varepsilon \text{ given at } t = 0; \ k|_\Gamma = k_\Gamma = 0, \quad \varepsilon|_\Gamma = \varepsilon_\Gamma \qquad (1.3.6)$$

however ε_Γ is not known so the model is not well posed near the solid walls. A coupling with a one equation model (unknown ε) near the walls can be done. Alternatively an attempt is usually made to remove the low Reynolds regions from the computational domain by applying *wall laws*

$$k|_\Gamma = u^{*2} c_\mu^{-\frac{1}{2}}, \quad \varepsilon|_\Gamma = \frac{u^{*3}}{\kappa \delta} \qquad (1.3.7)$$

$$u \cdot n = 0, \quad \alpha u \cdot \tau + \beta \frac{\partial u \cdot \tau}{\partial n} = \gamma \qquad (1.3.8)$$

where κ is the Von Karman constant ($\kappa = 0.41$), δ the grid size at the wall (an approximation of the boundary layer thickness), u^* (computed by (1.3.9)) the friction velocity, $\beta = c_\mu k^2 / \varepsilon$, $\alpha = \beta/[\kappa\delta(B + \kappa^{-1} \log(\delta/D))]$ where D is a roughtness constant, $\gamma = -u^*|u^*|$ and B is such that (1.3.8) matches the viscous sublayer. To compute u*, Reichard's law may be used:

$$u^* = \frac{u.\tau}{f(u^*)}; \quad f(u^*) = 2.5 log(1 + 0.4 y^+) + 7.8\left(1 - e^{-\frac{y^+}{11}} - \frac{y^+}{11} e^{-0.33 y^+}\right) \qquad (1.3.9)$$

with $y^+ = \delta u^*/\nu$. So α, β, γ in (1.3.8) are nonlinear functions of $u \cdot \tau$. For smooth walls an easier alternative is $\alpha = \gamma = 0$.

For physical and mathematical reasons it is essential that the system (1.3.1)–(1.3.5) yields positive values for k and ε.

We shall now show that if the system has a smooth solution for given positive initial data and positive Dirichlet conditions on the boundaries then k and ε stay positive and bounded at later times. For this purpose one looks at

$$\theta = \frac{k}{\varepsilon}. \qquad (1.3.10)$$

If D_t denotes the total derivative operator, $\partial/\partial t + u\nabla$ and E denotes $\frac{1}{2}|\nabla u + \nabla u^T|^2$, then

$$
\begin{aligned}
D_t \theta &= \frac{1}{\varepsilon} D_t k - \frac{k}{\varepsilon^2} D_t \varepsilon \\
&= \theta^2 E(c_\mu - c_1) - 1 + c_2 + c_\mu \nabla \cdot \frac{k^2}{\varepsilon} \nabla\theta + 2c_\mu \theta^2 \nabla\theta \cdot \nabla\left(\frac{k}{\theta}\right) \\
&\quad + (c_\mu - c_\varepsilon) \frac{k}{\varepsilon^2} \nabla \cdot \frac{k^2}{\varepsilon} \nabla\varepsilon \qquad (1.3.11)
\end{aligned}
$$

In this equation it is seen that when the viscous terms are removed it becomes autonomous (independent of k):

$$D_t\theta = \theta^2 E(c_\mu - c_1) - 1 + c_2. \qquad (1.3.12)$$

Since $E(c_\mu - c_1) < 0$ and $c_2 - 1 > 0$ the solution of (1.3.12) is positive and bounded.

Considering again (1.3.4), (1.3.5) without viscous terms one may search for another variable with nice properties; following Mohammadi [14] let

$$\sigma = k^{-\alpha}\varepsilon^\beta \qquad (1.3.13)$$

then

$$\frac{1}{\sigma}D_t\sigma = -\frac{\alpha}{k}D_t k + \frac{\beta}{\varepsilon}D_t\varepsilon = -\frac{E}{2}\theta(\beta c_1 - \alpha c_\mu) + \frac{1}{\theta}(\alpha - c_2\beta). \qquad (1.3.14)$$

Therefore if

$$\beta c_1 - \alpha c_\mu \leq 0, \quad \alpha - c_2\beta \leq 0, \quad \text{i.e.} \quad 1.44\beta \leq \alpha \leq 1.92\beta \qquad (1.3.15)$$

then σ is decreasing. One simple choice is $\beta = 2$, $\alpha = 3$:

$$\sigma = \frac{\varepsilon^2}{k^3} \qquad (1.3.16)$$

and if we let $c_3 = 0.0054$, $c_4 = 0.016$ its equation is

$$D_t\sigma = -\sigma[0.0054E\theta + 0.016\frac{1}{\theta}] = -\sigma(c_3 E\theta + \frac{c_4}{\theta}). \qquad (1.3.17)$$

A discretization of (1.3.12), (1.3.17) which preserves the positivity and boundedness of θ and σ could be

$$\frac{1}{\delta t}(\theta^{m+1} - \theta^m o X^m) = -\theta^m\theta^{m+1}E^m(c_1 - c_\mu) + c_2 - 1 \qquad (1.3.18)$$

$$\frac{1}{\delta t}(\sigma^{m+1} - \sigma^m o X^m) = -\sigma^{m+1}[c_3 E^m\theta^m + \frac{c_4}{\theta^m}] \qquad (1.3.19)$$

which is also

$$\theta^{m+1} = \frac{(c_2 - 1)\delta t + \theta^m o X^m}{1 + (c_1 - c_\mu)\delta E^m\theta^m} \qquad (1.3.20)$$

$$\sigma^{m+1} = \frac{\sigma^m o X^m}{1 + \delta t c_3 E^m\theta^m + \delta t c_4/\theta^m}. \qquad (1.3.21)$$

Notice that $\theta^{m+1}, \sigma^{m+1}$ are positive and bounded if $\theta^m \sigma^m$ are positive and bounded even of the time step δt is large.

Now we are in a position to add the viscous terms and we propose the following algorithm:

Algorithm 1 $(k - \varepsilon)$

1. Compute $k^m o X^m, \varepsilon^m o X^m$ (i.e. convect k^m and ε^m)

2. Compute $\bar{k}^m, \bar{\varepsilon}^m$ by (1.3.20)–(1.3.21), i.e.

$$\bar{k}^m = (\theta^{m+1})^{-2}(\sigma^{m+1})^{-1}; \bar{\varepsilon}^m = (\theta^{m+1})^{-3}(\sigma^{m+1})^{-1} \qquad (1.3.22)$$

where $\theta^{m+1}, \sigma^{m+1}$ are given by (1.3.20)–(1.3.21) with $\theta^m = k^m/\varepsilon^m$, $\sigma^m = (\varepsilon^m)^2(k^m)^{-3}$.

3. Solve

$$\frac{1}{\delta t} k^{m+1} - \nabla \cdot (c_\mu \frac{k^{m^2}}{\varepsilon^m} \nabla k^{m+1}) = \frac{1}{\delta t} \bar{k}^m, \qquad (1.3.23)$$

$$\frac{1}{\delta t} \varepsilon^{m+1} - \nabla \cdot (c_\varepsilon \frac{k^{m^2}}{\varepsilon^m} \nabla \varepsilon^{m+1}) = \frac{1}{\delta t} \bar{\varepsilon}^m. \qquad (1.3.24)$$

The discretization of (1.3.23), (1.3.24) by the FEM is straightforward; however the positivity and boundedness of k^{m+1} and ε^{m+1} is guaranteed only if the maximum principle applies and this can be proved only with the Finite Elements of degree 1 on triangles which have no obtuse angle. Furthermore mass lumping should be used. Under these hypothesis Algorithm 1 generates positive and bounded values for k and ε under all circumstances (all δt and all E) when Dirichlet conditions are applied to k and ε on Γ.

Algorithm 1 is hard to adapt to other methods than the characteristic Galerkin method; but one can always use equations (1.3.12), (1.3.17) for θ and σ to which viscous terms can be added so as to match as best as possible the $k - \varepsilon$ model.

1.4   Some Numerical Results

1.4.1   Efficiency of the $\theta - \sigma$ algorithm

Figures 1.1–1.6: Turbulent jet. A turbulent jet at $Re = 10^6$ computed with 600 nodes.

- By a classical $k - \varepsilon$ algorithm where each equation is solved for its unknown and the process is iterated. (Figure 1.1 shows k, Figure 1.2 shows ε)

- By the $\theta - \varphi$ algorithm (Figure 1.3 shows k, Figure 1.4 shows ε)

Figure 1.1: Turbulent jet computed by a classical $k - \varepsilon$ algorithm. k shown here.

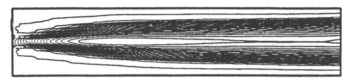

Figure 1.2: Turbulent jet computed by a classical $k - \varepsilon$ algorithm. ε shown here.

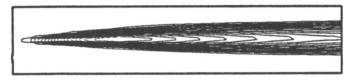

Figure 1.3: Turbulent jet computed by the $\theta - \sigma$ algorithm. k shown here.

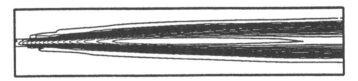

Figure 1.4: Turbulent jet computed by the $\theta - \sigma$ algorithm. ε shown here.

Figures 1.5–1.6 shows the convergence history of the algorithms. For the classical algorithm; it is seen on this configuration that convergence is obtained first but if the process is continued there is a divergence of ε. This does not happen with the second algorithm. Theoretically both methods should give the same result but a stability condition exists for the classical algorithm. This is probably why it diverges here.

1.4.2 Flow in a cavity with the $k - \varepsilon$ model

The domain is a rectangle tangent to a pipe of square section and 2D. The inflow conditions on u_h, k_h, ε_h, are taken from the experiment of Comte-Bellot [6]; Neumann conditions are given at the outflow. The $k - \varepsilon$ logarithmic wall laws are used on the horizontal walls with $Re = 57000$. Figures 1.7–1.9 display k, ε, u and show good

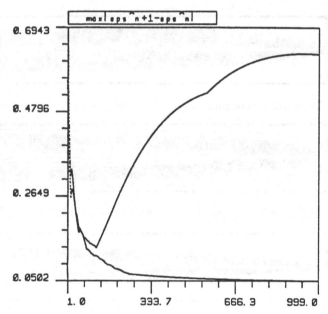

Figure 1.5: Convergence history of k for the classical $k - \varepsilon$ algorithm (top curve) and for the $\theta - \sigma$ algorithm.

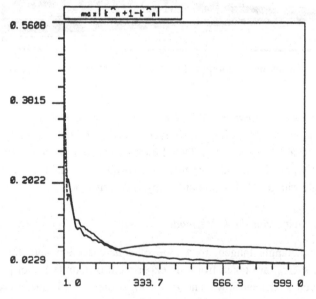

Figure 1.6: Convergence history of ε for the classical $k - \varepsilon$ algorithm (top curve) and for the $\theta - \sigma$ algorithm.

agreement with other similar computations (Viollet [24]). Convergence is obtained in 50 time steps approximatively.

1.4.3 Flow behind a cylinder with the $k - \varepsilon$ model at $Re = 10000$

Starting from a simulation of the Navier-Stokes equations at $Re = 500$ without turbulence modeling, the $k - \varepsilon$ model is activated with $k = \varepsilon = 10^{-7}$ at inflow boundaries. Figures 1.10–1.11 show $\nabla \times u$, k at t = 30". Reichard's wall law has been used on the cylinder. The flow does not converge to a stationary state. Higher Reynolds number are difficult to simulate because in the part of the flow which is not in the wake there is no turbulence and so the flow is governed by the Euler equation which is numerically unstable.

1.5 Discussion and Conclusion

Addition of a turbulence model into a computer program which was written for the Navier-Stokes equations can be deceptive because the system with the turbulence model may be much more unstable than the Navier-Stokes equations. It is not the case of algebraic stress models but it is the case of two equations models like $k - \varepsilon$.

Several researchers known to the authors have attemped a simulation of the flow behind a cylinder with the $k - \varepsilon$ model without success. The main difficulty being that classical algorithms (explicit or Newton steps) fail to converge or that they generate negative values for k or ε. A cosmetic cure is to truncate the values whenever they fall below a critical value, but our experience has shown us that the results depend upon the critical threshold chosen. We have shown in this paper that by carefully studying the positivity of the variables it is possible to construct semi-implicit L^{∞}-stable algorithms. Our algorithm is sufficiently robust to yield a solution to the difficult problem of simulating the flow behind a cylinder with the model.

Some engineer may object that the solution to the $k - \varepsilon$ model cannot by definition be transient. But from the mathematical stand point there is no reason to rule out time dependant solutions. From a theoretical point of view it is almost certain that both solutions exist, as for the Navier-Stokes equations without turbulence modeling. So the real question is the stability of the stationary solution. Suppose that by other methods one is able to compute a stationary solution of the model; if the solution is unstable then any small pertubation will induce a bifurcation toward the transient solution. Then is there any value in a turbulence model that produces unstable solutions?

For compressible fluids the problem of stability of algorithm is even more critical but the same method applies (see Mohammadi [14]) and similar success has been obtained.

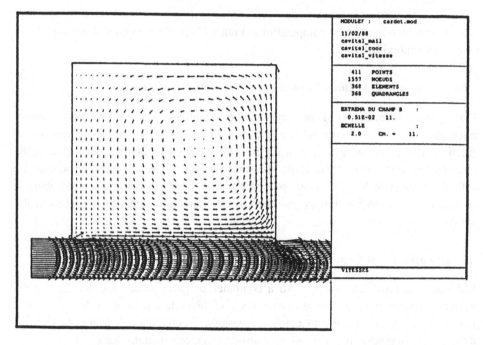

Figure 1.7: Flow in a cavity at $Re = 57000$. u shown here.

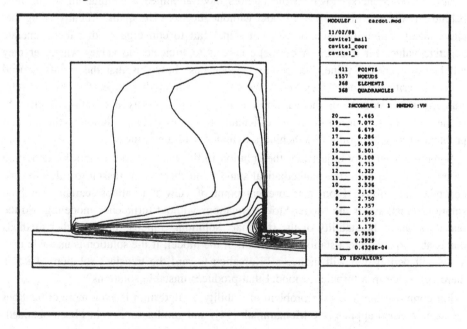

Figure 1.8: Flow in a cavity at $Re = 57000$. k shown here.

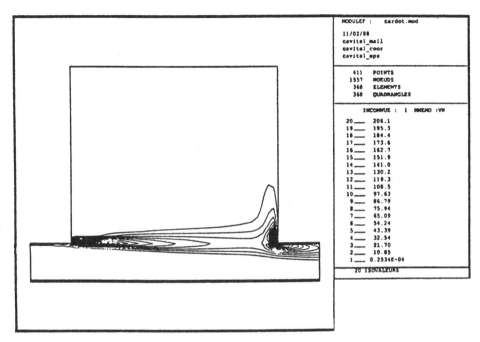

Figure 1.9: Flow in a cavity at $Re = 57000$. ε shown here.

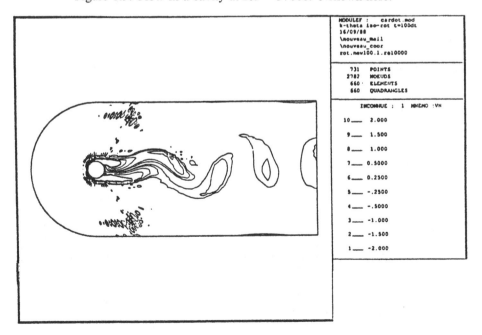

Figure 1.10: Flow behind a cylinder at $Re = 10000$. $\nabla \times u$ shown here.

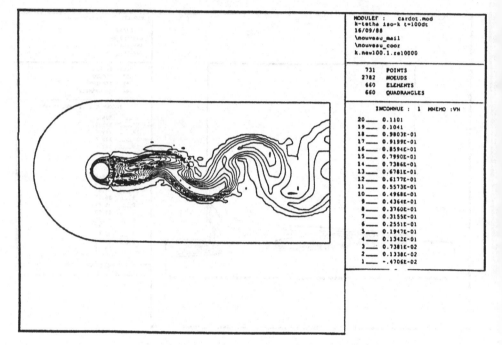

Figure 1.11: Flow behind a cylinder at $Re = 10000$; k shown here.

References

[1] D. Arnold, F. Brezzi, M. Fortin. (1984). "A stable finite element for the Stokes equations," *Calcolo* 21(**4**) 337–344.

[2] B. S. Baldwin, H. Lomax. (1978). "Thin layer approximation and algebraic model for separated turbulent flows," AIAA **78**-257, Huntsville.

[3] A. Baker. (1985). *Finite element computational fluid mechanics*, McGraw-Hill.

[4] C. Bègue, B. Cardot, C. Parès, O. Pironneau. (1990). "Simulation of turbulence with transient mean" *Int. J. Num. Meth. in Fluids* **11**.

[5] J.P. Benqué, O. Daubert, J. Goussebaile, H. Haugel. (1986). "Splitting up techniques for computations of industrial flows," In *Vistas in applied mathematics*, A. V. Balakrishnan ed. Optimization Software inc. Springer.

[6] G. Comte-Bellot. (1990). "Ecoulement turbulent entre deux parois planes," Publications scientifiques et techniques du ministère de l'air.

[7] J.W. Deardorff. (1970). "A numerical study of 3-d turbulent channel flow at large Reynolds numbers," J. Fluid Mech. **41**, 2, pp. 453–480.

[8] J. Douglas, T. F. Russell. (1982). "Numerical methods for convection dominated diffusion problems based on combining the method of characteristics with finite element methods or finite difference method," *SIAM J. Numer Anal.* 19, **5** 871–885.

[9] R. Glowinski. (1984). *Numerical Methods for Nonlinear Variational Methods.* Springer Series in Computational Physics.

[10] J. Goussebaïle, A. Jacomy. (1985). "Application à la thermo-hydrolique des méthodes d'éclatement d'opérateur dans le cadre éléments finis: traitement du modèle $k - \varepsilon$," Rapport EDF-LNH HE/41/85.11.

[11] A. G. Hutton, R. M. Smith, S. Hickmott. "The computation of turbulent flows of industrial complexity by the finite element method. Progress and prospects." *Finite Éléments in Fluids.* V7, Chap 15.

[12] B. E. Launder, D. B. Spalding. (1972). *Mathematical Models of Turbulence.* Academic Press.

[13] J. L. Lions(1968). *Quelques Methodes de Resolution des Problèmes aux Limites Nonlineaires,* Dunod.

[14] B. Mohamadi (Jan 1991). "A stable finite element algorithm for the $k - \varepsilon$ model for compressible turbulence," INRIA report.

[15] P. Moin, J. Kim. (1982). "Large eddy simulation of turbulent channel flow," *J. Fluid. Mech,* **118**, p. 341.

[16] C. Parès. (1988). "Un traitement faible par élément finis de la condition de glissement sur une paroi pour les équations de Navier-Stokes," Note C.R.A.S. **307** I 101-106.

[17] O. Pironneau. (1982). "On the transport-diffusion algorithm and its applications to the Navier-Stokes equations," *Numer. Math.* **38**, 309–332.

[18] O. Pironneau. (1989). *Finite Element Methods for Fluids,* Wiley.

[19] W. Rodi (1980). "Turbulence models and their application in hydraulics," Int. Ass. for Hydraulic Research, state-of-the-art paper Delft.

[20] V. Schumann. (1975). "Subgrid scale model for finite difference simulations of turbulent flows in plane channel and annuli," J. Comp. Physics. **18** pp. 376–404.

[21] J. S. Smagorinsky. (1963). "General circulation model of the atmosphere," *Mon. Weather Rev,* **91** pp. 99–164.

[22] C.G. Speziale. "Turbulence modeling in non-inertial frames of reference," ICASE report No 88.18.

[23] F. Thomasset. (1981). *Implementation of Finite Element Methods for Navier-Stokes Eq,* Springer series in Comp. Physics.

[24] P. L. Viollet. (1981). "On the modeling of turbulent heat and mass transfers for computations of buyoancy affected flows," *Proc. Int. Conf. Num. Meth. for Laminar and Turbulent Flows.* Venezia.

2 On Some Finite Element Methods for the Numerical Simulation of Incompressible Viscous Flow

Edward J. Dean and Roland Glowinski

Abstract

In this article we discuss the solution of the Navier-Stokes equations modelling unsteady incompressible viscous flow, by numerical methods combining operator splitting for the time discretization and finite elements for the space discretization.

The discussion includes the description of conjugate gradient algorithms which are used to solve the advection-diffusion and Stokes type problems produced at each time step by the operator splitting methods.

Introduction and Synopsis

The main goal of this article is to review several issues associated to the numerical solution of the Navier-Stokes equations modelling incompressible viscous flow. The methodology to be discussed relies systematically on variational priciples and is definitely oriented to Galerkin approximations. Also, we shall take advantage of time discretizations by operator splitting to decouple the two main difficulties occuring in the Navier-Stokes model, namely the incompressibility condition $\nabla \cdot \mathbf{u} = 0$ and the advection term $(\mathbf{u} \cdot \nabla)\mathbf{u}$, $\mathbf{u}$ being here the velocity field. The space approximation will be based on finite element methods and we shall discuss with some details the compatibility conditions existing between the velocity and pressure spaces; the practical implementation of these finite element methods will also be addressed.

This article relies heavily on [1]–[7] and does not have the pretention to cover the full field of finite element methods for the Navier-Stokes equations; concentrating on books only, pertinent references in this direction are [8]–[14] (see also the references therein).

This article is organized in sections whose list is given just below.

1. The Navier-Stokes equations for incompressible viscous flow

2. Operator splitting methods for initial value problems. Application to the Navier-Stokes equations

3. Iterative solution of the advection-diffusion sub-problems

17

4. Iterative solution of the Stokes type sub-problems

5. Finite element approximations of the Navier-Stokes equations

6. Comments on the Numerical Methodology.

2.1   The Navier-Stokes equations for incompressible viscous flow

Unsteady, isothermal flows of *incompressible, viscous, Newtonian* fluids are modelled by the following *Navier-Stokes equations*:

$$\frac{\partial \mathbf{u}}{\partial t} - \nu \nabla^2 \mathbf{u} + (\mathbf{u} \cdot \nabla)\mathbf{u} + \nabla p = \mathbf{f} \text{ in } \Omega \text{ (momentum equation)}, \qquad (2.1.1)$$

$$\nabla \cdot \mathbf{u} = 0 \text{ in } \Omega \text{ (incompressibility condition)}. \qquad (2.1.2)$$

In (2.1.1), (2.1.2), Ω ($\subset \Re^d$, $d = 2, 3$ in practice) denotes the flow region; its boundary will be denoted by Γ and we shall denote by $x = \{x_i\}_{i=1}^d$, the generic point of $\Re^d$. Also, in (2.1.1), (2.1.2) (and in the following)

(a) $\mathbf{u} = \{u_i\}_{i=1}^d$ is the *velocity* and p is the *pressure*;

(b) ν (> 0) is a *viscosity* coefficient;

(c) $\nabla = \left\{ \frac{\partial}{\partial x_i} \right\}_{i=1}^d$, $\nabla^2 = \Delta = \sum_{i=1}^d \frac{\partial^2}{\partial x_i^2}$, $\mathbf{u} \cdot \mathbf{v} = \sum_{i=1}^d u_i \cdot v_i$, $\forall \mathbf{u} = \{u_i\}_{i=1}^d$, $\forall \mathbf{v} = \{v_i\}_{i=1}^d$, $\nabla \mathbf{u} \cdot \nabla \mathbf{v} = \sum_{i=1}^d \sum_{j=1}^d \frac{\partial u_i}{\partial x_j} \frac{\partial v_i}{\partial x_j}$, $\forall \mathbf{u}, \mathbf{v}$, $|\mathbf{v}|^2 = \mathbf{v} \cdot \mathbf{v}$, $|\nabla \mathbf{v}|^2 = \nabla \mathbf{v} \cdot \nabla \mathbf{v}$;

(d) $\nabla \cdot \mathbf{v} = \sum_{i=1}^d \frac{\partial v_i}{\partial x_i}$, $\forall \mathbf{v}$, $(\mathbf{v} \cdot \nabla)\mathbf{w} = \left\{ \sum_{j=1}^d v_j \frac{\partial w_i}{\partial x_j} \right\}_{i=1}^d$, $\forall \mathbf{v}, \mathbf{w}$;

(e) $\mathbf{f} = \{f_i\}_{i=1}^d$ is a *density of external forces*.

Relations (2.1.1), (2.1.2) are not sufficient to define a flow; we have to consider further conditions, such as the *initial conditions*

$$\mathbf{u}(x, 0) = \mathbf{u}_0(x) \quad \text{(with } \nabla \cdot \mathbf{u}_0 = 0), \qquad (2.1.3)$$

and the *boundary condition*

$$\mathbf{u} = \mathbf{g} \text{ on } \Gamma \quad \text{(with } \int_\Gamma \mathbf{g} \cdot \mathbf{n} \, d\Gamma = 0), \qquad (2.1.4)$$

in (2.1.4), $\mathbf{n}$ denotes the unit vector of the outward normal at Γ. The boundary condition (2.1.4) is of Dirichlet type; more complicated boundary conditions are described in, e.g., [1], [3], [5], [13], among them, the following *mixed boundary conditions*

$$\mathbf{u} = \mathbf{g}_0 \text{ on } \Gamma_0, \quad \sigma \mathbf{n} = \mathbf{g}_1 \text{ on } \Gamma_1; \qquad (2.1.5)$$

where, in 2.1.5, Γ_0 and Γ_1 are two subsets of Γ satisfying $\Gamma_0 \cap \Gamma_1 = \emptyset$, closure of $\Gamma_0 \cup \Gamma_1 = \Gamma$, and where the (stress) tensor σ is defined by

$$\sigma = 2\nu\mathbf{D} - p\mathbf{I} \qquad (2.1.6)$$

with

$$D_{ij} = \frac{1}{2}\left(\frac{\partial u_i}{\partial x_j} + \frac{\partial u_j}{\partial x_i}\right). \qquad (2.1.7)$$

Another mixed boundary condition which occurs often in applications is given by

$$\mathbf{u} = \mathbf{g}_0 \text{ on } \Gamma_0, \qquad \nu\frac{\partial \mathbf{u}}{\partial n} - \mathbf{n}p = \mathbf{g}_1 \text{ on } \Gamma_1, \qquad (2.1.8)$$

with $\frac{\partial \mathbf{u}}{\partial n} = \left\{\frac{\partial u_i}{\partial n}\right\}_{i=1}^{N}$ ($= \{\nabla u_i \cdot \mathbf{n}\}_{i=1}^{N}$); (2.1.8) is less physical than (2.1.5), but like (2.1.5), it is quite useful to implement *downstream boundary conditions* for flow in unbounded regions.

The existence and possible uniqueness of solutions for problem (2.1.1)–(2.1.4) is discussed in, e.g., references [10, 11, 15, 16, 17, 18].

Solving numerically (2.1.1)–(2.1.4) (or (2.1.1)–(2.1.3), (2.1.5), or (2.1.1)–(2.1.3), (2.1.8)) is not trivial at all for the following reasons:

(i) The above problems are *nonlinear*;

(ii) The *incompressibility* condition (2.1.2);

(iii) The above problems are *systems* of partial differential equations, coupled through the nonlinear term $(\mathbf{u} \cdot \nabla)\mathbf{u}$, the incompressibility condition $\nabla \cdot \mathbf{u} = 0$, and sometimes through the boundary conditions (as it is the case in (2.1.5)).

In the following sections, we shall show that a time discretization by operator splitting will partly overcome the above difficulties; in particular, we shall be able to decouple those difficulties associated to the *nonlinearity* with those associated to the *incompressibility* condition.

2.2 Operator splitting methods for initial value problems

2.2.1 Generalities

We follow here the approach in [3–5] (see also references [6, 7]); therefore, let us consider the following initial value problem

$$\frac{d\varphi}{dt} + A(\varphi) = 0, \qquad (2.2.1)$$

$$\varphi(0) = \varphi_0, \qquad (2.2.2)$$

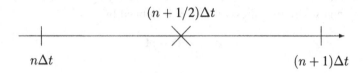

Figure 2.1: Dividing the time interval in the Peaceman-Rachford scheme.

where A is an operator (possibly nonlinear) from a Hilbert space H into itself, and where $\varphi_0 \in$ H. Suppose now that operator A has the following nontrivial decomposition

$$A = A_1 + A_2 \qquad (2.2.3)$$

(by nontrivial, we mean that A_1 and A_2 are individually simpler than A).

It is then quite natural to integrate the initial value problem (2.2.1), (2.2.2) by numerical methods taking advantage of the decomposition property (2.2.3); such a goal can be achieved by the *operator splitting schemes* discussed in the following paragraphs (for further information concerning operator splitting and related methods, see, e.g., [19–24], and the references therein).

2.2.2   The Peaceman-Rachford scheme

Let Δt (> 0) by a *time discretization step*, and denote by $\varphi^{n+\alpha}$ an approximation of $\varphi((n+\alpha)\Delta t)$, where φ is the solution of the initial value problem (2.2.1), (2.2.2). The fundamental idea behind the Peaceman-Rachford scheme (introduced in [25]) is quite simple:

Divide the time interval $[n\Delta t, (n + 1)\Delta t]$ into two subintervals, using the mid-point $(n+1/2)\Delta t$ (as shown in Figure 2.1) and then, the approximate solution φ^n being known at $n\Delta t$, compute first $\varphi^{n+1/2}$ using over $[n\Delta t, (n+1/2)\Delta t]$ a scheme of *backward Euler* type with respect to A_1 and of *forward Euler* type with respect to A_2; proceed similarly over $[(n + 1/2)\Delta t, (n + 1)\Delta t]$, switching the roles of A_1 and A_2.

The following scheme (due precisely to Peaceman and Rachford) realizes this program:

$$\varphi^0 = \varphi_0 ; \qquad (2.2.4)$$

then for $n \geq 0$, assuming that φ^n is known, compute successively $\varphi^{n+1/2}$ and φ^{n+1} via

$$\frac{\varphi^{n+1/2} - \varphi^n}{\Delta t/2} + A_1\left(\varphi^{n+1/2}\right) + A_2\left(\varphi^n\right) = 0, \qquad (2.2.5)$$

$$\frac{\varphi^{n+1} - \varphi^{n+1/2}}{\Delta t/2} + A_1\left(\varphi^{n+1/2}\right) + A_2\left(\varphi^{n+1}\right) = 0. \tag{2.2.6}$$

The convergence of scheme (2.2.4)–(2.2.6) has been proved in [26], [27] under quite general hypotheses concerning the properties of A_1 and A_2 (see also [28]); indeed, A_1 and/or A_2 can be nonlinear and even multivalued. To give a flavor of the accuarcy and stability properties of scheme (2.2.4)–(2.2.6), we shall consider the trivial situation where in (2.2.1), (2.2.2) we have $H = \Re^N$, $\varphi_0 \in \Re^N$ and A is a $N \times N$ matrix, symmetric and positive definite. The solution of (2.2.1), (2.2.2) is then

$$\varphi(t) = e^{-At}\varphi_0. \tag{2.2.7}$$

If one projects (2.2.7) over a vector basis of $\Re^N$ consisting of eigenvectors of A, we obtain – with obvious notation –

$$\varphi_i(t) = e^{-\lambda_i t}\varphi_{0i}, \quad i = 1, \dots, N, \tag{2.2.8}$$

where $0 < \lambda_1 \leq \lambda_2 \cdots \leq \lambda_N$ denote the eigenvalues of A.

In order to apply scheme (2.2.4)–(2.2.6), we consider the following decomposition of matrix A

$$A = \alpha A + \beta A, \tag{2.2.9}$$

with $\alpha + \beta = 1$, $0 < \alpha < 1$, $0 < \beta < 1$. Applying (2.2.4)–(2.2.6) with $A_1 = \alpha A$, $A_2 = \beta A$ yields

$$\varphi^{n+1} = \left(I + \frac{\Delta t}{2}\beta A\right)^{-1}\left(I - \frac{\Delta t}{2}\alpha A\right)\left(I + \frac{\Delta t}{2}\alpha A\right)^{-1}\left(I - \frac{\Delta t}{2}\beta A\right)\varphi^n. \tag{2.2.10}$$

The discrete analogues of (2.2.7), (2.2.8) are then

$$\varphi^n = \left(I + \frac{\Delta t}{2}\beta A\right)^{-n}\left(I - \frac{\Delta t}{2}\alpha A\right)^{n}\left(I + \frac{\Delta t}{2}\alpha A\right)^{-n}\left(I - \frac{\Delta t}{2}\beta A\right)^{n}\varphi_0, \tag{2.2.11}$$

$$\varphi_i^n = \left(\frac{1 - \frac{\Delta t}{2}\alpha\lambda_i}{1 + \frac{\Delta t}{2}\alpha\lambda_i}\right)^n \left(\frac{1 - \frac{\Delta t}{2}\beta\lambda_i}{1 + \frac{\Delta t}{2}\beta\lambda_i}\right)^n \varphi_{0i}, \tag{2.2.12}$$

respectively. Since $0 \leq |\frac{1-\xi}{1+\xi}| < 1$, $\forall \xi > 0$, we have

$$|\varphi_i^n| \leq |\varphi_{0i}|, \; \forall i = 1, \dots, N, \; \forall n \geq 1, \tag{2.2.13}$$

which implies the *stability* of the Peaceman-Rachford scheme (for the simple problem considered here, at least). We also have

$$\lim_{n \to +\infty} \varphi_i^n = 0, \; \forall i = 1, \dots, N, \tag{2.2.14}$$

which is the discrete analogue of $\lim_{t \to +\infty} \varphi_i(t) = 0$.

Let's study now the *accuracy* of the above scheme; motivated by (2.2.12) we introduce the *rational function*

$$R_1(\xi) = \left(\frac{1 - \alpha \frac{\xi}{2}}{1 + \alpha \frac{\xi}{2}} \right) \left(\frac{1 - \beta \frac{\xi}{2}}{1 + \beta \frac{\xi}{2}} \right). \tag{2.2.15}$$

We have, in the neighborhood of $\xi = 0$,

$$R_1(\xi) = 1 - \xi + \frac{\xi^2}{2} - (\alpha^2 + \beta^2 + \alpha\beta) \frac{\xi^3}{4} + \xi^4 O(1); \tag{2.2.16}$$

we have on the other hand

$$e^{-\xi} = 1 - \xi + \frac{\xi^2}{2} - \frac{\xi^3}{6} + \xi^4 O(1). \tag{2.2.17}$$

Comparing (2.2.16) and (2.2.17) shows that scheme (2.2.4)–(2.2.6) is *second-order accurate* for any pair $\{\alpha, \beta\}$ satisfying $\alpha + \beta = 1, 0 < \alpha < 1, 0 < \beta < 1$. Indeed if one takes $\alpha = \beta = 1/2$, we have $(\alpha^2 + \beta^2 + \alpha\beta)/4 = 3/16 = 1/6 + 1/48$, which shows that for $\alpha = \beta = 1/2$, scheme (2.2.4)–(2.2.6) is "almost" *third-order accurate* (3/16 is the smallest value of $(\alpha^2 + \beta^2 + \alpha\beta)/4$ compatible with $\alpha + \beta = 1$).

Let's discuss now the main drawback of the Peaceman-Rachford scheme:

Relation (2.2.8) shows that the larger λ_i, the faster $\varphi_i(t)$ converges to zero as $t \to +\infty$; considering now the discrete analogue of (2.2.8), namely (2.2.12), we observe that for *large values* of $\lambda_i \Delta t$ we have $R_1(\lambda_i \Delta t) \sim 1$, implying that, in (2.2.12), φ_i^n converges slowly to zero as $n \to +\infty$; from this property (which is also shared by the *Crank-Nicolson scheme*) we can expect scheme (2.2.4)–(2.2.6) not to be very well suited (unless Δt is very small) to simulate fast transient phenomena and to efficiently capture steady state solutions of (2.2.1), (2.2.2) (i.e. the solutions of $A(\varphi) = 0$), if operator A is *stiff*.

Remark 2.2.1 *We observe that operators A_1 and A_2 play essentially symmetric roles in scheme (2.2.4)–(2.2.6).*

2.2.3   The Douglas-Rachford scheme

The *Douglas-Rachford* scheme (cf. reference [29]) is a *predictor-corrector* variant of the Peaceman-Rachford scheme (2.2.4)–(2.2.6); applied to the numerical integration of the inital value problem (2.2.1), (2.2.2), it takes the following form

$$\varphi^0 = \varphi_0; \tag{2.2.18}$$

then, for $n \geq 0$, φ^n being known, we compute $\hat{\varphi}^{n+1}$ and φ^{n+1} as follows

$$\frac{\hat{\varphi}^{n+1} - \varphi^n}{\Delta t} + A_1(\hat{\varphi}^{n+1}) + A_2(\varphi^n) = 0\,, \qquad (2.2.19)$$

$$\frac{\varphi^{n+1} - \varphi^n}{\Delta t} + A_1(\hat{\varphi}^{n+1}) + A_2(\varphi^{n+1}) = 0\,. \qquad (2.2.20)$$

The convergence of scheme (2.2.18)–(2.2.20) has been proved in [26]–[28], under quite general hypotheses concerning A_1 and A_2.

Following Section 2.2.2 we consider the case when in (2.2.1), (2.2.2) we have H = $\Re^N$, $\varphi_0 \in \Re^N$ and A is a $N \times N$, symmetric and positive definite matrix. Using the decomposition (2.2.9) of A we obtain

$$\varphi^{n+1} = (I + \beta \Delta t A)^{-1}(I + \alpha \Delta t A)^{-1}(I + \alpha\beta|\Delta t|^2 A^2)\varphi^n\,. \qquad (2.2.21)$$

Relation (2.2.21) implies that

$$\varphi^n = (I + \beta \Delta t A)^{-n}(I + \alpha \Delta t A)^{-n}(I + \alpha\beta|\Delta t|^2 A^2)^n\varphi_0\,, \qquad (2.2.22)$$

$$\varphi_i^n = \frac{(1 + \alpha\beta|\Delta t|^2 \lambda_i^2)^n}{(1 + \alpha\Delta t\lambda_i)^n(1 + \beta\Delta t\lambda_i)^n}\varphi_{0i}\,. \qquad (2.2.23)$$

Introduce now R_2 defined by

$$R_2(\xi) = \frac{1 + \alpha\beta\xi^2}{(1 + \alpha\xi)(1 + \beta\xi)}\,; \qquad (2.2.24)$$

since $0 < R_2(\xi) < 1$, $\forall \xi > 0$, we have

$$|\varphi_i^n| \leq |\varphi_{0i}|\,, \forall i = 1, \ldots, N, \forall n \geq 1\,,$$

which implies the *stability* of the Douglas-Rachford scheme; we also have

$$\lim_{n \to +\infty} \varphi_i^n = 0, \forall i = 1, \ldots, N\,.$$

To study the *accuracy* of the Douglas-Rachford scheme we observe that in the neighborhood of $\xi = 0$ we have

$$R_2(\xi) = 1 - \xi + \xi^2 + \xi^3 O(1)\,. \qquad (2.2.25)$$

which implies, by comparing to (2.2.17) that scheme (2.2.18)–(2.2.20) is *first order accurate* only.

Finally, since

$$\lim_{\xi \to +\infty} R_2(\xi) = 1,$$ (2.2.26)

we can expect that scheme (2.2.18)–(2.2.20) will behave essentially as scheme (2.2.4)–(2.2.6), i.e., *poorly*, concerning the numerical integration of *stiff* differential systems. This prediction is confirmed by numerical experiments.

Remark 2.2.2 *Unlike the Peaceman-Rachford scheme (2.2.4)–(2.2.6), we observe that the roles played by operators A_1 and A_2 are nonsymmetric in scheme (2.2.18)–(2.2.20). As a rule of thumb we suggest to take for A_2 the operator with the best monotonicity properties (see [24] and [30] for more details).*

Remark 2.2.3 *Unlike scheme (2.2.4)–(2.2.6), scheme (2.2.18)–(2.2.20) is very easy to generalize to operator decompositions involving more than two operators. Consider therefore the numerical integration of (2.2.1), (2.2.2) when*

$$A = \sum_{i=1}^{q} A_i,$$ (2.2.27)

with $q \geq 2$. Following [31], we generalize scheme (2.2.18)–(2.2.20) by

$$\varphi^0 = \varphi_0;$$ (2.2.28)

and then for $n \geq 0$, φ^n being known compute $\varphi^{n+\frac{1}{q}}, \ldots, \varphi^{n+\frac{i}{q}}, \ldots, \varphi^{n+1}$ as follows

$$\frac{\varphi^{n+\frac{1}{q}} - \varphi^n}{\Delta t} + \frac{1}{q-1} A_1(\varphi^{n+\frac{1}{q}}) + \left(1 - \frac{1}{q-1}\right) A_1(\varphi^n)$$
$$+ \sum_{j=2}^{q} A_j(\varphi^n) = 0,$$ (2.2.29a)

$$\frac{\varphi^{n+\frac{i}{q}} - \varphi^n}{\Delta t} + \sum_{j=1}^{i-1} \left\{ \frac{1}{q-1} A_j(\varphi^{n+\frac{j}{q}}) + \left(1 - \frac{1}{q-1}\right) A_j(\varphi^n) \right\}$$
$$+ \frac{1}{q-1} A_i(\varphi^{n+\frac{i}{q}}) + \left(1 - \frac{1}{q-1}\right) A_i(\varphi^n) + \sum_{j=i+1}^{q} A_j(\varphi^n) = 0,$$ (2.2.29b)

$$\frac{\varphi^{n+1} - \varphi^n}{\Delta t} + \sum_{j=1}^{q-1} \left\{ \frac{1}{q-1} A_j(\varphi^{n+\frac{j}{q}}) + \left(1 - \frac{1}{q-1}\right) A_j(\varphi^n) \right\}$$
$$+ \frac{1}{q-1} A_q(\varphi^{n+1}) + \left(1 - \frac{1}{q-1}\right) A_q(\varphi^n) = 0.$$ (2.2.29c)

In relations (2.2.29), $\varphi^{n+i/q}$ denotes an approximate solution at step i of the computation process; it does not denote an approximation of $\varphi((n + i/q)/\Delta t)$.

2.2.4 A θ-scheme

This scheme (introduced by the author in [2], [3], [4]) is a variation of schemes discussed in [21], [22], [23] and is discussed with more details in [24]. It is in fact a variant of the Peaceman-Rachford scheme described in Section 2.2.2.

Let θ be a number on the open interval $(0, 1/2)$ (in practice $\theta \in (0, 1/3]$); the θ-scheme applied to the solution of the initial value problem (2.2.1), (2.2.2), when $A = A_1 + A_2$, is described as follows:

$$\varphi^0 = \varphi_0 \, ; \tag{2.2.30}$$

then for $n \geq 0$, φ^n being known, we compute $\varphi^{n+\theta}$, $\varphi^{n+1-\theta}$, φ^{n+1} as follows:

$$\frac{\varphi^{n+\theta} - \varphi^n}{\theta \Delta t} + A_1(\varphi^{n+\theta}) + A_2(\varphi^n) = 0 \, , \tag{2.2.31}$$

$$\frac{\varphi^{n+1-\theta} - \varphi^{n+\theta}}{(1 - 2\theta)\Delta t} + A_1(\varphi^{n+\theta}) + A_2(\varphi^{n+1-\theta}) = 0 \, , \tag{2.2.32}$$

$$\frac{\varphi^{n+1} - \varphi^{n+1-\theta}}{\theta \Delta t} + A_1(\varphi^{n+1}) + A_2(\varphi^{n+1-\theta}) = 0 \, . \tag{2.2.33}$$

We consider again the simple situation where $H = \Re^N$, $\varphi_o \in \Re^N$, where A is an $N \times N$ symmetric and positive definite matrix, and where $A_1 = \alpha A$, $A_2 = \beta A$ with $\alpha + \beta = 1$, $0 < \alpha < 1$, $0 < \beta < 1$. Introduce $\theta' = 1 - 2\theta$; we have then

$$\varphi^{n+1} = (I + \alpha\theta\Delta tA)^{-2}(I - \beta\theta\Delta tA)^2(I + \beta\theta'\Delta tA)^{-1}(I - \alpha\theta'\Delta tA)\varphi^n \, , \tag{2.2.34}$$

which implies

$$\varphi_i^n = \frac{(1 - \beta\theta\Delta t\lambda_i)^{2n}(1 - \alpha\theta'\Delta t\lambda_i)^n}{(1 + \alpha\theta\Delta t\lambda_i)^{2n}(1 + \beta\theta'\Delta t\lambda_i)^n} \varphi_{0i} \, . \tag{2.2.35}$$

Consider now the rational function R_3 defined by

$$R_3(\xi) = \frac{(1 - \beta\theta\xi)^2(1 - \alpha\theta'\xi)}{(1 + \alpha\theta\xi)^2(1 + \beta\theta'\xi)} \, . \tag{2.2.36}$$

Since

$$\lim_{\xi \to +\infty} |R_3(\xi)| = \beta/\alpha \, , \tag{2.2.37}$$

we should prescribe

$$\alpha \geq \beta \tag{2.2.38}$$

to obtain from (2.2.34), (2.2.35) the stability of the θ-scheme (2.2.30)–(2.2.33) for the large eigenvalues of A. Next, with respect to accuracy of scheme (2.2.30)–(2.2.33), we can show that

$$R_3(\xi) = 1 - \xi + \frac{\xi^2}{2}\left\{1 + (\beta^2 - \alpha^2)(2\theta^2 - 4\theta + 1)\right\} + \xi^3 O(1). \qquad (2.2.39)$$

It follows from (2.2.39) that scheme (2.2.30)–(2.2.33) is *second-order accurate* if either

$$\alpha = \beta \quad (= 1/2 \text{ from } \alpha + \beta = 1), \qquad (2.2.40)$$

or

$$\theta = 1 - 1/\sqrt{2} = .29289\ldots; \qquad (2.2.41)$$

scheme (2.2.30)–(2.2.33) is only *first-order* accurate if neither (2.2.40) nor (2.2.41) holds.

If one takes $\alpha = \beta = 1/2$, it follows from (2.2.35) that scheme (2.2.30)–(2.2.33) is *unconditionally stable* $\forall \theta \in (0, 1/2)$; however since (from (2.2.37)) we have

$$\lim_{\xi \to +\infty} |R_3(\xi)| = 1, \qquad (2.2.42)$$

the remark concerning schemes (2.2.4)–(2.2.6) and (2.2.18)–(2.2.20), with regard to the integration of stiff systems, still holds. In practice, we will choose α and β in order to have the same matrix for all the partial steps of the integration method, i.e., α, β, θ have to satisfy

$$\alpha\theta = \theta(1 - 2\theta), \qquad (2.2.43)$$

which implies

$$\alpha = (1 - 2\theta)/(1 - \theta), \qquad \beta = \theta/(1 - \theta). \qquad (2.2.44)$$

Combining (2.2.38), (2.2.44) yields

$$0 < \theta < 1/3. \qquad (2.2.45)$$

For $\theta = 1/3$, (2.2.44) implies $\alpha = \beta = 1/2$.

If $0 < \theta < 1/3$ and if α and β are given by (2.2.44) we have

$$\lim_{\xi \to +\infty} |R_3(\xi)| = \beta/\alpha = \theta/(1 - 2\theta) < 1. \qquad (2.2.46)$$

Indeed, we can prove that if $\theta \in [\theta^*, 1/3]$ (with $\theta^* = .087385580\ldots$) and if α and β are given by (2.2.44), then scheme (2.2.30)–(2.2.33) is *unconditionally stable*; moreover if $\theta \in (\theta^*, 1/3)$ (with α, β still given by (2.2.44)), property (2.2.46) gives to scheme (2.2.30)–(2.2.33) good asymptotic properties as $n \to +\infty$, making it well suited to compute steady state solutions.

If $\theta = 1 - 1/\sqrt{2}$ (resp., $\theta = .25$) we have $\alpha = 2 - \sqrt{2}$, $\beta = \sqrt{2} - 1$, $\beta/\alpha = 1/\sqrt{2}$ (resp., $\alpha = 2/3$, $\beta = 1/3$, $\beta/\alpha = 1/2$).

2.2.5 Application to the Navier-Stokes Equations

We discuss now the application of the time discretization schemes described in the above sections to the solution of the time-dependent Navier-Stokes equations (2.1.1), (2.1.2), with the initial-value condition (2.1.3); we suppose that the boundary conditions are of the following mixed type

$$\mathbf{u} = \mathbf{g}_0 \text{ on } \Gamma_0, \qquad \nu \frac{\partial \mathbf{u}}{\partial n} - \mathbf{n}p = \mathbf{g}_1 \text{ on } \Gamma_1, \qquad (2.2.47)$$

with $\Gamma_0 \cap \Gamma_1 = \emptyset$, closure of $\Gamma_0 \cap \Gamma_1 = \Gamma$; if $\Gamma_1 = \emptyset$ we need to have $\int_\Gamma \mathbf{g}_0 \cdot \mathbf{n} \, d\Gamma = 0$.

In this section we shall consider application of the θ-scheme only, since it is the one producing the best results with regard to accuracy and convergence to steady states. We obtain therefore the following scheme:

$$\mathbf{u}^0 = \mathbf{u}_0 ; \qquad (2.2.48)$$

then for $n \geq 0$ and starting from $\mathbf{u}^n$ we solve

$$\frac{\mathbf{u}^{n+\theta} - \mathbf{u}^n}{\theta \Delta t} - \alpha\nu\Delta\mathbf{u}^{n+\theta} + \nabla p^{n+\theta} = \mathbf{f}^{n+\theta} + \beta\nu\Delta\mathbf{u}^n - (\mathbf{u}^n \cdot \nabla)\mathbf{u}^n \text{ in } \Omega, \qquad (2.2.49a)$$

$$\nabla \cdot \mathbf{u}^{n+\theta} = 0 \text{ in } \Omega \qquad (2.2.49b)$$

$$\mathbf{u}^{n+\theta} = \mathbf{g}_0^{n+\theta} \text{ on } \Gamma_0, \qquad \alpha\nu\frac{\partial\mathbf{u}^{n+\theta}}{\partial n} - \mathbf{n}p^{n+\theta} = \mathbf{g}_1^{n+\theta} - \beta\nu\frac{\partial\mathbf{u}^n}{\partial n} \text{ on } \Gamma_1, \qquad (2.2.49c)$$

$$\frac{\mathbf{u}^{n+1-\theta} - \mathbf{u}^{n+\theta}}{(1-2\theta)\Delta t} - \beta\nu\mathbf{u}^{n+1-\theta} + (\mathbf{u}^{n+1-\theta} \cdot \nabla)\mathbf{u}^{n+1-\theta}$$
$$= \mathbf{f}^{n+1-\theta} + \alpha\nu\Delta\mathbf{u}^{n+\theta} - \nabla p^{n+\theta} \text{ in } \Omega, \qquad (2.2.50a)$$

$$\left.\begin{array}{l} \mathbf{u}^{n+1-\theta} = \mathbf{g}_0^{n+1-\theta} \text{ on } \Gamma_0, \\[2mm] \beta\nu\dfrac{\partial\mathbf{u}^{n+1-\theta}}{\partial n} = \mathbf{g}_1^{n+1-\theta} + \mathbf{n}p^{n+\theta} - \alpha\nu\dfrac{\partial\mathbf{u}^{n+\theta}}{\partial n} \text{ on } \Gamma_1, \end{array}\right\} \qquad (2.2.50b)$$

$$\frac{\mathbf{u}^{n+1} - \mathbf{u}^{n+1-\theta}}{\theta\Delta t} - \alpha\nu\Delta\mathbf{u}^{n+1} + \nabla p^{n+1}$$
$$= \mathbf{f}^{n+1} + \beta\nu\Delta\mathbf{u}^{n+1-\theta} - (\mathbf{u}^{n+1-\theta} \cdot \nabla)\mathbf{u}^{n+1-\theta} \text{ in } \Omega, \qquad (2.2.51a)$$

$$\nabla \cdot \mathbf{u}^{n+1} = 0 \text{ in } \Omega, \qquad (2.2.51b)$$

$$\mathbf{u}^{n+1} = \mathbf{g}_0^{n+1} \text{ on } \Gamma_0, \qquad \alpha\nu\frac{\partial\mathbf{u}^{n+1}}{\partial n} - \mathbf{n}p^{n+1} = \mathbf{g}_1^{n+1} - \beta\nu\frac{\partial\mathbf{u}^{n+1-\theta}}{\partial n} \text{ on } \Gamma_1. \qquad (2.2.51c)$$

The choice of α and β is discussed below. We observe that using the above θ-scheme we have been able to *decouple* the nonlinearity and the incompressibility in the Navier-Stokes equations (2.1.1), (2.1.2). In Sections 2.3 and 2.4, we will describe the specific treatment of the subproblems encountered at each step of (2.2.48)–(2.2.51). We note that $\mathbf{u}^{n+\theta}$ and $\mathbf{u}^{n+1}$ are obtained from the solution of linear problems very close to the steady Stokes problem. The good choice for α and β is given by (2.2.44); with such a choice many computer subprograms are common to both the linear and nonlinear subproblems, saving thereby quite a substantial amount of core memory. Concerning θ, numerical experiments show that $\theta = 1 - 1/\sqrt{2}$ seems to produce the best results, even in those situations where the Reynolds number is large.

To conclude this section we would like to mention that there is practically no loss in accuracy and stability by replacing $(\mathbf{u}^{n+1-\theta} \cdot \nabla)\mathbf{u}^{n+1-\theta}$ by $(\mathbf{u}^{n+\theta} \cdot \nabla)\mathbf{u}^{n+1-\theta}$ in (2.2.50a).

Remark 2.2.4 *For $\theta = 1/4$, the convergence of scheme (2.2.48)–(2.2.51) is proved in [32].*

2.3 Iterative solution of the advection-diffusion subproblems

2.3.1 Classical and variational formulations, synopsis

At each full step of the operator-splitting method (2.2.48)–(2.2.51) we have to solve a nonlinear elliptic system of the following type:

$$\alpha\mathbf{u} - \nu\Delta\mathbf{u} + (\mathbf{u} \cdot \nabla)\mathbf{u} = \mathbf{f} \text{ in } \Omega, \quad (2.3.1)$$

$$\mathbf{u} = \mathbf{g}_0 \text{ on } \Gamma_0, \quad \nu\frac{\partial\mathbf{u}}{\partial n} = \mathbf{g}_1 \text{ on } \Gamma_1, \quad (2.3.2)$$

where α and ν are two positive constants and where $\mathbf{f}$, $\mathbf{g}_0$ and $\mathbf{g}_1$ are three given functions defined on Ω, Γ_0 and Γ_1, respectively. We do not discuss here the existence and uniqueness of solutions for (2.3.1), (2.3.2). We now introduce the following functional spaces of *Sobolev* type:

$$H^1(\Omega) = \left\{ \varphi | \varphi \in L^2(\Omega), \ \frac{\partial\varphi}{\partial x_i} \in L^2(\Omega), \ \forall i = 1, \ldots, d \right\}, \quad (2.3.3)$$

$$V_0 = \left\{ \mathbf{v} | \mathbf{v} \in (H^1(\Omega))^d, \ \mathbf{v} = 0 \text{ on } \Gamma_0 \right\}, \quad (2.3.4)$$

$$V_g = \left\{ \mathbf{v} | \mathbf{v} \in (H^1(\Omega))^d, \ \mathbf{v} = \mathbf{g}_0 \text{ on } \Gamma_0 \right\}; \quad (2.3.5)$$

if $\mathbf{g}_0$ is sufficiently smooth, then V_g is nonempty. We use the notation of Section 2.1, completed by

$$dx = dx_1 \cdots dx_d.$$

Using *Green's formula*, we can prove that for sufficiently smooth functions **u** and **v** belonging to $(H^1(\Omega))^d$ and V_0, respectively, we have

$$\int_{\Gamma_1} \frac{\partial \mathbf{u}}{\partial n} \cdot \mathbf{v} \, d\Gamma = \int_\Omega \nabla \mathbf{u} \cdot \nabla \mathbf{v} \, dx + \int_\Omega \Delta \mathbf{u} \cdot \mathbf{v} \, dx. \tag{2.3.6}$$

It can also be proved that if **u** is a solution of (2.3.1), (2.3.2) belonging to V_g, it is also a solution of the following *nonlinear variational problem*

$$\left.\begin{array}{l} \mathbf{u} \in V_g; \quad \forall \mathbf{v} \in V_0 \text{ we have} \\[2mm] \alpha \displaystyle\int_\Omega \mathbf{u} \cdot \mathbf{v} \, dx + \nu \int_\Omega \nabla \mathbf{u} \cdot \nabla \mathbf{v} \, dx + \int_\Omega (\mathbf{u} \cdot \nabla)\mathbf{u} \cdot \mathbf{v} \, dx \\[4mm] \qquad = \displaystyle\int_\Omega \mathbf{f} \mathbf{v} \, dx + \int_{\Gamma_1} \mathbf{g}_1 \cdot \mathbf{v} \, d\Gamma. \end{array}\right\} \tag{2.3.7}$$

We observe that (2.3.1), (2.3.2), (2.3.7) is not equivalent to a problem of the Calculus of Variations since there is no functional of **v** with $(\mathbf{v} \cdot \nabla)\mathbf{v}$ as a differential; however, using a convenient *least squares formulation* we shall be able to solve (2.3.1), (2.3.2), (2.3.7) by iterative methods from *Nonlinear Programming* such as *conjugate gradient algorithms*.

2.3.2 Least-squares formulation of problem (2.3.1), (2.3.2), (2.3.7)

Let $\mathbf{v} \in V_g$; to **v** we associate the solution $\mathbf{y} = \mathbf{y}(\mathbf{v}) \in V_0$ of

$$\left.\begin{array}{l} \alpha \mathbf{y} - \nu \Delta \mathbf{y} = \alpha \mathbf{v} - \nu \Delta \mathbf{v} + (\mathbf{v} \cdot \nabla)\mathbf{v} - \mathbf{f} \text{ in } \Omega, \\[2mm] \mathbf{y} = 0 \text{ on } \Gamma_0, \quad \nu \dfrac{\partial \mathbf{y}}{\partial n} = \nu \dfrac{\partial \mathbf{v}}{\partial n} - \mathbf{g}_1 \text{ on } \Gamma_1. \end{array}\right\} \tag{2.3.8}$$

We observe that **y** is obtained from **v** via the solution of d uncoupled linear elliptic problems (one for each component of **y**); using (2.3.6), it is easily shown that (2.3.8) is equivalent to the linear variational problem

$$\left.\begin{array}{l} \mathbf{y} \in V_0; \quad \forall \mathbf{z} \in V_0 \text{ we have} \\[2mm] \alpha \displaystyle\int_\Omega \mathbf{y} \cdot \mathbf{z} \, dx + \nu \int_\Omega \nabla \mathbf{y} \cdot \nabla \mathbf{z} \, dx = \alpha \int_\Omega \mathbf{v} \cdot \mathbf{z} \, dx + \nu \int_\Omega \nabla \mathbf{v} \cdot \nabla \mathbf{z} \, dx \\[4mm] \qquad\qquad + \displaystyle\int_\Omega (\mathbf{v} \cdot \nabla)\mathbf{v} \cdot \mathbf{z} \, dx - \int_\Omega \mathbf{f} \cdot \mathbf{z} \, dx \\[4mm] \qquad\qquad - \displaystyle\int_{\Gamma_1} \mathbf{g}_1 \cdot \mathbf{z} \, d\Gamma, \end{array}\right\} \tag{2.3.9}$$

which has a *unique* solution.

Suppose now that $\mathbf{v}$ is a solution of the nonlinear problem (2.3.1), (2.3.2), (2.3.7); the corresponding $\mathbf{y}$ (obtained from the solution of (2.3.8), (2.3.9)) is clearly $\mathbf{y} = 0$; from this observation, it is quite natural to introduce the following (nonlinear) least-squares formulation of (2.3.1), (2.3.2), (2.3.7):

$$\text{Find } \mathbf{u} \in V_g \text{ such that } J(\mathbf{u}) \le J(\mathbf{v}), \quad \forall \mathbf{v} \in V_g, \tag{2.3.10}$$

where the functional $J : (\mathrm{H}^1(\Omega))^d \to \Re$ is defined by

$$J(\mathbf{v}) = \frac{1}{2} \int_\Omega \left\{ \alpha |\mathbf{y}|^2 + \nu |\nabla \mathbf{y}|^2 \right\} dx, \tag{2.3.11}$$

with $\mathbf{y}$ defined from $\mathbf{v}$ by (2.3.8), (2.3.9). Observe that if $\mathbf{u}$ is a solution of (2.3.10), such that $J(\mathbf{u}) = 0$, then it is also a solution of (2.3.1), (2.3.2), (2.3.7).

2.3.3   Conjugate-gradient solution of the least-squares problem (2.3.10)

Description of the algorithm.   We use the *Fletcher-Reeves* version (see [33]) of the conjugate-gradient method to solve the minimization problem (2.3.10); we have then (with $J'(\mathbf{v})$ the differential of J at $\mathbf{v}$):

Step 0: Initialization

$$\mathbf{u}^0 \in V_g \text{ is given;} \tag{2.3.12}$$

we then define $\mathbf{g}^0, \mathbf{w}^0 \in V_0$ *by*

$$\left. \begin{array}{l} \mathbf{g}^0 \in V_0, \\[2mm] \alpha \displaystyle\int_\Omega \mathbf{g}^0 \cdot \mathbf{z} \, dx + \nu \displaystyle\int_\Omega \nabla \mathbf{g}^0 \cdot \nabla \mathbf{z} \, dx = \langle J'(\mathbf{u}^0), \mathbf{z} \rangle, \quad \forall \mathbf{z} \in V_0, \end{array} \right\} \tag{2.3.13}$$

$$\mathbf{w}^0 = \mathbf{g}^0, \tag{2.3.14}$$

respectively.   □

Then for $m \ge 0$, *assuming that* $\mathbf{u}^m$, $\mathbf{w}^m$, $\mathbf{g}^m$ *are known we obtain* $\mathbf{u}^{m+1}$, $\mathbf{g}^{m+1}$, $\mathbf{w}^{m+1}$ *by*

Step 1: Descent

$$\left. \begin{array}{l} \lambda_m \in \Re, \\[2mm] J(\mathbf{u}^m - \lambda_m \mathbf{w}^m) \le J(\mathbf{u}^m - \lambda \mathbf{w}^m), \quad \forall \lambda \in \Re, \end{array} \right\} \tag{2.3.15}$$

$$\mathbf{u}^{m+1} = \mathbf{u}^m - \lambda_m \mathbf{w}^m. \tag{2.3.16}$$

Step 2: Calculation of the new descent direction
Solve

$$\left.\begin{array}{l} \mathbf{g}^{m+1} \in V_0\,, \\[2mm] \alpha \displaystyle\int_\Omega \mathbf{g}^{m+1} \cdot \mathbf{z}\, dx + \nu \int_\Omega \nabla \mathbf{g}^{m+1} \cdot \nabla \mathbf{z}\, dx = \langle J'(\mathbf{u}^{m+1}), \mathbf{z} \rangle, \quad \forall \mathbf{z} \in V_0\,, \end{array}\right\} \qquad (2.3.17)$$

and compute

$$\gamma_m = \frac{\alpha \int_\Omega |\mathbf{g}^{m+1}|^2 dx + \nu \int_\Omega |\nabla \mathbf{g}^{m+1}|^2 dx}{\alpha \int_\Omega |\mathbf{g}^m|^2 dx + \nu \int_\Omega |\nabla \mathbf{g}^m|^2 dx}, \qquad (2.3.18)$$

$$\mathbf{w}^{m+1} = \mathbf{g}^{m+1} + \gamma_m \mathbf{w}^m. \qquad \square \qquad (2.3.19)$$

Set $m = m + 1$ and go to (2.3.15).

As we shall see in Section 2.3.3, applying algorithm (2.3.12)–(2.3.19) to the solution of the least-squares problem (2.3.10), requires at each iteration the solution of three elliptic systems (i.e. $3N$ scalar elliptic problems), with *mixed* Dirichlet-Neumann boundary conditions, associated to the elliptic operator $\alpha I - \nu \Delta$.

Calculation of J'. A most important step when making use of (2.3.12)–(2.3.19) to solve (2.3.10) is the calculation of $\langle J'(\mathbf{u}^{m+1}), \mathbf{z} \rangle$ at each iteration; we can easily prove (see, e.g., [1, Chapter 7]) that $J'(\mathbf{v})$ can be identified with the linear functional from V_0 to $\Re$ defined by

$$\begin{aligned} \langle J'(\mathbf{v}), \mathbf{z} \rangle = {} & \alpha \int_\Omega \mathbf{y} \cdot \mathbf{z}\, dx + \nu \int_\Omega \nabla \mathbf{y} \cdot \nabla \mathbf{z}\, dx \\ & + \int_\Omega \mathbf{y} \cdot (\mathbf{v} \cdot \nabla) \mathbf{z}\, dx + \int_\Omega \mathbf{y} \cdot (\mathbf{z} \cdot \nabla) \mathbf{v}\, dx, \quad \forall \mathbf{z} \in V_0\,, \end{aligned} \qquad (2.3.20)$$

where $\mathbf{y}$ is obtained from $\mathbf{v}$ by solving (2.3.8), (2.3.9), and where

$$(\mathbf{v} \cdot \nabla) \mathbf{w} = \left\{ \sum_{j=1}^N \mathbf{v}_j \frac{\partial w_i}{\partial x_j} \right\}_{i=1}^d, \quad \forall \mathbf{v}, \mathbf{w}\,;$$

$\langle J'(\mathbf{v}), \mathbf{z} \rangle$ has therefore a *purely integral representation*, which is of major importance in view of finite element (or spectral, or wavelet) implementations of (2.3.12)–(2.3.19). From the above results, to obtain $\langle J'(\mathbf{u}^{m+1}), \mathbf{z} \rangle$ we should proceed as follows:

(a) Compute $\mathbf{y}^{m+1}$, associated to $\mathbf{u}^{m+1}$ by (2.3.8), (2.3.9), as indicated below.

(b) We then obtain $\langle J'(\mathbf{u}^{m+1}), \mathbf{z} \rangle$ by taking $\mathbf{v} = \mathbf{u}^{m+1}$ and $\mathbf{y} = \mathbf{y}^{m+1}$ in (2.3.20).

Calculation of λ_m. Comments on Algorithm (2.3.12)–(2.3.19). A problem of practical importance is the calculation of λ_m. Let's denote by $\mathbf{y}^m(\lambda)$ the solution of (2.3.8), (2.3.9) associated to $\mathbf{v} = \mathbf{u}^m - \lambda \mathbf{w}^m$. We clearly have

$$\mathbf{y}^m(0) = \mathbf{y}^m, \quad \mathbf{y}^m(\lambda_m) = \mathbf{y}^{m+1}, \tag{2.3.21}$$

and also

$$\mathbf{y}^m(\lambda) = \mathbf{y}^m - \lambda \mathbf{y}_1^m + \lambda^2 \mathbf{y}_2^m, \tag{2.3.22}$$

where $\mathbf{y}_1^m$, $\mathbf{y}_2^m$ are the solutions of

$$\left.\begin{aligned}
&\mathbf{y}_1^m \in V_0; \ \forall \mathbf{z} \in V_0 \text{ we have} \\
&\alpha \int_\Omega \mathbf{y}_1^m \cdot \mathbf{z} \, dx + \nu \int_\Omega \nabla \mathbf{y}_1^m \cdot \nabla \mathbf{z} \, dx \\
&\qquad = \alpha \int_\Omega \mathbf{w}^m \cdot \mathbf{z} \, dx + \nu \int_\Omega \nabla \mathbf{w}^m \cdot \nabla \mathbf{z} \, dx \\
&\qquad\quad + \int_\Omega (\mathbf{u}^m \cdot \nabla) \mathbf{w}^m \cdot \mathbf{z} \, dx + \int_\Omega (\mathbf{w}^m \cdot \nabla) \mathbf{u}^m \cdot \mathbf{z} \, dx,
\end{aligned}\right\} \tag{2.3.23}$$

$$\left.\begin{aligned}
&\mathbf{y}_2^m \in V_0; \ \forall \mathbf{z} \in V_0 \text{ we have} \\
&\alpha \int_\Omega \mathbf{y}_2^m \cdot \mathbf{z} \, dx + \nu \int_\Omega \nabla \mathbf{y}_2^m \cdot \nabla \mathbf{z} \, dx = \int_\Omega (\mathbf{w}^m \cdot \nabla) \mathbf{w}^m \cdot \mathbf{z} \, dx,
\end{aligned}\right\} \tag{2.3.24}$$

respectively. Since

$$J(\mathbf{u}^m - \lambda \mathbf{w}^m) = \frac{1}{2} \int_\Omega \left\{ \alpha |\mathbf{y}^m(\lambda)|^2 + \nu |\nabla \mathbf{y}^m(\lambda)|^2 \right\} dx, \tag{2.3.25}$$

the function $\lambda \to J(\mathbf{u}^m - \lambda \mathbf{w}^m)$ is, from (2.3.22), a *quartic polynomial* in λ, that we shall denote by $j_m(\lambda)$; λ_m is therefore a solution of the *cubic equation*

$$j_m'(\lambda) = 0. \tag{2.3.26}$$

We shall use the standard *Newton's method* to compute λ_m from (2.3.26), starting from $\lambda = 0$. The resulting algorithm is given by

$$\lambda^0 = 0, \tag{2.3.27}$$

then for $k \geq 0$, we obtain λ^{k+1} from λ^k by

$$\lambda^{k+1} = \lambda^k - \frac{j_m'(\lambda^k)}{j_m''(\lambda^k)}. \tag{2.3.28}$$

In our calculations, we always observed a *fast convergence* of (2.3.27), (2.3.28) (three iterations at most). Once λ_m is known, we know $\mathbf{y}^{m+1}$ since (from (2.3.21)) $\mathbf{y}^{m+1} = \mathbf{y}^m(\lambda_m)$.

If we now count the number of elliptic systems for $\alpha I - \nu\Delta$ to be solved at each iteration we observe that we have to solve three such systems, namely (2.3.23), (2.3.24), and (2.3.17) (to obtain $\mathbf{g}^{m+1}$); this number is optimal for a nonlinear problem, since the solution of a linear problem by a least-squares conjugate gradient method requires the solution at each iteration of two linear systems associated to the preconditioning operator.

From the above observations, it is clear that the practical implementation of (2.3.12)–(2.3.19) will require efficient (direct or iterative) elliptic solvers; we shall address this issue in Section 2.5.

Another important issue concerning algorithm (2.3.12)–(2.3.19) is its stopping criterion; we have used

$$J(\mathbf{u}^m)/J(\mathbf{u}^0) \leq \epsilon, \tag{2.3.29}$$

with ϵ of the order of 10^{-6}.

Algorithm (2.3.12)–(2.3.19) (in fact its finite dimensional variants) is quite efficient when used in combination with the operator splitting methods of Section 2.2; 3 to 5 iterations suffice to reduce the value of the cost function by a factor of 10^4 to 10^6 (assuming that we initialize with the velocity field computed at the previous partial step of the θ-scheme). Recently, we have been implementing GMRES type algorithms for solving problem (2.3.1), (2.3.2), (2.3.7) (see, e.g., [34], [35], [36] for details on GMRES algorithms, including applications to Computational Fluid Dynamics); using the same preconditioner as in algorithm (2.3.12)–(2.3.19) leads to an algorithm roughly twice as fast as the one presented here. Actually a similar improvement can be obtained using the algorithms described in Section 2.3.4, below.

2.3.4 On Newton's method for problem (2.3.1), (2.3.2), (2.3.7) and other linearization methods

Application of Newton's method to the solution of problem (2.3.1), (2.3.2), (2.3.7). The basic Newton's method applied to problem (2.3.1), (2.3.2) yields the following algorithm

$$\mathbf{u}^0 \in V_g \quad \text{is given;} \tag{2.3.30}$$

then for $m \geq 0$ we obtain $\mathbf{u}^{m+1}$ from $\mathbf{u}^m$ via the solution of

$$\left.\begin{array}{l} \alpha\mathbf{u}^{m+1} - \nu\Delta\mathbf{u}^{m+1} + (\mathbf{u}^{m+1}\cdot\nabla)\mathbf{u}^m + (\mathbf{u}^m\cdot\nabla)\mathbf{u}^{m+1} \\ \qquad\qquad = \mathbf{f} + (\mathbf{u}^m\cdot\nabla)\mathbf{u}^m \text{ in } \Omega, \\ \mathbf{u}^{m+1} = \mathbf{g}_0 \text{ on } \Gamma_0, \quad \nu\dfrac{\partial\mathbf{u}^{m+1}}{\partial n} = \mathbf{g}_1 \text{ on } \Gamma_1. \end{array}\right\} \tag{2.3.31}$$

The *linear* problem (2.3.31) can be solved (after an appropriate space discretization) by a GMRES algorithm (or by a variant of the least-squares conjugate gradient method of Section 2.3.3, taking of course advantage of the *linearity* of problem (2.3.31); see also following remark).

A remark concerning a linearized variant of scheme (2.2.48)–(2.2.51). As mentioned in Section 2.2.5 a natural variant of the θ-scheme (2.2.48)–(2.2.51) is obtained if one substitutes $(\mathbf{u}^{n+\theta} \cdot \nabla)\mathbf{u}^{n+1-\theta}$ for $(\mathbf{u}^{n+1-\theta} \cdot \nabla)\mathbf{u}^{n+1-\theta}$ in (2.2.50a). So doing, we are led to consider the solution of *linear* problems of the following type:

$$\left.\begin{aligned}\alpha\mathbf{u} - \nu\Delta\mathbf{u} + (\mathbf{U} \cdot \nabla)\mathbf{u} = \mathbf{f} \text{ in } \Omega, \\ \mathbf{u} = \mathbf{g}_0 \text{ on } \Gamma_0, \quad \nu\frac{\partial\mathbf{u}}{\partial n} = \mathbf{g}_1 \text{ on } \Gamma_1,\end{aligned}\right\} \tag{2.3.32}$$

where $\mathbf{f}$, $\mathbf{U}$, $\mathbf{g}_0$, $\mathbf{g}_1$ are given functions, with $\mathbf{U}$ satisfying $\nabla \cdot \mathbf{U} = 0$. Problem (2.3.32) after being appropriately discretized can be solved by either a least-squares conjugate gradient method close to the one described in Section 2.3.3, but cheaper to implement since from the linearity of (2.3.32), we have to solve only 2 elliptic systems associated to $\alpha I - \nu\Delta$ per iteration. An interesting alternative is clearly to use a preconditioned GMRES algorithm to solve (2.3.32), with $\alpha I - \nu\Delta$ as preconditioner.

2.4 Iterative solution of the Stokes type subproblems

2.4.1 Generalities

At each full step of the θ-scheme (2.2.48)–(2.2.51) we have to solve two linear problems of the following type

$$\left.\begin{aligned}\alpha\mathbf{u} - \nu\Delta\mathbf{u} + \nabla p = \mathbf{f} \text{ in } \Omega, \\ \nabla \cdot \mathbf{u} = 0 \text{ in } \Omega, \\ \mathbf{u} = \mathbf{g}_0 \text{ on } \Gamma_0, \quad \nu\frac{\partial\mathbf{u}}{\partial n} - \mathbf{n}p = \mathbf{g}_1 \text{ on } \Gamma_1,\end{aligned}\right\} \tag{2.4.1}$$

where α and ν are two positive constants ($\alpha \sim 1/\Delta t$) and $\mathbf{f}$, $\mathbf{g}_0$, $\mathbf{g}_1$ are given functions. Considering again the following space

$$V_g = \left\{\mathbf{v}|\mathbf{v} \in (H^1(\Omega))^d, \ \mathbf{v} = \mathbf{g}_0 \text{ on } \Gamma_0\right\},$$

it can be shown that if $\mathbf{f}$, $\mathbf{g}_0$, $\mathbf{g}_1$ are sufficiently smooth, then problem (2.4.1) has a *unique solution* in $V_g \times L^2(\Omega)$ (in $V_g \times (L^2(\Omega)/\Re)$ if $\Gamma_0 = \Gamma$ and $\int_\Gamma \mathbf{g}_0 \cdot \mathbf{n} \, d\Gamma = 0$; $p \in L^2(\Omega)/\Re$ means that p is defined only to within an arbitrary constant).

Due to the incompressibility condition $\nabla \cdot \mathbf{u} = 0$, problem (2.4.1) is a nontrival one. However, suppose that p is known in $L^2(\Omega)$, then we obtain $\mathbf{u}$ via the solution of an elliptic system whose variational formulation is given by

$$\left.\begin{array}{l} \mathbf{u} \in V_g \,;\, \forall \mathbf{v} \in V_0 \text{ we have} \\[2mm] \alpha \displaystyle\int_\Omega \mathbf{u} \cdot \mathbf{v}\, dx + \nu \int_\Omega \nabla \mathbf{u} \cdot \nabla \mathbf{v}\, dx \\[4mm] \qquad = \displaystyle\int_\Omega \mathbf{f} \cdot \mathbf{v}\, dx + \int_\Omega p \nabla \cdot \mathbf{v}\, dx + \int_{\Gamma_1} \mathbf{g}_1 \cdot \mathbf{v}\, d\Gamma . \end{array}\right\} \tag{2.4.2}$$

Problem (2.4.2) can be easily solved by finite element or finite difference methods. This obvious observation is at the basis of powerful iterative methods for solving the Stokes problem (2.4.1). One of these methods will be discussed in the following paragraphs. Indeed these methods are sophisticated variants of the following very simple algorithm:

$$p^0 \in L^2(\Omega) \text{ is given;} \tag{2.4.3}$$

then for $m \geq 0$, assuming that p^m is known, we compute $\mathbf{u}^m$ and p^{m+1} via

$$\left.\begin{array}{l} \mathbf{u}^m \in V_g \,;\, \forall \mathbf{v} \in V_0 \text{ we have} \\[2mm] \alpha \displaystyle\int_\Omega \mathbf{u}^m \cdot \mathbf{v}\, dx + \nu \int_\Omega \nabla \mathbf{u}^m \cdot \nabla \mathbf{v}\, dx \\[4mm] \qquad = \displaystyle\int_\Omega \mathbf{f} \cdot \mathbf{v}\, dx + \int_\Omega p^m \nabla \cdot \mathbf{v}\, dx + \int_{\Gamma_1} \mathbf{g}_1 \cdot \mathbf{v}\, d\Gamma , \end{array}\right\} \tag{2.4.4}$$

$$p^{m+1} = p^m - \rho \nabla \cdot \mathbf{u}^m . \qquad \square \tag{2.4.5}$$

Problem (2.4.4) is clearly equivalent to

$$\left.\begin{array}{l} \alpha \mathbf{u}^m - \nu \Delta \mathbf{u}^m = \mathbf{f} - \nabla p^m \text{ in } \Omega \\[3mm] \mathbf{u}^m = \mathbf{g}_0 \text{ on } \Gamma_0, \; \nu \dfrac{\partial \mathbf{u}^m}{\partial n} = \mathbf{g}_1 + \mathbf{n} p^m \text{ on } \Gamma_1 . \end{array}\right\} \tag{2.4.6}$$

Concerning now the convergence of algorithm (2.4.3)–(2.4.5), it follows from, e.g., [1, Appendix 3] that if

$$0 < \rho < 2\frac{\nu}{d} \tag{2.4.7}$$

then

$$\lim_{m \to +\infty} \{\mathbf{u}^m, p^m\} \to \{\mathbf{u}, p\} \text{ in } (H^1(\Omega))^d \times L^2(\Omega), \tag{2.4.8}$$

where $\{\mathbf{u}, p\}$ is a solution of (2.4.1).

Algorithm (2.4.3)–(2.4.6) may be slow in practice, particularly for flow at large Reynolds number where $\alpha \sim 1/\Delta t$ is taken very large (to follow the fast dynamics of such flow) and where ν is small.

To improve the speed of convergence, we shall use a *preconditioned conjugate gradient* version of algorithm (2.4.3)–(2.4.5).

2.4.2 A functional equation satisfied by the pressure

Let's define the space P as follows:
 If $\Gamma_0 = \Gamma$, one takes

$$P = L_0^2(\Omega) = \{q | q \in L^2(\Omega), \int_\Omega q \, dx = 0\} \, ; \qquad (2.4.9)$$

if $\int_{\Gamma_1} d\Gamma > 0$, one takes

$$P = L^2(\Omega) \, . \qquad (2.4.10)$$

Next, we define an operator $A: P \to P$ as follows:

(i) *For $q \in P$, solve*

$$\left. \begin{array}{l} \alpha \mathbf{u}_q - \nu \Delta \mathbf{u}_q = -\nabla q \text{ in } \Omega, \\[2mm] \mathbf{u}_q = 0 \text{ on } \Gamma_0, \; \nu \dfrac{\partial \mathbf{u}_q}{\partial n} = \mathbf{n}q \text{ on } \Gamma_1 \, . \end{array} \right\} \qquad (2.4.11)$$

The variational formulation of (2.4.11) is given by

$$\left. \begin{array}{l} \mathbf{u}_q \in V_0; \; \forall \mathbf{v} \in V_0, \text{ we have} \\[2mm] \alpha \displaystyle\int_\Omega \mathbf{u}_q \cdot \mathbf{v} \, dx + \nu \int_\Omega \nabla \mathbf{u}_q \cdot \nabla \mathbf{v} \, dx = \int_\Omega q \nabla \cdot \mathbf{v} \, dx \, . \end{array} \right\} \qquad (2.4.12)$$

(ii) *Define A by*

$$Aq = \nabla \cdot \mathbf{u}_q \, . \qquad (2.4.13)$$

Properties of operator A: Operator A is clearly linear from P into $L^2(\Omega)$. If $\Gamma_0 = \Gamma$, we have more, since

$$\int_\Omega Aq \, dx = \int_\Omega \nabla \cdot \mathbf{u}_q \, dx = \int_\Omega \mathbf{u}_q \cdot \mathbf{n} \, d\Gamma = 0$$

implies that $Aq \in L_0^2(\Omega) = P$.
 Now, taking $\mathbf{v} = \mathbf{u}_{q'}$ in (2.4.12), we observe that

$$\int_\Omega (Aq')q \, dx = \alpha \int_\Omega \mathbf{u}_q \cdot \mathbf{u}_{q'} \, dx + \nu \int_\Omega \nabla \mathbf{u}_q \cdot \nabla \mathbf{u}_{q'} \, dx \, . \qquad (2.4.14)$$

From (2.4.14), operator A is clearly *symmetric* and *positive semi-definite*.
Now, combining (2.4.11) and (2.4.14), we obtain that

$$\int_\Omega (Aq)q\,dx = 0 \Rightarrow q = 0, \qquad (2.4.15)$$

i.e., A is *positive definite* over P.
Indeed, it can be shown that there exists a constant $c > 0$ such that

$$\int_\Omega (Aq)q\,dx \geq c\|q\|^2_{L^2(\Omega)}, \quad \forall q \in P, \qquad (2.4.16)$$

i.e., *operator A is strongly elliptic over P*.
We observe that

$$Aq = -\nabla \cdot (\alpha I - \nu\Delta)^{-1}\nabla q, \qquad (2.4.17)$$

where the boundary conditions associated to the elliptic operator $\alpha I - \nu\Delta$ are those in (2.4.11). $\qquad\square$

Back to problem (2.4.1), we introduce $\mathbf{u}_0 \in V_g$ solution of

$$\left.\begin{array}{l} \alpha\mathbf{u}_0 - \nu\Delta\mathbf{u}_0 = \mathbf{f} \text{ in } \Omega, \\[2mm] \mathbf{u}_0 = \mathbf{g}_0 \text{ on } \Gamma_0, \, \nu\dfrac{\partial\mathbf{u}_0}{\partial n} = \mathbf{g}_1 \text{ on } \Gamma_1, \end{array}\right\} \qquad (2.4.18)$$

and then $\bar{\mathbf{u}} = \mathbf{u} - \mathbf{u}_0$.
We have then

$$\left.\begin{array}{l} \alpha\bar{\mathbf{u}} - \nu\Delta\bar{\mathbf{u}} = -\nabla p \text{ in } \Omega, \\[2mm] \bar{\mathbf{u}} = 0 \text{ on } \Gamma_0, \, \nu\dfrac{\partial\bar{\mathbf{u}}}{\partial n} = \mathbf{n}p \text{ on } \Gamma_1, \end{array}\right\} \qquad (2.4.19)$$

and

$$\nabla \cdot \bar{\mathbf{u}} = \nabla(\mathbf{u} - \mathbf{u}_0) = -\nabla \cdot \mathbf{u}_0. \qquad (2.4.20)$$

It follows from (2.4.19), (2.4.20), and from the definition of A (see (2.4.11), (2.4.13)) that p is the solution of

$$Ap = -\nabla \cdot \mathbf{u}_0. \qquad (2.4.21)$$

A *variational formulation* of (2.4.21) is given by

$$\left.\begin{array}{l} p \in P, \\[2mm] \displaystyle\int_\Omega (Ap)q\,dx = -\int_\Omega \nabla \cdot \mathbf{u}_0 q\,dx, \quad \forall q \in P. \end{array}\right\} \qquad (2.4.22)$$

From the properties of A (*symmetry* and *strong ellipticity*) problem (2.4.21), (2.4.22) can be solved by a conjugate gradient algorithm operating over the space P. Such an algorithm will be discussed in the following Section 2.4.3.

38 E. J. DEAN AND R. GLOWINSKI

Remark 2.4.1 *We should easily show that algorithm (2.4.3)–(2.4.5) is in fact equivalent to*

$$p^0 \ given; \tag{2.4.23}$$

then for $m \geq 0$ we obtain p^{m+1} from p^m by

$$p^{m+1} = p^m - \rho(Ap^m + \nabla \cdot \mathbf{u}_0). \tag{2.4.24}$$

Relation (2.4.24) shows that algorithm (2.4.3)–(2.4.5) is nothing but a gradient method, *with* fixed step ρ, *applied to the solution of problem (2.4.21), (2.4.22).*

2.4.3 Conjugate gradient solution of problem (2.4.21), (2.4.22)

Generalities. A first conjugate gradient algorithm. From the properties of operator A (see Section 2.4.2, above), problem (2.4.21), (2.4.22) is a particular case of the following family of *linear variational problems*

$$\text{Find } u \in V \text{ such that } a(u, v) = L(v), \quad \forall v \in V, \tag{2.4.25}$$

where, in (2.4.25), we have

(i) V is real Hilbert space for the scalar product $(\cdot, \cdot)$ and the corresponding norm $\| \cdot \|$.

(ii) $a : V \times V \to \Re$ is *bilinear, continuous, symmetric,* and *V-elliptic* (the last property means there exists $\beta > 0$ such that $a(v, v) \geq \beta \|v\|^2, \forall v \in V$).

(iii) $L : V \to \Re$ is *linear* and *continuous.*

It follows then from the *Lax-Milgram theorem* (see e.g. [1, Appendix 1]) that problem (2.4.25) has a *unique* solution (symmetry is not required for the existence and uniqueness of a solution of problem (2.4.25)).

Due to the symmetry of a $(\cdot, \cdot)$, problem (2.4.25) is equivalent to the *minimization* problem

$$u \in V,$$
$$J(u) \leq J(v), \forall v \in V, \tag{2.4.26}$$

with $J(v) = \frac{1}{2}a(v, v) - L(v)$.

From the properties of V, $a(\cdot, \cdot)$, $L(\cdot)$, problem (2.4.25), (2.4.26) can be solved by the following conjugate gradient algorithm:

$$u^0 \in V \text{ is given}; \tag{2.4.27}$$

then for $m \geq 0$, solve

$$g^0 \in V,$$
$$(g^0, v) = a(u^0, v) - L(v), \; \forall v \in V, \tag{2.4.28}$$

and set

$$w^0 = g^0. \qquad \square \tag{2.4.29}$$

Then for $m \geq 0$ assuming that u^m, g^m, w^m are known, we compute u^{m+1}, g^{m+1}, w^{m+1} as follows:

Compute

$$\rho_m = \frac{\|g^m\|^2}{a(w^m, w^m)}, \tag{2.4.30}$$

$$u^{m+1} = u^m - \rho_m w^m, \tag{2.4.31}$$

and solve

$$g^{m+1} \in V,$$
$$(g^{m+1}, v) = (g^m, v) - \rho_m a(w^m, v), \; \forall v \in V; \tag{2.4.32}$$

if $\|g^{m+1}\|/\|g^0\| \leq \epsilon$, take $u = u^{m+1}$; if not, compute

$$\gamma_m = \frac{\|g^{m+1}\|^2}{\|g^m\|^2}, \tag{2.4.33}$$

and then

$$w^{m+1} = g^{m+1} + \gamma_m w^m. \qquad \square \tag{2.4.34}$$

Do $m = m + 1$ and go to (2.4.30). $\qquad \square$

Concerning the convergence of algorithm (2.4.27)–(2.4.34), it follows from, e.g., [37] that

$$\|u^m - u\| \leq 2\sqrt{\nu_a}\|u - u^0\| \left(\frac{\sqrt{\nu_a} - 1}{\sqrt{\nu_a} + 1}\right)^m, \tag{2.4.35}$$

where, in (2.4.35), the *condition number* ν_a of the bilinear form $a(\cdot, \cdot)$ is defined by

$$\nu_a = \left(\underset{v \in V - \{0\}}{\mathrm{Sup}} \frac{a(v, v)}{\|v\|^2}\right) / \left(\underset{v \in V - \{0\}}{\mathrm{Inf}} \frac{a(v, v)}{\|v\|^2}\right); \tag{2.4.36}$$

it follows from (2.4.36) that the "closer" $a(\cdot, \cdot)$ is to $(\cdot, \cdot)$, the faster the convergence of algorithm (2.4.27)–(2.4.34).

Application to the solution of problem (2.4.21), (2.4.22):

Since problem (2.4.21), (2.4.22) is a particular case of (2.4.25), it is quite natural to use algorithm (2.4.27)–(2.4.34) to solve it. Since space P is a closed subspace of $L^2(\Omega)$ in the two cases that we are considering (P is then defined either by (2.4.9) or (2.4.10)), it is quite natural to equip P with the canonical scalar product of $L^2(\Omega)$, i.e., $\{q, q'\} \to \int_\Omega qq'\, dx$. With such a scalar product, algorithm (2.4.27)–(2.4.34), applied to the solution of problem (2.4.21), (2.4.22), takes the following form:

$$p^0 \in P \text{ is given;} \tag{2.4.37}$$

solve then

$$\left.\begin{aligned} \alpha \mathbf{u}^0 - \nu \mathbf{u}^0 &= \mathbf{f} - \nabla p^0 \text{ in } \Omega, \\ \mathbf{u}^0 = \mathbf{g}_0 \text{ on } \Gamma_0,\ \nu \frac{\partial \mathbf{u}^0}{\partial n} &= \mathbf{n} p^0 + \mathbf{g}_1 \text{ on } \Gamma_1, \end{aligned}\right\} \tag{2.4.38}$$

and set

$$g^0 = \nabla \cdot \mathbf{u}^0, \tag{2.4.39}$$

$$w^0 = g^0. \qquad \Box \tag{2.4.40}$$

Then, for $m \geq 0$ assuming that p^m, $\mathbf{u}^m$, g^m, w^m are known, compute p^{m+1}, $\mathbf{u}^{m+1}$, g^{m+1}, w^{m+1} as follows:

Solve

$$\left.\begin{aligned} \alpha \bar{\mathbf{u}}^m - \nu \bar{\mathbf{u}}^m &= -\nabla w^m \text{ in } \Omega, \\ \bar{\mathbf{u}}^m = 0 \text{ on } \Gamma_0,\ \nu \frac{\partial \bar{\mathbf{u}}^m}{\partial n} &= \mathbf{n} w^m \text{ on } \Gamma_1, \end{aligned}\right\} \tag{2.4.41}$$

and set

$$\bar{g}^m = \nabla \cdot \bar{\mathbf{u}}^m. \tag{2.4.42}$$

Compute now

$$\rho_m = \frac{\|g^m\|_{L^2(\Omega)}^2}{\int_\Omega w^m \bar{g}^m\, dx} \left(= \frac{\|g^m\|_{L^2(\Omega)}^2}{\int_\Omega (\alpha |\bar{\mathbf{u}}^m|^2 + \nu |\nabla \bar{\mathbf{u}}^m|^2)\, dx} \right), \tag{2.4.43}$$

$$p^{m+1} = p^m - \rho_m w^m, \tag{2.4.44}$$

$$\mathbf{u}^{m+1} = \mathbf{u}^m - \rho_m \bar{\mathbf{u}}^m, \tag{2.4.45}$$

$$g^{m+1} = g^m - \rho_m \bar{g}^m. \tag{2.4.46}$$

If $\|g^{m+1}\|_{L^2(\Omega)}/\|g^0\|_{L^2(\Omega)} \leq \epsilon$ take $\mathbf{u} = \mathbf{u}^{m+1}$, $p = p^{m+1}$; if not, compute

$$\gamma_m = \frac{\|g^{m+1}\|_{L^2(\Omega)}^2}{\|g^m\|_{L^2(\Omega)}^2}, \tag{2.4.47}$$

and update w^m by

$$w^{m+1} = g^{m+1} = \gamma_m w^m . \qquad \square \qquad (2.4.48)$$

Set $m = m + 1$ and go to (2.4.41).

The *implementation* and the *convergence* of algorithm (2.4.37)–(2.4.48) deserve several comments:

(i) The complicated part of the algorithm is step (2.4.41) where we have to solve an elliptic system for the operator $\alpha I - \nu\Delta$. For *large Reynolds number* calculations (where $\alpha \sim 1/\Delta t$ is large and ν small) the discrete analogues of $\alpha I - \nu\Delta$ are well-conditioned making the iterative solution of the associated linear systems fairly fast. We shall go back to this question in Section 2.5.

(ii) We have $\lim_{m \to +\infty}\{u^m, p^m\} = \{u, p\}$ where $\{u, p\}$ is the solution of the Stokes problem (2.4.1), (2.4.2) (such that $\int_\Omega p\,dx = 0$ if $\Gamma_0 = \Gamma$).

(iii) The convergence of algorithm (2.4.37)–(2.4.48) is quite fast if $\frac{\nu}{\alpha} \gg 1$. Unfortunately (because these are the most interesting situations) the performance of the above algorithm deteriorates as $\frac{\nu}{\alpha}$ decreases and is quite bad if $\frac{\alpha}{\nu} \gg 1$. To understand this behavior, let's recall that

$$Aq = -\nabla \cdot (\alpha I - \nu\Delta)^{-1}\nabla q . \qquad (2.4.49)$$

If $\frac{\nu}{\alpha} \gg 1$, we have

$$Aq \sim -\frac{1}{\nu}\nabla \cdot (-\Delta)^{-1}\nabla q ; \qquad (2.4.50)$$

assuming that the various operators in (2.4.50) *commute* (which is not strictly true) we obtain

$$Aq \sim -\frac{1}{\nu}\Delta(-\Delta)^{-1}q = \frac{1}{\nu}q ; \qquad (2.4.51)$$

relations (2.4.51) implies that A behaves spectrally over $L^2(\Omega)$ like a multiple of the identity operator. From this behavior we can expect a fast convergence of algorithm (2.4.37)–(2.4.48) if $\frac{\nu}{\alpha} \gg 1$; numerical experiments justify this prediction. Let's consider now the opposite case, i.e., $\frac{\alpha}{\nu} \gg 1$; the same kind of arguments will show that

$$Aq \sim -\frac{1}{\alpha}\Delta q . \qquad (2.4.52)$$

From (2.4.52) it is quite clear that for $\frac{\alpha}{\nu} \gg 1$, operator A is far from behaving like a multiple of the identity operator and therefore the very slow convergence of algorithm (2.4.37)–(2.4.48) is not surprising in this case (we can expect a condition number of the order of h^{-2} after space discretization). A cure to the deterioration of the convergence properties of algorithm (2.4.37)–(2.4.48) will be given in the next paragraph.

*A quasi-optimal preconditioning for the conjugate gradient solution of problem (2.4.21),
(2.4.22).* Let's define $B \in \mathcal{L}(P,P)$ by

$$Bq = \alpha\varphi_q + \nu q \tag{2.4.53}$$

where φ_q is the solution of

$$-\Delta\varphi_q = q \text{ in } \Omega, \quad \frac{\partial\varphi_q}{\partial n} = 0 \text{ on } \Gamma, \quad \int_\Omega \varphi_q \, dx = 0 \tag{2.4.54}$$

if $\Gamma_0 = \Gamma$, and of

$$-\Delta\varphi_q = q \text{ in } \Omega, \quad \frac{\partial\varphi_q}{\partial n} = 0 \text{ on } \Gamma_0, \quad \varphi_q = 0 \text{ on } \Gamma_1 \tag{2.4.55}$$

if $\int_{\Gamma_1} d\Gamma > 0$.

We prove now the following

Theorem 2.4.1 *Operator B is continuous, self-adjoint and strongly elliptic from P onto
P.*

Proof: To prove the above results, we can, for example, prove that the bilinear form
$b(\cdot,\cdot)$ defined by

$$b(q,q') = \int_\Omega (Bq)q' \, dx \tag{2.4.56}$$

is continuous, symmetric and P-elliptic. From (2.4.53), (2.4.54) or (2.4.55), and from
Green's formula, we have

$$\begin{aligned}
b(q,q') &= \alpha \int_\Omega \varphi_q q' \, dx + \nu \int_\Omega qq' \, dx \\
&= -\alpha \int_\Omega \varphi_q \Delta\varphi_{q'} \, dx + \nu \int_\Omega qq' \, dx \\
&= \alpha \int_\Omega \nabla\varphi_q \cdot \nabla\varphi_{q'} \, dx + \nu \int_\Omega qq' \, dx - \alpha \int_\Gamma \frac{\partial\varphi_{q'}}{\partial n}\varphi_q \, d\Gamma \\
&= \alpha \int_\Omega \nabla\varphi_q \cdot \nabla\varphi_{q'} \, dx + \nu \int_\Omega qq' \, dx, \quad \forall q,q' \in P.
\end{aligned} \tag{2.4.57}$$

Relation (2.4.57) shows that the bilinear form $b(\cdot,\cdot)$ is symmetric; its continuity is
obvious from the fact that the linear mapping $q \to \varphi_q$ is continuous from P into $H^1(\Omega)$
implying the continuity of the bilinear form

$$\{q,q'\} \to \int_\Omega \nabla\varphi_q \cdot \nabla\varphi_{q'} \, dx$$

Finally, $b(\cdot, \cdot)$ is P-elliptic since

$$b(q, q) \geq \nu \int_{\Omega} |q|^2 \, dx, \quad \forall q \in P. \qquad \square$$

The properties of operator B imply that it has an inverse, that we shall denote by S, which is also *self-adjoint symmetric* and *strongly-elliptic* over the space P. If we denote by $s(\cdot, \cdot)$ the bilinear form defined by

$$s(q, q') = \int_{\Omega} (Sq) q' \, dx \qquad (2.4.58)$$

it defines over P a scalar product equivalent to the canonical one. From now on, we shall equip space P with the bilinear form $s(\cdot, \cdot)$ as scalar product; in particular, we shall describe below the variant of the conjugate gradient algorithm (2.4.37)–(2.4.48) obtained by replacing the canonical scalar product of $L^2(\Omega)$ by $s(\cdot, \cdot)$.

From the results of various numerical experiments let us mention that the operator S (in fact, its discrete analogue) has excellent preconditioning properties making the convergence of the conjugate gradient algorithm described in the next section *uniformly fast*, for a given geometry, when $\frac{\alpha}{\nu}$ varies from 0 to $+\infty$.

The preconditioning operator S was introduced by J. Cahouet (see [38]) in the mid-eighties for the case where $\Gamma = \Gamma_0$ (pure Dirichlet boundary condition for the velocity); the extension to the case where $\int_{\Gamma_1} d\Gamma > 0$ is – to our knowledge – due to the present author.

To conclude this paragraph let's show why one could anticipate good preconditioning properties for operator S:

Suppose for simplicity that $\Omega = (0, 1)^d$; we consider the Stokes problem

$$\left. \begin{array}{l} \alpha \mathbf{u} - \nu \Delta \mathbf{u} + \nabla p = \mathbf{f} \text{ in } \Omega, \\[4pt] \nabla \cdot \mathbf{u} = 0 \text{ in } \Omega, \\[4pt] \mathbf{u}, \nabla \mathbf{u} \text{ and } p \text{ are periodic at } \Gamma. \end{array} \right\} \qquad (2.4.59)$$

In the present context we say that a function v is periodic at Γ if

$$\left. \begin{array}{l} v(x_1, \ldots, x_{i-1}, 0, x_{i+1}, \ldots, x_d) = v(x_1, \ldots, x_{i-1}, 1, x_{i+1}, \ldots, x_d), \\[4pt] \forall i = 1, \ldots, d, \quad \forall x_j \in (0, 1), \quad \forall j = 1, \ldots, d, \quad j \neq i. \end{array} \right\} \qquad (2.4.60)$$

Solving problem (2.4.59) is trivial here since we obviously have

$$\left. \begin{array}{l} \Delta p = \nabla \cdot \mathbf{f} \text{ in } \Omega, \\[4pt] p \text{ and } \nabla p \text{ perodic at } \Gamma; \end{array} \right\} \qquad (2.4.61)$$

it can be shown that problem (2.4.61) has a unique solution in $H^1(\Omega)/\Re$, if $\mathbf{f}$ is sufficiently smooth. Next we solve

$$\left.\begin{array}{l} \alpha\mathbf{u} - \nu\Delta\mathbf{u} = \mathbf{f} - \nabla p \text{ in } \Omega, \\[4pt] \mathbf{u} \text{ and } \nabla\mathbf{u} \text{ perodic at } \Gamma, \end{array}\right\} \tag{2.4.62}$$

which has a unique solution. Finally, if we denote $\nabla \cdot \mathbf{u}$ by φ, then φ satisfies (from (2.4.61), (2.4.62))

$$\left.\begin{array}{l} \alpha\varphi - \nu\Delta\varphi = 0 \text{ in } \Omega, \\[4pt] \varphi \text{ and } \nabla\varphi \text{ perodic at } \Gamma; \end{array}\right\} \tag{2.4.63}$$

whose unique solution is $\varphi = 0 \Rightarrow \nabla \cdot \mathbf{u} = 0$ on Ω. Consider now the operator A associated to problem (2.4.59); it is defined by

$$Aq = \nabla \cdot \mathbf{u}_q, \tag{2.4.64}$$

with

$$\left.\begin{array}{l} \alpha\mathbf{u}_q - \nu\Delta\mathbf{u}_q = -\nabla q \text{ in } \Omega, \\[4pt] \mathbf{u}_q \text{ and } \nabla\mathbf{u}_q \text{ perodic at } \Gamma; \end{array}\right\} \tag{2.4.65}$$

i.e.,

$$Aq = -\nabla \cdot (\alpha I - \nu\Delta)^{-1}\nabla q. \tag{2.4.66}$$

Now we define operator B by

$$Bq = \alpha\varphi_q + \nu q, \tag{2.4.67}$$

where

$$\left.\begin{array}{l} -\Delta\varphi_q = q \text{ in } \Omega, \\[4pt] \varphi_q, \nabla\varphi_q \text{ periodic at } \Gamma; \quad \displaystyle\int_\Omega \varphi_q \, dx = 0. \end{array}\right\} \tag{2.4.68}$$

Evaluate now ABq; we have

$$ABq = -\nabla \cdot ((\alpha I - \nu\Delta)^{-1}\nabla(\alpha\varphi_q + \nu q)). \tag{2.4.69}$$

From the *periodic boundary conditions*, the various operators involved in the right hand side of (2.4.69) *commute* implying (from (2.4.68)) that, $\forall q$, we have

$$\begin{aligned} ABq &= -\nabla \cdot ((\alpha I - \nu\Delta)^{-1}\nabla(\alpha I - \nu\Delta)\varphi_q) \\ &= (\alpha I - \nu\Delta)^{-1}(\alpha I - \nu\Delta)(-\Delta)\varphi_q = -\Delta\varphi_q = q, \end{aligned} \tag{2.4.70}$$

implying in turn that

$$B = A^{-1} \tag{2.4.71}$$

it is clear that in that case $S = B^{-1} = A$ is the *optimal preconditioner* (for those readers which are not convinced by (2.4.70), we mention that the above result can be proved by *Fourier Analysis*, as shown in [38]).

Preconditioned conjugate gradient solution of problem (2.4.21), (2.4.22). In this paragraph we discuss the solution of problem (2.4.21), (2.4.22) (in fact of problem (2.4.1), (2.4.2)) by a conjugate gradient algorithm preconditioned by the operator S defined in the previous paragraphs.

We consider the case where $\int_{\Gamma_1} d\Gamma > 0$.

Description of the algorithm:

$$p^0 \in P \text{ is given}; \tag{2.4.72}$$

solve

$$\left. \begin{aligned} \alpha \mathbf{u}^0 - \nu \Delta \mathbf{u}^0 &= \mathbf{f} - \nabla p^0 \text{ in } \Omega, \\ \mathbf{u}^0 = \mathbf{g}_0 \text{ on } \Gamma_0, \quad \nu \frac{\partial \mathbf{u}}{\partial n} &= \mathbf{g}_1 + \mathbf{n} p^0 \text{ on } \Gamma_1, \end{aligned} \right\} \tag{2.4.73}$$

and set

$$r^0 = \nabla \cdot \mathbf{u}^0. \tag{2.4.74}$$

Solve now

$$\left. \begin{aligned} - \Delta \varphi^0 &= r^0 \text{ in } \Omega, \\ \frac{\partial \varphi^0}{\partial n} = 0 \text{ on } \Gamma_0, \quad \varphi^0 &= 0 \text{ on } \Gamma_1, \end{aligned} \right\} \tag{2.4.75}$$

and set

$$g^0 = \nu r^0 + \alpha \varphi^0, \tag{2.4.76}$$

$$w^0 = g^0. \qquad \square \tag{2.4.77}$$

Then for $m \geq 0$ assuming that p^m, $\mathbf{u}^m$, r^m, g^m, w^m are known, compute p^{m+1}, $\mathbf{u}^{m+1}$, r^{m+1}, g^{m+1}, w^{m+1} as follows:

Solve

$$\left. \begin{aligned} \alpha \bar{\mathbf{u}}^m - \nu \Delta \bar{\mathbf{u}}^m &= -\nabla w^m \text{ in } \Omega, \\ \bar{\mathbf{u}}^m = 0 \text{ on } \Gamma_0, \quad \nu \frac{\partial \bar{\mathbf{u}}^m}{\partial n} &= \mathbf{n} w^m \text{ on } \Gamma_1, \end{aligned} \right\} \tag{2.4.78}$$

and set

$$\bar{r}^m = \nabla \cdot \bar{\mathbf{u}}^m. \tag{2.4.79}$$

Compute

$$\rho_m = \frac{\int_\Omega r^m g^m \, dx}{\int_\Omega \bar{r}^m w^m \, dx} \tag{2.4.80}$$

and then

$$p^{m+1} = p^m - \rho_m w^m, \tag{2.4.81}$$

$$\mathbf{u}^{m+1} = \mathbf{u}^m - \rho_m \bar{\mathbf{u}}^m, \tag{2.4.82}$$

$$r^{m+1} = r^m - \rho_m \bar{r}^m. \tag{2.4.83}$$

Solve, next,

$$\left. \begin{array}{l} -\Delta\bar{\varphi}^m = \bar{r}^m \text{ in } \Omega, \\[2mm] \dfrac{\partial\bar{\varphi}^m}{\partial n} = 0 \text{ on } \Gamma_0, \quad \bar{\varphi}^m = 0 \text{ on } \Gamma_1, \end{array} \right\} \tag{2.4.84}$$

and compute

$$g^{m+1} = g^m - \rho_m(\nu\bar{r}^m + \alpha\bar{\varphi}^m). \tag{2.4.85}$$

If $\int_\Omega r^{m+1} g^{m+1}\, dx / \int_\Omega r^0 g^0\, dx \leq \epsilon$, *take* $p = p^{m+1}$, $\mathbf{u} = \mathbf{u}^{m+1}$; *if not compute*

$$\gamma_m = \frac{\int_\Omega r^{m+1} g^{m+1}\, dx}{\int_\Omega r^m g^m\, dx} \tag{2.4.86}$$

and then

$$w^{m+1} = g^{m+1} + \gamma_m w^m. \qquad \square \tag{2.4.87}$$

Set $m = m + 1$ *and go back to* (2.4.78).

Algorithm (2.4.72)–(2.4.87) has proved to be quite efficient for solving Navier-Stokes problems on a quite large range of Reynolds numbers. To conclude this paragraph we shall make the following remarks:

Remark 2.4.2 *In the case where* $\Gamma_0 = \Gamma$, *we should replace* (2.4.73), (2.4.75), (2.4.78), (2.4.84) *by*

$$\left. \begin{array}{l} \alpha\mathbf{u}^0 - \nu\Delta\mathbf{u}^0 = \mathbf{f} - \nabla p^0 \text{ in } \Omega, \\[2mm] \mathbf{u}^0 = \mathbf{g}_0 \text{ on } \Gamma, \end{array} \right\} \tag{2.4.88}$$

$$\left. \begin{array}{l} -\Delta\varphi^0 = r^0 \text{ in } \Omega, \\[2mm] \dfrac{\partial\varphi^0}{\partial n} = 0 \text{ on } \Gamma, \quad \displaystyle\int_\Omega \varphi^0\, dx = 0, \end{array} \right\} \tag{2.4.89}$$

$$\left. \begin{array}{l} \alpha\bar{\mathbf{u}}^m - \nu\Delta\bar{\mathbf{u}}^m = -\nabla w^m \text{ in } \Omega, \\[2mm] \bar{\mathbf{u}}^m = 0 \text{ on } \Gamma, \end{array} \right\} \tag{2.4.90}$$

$$\left. \begin{array}{l} -\Delta\bar{\varphi}^m = \bar{r}^m \text{ in } \Omega, \\[2mm] \dfrac{\partial\bar{\varphi}^m}{\partial n} = 0 \text{ on } \Gamma, \quad \displaystyle\int_\Omega \bar{\varphi}^m\, dx = 0, \end{array} \right\} \tag{2.4.91}$$

respectively. $\square$

Remark 2.4.3 *Each iteration of algorithm (2.4.72)–(2.4.87) requires the solution of one elliptic system for the operator $\alpha I - \nu\Delta$. As already mentioned, for flow at large Reynolds number where $\alpha \sim 1/\Delta t$ is large and ν is small the discrete analogues of operator $\alpha I - \nu\Delta$ are fairly well conditioned matrices making the iterative solution of these elliptic systems quite inexpensive. We also have to solve the Poisson problems (2.4.75), (2.4.84) (or (2.4.89), (2.4.91)); we shall discuss this aspect of the numerical implementation in the following Section 2.5.*

2.5 Finite element approximation of the Navier-Stokes equations

2.5.1 Generalities, Synopsis

We have discussed in Section 2.2 the time discretization of the Navier-Stokes equations (2.1.1), (2.1.2) coupled to convenient initial and boundary conditions. To achieve the computer implementation of the solution methods described in the above sections we still have to address the *space discretization issue*; in this article, we shall focus on *finite element methods*. There exists quite a large literature concerning the finite element approximation of the Navier-Stokes equations (indeed, every issue of the *International Journal of Numerical Methods in Fluids* contains at least one article on these topics); concentrating on books we shall mention [1], [8]–[14].

It is a fairly general opinion that the main difficulty related to the space approximation of the Navier-Stokes equations, in the *pressure-velocity formulation*, is the treatment of the *incompressibility condition* $\nabla \cdot \mathbf{u} = 0$. Actually, we have seen in Section 2.4.3 that solving the *periodic Stokes problem*, namely

$$\left.\begin{array}{l} \alpha\mathbf{u} - \nu\Delta\mathbf{u} + \nabla p = \mathbf{f} \text{ in } \Omega, \\ \nabla \cdot \mathbf{u} = 0 \text{ in } \Omega, \\ \mathbf{u}, \nabla\mathbf{u}, p \text{ periodic at } \Gamma, \end{array}\right\} \tag{2.5.1}$$

when $\Omega = (0,1)^N$, is quite easy; we recall that we compute first the *pressure* p from

$$\left.\begin{array}{l} \Delta p = \nabla \cdot \mathbf{f} \text{ in } \Omega, \\ p, \nabla p \text{ periodic at } \Gamma, \end{array}\right\} \tag{2.5.2}$$

and then, the *velocity* $\mathbf{u}$ from

$$\left.\begin{array}{l} \alpha\mathbf{u} - \nu\Delta\mathbf{u} = \mathbf{f} - \nabla p \text{ in } \Omega, \\ \mathbf{u}, \nabla\mathbf{u} \text{ periodic at } \Gamma. \end{array}\right\} \tag{2.5.3}$$

Assuming that $\mathbf{f}$ is sufficiently smooth, problems (2.5.2), (2.5.3) are well-posed in $H^1(\Omega)/\Re$ and $(H^1(\Omega))^d$; the vector field $\mathbf{u}$ is *automatically divergence-free* as shown in

Section 2.4.3. *Variational formulations* of problems (2.5.2), (2.5.3) are given by

$$\left. \begin{aligned} & p \in H^1_P(\Omega) \\ & \int_\Omega \nabla p \cdot \nabla q \, dx = \int_\Omega \mathbf{f} \cdot \nabla q \, dx, \quad \forall q \in H^1_P(\Omega), \end{aligned} \right\} \tag{2.5.4}$$

$$\left. \begin{aligned} & \mathbf{u} \in (H^1_P(\Omega))^d; \quad \forall \mathbf{v} \in (H^1_P(\Omega))^d, \text{ we have} \\ & \alpha \int_\Omega \mathbf{u} \cdot \mathbf{v} \, dx + \nu \int_\Omega \nabla \mathbf{u} \cdot \nabla \mathbf{v} \, dx = \int_\Omega \mathbf{f} \cdot \mathbf{v} \, dx + \int_\Omega p \nabla \cdot \mathbf{v} \, dx, \end{aligned} \right\} \tag{2.5.5}$$

respectively, with in (2.5.5), (2.5.4), $H^1_P(\Omega)$ defined by

$$H^1_P(\Omega) = \{ q | q \in H^1(\Omega), \; q \text{ periodic at } \Gamma \} . \tag{2.5.6}$$

Solving problem (2.5.1), by *Galerkin type methods*, via the equivalent variational formulation (2.5.4), (2.5.5) is quite easy. We introduce first two families $\{P_h\}_h$ and $\{V_h\}_h$ of *finite dimensional spaces*; we suppose that these families satisfy

$$P_h \subset H^1_P(\Omega), \quad \forall h, \quad V_h \subset (H^1_P(\Omega))^d, \quad \forall h, \tag{2.5.7}$$

$$\forall q \in H^1_P(\Omega), \exists \{q_h\}_h, \text{ such that } q_h \in P_h, \forall h \text{ and } \lim_{h \to 0} \| q_h - q \|_{H^1(\Omega)} = 0, \tag{2.5.8}$$

$$\forall \mathbf{v} \in (H^1_P(\Omega))^d, \exists \{\mathbf{v}_h\}_h, \text{ such that } \mathbf{v}_h \in V_h, \forall h \text{ and } \lim_{h \to 0} \| \mathbf{v}_h - \mathbf{v} \|_{(H^1_P(\Omega))^d} = 0. \tag{2.5.9}$$

Starting from the variational formulation (2.5.4), (2.5.5), we approximate problem (2.5.1) by

$$\left. \begin{aligned} & p_h \in P_h, \\ & \int_\Omega \nabla p_h \cdot \nabla q_h \, dx = \int_\Omega \mathbf{f}_h \cdot \nabla q_h \, dx, \quad \forall q_h \in P_h, \end{aligned} \right\} \tag{2.5.10}$$

$$\left. \begin{aligned} & \mathbf{u}_h \in V_h, \\ & \alpha \int_\Omega (\alpha \mathbf{u}_h \cdot \mathbf{v}_h + \nu \nabla \mathbf{u}_h \cdot \nabla \mathbf{v}_h) \, dx \\ & \qquad\qquad = \int_\Omega \mathbf{f}_h \cdot \mathbf{v}_h \, dx + \int_\Omega p_h \nabla \cdot \mathbf{v}_h \, dx, \quad \forall \mathbf{v}_h \in V_h, \end{aligned} \right\} \tag{2.5.11}$$

where, in (2.5.10), (2.5.11), $\mathbf{f}_h$ is an approximation of $\mathbf{f}$ such that $\lim_{h \to 0} \| \mathbf{f} - \mathbf{f}_h \|_{(L_2(\Omega))^d} = 0$.

It is a fairly easy exercise to prove that problems (2.5.10) and (2.5.11) are *well-posed* in $P_h/\Re$ and V_h, respectively, and also that $\lim_{h \to 0} \{\mathbf{u}_h, p_h\} = \{\mathbf{u}, p\}$ in $(H^1(\Omega))^{d+1}$, where $\{\mathbf{u}, p\}$ is a solution of (2.5.1).

From the above results it appears that approximating the periodic Stokes problem (2.5.1) is a rather simple issue. Indeed, we can combine any pressure approximation to any velocity one, as long as properties (2.5.8), (2.5.9) are satisfied; thus pressure and velocity approximations can be of different nature, use different grid and/or basis functions, etc.... On the other hand, approximating the Stokes-Neumann and/or Dirichlet problem (2.4.1) is a much more complicated matter, since compatibility conditions between the velocity and pressure approximations seem to be required if one wants to avoid spurious oscillations. In the next paragraph (Section 2.5.2) we shall take advantage of a particular case with a simple geometry to study, via Fourier Analysis, the mechanism producing numerical instabilities, and then suggest some cures for this unwanted phenomenon. In Section 2.5.3 we shall apply the results of Section 2.5.2 to the full discretization of the Navier-Stokes equations.

2.5.2 A Fourier Analysis of the numerical instability mechanism

In order to analyze the instability mechanism associated to the "naive" approximations of the Stokes problem we shall focus our attention on the *Stokes-Dirichlet problem*, namely

$$\left.\begin{array}{l} \alpha \mathbf{u} - \nu \Delta \mathbf{u} + \nabla p = \mathbf{f} \text{ in } \Omega, \\[4pt] \nabla \cdot \mathbf{u} = 0 \text{ in } \Omega, \\[4pt] \mathbf{u} = \mathbf{g} \text{ on } \Gamma, \text{ with } \int_\Gamma \mathbf{g} \cdot \mathbf{n} \, d\Gamma = 0, \end{array}\right\} \tag{2.5.12}$$

in the particular case where $\Omega = (0,1) \times (0,1)$.

Following Section 2.4.2, we observe that the *unique* pressure p in (2.5.12), such that $\int_\Omega p \, dx = 0$, is the solution of the functional equation

$$Ap = -\nabla \cdot \mathbf{u}_0, \tag{2.5.13}$$

where $\mathbf{u}_0$ is the solution of the following Dirichlet system

$$\left.\begin{array}{l} \alpha \mathbf{u}_0 - \nu \Delta \mathbf{u}_0 = \mathbf{f} \text{ in } \Omega, \\[4pt] \mathbf{u}_0 = \mathbf{g} \text{ on } \Gamma, \end{array}\right\} \tag{2.5.14}$$

and where A is defined by

$$Aq = -\nabla \cdot \left((\alpha I - \nu \Delta)^{-1} \nabla q \right) ; \tag{2.5.15}$$

the boundary conditions associated to $\alpha I - \nu \Delta$ in (2.5.15) are the *homogeneous Dirichlet* ones.

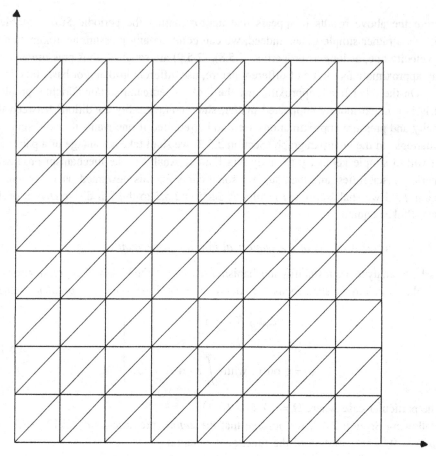

Figure 2.2: Regular triangulations obtained by the finite element method.

We shall discuss now the approximation of operator A:

To the integer I we associate $h = 1/(I+1)$ and introduce the grid points $M_{ij} = \{ih, jh\}$, $0 \le i, j \le I+1$; the points M_{ij} can be used to define either *finite difference* or *finite element* approximations of problems (2.5.12) and (2.5.13). For further simplicity we shall consider finite difference approximations but the following discussion could have been done in a *finite element* framework, using *piecewise linear* approximations associated to the triangulation of Figure 2.2, and the *trapezoidal rule* to evaluate integrals like $\alpha \int_\Omega vw\, dx$ in the variational formulations of the elliptic systems – associated to $\alpha I - \nu\Delta$ – encountered in the definition of $\mathbf{u}_0$ and A.

The *pressure* p will be approximated by $p_h = \{p_{ij}\}_{0 \le i,j \le I+1}$ and those *velocity* fields

$\mathbf{v}$ vanishing on Γ by $\mathbf{v}_h = \{\mathbf{v}_{ij}\}_{1 \le i,j \le I}$, with $\mathbf{v}_{ij} \in \Re^2$. We define the discrete pressure and velocity spaces P_h and V_{0h} by

$$P_h = \{q_h | q_h = \{q_{ij}\}, \quad 0 \le i,j \le I+1\}, \tag{2.5.16}$$

and

$$V_{0h} = \{\mathbf{v}_h | \mathbf{v}_h = \{\mathbf{v}_{ij}\}, \quad \mathbf{v}_{ij} \in \Re^2, \quad 1 \le i,j \le I\}. \tag{2.5.17}$$

To study the *kernel* and the *damping* properties of the discrete analogue of operator A it is convenient to introduce the following *vector bases* of P_h and V_{0h}:

$$\mathcal{B}_{ph} = \{\varphi_{mn} | \varphi_{mn} = \{\cos mi\pi h \times \cos nj\pi h\}_{0 \le i,j \le I+1}, 0 \le m,n \le I+1\}, \tag{2.5.18}$$

$$\begin{aligned} \mathcal{B}_{vh} = &\{\{\sin mi\pi h \times \sin nj\pi h, 0\}_{1 \le i,j \le I}, 1 \le m,n \le I\} \cup \\ &\{\{0, \sin ki\pi h \times \sin lj\pi h\}_{1 \le i,j \le I}, 1 \le k,l \le I\}, \end{aligned} \tag{2.5.19}$$

respectively.

The convenience of the above bases is due to the fact that their elements are the *eigenvalues* of the matrices which approximate (via finite difference discretizations) the elliptic operator $-\Delta$ for the homogeneous Neumann and Dirichlet boundary conditions, respectively. The finite difference method to be described below is not used in practice (until recently, at least) since it is known to be *unstable*. However, since the corresponding discretization is very close to the one obtained by finite element methods using regular triangulations such as the one in Figure 2.2, and on these triangulations piecewise linear approximations for both pressure and velocity, we shall consider it in detail. Indeed, the crucial part is the way $(\alpha I - \nu\Delta)^{-1}\nabla p$ is approximated:

Consider $p_h \in P_h$; we approximate ∇p at M_{ij} by

$$(\delta p)_{ij} = \left\{ \frac{p_{i+1j} - p_{i-1j}}{2h}, \frac{p_{ij+1} - p_{ij-1}}{2h} \right\}, \quad 1 \le i,j \le I. \tag{2.5.20}$$

If we denote by $\mathbf{w}_h$ the element of V_{0h} approximating $(\alpha I - \nu\Delta)^{-1}\nabla p$ we shall obtain it via the solution of the following linear system (with $\mathbf{w}_{kl} = 0$ if $M_{kl} \in \Gamma$)

$$\alpha\mathbf{w}_{ij} - \frac{\nu}{h^2}(\mathbf{w}_{i+1j} + \mathbf{w}_{i-1j} + \mathbf{w}_{ij+1} + \mathbf{w}_{ij-1} - 4\mathbf{w}_{ij}) = (\delta p)_{ij}, \quad 1 \le i,j \le I. \tag{2.5.21}$$

To study the properties of the mapping

$$p_h \rightarrow \mathbf{w}_h : P_h \rightarrow V_{0H}, \tag{2.5.22}$$

we consider the particular case where $p_h = \varphi_{mn} \in \mathcal{B}_{ph}$; the corresponding value of $(\delta p)_{ij}$, denoted by $(\delta\varphi_{mn})_{ij}$ is then given by

$$(\delta\varphi_{mn})_{ij} = -\left\{ \frac{\sin m\pi h}{h} \sin mi\pi h \times \cos nj\pi h, \frac{\sin n\pi h}{h} \cos mi\pi h \times \sin nj\pi h \right\}.$$

(2.5.23)

If $m = n = I + 1$, we clearly have $(\delta\varphi_{mn})_{ij} = 0$, $\forall 1 \leq i,j \leq I$; indeed, relation (2.5.23) tells us more since it follows from Figure 2.3, below (where we have visualized the function $m\pi \to \frac{\sin m\pi h}{h}$), that for $m, n > \frac{I+1}{2}$, the vectors φ_{mn} are *strongly damped* by the finite difference approximation of ∇ defined by (2.5.20). If we consider now the matrix in the left hand side of (2.5.21), it is quite easy to check that the *eigenvectors* of this matrix are either

$$\{\sin mi\pi h \times \sin nj\pi h, 0\}_{1 \leq i,j \leq I}, \quad 1 \leq m,n \leq I,$$

(2.5.24)

or

$$\{0, \sin mi\pi h \times \sin nj\pi h\}_{1 \leq i,j \leq I}, \quad 1 \leq m,n \leq I,$$

(2.5.25)

the corresponding eigenvalues being

$$\alpha + \frac{4\nu}{h^2}(\sin^2 m\frac{\pi}{2}h + \sin^2 n\frac{\pi}{2}h).$$

(2.5.26)

Since $\mathbf{w}_h$ is obtained by multiplying the right hand side of (2.5.21) by the inverse of the above matrix, we observe from (2.5.26) that the damping of the high wave number modes of p_h associated to the discretization of ∇ is further amplified; actually, the traditional finite difference discretizations of the divergence operator have a similar behavior.

To summarize the above analysis we can say that the pressure modes such that $m, n > \frac{I+1}{2}$ are strongly damped by the discrete analogue of the operator A defined by (2.5.15); this property implies that spurious pressure and velocity oscillations are produced if one relies on the above approach to solve the Stokes problem (2.5.12), via the pressure equation (2.5.13). Actually, the finite element approximations of (2.5.12) which use the same mesh and type of finite elements for pressure and velocity suffer from the same drawback that the finite difference method which has been described above (we insist on the fact that this method is essentially equivalent to a finite element one using piecewise linear approximations for both pressure and velocity on triangulations such as the one in Figure 2.3). To overcome the above instability we can either

(a) *Use different types of approximations for pressure and velocity*

or

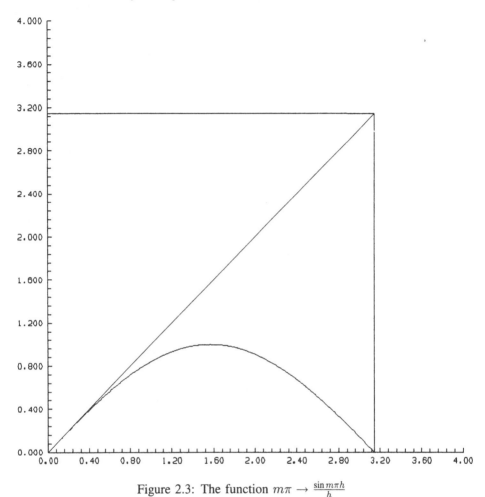

Figure 2.3: The function $m\pi \rightarrow \frac{\sin m\pi h}{h}$

(b) *Use the same type of approximation for pressure and velocity, combined with a regularization procedure.*

Approach (a) is well known and is related to the so-called *inf-sup condition*; finite element approximations which satisfy it are discussed in e.g. [1]–[8], [11], [13], [14]; the main idea here is to construct pressure spaces which are "poor" in high frequency modes, compared to the velocity space. Figure 2.3 suggests an obvious remedy which is to use a pressure grid which is *twice coarser* than the velocity one, and then use approximations of the same type on both grids. This observation makes sense for finite difference,

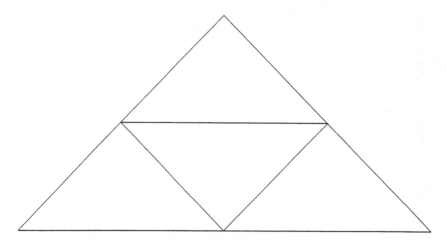

Figure 2.4: $\mathcal{T}_{h/2}$ obtained by joining midpoints in $T \in \mathcal{T}_h$.

finite element, spectral, pseudo-spectral, and wavelet approximation of problem (2.5.12); the well-known (and converging) finite element method obtained by using a *continuous piecewise linear* approximation of the *pressure* (resp. of the *velocity*) on a triangulation $\mathcal{T}_h$ (resp. $\mathcal{T}_{h/2}$, obtained from $\mathcal{T}_h$ by joining as shown in Figure 2.4 the midpoints in any $T \in \mathcal{T}_h$) definitely follows the above rule. This method is discussed in e.g. [1]–[7], [11], [13], [14] (some of the above references show numerical results obtained with it).

Approach (b) which has been recently strongly advocated by several authors (see, e.g., [39]), leads essentially to *Tychonoff regularization procedures*, an obvious one being to "regularize" equation (2.5.13) by the following problem (written in *variational* form):

$$\left. \begin{aligned} &p_\epsilon \in \mathrm{H}^1(\Omega)\,, \\ &\epsilon \int_\Omega \nabla p_\epsilon \cdot \nabla q \, dx + \int_\Omega (A p_\epsilon) q \, dx = \int_\Omega \nabla \cdot \mathbf{u}_0 q \, dx, \quad \forall q \in \mathrm{H}^1(\Omega)\,, \end{aligned} \right\} \quad (2.5.27)$$

where, in (2.5.27), ϵ is a *positive* parameter.

Very good results have been obtained with approach (b) (see [39]), however, we prefer approach (a), for the following reasons:

(i) It is *parameter free*, unlike the second approach which requires the adjustment of the regularization parameter.

(ii) In general, the mesh size is adjusted, globally or locally, on the basis of the velocity behavior (boundary and shear layer thicknesses, for example). Therefore, compared to approach (a), approach (b) will be 4 times more costly (8 times for three dimensional problems) from the pressure point of view, without further gains in accuracy.

(iii) Multilevel solvers have been recently developed for Stokes problems; since methods of type (a) have also a multilevel structure concerning the approximation of pressure and velocity, we think that they are better suited than those of type (b) for multilevel solution methods such as multigrid.

(iv) Tychonoff regularization procedures are systematic methods for stabilizing ill-posed problems; in most cases, the adjustment of the regularization parameter is a delicate problem in itself, therefore, if there exist alternatives which are parameter free, we definitely think that the latter are preferable, particularly if they are based on an analysis of the mechanism producing the unwanted oscillations. Actually, we have nothing against regularization procedures since we have been using them, and are still using them, to solve *control problems* for the *wave equation* (see [6], [40]) and for the *Euler-Bernouilli equation* modelling beam and plate motions. However, as a result of the present analysis, we have introduced in [41], new solution methods for some of the above control problems, which are more efficient than those discussed in [6], [40].

2.5.3 *Finite element methods for the time dependent Navier-Stokes equations*

We shall describe in this paragraph a specific class of finite element approximations for the time dependent Navier-Stokes equations (2.1.1)–(2.1.3) completed by convenient boundary conditions. Actually, these methods, which lead to continuous approximations for pressure and velocity, are fairly simple; most of them have been known for years and are quite popular in the finite element oriented part of the Computational Fluid Dynamics community. They have been advocated, for example, by Hood and Taylor (see [42]) and follow the general approximation principles discussed in Section 2.5.2. Other finite element approximations for the incompressible Navier-Stokes equations can be found in, e.g., [1], [8], [10], [11], [13], [14], [39], [43] (see also the references therein).

Basic hypotheses. Fundamental discrete spaces. We suppose that Ω is a bounded polygonal domain of $\Re^2$. With $\mathcal{T}_h$ a standard finite element *triangulation* of Ω (see, e.g. [44], [45] for this notion), and h the maximal length of the edges of $\mathcal{T}_h$, we introduce the following discrete spaces (with P_k =space of the polynomials in two variables of degree $\leq k$):

$$P_h = \left\{ q_h | q_h \in C^0(\overline{\Omega}), \quad q_h|_T \in P_1, \quad \forall T \in \mathcal{T}_h \right\}, \tag{2.5.28}$$

$$V_h = \left\{ \mathbf{v}_h | \mathbf{v}_h \in C^0(\overline{\Omega}) \times C^0(\overline{\Omega}), \quad \mathbf{v}_h|_T \in P_2 \times P_2, \quad \forall T \in \mathcal{T}_h \right\}. \tag{2.5.29}$$

If the boundary conditions imply $\mathbf{u} = \mathbf{g}_0$ on Γ_0, we shall need the space V_{0h} defined by

$$V_{0h} = \{\mathbf{v}_h | \mathbf{v}_h \in V_h, \quad \mathbf{v}_h = 0 \text{ on } \Gamma\}, \text{ if } \Gamma_0 = \Gamma, \tag{2.5.30a}$$

and by

$$V_{0h} = \{\mathbf{v}_h | \mathbf{v}_h \in V_h, \quad \mathbf{v}_h = 0 \text{ on } \Gamma_0\}, \text{ if } \int_{\Gamma_1} d\Gamma > 0; \tag{2.5.30b}$$

if we are in the situation associated to (2.5.30b) it is of fundamental importance to have the points at the interface of Γ_0 and $\Gamma_1 (= \Gamma/\Gamma_0)$ as vertices of $\mathcal{T}_h$.

Two useful variants of V_h (and V_{0h}) are obtained as follows: either

$$V_h = \{\mathbf{v}_h | \mathbf{v}_h \in C^0(\overline{\Omega}) \times C^0(\overline{\Omega}), \quad \mathbf{v}_h|_T \in P_1 \times P_1, \quad \forall T \in \mathcal{T}_{h/2}\}, \tag{2.5.31}$$

or (this space has been introduced in [46])

$$V_h = \{\mathbf{v}_h | \mathbf{v}_h \in C^0(\overline{\Omega}) \times C^0(\overline{\Omega}), \quad \mathbf{v}_h|_T \in P_{1T}^* \times P_{1T}^*, \quad \forall T \in \mathcal{T}_h\}, \tag{2.5.32}$$

In (2.5.31), $\mathcal{T}_{h/2}$ is the triangulation of Ω obtained from $\mathcal{T}_h$ by joining the midpoints of the edges of $T \in \mathcal{T}_h$, as already shown in Figure 2.4; for the same triangulation $\mathcal{T}_h$, we have the same global number of unknowns if we use V_h defined by either (2.5.29) or (2.5.31); however, the matrices encountered in the latter case are more compact and sparse. In (2.5.32), P_{1T}^* is the subspace of P_3 defined as follows

$$\begin{aligned} P_{1T}^* = \{q | q = q_1 + \lambda \varphi_T, \text{ with } q_1 \in P_1, \lambda \in \Re \\ \text{and } \varphi_T \in P_3, \varphi_T = 0 \text{ on } \partial T, \varphi_T(G_T) = 1\}, \end{aligned} \tag{2.5.33}$$

where, in (2.5.33), G_T is the *centroid* of T; a function like φ_T is usually called a *bubble function*.

Approximation of the boundary conditions. If the boundary conditions are defined by

$$\mathbf{u} = \mathbf{g} \text{ on } \Gamma, \text{ with } \int_\Gamma \mathbf{g} \cdot \mathbf{n} \, d\Gamma = 0, \tag{2.5.34}$$

it is of fundamental importance to approximate $\mathbf{g}$ by $\mathbf{g}_h$ such that

$$\int_\Gamma \mathbf{g}_h \cdot \mathbf{n} \, d\Gamma = 0. \tag{2.5.35}$$

Let us discuss the construction of such a $\mathbf{g}_h$ (we follow here [1, Appendix 3]). For simplicity, we shall suppose that $\mathbf{g}$ is continuous over Γ. We first define the space γV_h as

$$\gamma V_h = \{\boldsymbol{\mu}_h | \boldsymbol{\mu}_h = \mathbf{v}_h|_\Gamma, \quad \mathbf{v}_h \in V_h\}, \tag{2.5.36}$$

i.e., γV_h is the space of the traces on Γ of those functions $\mathbf{v}_h$ belonging to V_h. Actually, if V_h is defined by (2.5.29), γV_h is also the space of those functions defined over Γ, taking their values in $\Re^2$, continuous over Γ and piecewise quadratic over the edges of $\mathcal{T}_h$ contained in Γ.

Our problem is to construct an approximation $\mathbf{g}_h$ of $\mathbf{g}$ such that

$$\mathbf{g}_h \in \gamma V_h, \qquad \int_{\Gamma} \mathbf{g}_h \cdot \mathbf{n} \, d\Gamma = 0. \tag{2.5.37}$$

If $\pi_h \mathbf{g}$ is the unique element of γV_h, obtained by piecewise quadratic interpolation of $\mathbf{g}$ over Γ, i.e., obtained from the values taken by $\mathbf{g}$ at those nodes of $\mathcal{T}_h$ belonging to Γ, we usually have $\int_{\Gamma} \pi_h \mathbf{g} \cdot \mathbf{n} \, d\Gamma \neq 0$. To overcome this difficulty we may proceed as follows:

(i) We define an approximation $\mathbf{n}_h$ of $\mathbf{n}$ as the solution of the following linear variational problem in γV_h:

$$\mathbf{n}_h \in \gamma V_h; \qquad \int_{\Gamma} \mathbf{n}_h \cdot \boldsymbol{\mu}_h \, d\Gamma = \int_{\Gamma} \mathbf{n}_h \cdot \boldsymbol{\mu}_h \, d\Gamma, \quad \forall \boldsymbol{\mu}_h \in \gamma V_h. \tag{2.5.38}$$

Problem (2.5.38) is equivalent to a linear system whose matrix is sparse, symmetric positive definite, extremely well-conditioned and easy to compute (also problem (2.5.38) needs to be solved only once).

(ii) Define $\mathbf{g}_h$ by

$$\mathbf{g}_h = \pi_h \mathbf{g} - \left(\int_{\Gamma} \pi_h \mathbf{g} \cdot \mathbf{n} \, d\Gamma \Big/ \int_{\Gamma} \mathbf{n} \cdot \mathbf{n}_h \, d\Gamma \right) \mathbf{n}_h. \tag{2.5.39}$$

It is easy to check that (2.5.38), (2.5.39) imply (2.5.37). $\qquad \square$

Now if the boundary conditions are defined by either

$$\mathbf{u} = \mathbf{g}_0 \text{ on } \Gamma_0, \qquad \sigma \mathbf{n} = \mathbf{g}_1 \text{ on } \Gamma_1, \tag{2.5.40}$$

or

$$\mathbf{u} = \mathbf{g} \text{ on } \Gamma_0, \qquad \nu \frac{\partial \mathbf{u}}{\partial n} - \mathbf{n}p = \mathbf{g}_1 \text{ on } \Gamma_1, \tag{2.5.41}$$

a well chosen variational formulation will take care *automatically* of the boundary condition on Γ_1; as we shall see in the next paragraph, there is no difficulty associated to the Dirichlet condition on Γ_0.

Space approximation of the time dependent Navier-Stokes equations.

A. *The pure Dirichlet case.* The problem that we consider is defined by (2.1.1)–(2.1.4);
using the spaces P_h, V_h and V_{0h} defined by (2.5.28), (2.5.29) (or (2.5.31), (2.5.32)) and
(2.5.30a), respectively, we approximate (2.1.1)–(2.1.4) by:
Find $\{\mathbf{u}_h(t), p_h(t)\} \in V_h \times P_h$, $\forall t > 0$, such that

$$
\begin{aligned}
\int_\Omega \dot{\mathbf{u}}_h \cdot \mathbf{v}_h \, dx + \nu \int_\Omega \nabla \mathbf{u}_h \cdot \nabla \mathbf{v}_h \, dx + \int_\Omega (\mathbf{u}_h \cdot \nabla)\mathbf{u}_h \cdot \mathbf{v}_h \, dx \\
+ \int_\Omega \nabla p_h \cdot \mathbf{v}_h \, dx = \int_\Omega \mathbf{f}_h \cdot \mathbf{v}_h \, dx, \quad \forall \mathbf{v}_h \in V_{0h},
\end{aligned}
\tag{2.5.42}
$$

$$
\int_\Omega \nabla \cdot \mathbf{u}_h q_h \, dx = 0, \quad \forall q_h \in P_h,
\tag{2.5.43}
$$

$$
\mathbf{u}_h = \mathbf{g}_h \text{ on } \Gamma, \text{ (with } \mathbf{g}_h \in \gamma V_h),
\tag{2.5.44}
$$

$$
\mathbf{u}_h(x, 0) = \mathbf{u}_{0h}(x) \text{ (with } \mathbf{u}_{0h} \in V_h);
\tag{2.5.45}
$$

in (2.5.42)–(2.5.45), we have used the notation $\dot{\mathbf{u}}_h = \frac{\partial \mathbf{u}_h}{\partial t}$, and $\mathbf{f}_h$, $\mathbf{u}_{0h}$, $\mathbf{g}_h$ are convenient
approximations of $\mathbf{f}$, $\mathbf{u}_0$, $\mathbf{g}$, respectively.

B. *Case of the Navier-Stokes equations (2.1.1), (2.1.2), (2.1.3), (2.1.5).* We recall that
in this case, the boundary conditions are given by

$$
\mathbf{u} = \mathbf{g}_0 \text{ on } \Gamma_0, \quad \sigma \mathbf{n} = \mathbf{g}_1 \text{ on } \Gamma_1.
\tag{2.5.46}
$$

The reader courageous enough to do some integration by parts will verify that the
variational formulation of the finite dimensional problem approximating (2.1.1)–(2.1.3),
(2.1.5) is given by
Find $\{\mathbf{u}_h(t), p_h(t)\} \in V_h \times P_h$, $\forall t > 0$, such that

$$
\begin{aligned}
\int_\Omega \dot{\mathbf{u}}_h \cdot \mathbf{v}_h \, dx + 2\nu \int_\Omega \mathbf{D}(\mathbf{u}_h) \cdot \mathbf{D}(\mathbf{v}_h) \, dx + \int_\Omega (\mathbf{u}_h \cdot \nabla)\mathbf{u}_h \cdot \mathbf{v}_h \, dx \\
- \int_\Omega p_h \nabla \cdot \mathbf{v}_h \, dx = \int_\Omega \mathbf{f}_h \cdot \mathbf{v}_h \, dx + \int_{\Gamma_1} \mathbf{g}_{1h} \cdot \mathbf{v}_h \, d\Gamma, \quad \forall \mathbf{v}_h \in V_{0h},
\end{aligned}
\tag{2.5.47}
$$

$$
\int_\Omega \nabla \cdot \mathbf{u}_h q_h \, dx = 0, \quad \forall q_h \in P_h,
\tag{2.5.48}
$$

$$
\mathbf{u}_h = \mathbf{g}_{0h} \text{ on } \Gamma,
\tag{2.5.49}
$$

$$
\mathbf{u}_h(x, 0) = \mathbf{u}_{0h}(x) \text{ (with } \mathbf{u}_{0h} \in V_h).
\tag{2.5.50}
$$

In (2.5.47)–(2.5.50), $\mathbf{f}_h$, $\mathbf{g}_{1h}$, $\mathbf{g}_{0h}$, $\mathbf{u}_{0h}$ are approximations of $\mathbf{f}$, $\mathbf{g}_1$, $\mathbf{g}_0$, $\mathbf{u}_0$, respectively, and the tensor $\mathbf{D}(\mathbf{w})$ is defined from

$$\mathbf{D} = \frac{1}{2}(\nabla + \nabla^t),$$

implying that $D_{ij}(\mathbf{w}) = \frac{1}{2}\left(\frac{\partial w_i}{\partial x_j} + \frac{\partial w_j}{\partial x_i}\right)$.

C. Case of the Navier-Stokes equations (2.1.1), (2.1.2), (2.1.3), (2.1.8). The boundary conditions are given by

$$\mathbf{u} = \mathbf{g}_0 \text{ on } \Gamma_0, \quad \nu\frac{\partial \mathbf{u}}{\partial n} - np = \mathbf{g}_1 \text{ on } \Gamma_1, \tag{2.5.51}$$

leading to the following approximate problem:

Find $\{\mathbf{u}_h(t), p_h(t)\} \in V_h \times P_h$, $\forall t > 0$, such that

$$\int_\Omega \dot{\mathbf{u}}_h \cdot \mathbf{v}_h \, dx + \nu \int_\Omega \nabla \mathbf{u}_h \cdot \nabla \mathbf{v}_h \, dx + \int_\Omega (\mathbf{u}_h \cdot \nabla)\mathbf{u}_h \cdot \mathbf{v}_h \, dx$$
$$- \int_\Omega p_h \nabla \cdot \mathbf{v} \, dx = \int_\Omega \mathbf{f}_h \cdot \mathbf{v}_h \, dx + \int_{\Gamma_1} \mathbf{g}_{1h} \cdot \mathbf{v}_h \, d\Gamma, \quad \forall \mathbf{v}_h \in V_{0h}, \tag{2.5.52}$$

completed by (2.5.48), (2.5.49), (2.5.50). ☐

Expanding p_h and $\mathbf{u}_h$ on vector bases of P_h and V_h, respectively, and taking for the test functions $\mathbf{v}_h$ and q_h *all* the elements of the vector bases of V_{0h} and P_h, the above formulations will produce a system of *ordinary differential equations* with respect to t, coupled to the linear relations associated to the discrete incompressibility conditions. Applying to these algebraic-differential problems the time discretization methods by operator splitting of Section 2.2 (see particularly Section 2.2.5) is straightforward (see [1]–[7] for more details).

As an example, we shall describe the time discretization of the pure Dirichlet problem (2.5.42)–(2.5.45); following Section 2.2.5, we obtain

$$\mathbf{u}_h^0 = \mathbf{u}_{0h}; \tag{2.5.53}$$

then for $n \geq 0$ compute (from $\mathbf{u}_h^n$) $\{\mathbf{u}_h^{n+\theta}, p_h^{n+\theta}\} \in V_h \times P_h$, then $\mathbf{u}_h^{n+1-\theta} \in V_h$, and finally $\{\mathbf{u}_h^{n+1}, p_h^{n+1}\} \in V_h \times P_h$ by solving the following elliptic systems:

$$\left.\begin{aligned}
\int_\Omega \frac{\mathbf{u}_h^{n+\theta} - \mathbf{u}_h^n}{\theta \Delta t} \cdot \mathbf{v}_h \, dx &+ \alpha\nu \int_\Omega \nabla \mathbf{u}_h^{n+\theta} \cdot \nabla \mathbf{v}_h \, dx + \int_\Omega \nabla p_h^{n+\theta} \cdot \mathbf{v}_h \, dx \\
&= \int_\Omega \mathbf{f}_h^{n+\theta} \cdot \mathbf{v}_h \, dx - \beta\nu \int_\Omega \nabla \mathbf{u}_h^n \cdot \nabla \mathbf{v}_h \, dx - \int_\Omega (\mathbf{u}_h^n \cdot \nabla)\mathbf{u}_h^n \cdot \mathbf{v}_h \, dx, \\
&\hspace{6cm} \forall \mathbf{v}_h \in V_{0h},
\end{aligned}\right\} \tag{2.5.54a}$$

$$\int_\Omega \nabla \cdot \mathbf{u}_h^{n+\theta} q_h \, dx = 0, \quad \forall q_h \in P_h, \tag{2.5.54b}$$

$$\mathbf{u}_h^{n+\theta} \in V_h, \quad p_h^{n+\theta} \in P_h, \quad \mathbf{u}_h^{n+\theta} = \mathbf{g}_h^{n+\theta} \text{ on } \Gamma, \tag{2.5.54c}$$

then

$$\left.\begin{aligned}
&\int_\Omega \frac{\mathbf{u}_h^{n+1-\theta} - \mathbf{u}_h^{n+\theta}}{(1-2\theta)\Delta t} \cdot \mathbf{v}_h \, dx + \beta\nu \int_\Omega \nabla \mathbf{u}_h^{n+1-\theta} \cdot \nabla \mathbf{v}_h \, dx \\
&\qquad\qquad + \int_\Omega (\mathbf{u}_h^{n+1-\theta} \cdot \nabla)\mathbf{u}_h^{n+1-\theta} \cdot \mathbf{v}_h \, dx \\
&= \int_\Omega \mathbf{f}_h^{n+1-\theta} \cdot \mathbf{v}_h \, dx - \alpha\nu \int_\Omega \nabla \mathbf{u}_h^{n+\theta} \cdot \nabla \mathbf{v}_h \, dx - \int_\Omega \nabla p_h^{n+\theta} \cdot \mathbf{v}_h \, dx, \\
&\hspace{8cm} \forall \mathbf{v}_h \in V_{0h},
\end{aligned}\right\} \tag{2.5.55a}$$

$$\mathbf{u}_h^{n+1-\theta} \in V_h, \quad \mathbf{u}_h^{n+1-\theta} = \mathbf{g}_h^{n+1-\theta} \text{ on } \Gamma, \tag{2.5.55b}$$

and finally

$$\left.\begin{aligned}
&\int_\Omega \frac{\mathbf{u}_h^{n+1} - \mathbf{u}_h^{n+1-\theta}}{\theta\Delta t} \cdot \mathbf{v}_h \, dx + \alpha\nu \int_\Omega \nabla \mathbf{u}_h^{n+1} \cdot \nabla \mathbf{v}_h \, dx + \int_\Omega \nabla p_h^{n+1} \cdot \mathbf{v}_h \, dx \\
&\qquad = \int_\Omega \mathbf{f}_h^{n+1} \cdot \mathbf{v}_h \, dx - \beta\nu \int_\Omega \nabla \mathbf{u}_h^{n+1-\theta} \cdot \nabla \mathbf{v}_h \, dx \\
&\qquad\qquad - \int_\Omega (\mathbf{u}_h^{n+1-\theta} \cdot \nabla)\mathbf{u}_h^{n+1-\theta} \cdot \mathbf{v}_h \, dx, \\
&\hspace{8cm} \forall \mathbf{v}_h \in V_{0h},
\end{aligned}\right\} \tag{2.5.56a}$$

$$\int_\Omega \nabla \cdot \mathbf{u}_h^{n+1} q_h \, dx = 0, \quad \forall q_h \in P_h, \tag{2.5.56b}$$

$$\mathbf{u}_h^{n+1} \in V_h, \quad p_h^{n+1} \in P_h, \quad \mathbf{u}_h^{n+1} = \mathbf{g}_h^{n+1} \text{ on } \Gamma, \tag{2.5.56c}$$

respectively.

We shall take again $\alpha = (1-2\theta)/(1-\theta)$ and $\beta = \theta/(1-\theta)$. Concerning the value of θ, we suggest $\theta = 1 - \frac{1}{\sqrt{2}}$.

2.5.4 Remarks concerning the computer solution of the discrete problems

The solution of the subproblems encountered at each step of the above operator splitting methods can be obtained by iterative methods which are the discrete analogues of the methods discussed in Sections 2.3 and 2.4. In particular, we shall have to solve quite systematically linear systems approximating elliptic systems associated to elliptic opera-tors such as $\alpha I - \nu \Delta$; for example, if the boundary conditions are of the Dirichlet type only (i.e., if $\Gamma_0 = \Gamma$), we shall have to solve problems like

$$\alpha \mathbf{u} - \nu \Delta \mathbf{u} = \mathbf{f} \text{ in } \Omega, \quad \mathbf{u} = \mathbf{g} \text{ on } \Gamma. \tag{2.5.57}$$

Paradoxically, solving (2.5.57) is not very expensive for flows at high Reynold num-bers. For such flows, the viscosity ν is small, and their fast dynamics require small Δt, i.e., large values of α. Suppose for simplicity that $\Omega = (0,1)^2$, and also that one uses over Ω a regular triangulation like the one in Figure 2.2 where $h = 1/(I+1)$ (I a positive integer). Suppose also that one uses continuous and piecewise linear approxima-tions of the velocity over the above triangulation, and that integrals like $\int_\Omega \mathbf{v} \cdot \mathbf{w}\, dx$ are approximated using the trapezoidal rule. One obtains then the approximation of (2.5.57) associated to a traditional *five point formula*, i.e. (with obvious notation)

$$\alpha \mathbf{u}_{ij} + \frac{\nu}{h^2}\left(4\mathbf{u}_{ij} - \mathbf{u}_{i+1j} - \mathbf{u}_{i-1j} - \mathbf{u}_{ij+1} - \mathbf{u}_{ij-1}\right) = \mathbf{f}_{ij},$$
$$1 \le i,j \le I, \quad \mathbf{u}_{kl} = \mathbf{g}_{kl} \text{ if } \{kh, lh\} \in \Gamma. \tag{2.5.58}$$

It is well known that the matrix in (2.5.58) has for smallest and largest eigenvalues

$$\lambda_{\min} = \alpha + \frac{8\nu}{h^2}\sin^2\frac{\pi h}{2}$$

and

$$\lambda_{\max} = \alpha + \frac{8\nu}{h^2}\sin^2\frac{I\pi h}{2}$$

respectively.

For small values of h, we clearly have

$$\lambda_{\min} \approx \alpha + 2\pi^2\nu, \quad \lambda_{\max} \approx \alpha + 8\nu/h^2,$$

implying for the condition number $\mathcal{N}$ of the above matrix

$$\mathcal{N} = \frac{\lambda_{\max}}{\lambda_{\min}} \approx \left(\alpha + 8\nu/h^2\right)/\left(\alpha + 2\pi^2\nu\right).$$

Suppose now that $\nu = 10^{-3}, h = 10^{-2}, \Delta t = 10^{-2} (\Rightarrow \alpha = 10^2)$; we have then

$$\mathcal{N} \approx 1.8 \tag{2.5.59}$$

Suppose now that we solve the linear system (2.5.58) by a *nonpreconditioned* conjugate gradient algorithm. It follows then from (2.5.59), and from Section 2.4.3, that the distance between the solution of (2.5.58) and the nth iterate converges to zero at least as fast as

$$\left(\frac{\sqrt{1.8} - 1}{\sqrt{1.8} + 1}\right)^n = (.145898\ldots)^n$$

which corresponds to a quite high speed of convergence. A similar conclusion would hold for successive over-relaxation with optimal parameter. Actually, the convergence of the above methods is sufficiently fast so that further speeding up (by a multigrid method, for example) would be useless. Indeed the above methods are quite simple and are fairly easy to vectorize. The above observation holds for the various problems of type (2.5.58) encountered at each step of (2.5.53)–(2.5.56). Concentrating on the discrete Stokes problems (2.5.54) and (2.5.56), their solution by a variant of algorithm (2.4.72)–(2.4.87) requires at each iteration the solution of a linear system approximating a Neumann problem of the following type

$$- \Delta\varphi = f \text{ in } \Omega, \quad \frac{\partial\varphi}{\partial n} = 0 \text{ on } \Gamma; \quad \int_\Omega \varphi \, dx = 0 \qquad (2.5.60)$$

(with $\int_\Omega f \, dx = 0$). The matrix approximating the Neumann-Laplace operator occurring in (2.5.60) does not enjoy the nice properties of $\alpha I - \nu\Delta$ and therefore the approximate solution of (2.5.60) may be costly (at least for three dimensional problems). For two dimensional problems, we use a *direct method* (à la Cholesky) for solving these discrete Neumann problems (after deleting one equation and setting to zero the corresponding unknown). Actually, we have to remember that the above discrete Neumann problems have to be solved in the pressure space P_h, and therefore if one uses the approximations defined by (2.5.28), (2.5.29) or (2.5.31), we have 8 times (resp. 16 times) more unknowns for velocity than for pressure if $\Omega \subset \Re^2$ (resp. $\Omega \subset \Re^3$). For three dimensional flows, multigrid methods seem to be well suited to solve problem (2.5.60).

The above observations still hold if one has to solve, instead of (2.5.57) and (2.5.60) elliptic systems with mixed Dirichlet-Neumann boundary conditions, like those associated to the solution of the Navier-Stokes equations (2.1.1)–(2.1.3), (2.1.5) or (2.1.1)–(2.1.3), (2.1.8).

2.6   Comments on the Numerical Methodology

It is not very easy to do comparisons between solution methods for a problem of the complexity of the Navier-Stokes equations. However, compared to most solution methods, the ones described in this paper enjoy the following properties:

(i) Since they use a finite element approach for the space approximation, they can handle complicated geometries. Moreover, since *isoparametric* variants are available, they can treat accurately a curved boundary.

(ii) They are well-suited to accurate time dependent and steady state calculations, because of the θ-scheme operator splitting.

(iii) They are quite robust because the least squares treatment of the nonlinearities and the (practically) unconditional stability of the θ-scheme.

(iv) Once h and Δt are given, and θ taken equal to its optimal experimental value, namely $1 - 1/\sqrt{2}$, our methods are parameter free, being based on conjugate gradient algorithms.

(v) They do not introduce artificial viscosity.

(vi) They rely on the solution of well conditioned elliptic problems for which very efficient methods exist.

Acknowledgment

We would like to acknowledge the help of the following individuals for friendly discussions and/or collaboration: J. P. Benque, M. O. Bristeau, J. Cahouet, J. P. Chabard, J. Goussebaile, P. Gubernatis, A. Haugel, F. Hussein, G. Labadie, W. Lawton, J. L. Lions, B. Mantel, J. Periaux, P. Perrier, O. Pironneau, H. Resnikoff, R. Sanders, T. Tezduyar, J. Weiss, M. F. Wheeler.

The support of the following corporations or institutions is also acknowledged: AWARE, Dassault Industries, INRIA, University of Colorado, University of Houston, Université Pierre et Marie Curie. We also benefited from the support of DARPA (Contracts AFOSR F49620-89-C-0125 and AFOSR-90-0334), DRET (Grants 83/403 and 86/34) and NSF (Grant INT 8612680).

Finally, special thanks are due to R. A. Nicolaides for suggesting that we write this article, and to J. A. Wilson for diligently processing it.

References

[1] R. Glowinski, *Numerical Methods for Nonlinear Variational Problems*, Springer, New York, 1984.

[2] R. Glowinski, Viscous flow simulation by finite element methods and related numerical techniques, in *Progress and Supercomputing in Computational Fluid Dynamics*, E. M. Murman, S. S. Abarbanel eds., Birkhauser, Boston, 1985, pp. 173–210.

[3] M. O. Bristeau, R. Glowinski, B. Mantel, J. Periaux, P. Perrier, Numerical Methods for Incompressible and Compressible Navier-Stokes Problems, in *Finite Elements in Fluids*, 6, R. H. Gallagher, G. Carey, J. T. Oden, O. C. Zienkiewicz eds., J. Wiley, Chichester, 1985, pp. 1–40.

[4] R. Glowinski, Splitting methods for the numerical solution of the incompressible Navier-Stokes equations, in *Vistas in Applied Mathematics*, A. V. Balakrishnan, A. A. Doronitsyn, J. L. Lions eds., Optimisation Software, New York, 1986, pp. 57–95.

[5] M. O. Bristeau, R. Glowinski, J. Periaux, Numerical Methods for the Navier-Stokes equations, *Computer Physics Reports*, 6, 1987, pp. 73–187.

[6] E. J. Dean, R. Glowinski, C. H. Li, Supercomputer solution of partial differential equation problems in Computational Fluid Dynamics and in Control, *Computer Physics Communications*, 53, 1989, pp. 401–439.

[7] R. Glowinski, Supercomputing and the finite element approximation of the Navier-Stokes equations for incompressible viscous fluids, in *Recent Advances in Computational Fluid Dynamics*, C. C. Chao, S. A. Orszag, W. Shyy eds., Lecture Notes in Engineering, 43, Springer, Berlin, 1989, pp. 277–315.

[8] F. Thomasset, *Implementation of Finite Element Methods for Navier-Stokes Equations*, Springer, New York, 1981.

[9] R. Peyret, T. D. Taylor, *Computational Methods for Fluid Flow*, Springer, New York, 1982.

[10] R. Temam, *Navier-Stokes Equations*, North-Holland, Amsterdam, 1977.

[11] V. Girault, P. A. Raviart, *Finite Element Methods for Navier-Stokes Equations*, Springer, Berlin, 1986.

[12] C. Cuvelier, A. Segal, A. van Steenhoven, *Finite Element Methods and Navier-Stokes Equations*, Reidel, Dordrecht, 1986.

[13] O. Pironneau, *Finite Element Methods for Fluids*, J. Wiley, Chichester, 1989.

[14] M. D. Gunzburger, *Finite Element Methods for Viscous Incompressible Flows*, Academic Press, Boston, 1989.

[15] O. A. Ladysenskaya, *The Mathematical Theory of Viscous Incompressible Flows*, Gordon and Breach, New York, 1969.

[16] J. L. Lions, *Quelques Méthodes de Résolution des Problèmes aux Limites Non Linéaires*, Dunod, Paris, 1969.

[17] L. Tartar, *Topics in Nonlinear Analysis*, Université Paris-Sud Orsay, Paris, 1978.

[18] H. O. Kreiss, J. Lorenz, *Initial-Boundary Value Problems and the Navier-Stokes Equations*, Academic Press, Boston, 1989.

[19] N. Yanenyo, *The Method of Fractional Steps*, Springer, Berlin, 1971.

[20] G. I. Marchuk, *Methods of Numerical Mathematics*, Springer, New York, 1975.

[21] G. Strang, On the construction and comparison of difference schemes, *SIAM J. Num. Anal.*, 5, 1968, pp. 506–517.

[22] J. T. Beale, A. Majda, Rates of Convergence for viscous splitting of the Navier-Stokes equations, *Math. Comp.*, 37, 1981, pp. 243–260.

[23] R. Leveque, J. Oliger, Numerical methods based on additive splitting for hyperbolic partial differential equations, *Math. Comp.*, 40, 1983, pp. 469–497.

[24] R. Glowinski, P. Le Tallec, *Augmented Lagrangian and Operator Splitting Methods in Nonlinear Mechanics*, SIAM, Philadelphia, 1989.

[25] D. Peaceman, M. Rachford, The numerical solution of parabolic and elliptic differential equations, *J. SIAM*, 3, 1955, pp. 28–41.

[26] P. L. Lions, B. Mercier, Splitting algorithms for the sum of two nonlinear operators, *SIAM J. Num. Anal.*, 16, 1979, pp. 964–979.

[27] E. Godlewski, *Méthodes à pas multiples et de directions alternées pour la discretisation d' équations d'évolution, Thèse de 3è cycle*, Université P et M. Curie, Paris, 1980.

[28] D. Gabay, Application of the method of multipliers to variational inequalities, in *Augmented Lagrangian Methods*, M. Fortin and R. Glowinski eds., North-Holland, Amsterdam, 1983, pp. 299–331.

[29] J. Douglas, H. Rachford, On the numerical solution of the heat conduction problem in two and three space variables, *Trans. Amer. Math. Soc.*, 82, 1956, pp. 421–439.

[30] M. Fortin, R. Glowinski (eds.), *Augmented Lagrangian Methods*, North-Holland, Amsterdam, 1983.

[31] J. Douglas, Alternating Direction Methods for 3 space variables, *Numerische Mathematik*, 4, 1962, pp. 41–63.

[32] E. Fernandez-Cara, M. M. Beltran, The convergence of two numerical schemes for the Navier-Stokes equations, *Numerische Mathematik*, 55, 1989, 1, pp. 33–60.

[33] E. Polak, *Computational Methods in Optimization*, Academic Press, New York, 1971.

[34] P. N. Brown, Y. Saad, Hybrid Krylov methods for nonlinear systems of equations, *SIAM J. Sci. Stat. Comp.*, 11, 1990, 3, pp. 450–481.

[35] D. P. Young, R. G. Melvin, J. T. Johnson, J. E. Bussoletti, L. B. Wigton, S. S. Samant, Application of sparse matrix solvers as effective preconditioners, *SIAM J. Sci. Stat. Comp.*, 10, 1989, 6, pp. 1186–1199.

[36] M. O. Bristeau, R. Glowinski, J. Periaux, Acceleration procedures for the numerical simulation of compressible and incompressible viscous flows, Chapter 6 of *Advances in Computational Nonlinear Mechanics*, I. S. Doltsinis ed., Springer, Wien, 1989, pp. 197–243.

[37] J. Daniel, *The Approximate Minimization of Functionals*, Prentice-Hall, Englewood Cliffs, N.J., 1970.

[38] J. Cahouet, J. P. Chabard, Some fast 3D solvers for the generalized Stokes problem, *Int. J. Num. Meth. in Fluids*, 8, 1988, pp. 269–295.

[39] T. J. R. Hughes, L. P. Franca, M. Balestra, A new finite element formulation for computational fluid dynamics: V. Circumventing the Babuska-Brezzi condition: A stable Petrov-Galerkin formulation of the Stokes problem accomodating equal-order interpolations, *Comp. Meth. Appl. Mech. Eng.*, 59, 1986, 1, pp. 85–100.

[40] R. Glowinski, C. H. Li, J. L. Lions, A numerical approach to the exact controllability of the wave equation (I) Dirichlet controls: Description of the numerical methods, *Japan J. Appl. Math.*, 7, 1990, 1, pp. 1–76.

[41] R. Glowinski, C. H. Li, On the numerical implementation of the Hilbert Uniqueness Method for the exact boundary controllability of the wave equation, *C. R. Acad. Sc., Paris*, T. 311, Série I, 1990, pp. 135–142.

[42] P. Hood, C. Taylor, A numerical solution of the Navier-Stokes equations using the finite element technique, *Computers and Fluids*, 1, 1973, pp. 73–100.

[43] T. J. R. Hughes, *The Finite Element Method*, Prentice-Hall, Englewood Cliffs, N.J., 1987.

[44] P. G. Ciarlet, *The Finite Element Method for Elliptic Problems*, North-Holland, Amsterdam, 1978.

[45] P. A. Raviart, J. M. Thomas, *Introduction à l'Analyse Numérique des Equations aux Dérivées Partielles*, Masson, Paris, 1983.

3   CFD – An Industrial Perspective

Michael S. Engelman

3.1 Introduction

This book represents a timely review of state-of-the-art algorithms in the many branches of Computational Fluid Dynamics (CFD). Algorithms for a wide range of incompressible flow phenomena are presented ranging from highly turbulent flows to low-speed non-Newtonian flows. Despite the enormous advances of the past ten years it is clear that there are still many interesting new Algorithmic directions evolving.

Much of the material presented herein is from the perspective of the algorithm researcher whose ultimate goal is the development of viable algorithms for the accurate and efficient simulation of real-world CFD problems. In this chapter, I would like to take the opportunity to discuss CFD from a different perspective – that of a commercial CFD package developer and more importantly that of the user of CFD codes. For, in the final analysis, it is the user who must apply the algorithms developed by researchers and packaged in commercial CFD codes. Over the last ten years CFD has evolved to the point where commercial CFD codes have been developed and introduced into the marketplace. These codes are being used with increasing frequency and it is probably fair to state that today the use of CFD in industry is increasingly becoming an accepted analysis tool.

Over the last decade the application of CFD has spread from its original aerospace beginnings to increasing application in a broad spectrum of industries ranging from the more traditional automotive and electronics industries to exciting new applications in the biomedical and food industries. In these industries CFD is still not used in a truly design environment; however, the real payback from the use of CFD will come when it is used on a daily basis as a design tool by design engineers who are not CFD specialists. In recent years most of the energies and resources of both academic researchers and commercial developers has been expended in algorithm development and in enhancing the computational capabilities and expanding the areas of applicability of CFD. Now that the algorithms and commercial codes have reached a reasonable level of maturity and acceptance, attention must be turned to those issues which need to be addressed in order for CFD to achieve greater acceptance at the design level. In the following sections of the chapter I would like to present a commercial CFD code developer's point of view as to just what these issues are, the current state of "commercial CFD" and some of the issues unique to the "selling" of CFD.

3.2 Application Areas

Typically the words Computational Fluid Dynamics conjure up visions of aerodynamics and simulation of flows over aircraft wings, fuselages and jet engines. Certainly
the aerospace industry pioneered the use of CFD and is still at the forefront of CFD
development. It is also an area where the use of CFD and its benefits have been long
realized and are not questioned. Oftentimes, however, it is not realized just how perva-
sive "low-speed" (mostly incompressible) flow applications are in the engineering world.
The following sections summarize a range of applications arbitrarily categorized by the
industry in which they arise – although many of these applications are cross-industry in
nature.

3.2.1 Automotive Industry

The automotive industry is the source for a wide range of CFD applications including:

- Vehicle aerodynamics – drag computations
- Climate control
- Engine block cooling
- Flow through valves, filters and control devices
- Sloshing of fuel in fuel tanks

Figure 3.1 shows the results of a simulation of air flow through a truck air filter; the
filter is modelled using a porous medium representation. The plots are clockwise from
the top left: mesh plot of the filter, pressure field, velocity field and particle traces. See
Aoyogi et al. [1988].

3.2.2 Electronics Industry

With the diminishing size of electronic devices and the emergence of desktop computers
which use natural convection rather than fans for cooling, there is an ever-growing need
for accurate predictions of heat transfer and cooling in such devices (see Mansingh and
Misegades [1990] and Torok [1985]). Applications include:

- Flow and thermal distribution in cabinets and devices, i.e., electronic packaging
- Component cooling
- Air flow in disk drives
- Conjugate heat transfer problems

Figure 3.2 shows the results of a simulation of air flow in a complete computer cabinet.
The plots are clockwise from the top left: edge plot of the cabinet and components,
velocity field, pressure field and mesh plot.

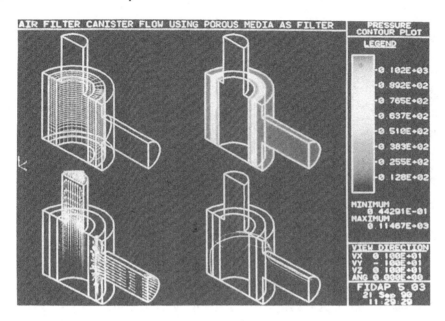

Figure 3.1: Flow through a truck air filter.

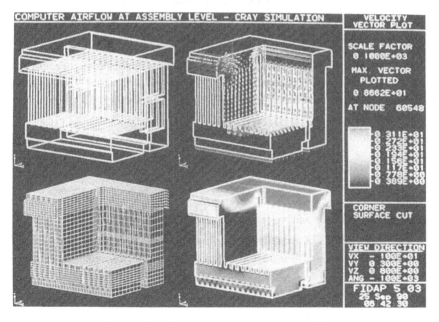

Figure 3.2: Flow through a computer cabinet.

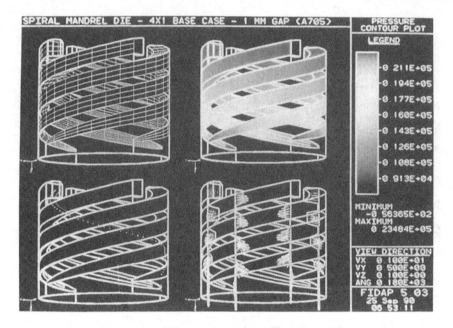

Figure 3.3: Extrusion from a spiral Mandrel die.

3.2.3 Chemical/Process Industry

The chemical and process industries provide many complex applications for CFD which lie in the very low Reynolds number regime (< 1) but which are highly nonlinear due to material nonlinearities (see Fraser et al. [1989] and Murnane et al. [1988]). These include:

- Plastic flows
- Glass flows
- Die and extrusion flows
- Filling of molds
- Reacting flows – heat and mass transfer in chemical reactors

Figure 3.3 shows the results of a simulation of plastic flow in a four headed spiral mandrel die. The plots are clockwise from the top left: mesh plot, pressure field, velocity field, and particle traces.

3.2.4 Metal-Forming Industry

Many challenging applications exist in the metal-forming industry including:

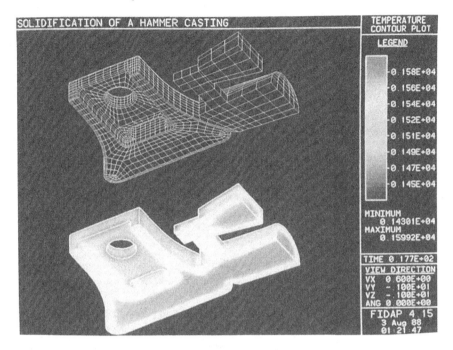

Figure 3.4: Solidification of a hammer casting.

- Continuous casting
- Strip casting
- Metal extrusion
- Hall cell simulation (aluminum manufacture)
- Solidification

Figure 3.4 shows the results of a simulation of the solidification of a hammer casting. The plots are for a single time step and show the mesh and temperature distribution. See Dantzig and Chao [1986].

3.2.5 Nuclear Industry

The nuclear industry has been a longstanding user of CFD partly due to the difficulty associated with building prototypes and performing experiments for the many scenarios that must be explored (see Lasbleiz et al. [1986]). Applications include:

- Flow through piping systems

- Reactor cooling
- Flow in reactors
- Storage of nuclear waste
- Cooling towers and thermal plumes

3.2.6 Thin Film/Coating Industry

An area rich in CFD applications is the thin/film coating industry (see Coyle et al. [1988]). Applications include:

- Coating of magnetic tape
- Coating of photographic film
- Coating of adhesives
- Manufacture of paper
- Optical fiber coating

Figure 3.5 shows the results of the simulation of the curtain coating of a moving substrate. The interesting feature of this simulation is that in addition to the flow field the shape of all surfaces, the height of coated liquid and the location of the contact point of the liquid on the substrate has been computed as part of the simulation.

3.2.7 Biomedical/Pharmaceutical industry

Increasing use is being made of CFD in the biomedical/pharmaceutical industry (see Engelman and Bercovier [1980]) where applications include:

- Blood flow in veins and arteries
- Flow past prostheses
- Flow in the heart
- Centrifugal processing
- Design of injection delivery systems

3.2.8 Food and Beverage Industry

The food and beverage industry represents an area with a wealth of applications for CFD (see Engelman and Sani [1983]). This will be one of the growth areas for CFD in coming years. Applications include:

- Pasteurization processes

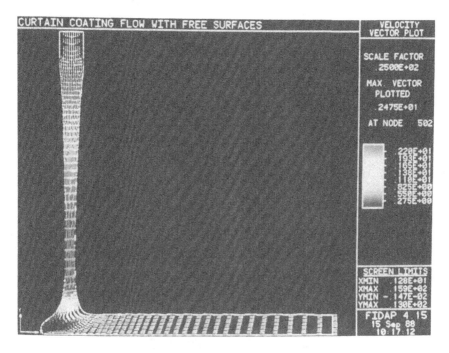

Figure 3.5: Curtain coater simulation.

- Flow, temperature and concentration distributions in process equipment
- Extrusion of dough-like fluids
- Convection ovens

Figure 3.6 shows the results of the simulation of the pasteurization of beer. The plot shows the temperature and velocity field distributions at one instant of the pasteurization process.

3.2.9 Aerospace/Defense Industry

The aerospace/defense industry has been at the forefront of the development and use of CFD over the past decade. Applications for "low-speed" CFD include:

- Flow around submerged bodies
- Variable and micro gravity effects
- Cabin ventilation
- Flows in fuel lines and fuel tanks

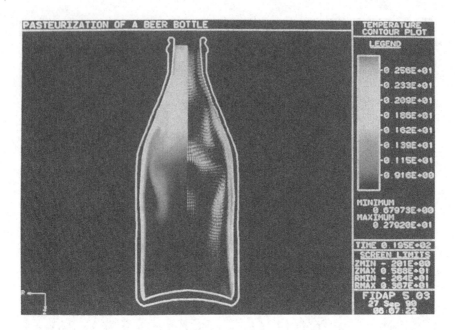

Figure 3.6: Pasteurization of beer.

Figure 3.7 shows the results of the simulation of the viscous turbulent flow past a submerged body. The plot shows the velocity and pressure fields on the surface of the body. See Givler [1991].

3.2.10 Miscellaneous Applications

There are numerous other applications (cf. Fontaine et al. [1988]) for CFD including:

- Crystal growth
- Chemical vapor deposition
- Ventilation and ducting flow
- Turbomachinery
- Environmental flows and flows around buildings

The list of above applications is only a small subset of the potential applications for CFD in industry. CFD is already heavily used in many of these areas while in many other areas its use is embryonic. The challenge for the future is make CFD more accessible to these industries and application areas.

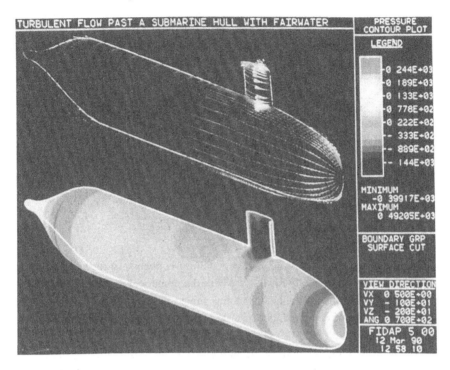

Figure 3.7: Flow past a submerged body.

3.3 CFD in Industry

The benefits to be gained from the computer simulation of the various flows described above are easy to enumerate. Experimental fluid dynamics is time-consuming and costly; with CFD many possibilities can be explored in a relatively short time span. Measurement devices often introduce disturbances into the flow fields they are measuring; with CFD there are no such disturbances. The CFD simulation provides comprehensive information on the velocity, temperature, concentration and pressure fields throughout the flow domain while experimental testing provides data only at those locations where instrumentation is installed. Severe operating conditions or design changes, whose simulation cannot be justified on practical grounds, can be studied.

However, it should be stressed that CFD codes will not eliminate the need for experimental testing, but rather will enhance the effectiveness of testing. The number of prototypes required can be reduced – CFD can be used to design the prototype for testing – thus speeding up the entire design cycle. Physical and numerical modelling are and will

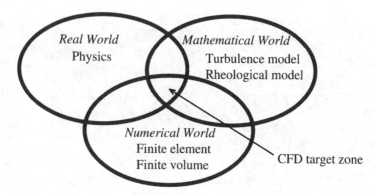

Figure 3.8: Worlds of CFD.

remain strongly complementary; physical modelling is necessary to verify and validate numerical solutions – especially for flows where the numerical models rely on numerical data. However, given the better understanding gained through the comprehensive nature of numerical simulations, the number of tests can be reduced and instrumentation better located.

Despite the wide ranging applications and clear well-defined benefits it cannot be said that the use of CFD as a *design tool* in the industries described above is widespread or even that the simulation of fluid flows is the regular recourse of the engineer faced with fluid flow problems. By contrast, in the area of structural and solid mechanics analysis, the use of general purpose finite element codes has revolutionized structural design and integrity. The development of similar codes for the problems of fluid flow and heat and mass transfer has lagged far behind largely due to the complexity of the physical processes involved and the complexity of the resultant numerical problem. Nevertheless, over the last decade commercial CFD codes have evolved to the point where the use of CFD in industry is an accepted analysis tool. Before looking forward to the issues facing CFD in the future, I believe it is important to review in more detail some of the reasons why "commercial CFD" has lagged so far behind structural analysis as well as some of the issues confronting the designer of a commercial CFD program.

3.4   The Worlds Of CFD

To understand the difficulties involved in simulating industrial fluid flow problems it is instructive to review the steps required to achieve a CFD capability. I like to refer to these steps as the "Worlds of CFD" – these worlds are depicted in Figure 3.8.

The starting point for any CFD simulation is of course the real problem being studied, e.g., flow through a valve, extrusion from a die, etc. Each such flow is governed by physical laws of nature – this is the Real World and this is the world that we wish to reproduce in the simulation as accurately as possible. The first step in the modelling process is to derive a set of mathematical equations which describes the physics of the flow. In the "simplest" case of laminar isothermal Newtonian incompressible flow, these are the well-known Navier-Stokes equations. Often, however, the flow is turbulent and although the same Navier-Stokes equations could be used, the number of grid points required to resolve the chaotic nature of turbulent flows far exceeds the capacities of the current, and even the next, generation of supercomputers. Thus, in practice, a turbulence model must be introduced to allow the simulation of such problems. Unfortunately, most turbulence models are empirical in nature and each has its advantages and disadvantages – no one turbulence model is applicable to all flow situations. Nor is there complete, or in many cases, even pa tial agreement as to the best turbulence model to use for a given application. Almost identical comments apply when dealing with non-Newtonian fluids, e.g. plastics; there are many rheological models available to describe the behavior of non-Newtonian fluids – almost all are empirical in nature. Moreover, similar modeling constraints apply to many other aspects of fluid flows. Our knowledge of the physics of some the fundamental aspects of fluid flows and our ability to model them accurately mathematically is still evolving. Thus the Mathematical World is truly a model of the Real World, i.e., it is not an exact representation. Each model employed, be it a turbulence model, a rheological model, or any other model, introduces the possibility of errors into the final CFD simulation results. Thus, as shown in Figure 3.8, although the Real and Mathematical Worlds do overlap the intersection is not a complete one, i.e. there will be times when the mathematical model does not accurately model the real flow. How closely these worlds coincide is largely dependent on the assumptions made by the mathematical models employed.

Let us assume for the moment that a mathematical model has been agreed upon. If this model could be solved analytically and a mathematical closed form of the solution derived, we would be able to evaluate any of the flow variables at any point in the fluid (and then, of course, there would be no need for CFD programs). Unfortunately, the equations of fluid flow are highly nonlinear and do not admit a mathematical solution except in the most trivial of cases. This leads us to the computer and the Numerical World where we seek a discretized solution at a finite number of points in the flow domain. The transformation from the mathematical model to the discrete problem to be solved on the computer is the main activity performed in the Numerical World. Many approaches are available to discretize the equations of fluid flow; some of the more commonly employed approaches include finite difference, finite volume, finite element and spectral methods. Once again, there is very little agreement as to the best method to use – all that can be said is that the accuracy and validity of the results obtained from a CFD simulation can be strongly dependent on the numerical techniques employed. Thus, once again, as shown in Figure 3.8, although the Mathematical and Numerical Worlds do overlap the intersection is not a complete one, i.e., there will be times that the numerical solution is not an accurate solution of the mathematical model. How closely these worlds coincide is largely dependent on the numerical algorithms employed.

The aim of any CFD code is to provide solutions that lie in the intersection of all three worlds, Real, Mathematical and Numerical – the so-called "CFD Target Zone" of Figure 3.8. It is crucial for the user of any CFD code, commercial or otherwise, to realize that the computed solution may not lie in this target zone. CFD codes are not "black boxes" that unerringly reproduce the complete physics. Rather, they are simulation tools which have a definable capability to simulate physical phenomena that may play an important role in a given engineering problem. Their potential can effectively and fully be tapped only if the user has a basic understanding of the mathematical and numerical assumptions that have been made in the code.

3.5 Components of a CFD Code

As alluded to in the previous section, there is very little agreement and in the past there has often been a great deal of emotionalism surrounding the question of numerical algorithms for the simulation of fluid flows. It is worthwhile examining the components that combine to form a functioning CFD code in order to dispel some of the misstatements that are often made regarding different CFD codes and CFD numerical techniques in general.

Three major components can be identified in any CFD code – these are a discretization scheme, a numerical algorithm and an equation solver. The function served by each of these components is as follows:

3.5.1 Discretization Scheme

As discussed previously, once a mathematical model of the flow under consideration has been agreed upon, the mathematical model must be transformed to a discrete problem so that it may be solved on the computer. This step is achieved using a discretization (or approximation) scheme which transforms the continuum set of equations to a discrete set of equations. A multitude of schemes abound in the literature for achieving this transformation including finite difference, finite volume, finite element, spectral, panel and boundary element methods – numerous references to this work can be found in the chapters of this book. In addition to these methods there are many others as well as many variants within each approach.

3.5.2 Numerical Algorithm

Once a discretized set of equations has been derived the next step is to develop a numerical algorithm for their solution. This is a non-trivial step as in general the equations of fluid flow are highly nonlinear. As with discretization schemes a host of possibilities exists and the literature is replete with algorithms bearing such names as SIMPLE, SIMPLER, TEACH, etc (cf. Roache [1972]). Many of these algorithms fall into a class of algorithms which can be categorized as "segregated" algorithms. In the segregated approach, a pressure or pressure update step is performed separately from the solution of the momentum equations; each momentum may also be solved in a segregated method. An alternate approach to a segregated approach is to solve all equations simultaneously – a fully coupled approach. The fully coupled approach tends to be more effective when there is strong coupling between the various solution fields. Other numerical algorithms are based on explicit time-marching procedures, the method of characteristics, the vortex method, to name but a few.

3.5.3 Equation Solver

Almost all numerical algorithms for the solution of the discretized equations of fluid flow require that a linear system of equations be solved at least once (and in many case multiple times) at each iteration or time step. These systems tend to be large and sparse and the structure of the coefficient matrix usually influences the choice of solver. Commonly used solvers include direct Gaussian elimination solvers, tridiagonal solvers and conjugate gradient based iterative solvers (cf. Hasbani and Engelman [1979], Saad [1981] and [1986]). As before there are advantages and disadvantages to the different choices: direct solvers always provide a solution in a single pass but can become prohibitively expensive as the problem size increases; iterative solvers are less expensive but convergence and the rate of convergence can be major issues, tridiagonal solvers are very efficient but place restrictions on mesh connectivity, etc.

The union of a discretization scheme, a numerical algorithm and an equation solver is a CFD code. Of course, the above representation is somewhat simplistic as one component is usually tightly coupled with another component, or at least limits the choices one has for the other components, e.g. an unstructured finite volume approach will not allow the use of a tridiagonal equation solver. Nevertheless, it is possible to identify these three components in any CFD code.

In the past there has been a tendency to categorize CFD codes based solely on the discretization scheme employed – this has been especially true in the competitive commercial CFD environment. However, this is a gross simplification – a finite element based code can use a SIMPLER type algorithm with an iterative equation solver just as a finite difference code can use a fully coupled approach with a direct solver. Historically, particular discretization schemes have tended to use certain discretization/algorithm/solver combinations, e.g. finite element/fully-coupled/direct solver or finite volume/segregated/iterative solver, partly, I believe, due to the lack of communication between researchers in the different areas. As a result much work has been duplicated and a great deal of misconception and folklore generated between practitioners of the different approaches. There are merits to all approaches and CFD codes in the future will blend the best features of the different discretization schemes, numerical algorithms and equation solvers. It is also important to realize that CFD applications are not a homogeneous group of applications – plastic flows are low Reynolds numbers flows, steel flows are highly turbulent, electronics applications range from high Peclet number forced convection flows to high Rayleigh number natural convection dominated flows, etc. No one numerical approach is best suited to this wide range of flow conditions.

3.6 Future Directions And Issues

In my opinion, the evolution of commercial CFD is also at a definite turning point in its short history. As mentioned earlier, over the last decade the application of CFD has broadened from its original aerospace beginnings to increasing application in a broad spectrum of industries. In this same period commercial CFD codes have evolved to the point where the use of CFD in industry is an accepted analysis tool. During this time, CFD vendors have expended most of their resources in enhancing the computational capabilities and expanding the areas of applicability of their software. However, the real payback from the use of CFD will come when it is used on a daily basis as a design tool by design engineers who are not CFD specialists. Now that these codes have reached a reasonable level of maturity and acceptance, attention must be directed to those issues which need to be addressed in order for CFD to achieve greater acceptance at the design level. In the remainder of this paper I will review those issues I believe will be key issues for commercial CFD in the 90's. The evolving needs of users of current CFD codes are largely responsible for the emergence of these issues.

3.6.1 Computational Efficiency / Accuracy

The characteristics of CFD problems are well-known. Almost all flows are highly nonlinear in nature: the source of the nonlinearity may be diverse including strong advection effects, turbulence, non-Newtonian effects, variable fluid properties, nonlinear boundary conditions such as radiative heat transfer, or combinations of these and other effects. Very often disparate length and/or time scales are present in the same problem. In more complex problems, the number of unknowns at each grid point can be large including velocity components, pressure, temperature, two or more turbulence variables and multiple species concentrations; thus the number of unknowns per grid point may range from four for a 3-D laminar isothermal problem to greater than twenty for a problem involving all the possible variables. Coupled with the fact that complex 3-D geometries can rarely be adequately modelled, even on a geometric level, with less than 50,000 – 100,000 mesh points leads to simulations involving from 200,000 to millions of nonlinear equations – a formidable challenge for even today's supercomputers.

Nevertheless, the continuing revolution in computer hardware is making CFD accessible to an ever widening range of companies. However, the appetite of CFD is such that it will always be computationally bound. Large-scale 3-D simulations of increasingly complex applications will continue to tax even future generations of supercomputers. Therefore computational efficiency will remain a crucial issue for years to come. However, computational efficiency goes hand in hand with computational accuracy; cost-effectiveness – NOT computational speed – is the key issue. The so-called "cost per grid point" measurement of a CFD code's speed is meaningless if it is not correlated with the accuracy achieved per grid point. The issue of accuracy is a critical one as accurate solutions are mandatory if CFD is to reach the design level. Algorithms are required which provide the user with a measure of a solution's accuracy and do not require the user to perform grid refinement and validation studies to verify a solution's accuracy. However, it must always be remembered that the solution provided by a CFD code, no matter how accurate numerically, is only as accurate as the underlying mathematical model.

An important area for algorithm researchers in CFD is the development of algorithms suited to emerging new computer architectures. Already most CFD codes make effective use of vector architectures. The trend in the future is clearly to multi-processor and parallel architectures. This will require the development of new algorithms which exploit the features of these new architectures. Massively parallel architectures appear to hold more promise for speeding up CFD calculations than the speedup of single CPU speeds. The blending of solution algorithms with hardware architectures without programming specifically for a particular machine is a difficult task and it is incumbent on computer manufacturers to provide the necessary tools to achieve these goals.

An area of considerable promise for improved cost-effectiveness is that of mesh and solution adaptation. This is the capability of numerical algorithms to modify themselves during a calculation to accommodate changing properties of numerical solutions. These modifications can be manifested in changes in the structure of the mesh, the order of the approximating functions, the location of nodes, the form of the solution algorithm, or in combinations of all these features. For example, in nearly all advection-dominated problems discontinuities or regions with sharp gradients appear. The regions in which the flow variables vary abruptly are usually small and are surrounded by large portions of the field in which the flow varies smoothly. It is therefore attractive to locally and adaptively refine the mesh where needed, until a preset tolerance for the error has been achieved.

The aim of any of the above modifications is to provide optimal automated control of the computational process: to control the computation to produce the "best" results for the least effort, i.e., reduce the cost and enhance the accuracy of the overall solution. A key issue is how "optimal" and "effort" are defined and measured; typically, it is the numerical error in the solution so that one attempts to control the accuracy of the computation. There is a valuable by-product of such an approach: if the error in a computation can indeed be estimated, then the user has a measure of the reliability of the calculation. The best adaptive procedures will function independently of the user, who will merely prescribe a level of error he can tolerate or a cost he is willing to endure to complete the flow simulation. Thereafter, the adaptive algorithm will make the decisions necessary to produce a solution within user-specified limits. For example, this control of error may be achieved by refining the mesh in areas of the solution domain where errors are too large and by coarsening the mesh where the error is small. Clearly this capability is tied to automated and unstructured grid capabilities – these topics will be discussed in the following section. Because of the obvious advantages of adaptive schemes, this field is receiving increased attention by researchers (cf. Demkowicz et al [1989] and Oden and Demkowicz [1988]).

3.6.2   Model (Mesh) Generation

Model generation continues to a be a frustration in computational fluid dynamics. For complex 3-D simulations it is not uncommon for 80%+ of the man-hours required to perform a simulation to be spent on model generation. Increased use of CFD is dependent on improved model generation techniques. Since automation of mesh generation has a potentially large payoff, extensive research is continuing in this area (cf. Blacker and Stephenson [1990], Ho-Le [1988], Yerri and Shephard [1984]). Unfortunately much of this research focuses on applications in the structural and solids mechanics fields and is often not optimal for fluid mechanics applications – for example, control of mesh density and structure in boundary layer regions is often difficult.

The accurate representation of arbitrary domains represents a challenging problem in any computational mechanics application. The magnitude of this problem does not become readily apparent in two dimensions, however, anyone trying to mesh complicated geometries in three dimensions with structured grids will encounter singular lines and the avoidable cost of overmeshing can no longer be ignored. This issue is of particular concern in CFD applications where boundary layer and other localized phenomena can be critical in complex geometries while other areas of the domain do not require the same resolution. It is my opinion that the ability to handle unstructured meshes will be mandatory for a CFD code to remain viable in an industrial environment.

Another consideration is that most real-world problems begin their life as solid model data in a CAD package. In the optimal environment the analyst should be able to generate a mesh discretization directly from the solid model database. This will require automated mesh generation techniques which can extract the necessary geometric data from the solid model database and with additional user information regarding mesh density and grading generate the appropriate mesh.

Finally, an automated mesh generation capability is a key component of any adaptive solution procedure as described in the previous section. The integration of automated and adaptive meshing techniques for unstructured meshes with geometry based CAD tools would relieve the analyst of the drudgery of the mesh generation and eliminate a great deal of the skill and art currently required to design adequate meshes and solution strategies for CFD simulations. This in turn would allow more time to be spent on the analysis of the results and the design process itself.

3.6.3   User Interface / Usability

The demographics of the average CFD user is changing – the typical user in the future will not be a CFD specialist. For CFD to reach the design environment ease of use is a vital consideration. The new generation of graphics workstations requires new Graphical User Interfaces – simplistic menus are no longer adequate. Moreover, the needs of both expert and novice users must be addressed – the current trend is clearly towards mouse-driven data entry with simplified point and click operation. Context sensitive on-line help should be available for all data entry. For those users more comfortable with the command-driven style of input, User Interfaces will need to support both modes of input interchangeably.

However, there is a potential danger underlying fancy easy-to-use user interfaces. Unfortunately, CFD is, and for the foreseeable future will be, a complex nonlinear application; an effective CFD user will always need a strong foundation in fluid mechanics in order to be able to relate the theoretical model to the physical model being addressed and to effectively evaluate the numerical results. And although an in depth knowledge of CFD may not be required, an effective CFD user should have a conceptual understanding of the numerical techniques used for the successful and cost-effective use of the program. "Smarter" software that provides automated tuning of solution methods and draws from the developer's and user's experience (i.e., some form of artificial intelligence) must be developed and will alleviate this need to a certain degree. The adaptive approach described earlier holds the greatest promise for achieving some of these goals.

3.6.4 Visualization

Rapid, user-friendly, flexible graphical display of data is a basic and vital tool during all stages of CFD: grid generation, debugging, evaluation and interpretation of results are rendered impossible without it. The visualization of CFD results, especially for 3-D internal flows, is particularly challenging and the problem of extracting information and graphically displaying this information in a meaningful way is acute. Matching these requirements to the capabilities offered by today's 3-D graphics workstations is a fascinating challenge. The use of transparency, translucency, dynamic manipulation of cutting surfaces and animation are but a few of the potential tools. Visualization tools need to be developed which mirror the physical techniques of dye injection, laser Doppler velocimetry and streaking to enable comparison of experiment and simulation. Animation will be essential to visualize transient analyses.

As larger simulations involving larger numbers of degrees of freedoms are performed the issue of visualizing results will continue to grow in importance and complexity. Already datasets from large-scale 3-D transient simulations can easily exceed a gigabyte in size and it is not inconceivable that post-processing operations can require more CPU resources than the computation itself. Particle tracing and interactive cutting surface manipulation on unstructured meshes of hundreds of thousands of grid points are highly compute intensive tasks. Clearly the advent of 3-D high-performance graphics workstations is revolutionizing CFD post-processing (and scientific visualization in general). This, together with the growth in size of CFD simulations, requires that greater attention be focussed on the entire computing environment. When a simulation is performed on a supercomputer it may not be viable to transfer to the workstation the datasets necessary for post-processing – the impracticability of downloading the data requires networked distributed file systems. Distributed graphics postprocessing, where the 3-D graphics is displayed and performed on the work-station while associated compute-intensive tasks are performed on the supercomputer (e.g., particle tracing computations), is another exciting possibility. Such systems are already in use today (see Grimsrud [1990]).

3.7 Concluding Comments

The potential for CFD in the 90's is unlimited. For this potential to be realized greater attention must be focused on the issues of computational efficiency, model generation, user interface and visualization. Although all these issues are important for the success and growth of CFD in the 90's, the development of viable algorithms for the simulation of real-world applications must always be the number one priority of CFD researchers and developers. There are still many application areas which cannot be addressed adequately, e.g. multiphase flows.

Thus there still remains a multitude of challenges facing CFD. The ultimate challenge for the commercial code developer is the unifying of techniques from a wide diversity of disciplines (graphical user interfaces, mesh generation, visualization, physical modelling and numerical algorithms, etc.) into a viable tool for the simulation of complex physical phenomenon. All these developments, issues and challenges promise to make the 90's a dynamic and exciting time to be involved in the field of Computational Fluid Dynamics.

References

[1] Aoyagi, Y., Takenaka, Y., Niino, S., Watanabe, A., and Joko, I. [1988] Numerical Simulation and Experimental Observation of Coolant Flow Around Cylinder Liners in V-8 Engine, SAE, 880109.

[2] Blacker, T. D. and Stepheson, M. D. [1990] "Paving, a new approach to automated quadrilateral mesh generation," Sandia Report, SAND90-0249, UC-705.

[3] Coyle, D. J., Macosko, C. W., and Scriven, L. E., [1988] "Film-splitting flows in forward roll coating," *J. Fluid Mechanics*, 171, 183–207.

[4] Dantzig, J. A. and Chao, L. S. [1986] The Effect of Shear Flows on Solidification Microstructures, Proceedings of 10th National Congress on Applied Mechanics.

[5] Demkowicz, L., Oden, J. T., Rachowicz, W., and Hardy, O. [1989] "Toward a Universal h-p adaptive finite element method," Parts 1 and 2, *Comp. Meth. in Appl. Mech. and Eng.*, 77, 77-180.14. Oden 1

[6] Engelman, M. S. and Bercovier, M. [1980] "Numerical simulation of Non-Newtonian blood flow models by a finite element method," *Computing Methods in Applied Sciences and Engineering*, Ed. R. Glowinski and J. L. Lions, North-Holland.

[7] Engelman, M. S. and Sani, R. L. [1983] FEM Simulation of an In-Package Pasteurization Process, *Numerical Heat Transfer*.

[8] Fontaine, J. P., Randriamampianuna, A., Extremet, G. P., and Bontoux, P. [1988] "Simulation of steady and time- dependent rotation-driven regimes in a liquid-encapsulated Czochralski configuration," *Journal of Crystal Growth* 97.

[9] Fraser, K., Coyle, D. J., and Bruker, I. [1989] "Evaluation of an injection molding mixing screw," *ANTEC*, 214.

[10] Givler, R., Gartling, D. K., Engelman, M. S., and Harotounian, V. [1991] "Navier-Stokes simulation of flow past three-dimensional submerged bodies," to appear in *Comp. Meth. in Appl. Mech. and Eng.*

[11] Grimsrud, A. [1990] MPGS 3.0 Users Manual, Cray Research, Minneapolis.

[12] Hasbani, Y. and Engelman, M. S. [1979] "An Out-of-Core Solution of Linear Equations with Non-Symmetric Coefficient Matrix," *Computers and Fluids*, 7, 13–31.

[13] Ho-Le, K. [1988] "Finite element mesh generation methods: A review and classification," *Computer Aided Design*, 20:1, 27–38.

[14] Lasbleiz, P., Chabard, J. P., and Hauguel, A., [1986] "Efficient computation of 3-D industrial turbulent flow by the finite element method," Proceedings of the Sixth Int. Symp. on FEM in Flow Problems, 189–194.

[15] Mansingh, V. and Misegades, K. [1990] "System level airflow analysis for a computer sytem processing unit," *Hewlett-Packard Journal*, October.

[16] Murnane, R. A., Johnson, W. W., and Moreland, N. J. [1988] "The Analysis of Glass melting processes using three-dimensional finite elements," *Int. J. Num. Meth. Fluid*, 8, 1491.

[17] Oden, J. T. and Demkowicz, L. [1988] "Advances in Adaptive Improvements: A survey of Adaptive Methods in Computational Fluid Mechanics," *State of the Art Surveys in Computational Mechanics*, Eds. A. K. Noor and J. T. Oden, ASME, N.Y.

[18] Roache, P. J. [1972] *Computational Fluid Dynamics*, Hermosa Publishers, Albuquerque, New Mexico.

[19] Saad, Y. [1981] "Krylov Subspace Methods for Solving Large Unsymmetric Linear Systems," *Math. Comp.*, 37, 105.

[20] Saad, Y., [1986] "GMRES: A Generalized Minimal Residual algorithm," *SIAM J. Sci. Stat. Comp.*, 7, 856–869.

[21] Torok, D. [1985] "Augmenting Experimental Methods for Flow and Thermal Performance Prediction in Electronic Packaging using Finite Elements," Fundamentals of Natural Convection/Electronic Cooling, ASME HTD-Vol. 32, 49–57.

[22] Yerry, M. A. and Shephard, M. S. [1984] "Automatic three-dimensional mesh generation by the modified-octree technique," *Int. J. Num. Meth. Eng.*, 20, 1965–1990.

4   Stabilized Finite Element Methods

Leopoldo P. Franca, Thomas J. R. Hughes, and Rolf Stenberg

4.1   Introduction

Physically the Stokes equations model "slow" flows of incompressible fluids or alternatively isotropic incompressible elastic materials. In Computational Fluid Dynamics, however, the Stokes equations have become an important model problem for designing and analyzing finite element algorithms. The reason being, that some of the problems encountered when solving the full Navier-Stokes equations are already present in the more simple Stokes equations. In particular, it gives the right setting for studying the stability problem connected with the choice of finite element spaces for the velocity and the pressure. It is well known that these spaces cannot be chosen independently when the discretization is based on the "Galerkin" variational form. This method belongs to the class of saddle-point problems for which an abstract theory has been developed by Brezzi [1974] and Babuška [1973]. The theory shows that the method is optimally convergent if the finite element spaces for velocity and pressure satisfy the "Babuška-Brezzi" or "inf-sup" condition. In computations the violation of this condition often leads to unphysical pressure oscillations and a "locking" of the velocity field, cf. Hughes [1987]. During the last decade this problem has been studied thoroughly and various velocity-pressure combinations have been shown to satisfy the Babuška-Brezzi condition. Unfortunately, however, it has turned out that many seemingly natural combinations do not satisfy it. (See Girault and Raviart [1986], Brezzi and Fortin [1991], and references therein.)

In this chapter we will review a recent technique of "stabilizing" mixed methods. In this approach the standard Galerkin form is modified by the addition of mesh-dependent terms which are weighted residuals of the differential equations. By this technique it is possible to avoid the stability problem connected with the classical mixed methods and hence the convergence can be established for a wide family of simple interpolations.

This methodology was first used in connection with advective flows in the works of Brooks and Hughes [1982], Hughes et al. [1979,1989] and Johnson et al. [1981,1984]. Developments of these formulations to mixed methods started in Hughes et al. [1986], motivated by the stabilization procedure proposed by Brezzi and Pitkäranta [1984] to the Stokes problem employing linear elements for velocity and pressure. Different from the formulation proposed by Brezzi and Pitkäranta, the work in Hughes et al. [1986]

87

presented a consistent formulation which allowed the construction of higher order approximations with optimal accuracy. Later numerous variants and extensions have been proposed and analyzed, cf. Hughes and Franca [1987], Franca et al. [1988b], Brezzi and Douglas [1988], Karam Filho and Loula [1988], Pierre [1988,1989], Douglas and Wang [1989], Durán and Notchetto [1989], Franca and Stenberg [1991], Silvester and Kechkar [1990], and Franca et al. [1990b]. The first papers, Hughes et al. [1986] and Hughes and Franca [1987], contained an error analysis which was improved in Brezzi and Douglas [1988] and Pierre [1989]. In Franca and Stenberg [1991] we presented a unified error analysis technique which can be used for all formulations.

Let us here also remark that these stabilization techniques are not restricted to problems in fluid mechanics. They can be applied to design stable finite element methods for a number of other problems in continuum mechanics, such as beams, plates and arches, see Franca and Hughes [1988a] and references therein.

The plan of the chapter is as follows. In the next section we introduce our notation and review the classical mixed formulation of the Stokes problem. In Section 4.3 we consider stabilization techniques. First, we review our technique by giving a self-contained analysis of the method of Hughes and Franca [1987] as modified in Franca and Stenberg [1991]. This method appears particularly attractive since it preserves the symmetry of the original Stokes operator. Next, we consider the method by Douglas and Wang [1989] in which the symmetry of the discretization is abandoned in favour of better stability characteristics for higher order interpolations. Finally, we consider two methods (proposed in Franca and Stenberg [1991]) which give rise to symmetric discretizations and in addition have the same stability robustness as the method of Douglas and Wang [1989].

4.2 Preliminaries

The purpose of this chapter is to discuss the stabilization of finite element methods for the Stokes problem. Hence, we will without loss of generality consider the Stokes equations with viscosity equal to unity and with homogeneous Dirichlet boundary conditions

$$
\begin{aligned}
-\Delta \mathbf{u} + \nabla p &= \mathbf{f} && \text{in } \Omega, \\
\operatorname{div} \mathbf{u} &= 0 && \text{in } \Omega, \\
\mathbf{u} &= \mathbf{0} && \text{on } \Gamma.
\end{aligned} \tag{4.2.1}
$$

Here $\mathbf{u} = (u_1, u_2, \ldots, u_N)$ is the velocity of the fluid, p is the pressure, and $\mathbf{f}$ is the body force. The domain $\Omega \subset \mathbb{R}^N$, $N = 2$ or 3, is assumed to be bounded with a polygonal or polyhedral boundary Γ.

Using the notation

$$B(\mathbf{w}, r; \mathbf{v}, q) = (\nabla \mathbf{w}, \nabla \mathbf{v}) - (\text{div } \mathbf{v}, r) - (\text{div } \mathbf{w}, q) \qquad (4.2.2)$$

and

$$F(\mathbf{v}, q) = (\mathbf{f}, \mathbf{v}), \qquad (4.2.3)$$

the variational formulation of the problem is the following. Given $\mathbf{f} \in [H^{-1}(\Omega)]^N$, find $(\mathbf{u}, p) \in [H_0^1(\Omega)]^N \times L_0^2(\Omega)$ such that

$$B(\mathbf{u}, p; \mathbf{v}, q) = F(\mathbf{v}, q) \qquad \forall (\mathbf{v}, q) \in [H_0^1(\Omega)]^N \times L_0^2(\Omega). \qquad (4.2.4)$$

Here and below we use standard notation (see Ciarlet [1978] for the notation not explicitly defined below). In particular, we denote by $(\cdot, \cdot)_D$ the inner product in $L^2(D)$, $[L^2(D)]^N$ or $[L^2(D)]^{N \times N}$ with the subscript D dropped for $D = \Omega$. Further, we denote the space of functions continuous on Ω by $C^0(\Omega)$ and

$$L_0^2(\Omega) = \{ q \in L^2(\Omega) \mid \int_\Omega q \, d\Omega = 0 \}.$$

C, C_j, $j \in \mathbb{N}$, and C_I stand for various positive constants independent of the mesh parameter h.

Remark. In the case of various "outflow" boundary conditions it is probably more correct to use $(\varepsilon(\mathbf{u}), \varepsilon(\mathbf{v}))$ (here $\varepsilon(\mathbf{u})$ is the symmetric part of the velocity gradient) instead of $(\nabla \mathbf{u}, \nabla \mathbf{v})$ in the variational and finite element formulations. For the present discussion this is an irrelevant matter since all results are trivially valid for this case as well. ☐

The problem 4.2.4 is well posed, i.e. there is a positive constant C such that

$$\sup_{\substack{(\mathbf{v}, q) \in [H_0^1(\Omega)]^N \times L_0^2(\Omega) \\ (\mathbf{v}, q) \neq (\mathbf{0}, 0)}} \frac{B(\mathbf{u}, p; \mathbf{v}, q)}{\|\mathbf{v}\|_1 + \|q\|_0} \geq C(\|\mathbf{u}\|_1 + \|p\|_0) \qquad \forall (\mathbf{u}, p) \in [H_0^1(\Omega)]^N \times L_0^2(\Omega).$$

$$(4.2.5)$$

This inequality is a simple consequence of the Poincaré inequality (or Korn's inequality when $(\varepsilon(\mathbf{u}), \varepsilon(\mathbf{v}))$ is used in the variational formulation) and the condition

$$\sup_{\mathbf{0} \neq \mathbf{v} \in [H_0^1(\Omega)]^N} \frac{(\text{div } \mathbf{v}, p)}{\|\mathbf{v}\|_1} \geq C\|p\|_0 \qquad \forall p \in L_0^2(\Omega). \qquad (4.2.6)$$

In the traditional mixed method the same variational problem is solved in some finite element subspaces. Let $\mathbf{V}_h \subset [H_0^1(\Omega)]^N$ and $P_h \subset L_0^2(\Omega)$ be defined as piecewise polynomials on a regular partitioning $\mathcal{C}_h$ of $\overline{\Omega}$ into elements consisting of triangles (tetrahedrons in $\mathbb{R}^3$) or convex quadrilaterals (hexahedrons). We then get the so called

Galerkin Method: Find $\mathbf{u}_h \in \mathbf{V}_h$ *and* $p_h \in P_h$ *such that*

$$B(\mathbf{u}_h, p_h; \mathbf{v}, q) = F(\mathbf{v}, q) \qquad \forall(\mathbf{v}, q) \in \mathbf{V}_h \times P_h. \qquad \qquad \square$$

In the discrete case the same steps used for proving that the Poincaré inequality and (4.2.6) implies (4.2.5) can be repeated. Hence, we have the following famous result.

Proposition 4.2.1 (Babuška [1973], Brezzi [1974]) *If the finite element spaces* $\mathbf{V}_h$ *and* P_h *satisfy the condition*

$$\sup_{0 \neq \mathbf{v} \in \mathbf{V}_h} \frac{(\operatorname{div} \mathbf{v}, p)}{\|\mathbf{v}\|_1} \geq C\|p\|_0 \qquad \forall p \in P_h, \qquad (4.2.7)$$

then the inequality

$$\sup_{\substack{(\mathbf{v},q) \in \mathbf{V}_h \times P_h \\ (\mathbf{v},q) \neq (\mathbf{0},0)}} \frac{B(\mathbf{u}, p; \mathbf{v}, q)}{\|\mathbf{v}\|_1 + \|q\|_0} \geq C(\|\mathbf{u}\|_1 + \|p\|_0) \qquad \forall(\mathbf{u}, p) \in \mathbf{V}_h \times P_h$$

is valid. $\qquad \qquad \square$

It is evident that this excludes many finite element spaces and, as already noted in the introduction, many natural choices cannot be used. For a survey of methods which have been proven to satisfy this condition we refer to Girault and Raviart [1986] and Brezzi and Fortin [1991].

In the case that Proposition 4.2.1 is valid, the method will converge optimally. If piecewise polynomials of degree k and l are used for the velocity and pressure, respectively, then we have the following error estimate

$$\|\mathbf{u} - \mathbf{u}_h\|_1 + \|p - p_h\|_0 \leq C(h^k|\mathbf{u}|_{k+1} + h^{l+1}|p|_{l+1}),$$

provided that $\mathbf{u} \in [H^{k+1}(\Omega)]^N$ and $p \in H^{l+1}(\Omega)$.

4.3 Stabilized Methods

Let us consider approximation by Lagrange elements which are without doubt the most popular in practice. For convenience we will adopt the following notation

$$R_m(K) = \begin{cases} P_m(K) & \text{if } K \text{ is a triangle or tetrahedron,} \\ Q_m(K) & \text{if } K \text{ is a quadrilateral or hexahedron,} \end{cases}$$

where for each integer $m \geq 0$, $P_m(K)$ and $Q_m(K)$ are the usual polynomial spaces on K (see Ciarlet [1978]).

The finite element spaces $\mathbf{V}_h$ and P_h for approximating the velocity and the pressure, respectively, are then defined as

$$\mathbf{V}_h = \{\mathbf{v} \in [H_0^1(\Omega)]^N \mid \mathbf{v}_{|K} \in [R_k(K)]^N \ \forall K \in \mathcal{C}_h\}, \qquad (4.3.1)$$

$$P_h = \{p \in \mathcal{C}^0(\Omega) \cap L_0^2(\Omega) \mid p_{|K} \in R_l(K) \ \forall K \in \mathcal{C}_h\}, \qquad (4.3.2)$$

or

$$P_h = \{p \in L_0^2(\Omega) \mid p_{|K} \in R_l(K) \ \forall K \in \mathcal{C}_h\}. \qquad (4.3.3)$$

Here we have two alternatives for the pressure depending on if it is approximated continuously or not. Let us here also remark that for a two-dimensional domain one could mix triangles and quadrilaterals.

For stating our results it is convenient to use the following notation

$$S(K) = \begin{cases} P_N(K) & \text{if } K \text{ is a triangle or tetrahedron,} \\ Q_2(K) & \text{if } K \text{ is a quadrilateral or hexahedron,} \end{cases}$$

and

$$\mathbf{S}_h = \{\, \mathbf{v} \in [H_0^1(\Omega)]^N \mid \mathbf{v}_{|K} \in [S(K)]^N \ \forall K \in \mathcal{C}_h\}. \qquad (4.3.4)$$

Next, we recall that since $\mathbf{V}_h$ is assumed to consist of piecewise polynomials, a simple scaling argument shows that there is a positive constant C_I such that

$$C_I \sum_{K \in \mathcal{C}_h} h_K^2 \|\Delta \mathbf{v}\|_{0,K}^2 \le \|\nabla \mathbf{v}\|_0^2 \qquad \forall \mathbf{v} \in \mathbf{V}_h. \qquad (4.3.5)$$

Now let us define:

Method I (Hughes and Franca [1987], Franca and Stenberg [1991]). *Find $\mathbf{u}_h \in \mathbf{V}_h$ and $p_h \in P_h$ such that*

$$B_h(\mathbf{u}_h, p_h; \mathbf{v}, q) = F_h(\mathbf{v}, q) \qquad \forall(\mathbf{v}, q) \in \mathbf{V}_h \times P_h,$$

with

$$B_h(\mathbf{w}, r; \mathbf{v}, q) = B(\mathbf{w}, r; \mathbf{v}, q) - \alpha \sum_{K \in \mathcal{C}_h} h_K^2(-\Delta\mathbf{w} + \nabla r, -\Delta\mathbf{v} + \nabla q)_K$$

and

$$F_h(\mathbf{v}, q) = F(\mathbf{v}, q) - \alpha \sum_{K \in \mathcal{C}_h} h_K^2(\mathbf{f}, -\Delta\mathbf{v} + \nabla q)_K. \qquad \Box$$

The first observation concerning this method (and all the methods to follow) is that it is consistent, i.e. if we (e.g.) assume that $\mathbf{f} \in [L^2(\Omega)]^N$, then the exact solution $(\mathbf{u}, p)$ satisfies the discrete equation

$$B_h(\mathbf{u}, p; \mathbf{v}, q) = F_h(\mathbf{v}, q) \qquad \forall(\mathbf{v}, q) \in \mathbf{V}_h \times P_h. \qquad (4.3.6)$$

The next result, when compared with Proposition 4.2.1, shows the superiority of this formulation compared with the Galerkin method.

Theorem 4.3.1 *Assume that either* $\mathbf{S}_h \subset \mathbf{V}_h$ *or* $P_h \subset C^0(\Omega)$ *and that* $0 < \alpha < C_I$. *Then the bilinear form of Method I satisfies*

$$\sup_{\substack{(\mathbf{v},q)\in \mathbf{V}_h\times P_h \\ (\mathbf{v},q)\neq(\mathbf{0},0)}} \frac{B_h(\mathbf{u}, p; \mathbf{v}, q)}{\|\mathbf{v}\|_1 + \|q\|_0} \geq C(\|\mathbf{u}\|_1 + \|p\|_0) \qquad \forall(\mathbf{u}, p) \in \mathbf{V}_h \times P_h. \qquad \square$$

Before going into the details of proving this result, let us note that it implies the following optimal error estimate.

Theorem 4.3.2 *Assume that either* $\mathbf{S}_h \subset \mathbf{V}_h$ *or* $P_h \subset C^0(\Omega)$ *and that* $0 < \alpha < C_I$. *For the approximate solution obtained with Method I we then have*

$$\|\mathbf{u} - \mathbf{u}_h\|_1 + \|p - p_h\|_0 \leq C(h^k|\mathbf{u}|_{k+1} + h^{l+1}|p|_{l+1}),$$

provided the exact solution satisfies $\mathbf{u} \in [H^{k+1}(\Omega)]^N$ *and* $p \in H^{l+1}(\Omega)$.

Proof: Let $\tilde{\mathbf{u}} \in \mathbf{V}_h$ be the interpolant of $\mathbf{u}$ and let $\tilde{p} \in P_h$ be the interpolant of p. Theorem 4.3.1 now implies the existence of $(\mathbf{v}, q) \in \mathbf{V}_h \times P_h$ such that

$$\|\mathbf{v}\|_1 + \|q\|_0 \leq C \qquad (4.3.7)$$

and

$$\|\tilde{\mathbf{u}} - \mathbf{u}_h\|_1 + \|\tilde{p} - p_h\|_0 \leq B_h(\mathbf{u}_h - \tilde{\mathbf{u}}, p_h - \tilde{p}; \mathbf{v}, q). \qquad (4.3.8)$$

The consistency (4.3.6) now gives

$$B_h(\mathbf{u}_h - \tilde{\mathbf{u}}, p_h - \tilde{p}; \mathbf{v}, q) = B_h(\mathbf{u} - \tilde{\mathbf{u}}, p - \tilde{p}; \mathbf{v}, q). \qquad (4.3.9)$$

For the right hand side above the Schwarz inequality gives

$$B_h(\mathbf{u} - \tilde{\mathbf{u}}, p - \tilde{p}; \mathbf{v}, q)$$
$$\leq C\Big(\|\mathbf{u} - \tilde{\mathbf{u}}\|_1^2 + \sum_{K\in\mathcal{C}_h} h_K^2\|\Delta(\mathbf{u} - \tilde{\mathbf{u}})\|_{0,K}^2 + \|p - \tilde{p}\|_0^2 + \sum_{K\in\mathcal{C}_h} h_K^2\|\nabla(p - \tilde{p})\|_{0,K}^2\Big)^{1/2}$$
$$\cdot \Big(\|\mathbf{v}\|_1^2 + \sum_{K\in\mathcal{C}_h} h_K^2\|\Delta\mathbf{v}\|_{0,K}^2 + \|q\|_0^2 + \sum_{K\in\mathcal{C}_h} h_K^2\|\nabla q\|_{0,K}^2\Big)^{1/2}.$$

$$(4.3.10)$$

Hence, the triangle inequality, the inverse inequalities (4.3.5) and

$$\Big(\sum_{K \in \mathcal{C}_h} h_K^2 \|\nabla q\|_{0,K}^2 \Big)^{1/2} \leq C \|q\|_0,$$

(4.3.7) to (4.3.10) give

$$\begin{aligned}
\|\mathbf{u} - \mathbf{u}_h\|_1 + \|p - p_h\|_0 \;\leq\; & C\Big(\|\mathbf{u} - \tilde{\mathbf{u}}\|_1 + \big(\sum_{K \in \mathcal{C}_h} h_K^2 \|\Delta(\mathbf{u} - \tilde{\mathbf{u}})\|_{0,K}^2 \big)^{1/2} \\
& + \|p - \tilde{p}\|_0 + \big(\sum_{K \in \mathcal{C}_h} h_K^2 \|\nabla(p - \tilde{p})\|_{0,K}^2 \big)^{1/2}\Big).
\end{aligned}$$

The following interpolation estimate is standard

$$\begin{aligned}
\|\mathbf{u} - \tilde{\mathbf{u}}\|_1 + \big(\sum_{K \in \mathcal{C}_h} h_K^2 \|\Delta(\mathbf{u} - \tilde{\mathbf{u}})\|_{0,K}^2 \big)^{1/2} + \|p - \tilde{p}\|_0 + \big(\sum_{K \in \mathcal{C}_h} h_K^2 \|\nabla(p - \tilde{p})\|_{0,K}^2 \big)^{1/2} \\
\leq C(h^k |\mathbf{u}|_{k+1} + h^{l+1} |p|_{l+1}).
\end{aligned}$$

and the theorem is thus proved. $\qquad\square$

Remarks

1. With the regularity assumption (valid in the two-dimensional case for a convex polygon Ω)

$$\|\mathbf{u}\|_2 + \|p\|_1 \leq C \|\mathbf{f}\|_0 \tag{4.3.11}$$

   the usual duality argument gives the optimal L^2-estimate (cf. Franca and Stenberg [1991])

$$\|\mathbf{u} - \mathbf{u}_h\|_0 \leq C(h^{k+1}|\mathbf{u}|_{k+1} + h^{l+2}|p|_{l+1}).$$

2. It is important to note that for piecewise linear approximations to the velocity the bilinear and linear form reduce to

$$B_h(\mathbf{w}, r; \mathbf{v}, q) = B(\mathbf{w}, r; \mathbf{v}, q) - \alpha \sum_{K \in \mathcal{C}_h} h_K^2 (\nabla r, \nabla q)_K$$

   and

$$F_h(\mathbf{v}, q) = F(\mathbf{v}, q) - \alpha \sum_{K \in \mathcal{C}_h} h_K^2 (\mathbf{f}, \nabla q)_K,$$

   and in this case no upper limit has to be imposed on α. If we additionally make the simplification $F_h(\mathbf{v}, q) = F(\mathbf{v}, q)$, we get the method of Brezzi and Pitkäranta [1984]. Let us also remark that for isoparametric bilinear (in $\mathbb{R}^2$) or trilinear (in $\mathbb{R}^3$) B_h and F_h can be defined as above without a loss of accuracy.

3. For continuous linear approximations (on triangles or tetrahedrons) for both the velocity and pressure it was noted in Pierre [1989] that there is a close connection between this method and a well known Galerkin method, the MINI element of Arnold, Brezzi and Fortin (cf. Girault and Raviart [1986] or Brezzi and Fortin [1991]). In the MINI element the stability is achieved by adding local "bubble functions" to the velocity space. When these bubble degrees of freedom are condensed, one gets the above stabilized method with a particular choice for the stabilization parameter α, see Pierre [1989] for the details.

4. From the error estimate above it is clear that the optimal choice for the finite element spaces is obtained with $l = k - 1$. We have, however, preferred not to do this choice when presenting our result, since we want to emphasize how arbitrary one can choose the spaces and still obtain a stable method. Also, for the full Navier-Stokes equations it might be good to use other combinations than $l = k - 1$ (such as equal order approximations). □

Next, let us review the technique given in Franca and Stenberg [1991] for verifying the stability. The following lemma turns out to be very useful for analyzing both traditional Galerkin methods (cf. Stenberg [1990]) and the present stabilized methods. For Galerkin methods this idea was first used by Verfürth [1984] for the analysis of "Taylor-Hood" methods, i.e. methods with a continuous pressure approximation.

Lemma 4.3.1 *Suppose that either one of the assumptions* $\mathbf{S}_h \subset \mathbf{V}_h$ *or* $P_h \subset C^0(\Omega)$ *is valid. Then there exist positive constants* C_1 *and* C_2 *such that*

$$\sup_{0 \neq \mathbf{v} \in \mathbf{V}_h} \frac{(\operatorname{div} \mathbf{v}, p)}{\|\mathbf{v}\|_1} \geq C_1 \|p\|_0 - C_2 \Big(\sum_{K \in \mathcal{C}_h} h_K^2 \|\nabla p\|_{0,K}^2 \Big)^{1/2} \qquad \forall p \in P_h .$$

Proof: Consider first the case $\mathbf{S}_h \subset \mathbf{V}_h$. Denote by Π_h the L^2-projection onto the space of piecewise constants, i.e.

$$\Pi_h q_{|K} = \frac{1}{\operatorname{meas}(K)} \int_K q \, d\Omega , \qquad \forall K \in \mathcal{C}_h .$$

Since $\mathbf{S}_h \subset \mathbf{V}_h$ the pair $(\mathbf{V}_h, \Pi_h P_h)$ satisfies the stability inequality (4.2.7) (see e.g. Girault and Raviart [1986]). Hence there is a constant C_1 such that for every $p \in P_h$ there exists $\mathbf{v} \in \mathbf{V}_h$, with $\|\mathbf{v}\|_1 = 1$, such that

$$(\operatorname{div} \mathbf{v}, \Pi_h p) \geq C_1 \|\Pi_h p\|_0 .$$

Using the interpolation estimate

$$\|(I - \Pi_h)p\|_{0,K} \leq C_3 h_K \|\nabla p\|_{0,K} , \qquad \forall K \in \mathcal{C}_h ,$$

we now get

$$
\begin{aligned}
(\operatorname{div} \mathbf{v}, p) &= (\operatorname{div} \mathbf{v}, \Pi_h p) + (\operatorname{div} \mathbf{v}, (I - \Pi_h)p) \\
&\geq C_1 \|\Pi_h p\|_0 - \|\mathbf{v}\|_1 \|(I - \Pi_h)p\|_0 \\
&= C_1 \|\Pi_h p\|_0 - \|(I - \Pi_h)p\|_0 \\
&\geq C_1 \|p\|_0 - (1 + C_1)\|(I - \Pi_h)p\|_0 \\
&\geq C_1 \|p\|_0 - (1 + C_1)C_3 \Big(\sum_{K \in \mathcal{C}_h} h_K^2 \|\nabla p\|_{0,K}^2\Big)^{1/2}
\end{aligned}
$$

and the asserted estimate with the first assumption is proved.

Next, let us consider the case $P_h \subset \mathcal{C}^0(\Omega)$. Since $p \in P_h \subset L_0^2(\Omega)$, by the continuous version of the "inf-sup" condition, i.e (4.2.6), there is a non-vanishing $\mathbf{w} \in [H_0^1(\Omega)]^N$ such that

$$
(\operatorname{div} \mathbf{w}, p) \geq C_4 \|p\|_0 \|\mathbf{w}\|_1 .
$$

Further, one can show (e.g. Girault and Raviart [1986], pp. 109-111) that there is an interpolant $\tilde{\mathbf{w}} \in \mathbf{V}_h$ to $\mathbf{w}$ such that

$$
\Big(\sum_{K \in \mathcal{C}_h} h_K^{-2} \|\mathbf{w} - \tilde{\mathbf{w}}\|_{0,K}^2\Big)^{1/2} \leq C_5 \|\mathbf{w}\|_1
$$

and

$$
\|\tilde{\mathbf{w}}\|_1 \leq C_6 \|\mathbf{w}\|_1 . \tag{4.3.12}
$$

Integrating by parts on each $K \in \mathcal{C}_h$, and using the above estimates we get

$$
\begin{aligned}
(\operatorname{div} \tilde{\mathbf{w}}, p) &= (\operatorname{div}(\tilde{\mathbf{w}} - \mathbf{w}), p) + (\operatorname{div} \mathbf{w}, p) \\
&\geq (\operatorname{div}(\tilde{\mathbf{w}} - \mathbf{w}), p) + C_4 \|\mathbf{w}\|_1 \|p\|_0 \\
&= \sum_{K \in \mathcal{C}_h} (\mathbf{w} - \tilde{\mathbf{w}}, \nabla p)_K + C_4 \|\mathbf{w}\|_1 \|p\|_0 \\
&\geq -\Big(\sum_{K \in \mathcal{C}_h} h_K^{-2} \|\mathbf{w} - \tilde{\mathbf{w}}\|_{0,K}^2\Big)^{1/2} \cdot \Big(\sum_{K \in \mathcal{C}_h} h_K^2 \|\nabla p\|_{0,K}^2\Big)^{1/2} + C_4 \|\mathbf{w}\|_1 \|p\|_0 \\
&\geq \Big\{-C_5\Big(\sum_{K \in \mathcal{C}_h} h_K^2 \|\nabla p\|_{0,K}^2\Big) + C_4 \|p\|_0\Big\} \|\mathbf{w}\|_1 .
\end{aligned}
$$

Dividing by $\|\mathbf{w}\|_1$ we obtain

$$
\frac{(\operatorname{div} \tilde{\mathbf{w}}, p)}{\|\mathbf{w}\|_1} \geq C_4 \|p\|_0 - C_5 \Big(\sum_{K \in \mathcal{C}_h} h_K^2 \|\nabla p\|_{0,K}^2\Big)^{1/2}
$$

and (4.3.12) then gives

$$\frac{(\operatorname{div}\tilde{\mathbf{w}}, p)}{\|\tilde{\mathbf{w}}\|_1} \geq \frac{(\operatorname{div}\tilde{\mathbf{w}}, p)}{C_6\|\mathbf{w}\|_1} \geq C_6^{-1}\{C_4\|p\|_0 - C_5(\sum_{K\in\mathcal{C}_h} h_K^2\|\nabla p\|_{0,K}^2)^{1/2}\}.$$

The assertion is thus proved. $\qquad\qquad\qquad\qquad\qquad\qquad\qquad\qquad\qquad$ □

Let us now use this lemma for the

Proof of Theorem 4.3.1. Let $(\mathbf{u}, p) \in \mathbf{V}_h \times P_h$ be given and let $\mathbf{w} \in \mathbf{V}_h$ be a function for which the supremum of Lemma 4.3.1 is obtained. We scale $\mathbf{w}$ so that $\|\mathbf{w}\|_1 = \|p\|_0$. Using the stability estimate of Lemma 4.3.1, the inverse inequality (4.3.5) and the arithmetic-geometric mean inequality we get

$$B_h(\mathbf{u}, p; -\mathbf{w}, 0)$$
$$= -(\nabla\mathbf{u}, \nabla\mathbf{w}) + (\operatorname{div}\mathbf{w}, p) - \alpha\sum_{K\in\mathcal{C}_h} h_K^2(-\Delta\mathbf{u} + \nabla p, \Delta\mathbf{w})_K$$
$$\geq -\|\nabla\mathbf{u}\|_0\|\nabla\mathbf{w}\|_0 + C_1\|p\|_0^2 - C_2(\sum_{K\in\mathcal{C}_h} h_K^2\|\nabla p\|_{0,K}^2)^{1/2}\|p\|_0$$
$$+ \alpha\sum_{K\in\mathcal{C}_h} h_K^2(\Delta\mathbf{u}, \Delta\mathbf{w})_K - \alpha\sum_{K\in\mathcal{C}_h} h_K^2(\nabla p, \Delta\mathbf{w})_K$$
$$\geq -\|\nabla\mathbf{u}\|_0\|\nabla\mathbf{w}\|_0 + C_1\|p\|_0^2 - C_2(\sum_{K\in\mathcal{C}_h} h_K^2\|\nabla p\|_{0,K}^2)^{1/2}\|p\|_0$$
$$- \alpha(\sum_{K\in\mathcal{C}_h} h_K^2\|\Delta\mathbf{u}\|_{0,K}^2)^{1/2}(\sum_{K\in\mathcal{C}_h} h_K^2\|\Delta\mathbf{w}\|_{0,K}^2)^{1/2}$$
$$- \alpha(\sum_{K\in\mathcal{C}_h} h_K^2\|\nabla p\|_{0,K}^2)^{1/2}(\sum_{K\in\mathcal{C}_h} h_K^2\|\Delta\mathbf{w}\|_{0,K}^2)^{1/2} \qquad (4.3.13)$$
$$\geq -\|\nabla\mathbf{u}\|_0\|\nabla\mathbf{w}\|_0 + C_1\|p\|_0^2 - C_2(\sum_{K\in\mathcal{C}_h} h_K^2\|\nabla p\|_{0,K}^2)^{1/2}\|p\|_0$$
$$- \alpha C_I^{-1}\|\nabla\mathbf{u}\|_0\|\nabla\mathbf{w}\|_0 - \alpha C_I^{-1/2}(\sum_{K\in\mathcal{C}_h} h_K^2\|\nabla p\|_{0,K}^2)^{1/2}\|\nabla\mathbf{w}\|_0$$
$$\geq -C_3\|\nabla\mathbf{u}\|_0\|\nabla\mathbf{w}\|_0 + C_1\|p\|_0^2 - C_4(\sum_{K\in\mathcal{C}_h} h_K^2\|\nabla p\|_{0,K}^2)^{1/2}\|p\|_0$$
$$\geq -C_3\|\nabla\mathbf{u}\|_0\|p\|_0 + C_1\|p\|_0^2 - C_4(\sum_{K\in\mathcal{C}_h} h_K^2\|\nabla p\|_{0,K}^2)^{1/2}\|p\|_0$$
$$\geq -\frac{C_3}{2\epsilon}\|\nabla\mathbf{u}\|_0^2 + (C_1 - \frac{\epsilon}{2}(C_3 + C_4))\|p\|_0^2 - \frac{C_4}{2\epsilon}\sum_{K\in\mathcal{C}_h} h_K^2\|\nabla p\|_{0,K}^2$$

$$\geq \quad -C_5\|\nabla\mathbf{u}\|_0^2 + C_6\|p\|_0^2 - C_7 \sum_{K\in\mathcal{C}_h} h_K^2 \|\nabla p\|_{0,K}^2,$$

when choosing $0 < \epsilon < 2C_1(C_3 + C_4)^{-1}$. Next, we note that Poincaré's inequality and the assumption $0 < \alpha < C_I$ give

$$
\begin{aligned}
B_h(\mathbf{u}, p; \mathbf{u}, -p) &= \|\nabla\mathbf{u}\|_0^2 + \alpha \sum_{K\in\mathcal{C}_h} h_K^2 \|\nabla p\|_{0,K}^2 - \alpha \sum_{K\in\mathcal{C}_h} h_K^2 \|\Delta\mathbf{u}\|_{0,K}^2 \\
&\geq (1 - \alpha C_I^{-1})\|\nabla\mathbf{u}\|_0^2 + \alpha \sum_{K\in\mathcal{C}_h} h_K^2 \|\nabla p\|_{0,K}^2 \qquad (4.3.14) \\
&\geq C_8(\|\mathbf{u}\|_1^2 + \sum_{K\in\mathcal{C}_h} h_K^2 \|\nabla p\|_{0,K}^2),
\end{aligned}
$$

Denote $(\mathbf{v}, q) = (\mathbf{u} - \delta\mathbf{w}, -p)$. Combining (4.3.14) and (4.3.15) gives

$$
\begin{aligned}
B_h(\mathbf{u}, p; \mathbf{v}, q) &= B_h(\mathbf{u}, p; \mathbf{u} - \delta\mathbf{w}, -p) \\
&= B_h(\mathbf{u}, p; \mathbf{u}, -p) + \delta B_h(\mathbf{u}, p; -\mathbf{w}, 0) \qquad (4.3.15) \\
&\geq (C_8 - \delta C_5)\|\nabla\mathbf{u}\|_0^2 + \delta C_6\|p\|_0^2 + (C_8 - \delta C_7) \sum_{K\in\mathcal{C}_h} h_K^2 \|\nabla p\|_{0,K}^2 \\
&\geq C(\|\nabla\mathbf{u}\|_0^2 + \|p\|_0^2)
\end{aligned}
$$

when choosing $0 < \delta < \min\{C_8 C_5^{-1}, C_8 C_7^{-1}\}$. On the other hand we have

$$
\begin{aligned}
\|\nabla\mathbf{v}\|_0 + \|q\|_0 &\leq \|\nabla\mathbf{u}\|_0 + \delta\|\nabla\mathbf{w}\|_0 + \|p\|_0 \\
&\leq C(\|\nabla\mathbf{u}\|_0 + \|p\|_0),
\end{aligned}
$$

which combined with (4.3.16) (and the Poincaré inequality) proves the stability estimate. □

For continuous approximations for the pressure the above method was introduced in Hughes and Franca [1987]. For discontinuous pressures, however, the original formulation contained "jump terms" on the pressure, i.e. the bilinear form was defined as

$$
\begin{aligned}
B_h(\mathbf{w}, r; \mathbf{v}, q) &= B(\mathbf{w}, r; \mathbf{v}, q) - \alpha \sum_{K\in\mathcal{C}_h} h_K^2 (-\Delta\mathbf{w} + \nabla r, -\Delta\mathbf{v} + \nabla q)_K \\
&\quad - \beta \sum_{T\in\Gamma_h} h_T (\llbracket r \rrbracket, \llbracket q \rrbracket)_T,
\end{aligned}
$$

where Γ_h denotes the collection of all element interfaces, $\llbracket s \rrbracket$ is the jump in s along the interface and h_T is the length or diameter of T. β is a positive parameter. With these jump terms there does not seem to be any reason to use a discontinuous pressure instead of a continuous one. (Some arguments in favour of a modified jump formulation are, however, given in Silvester and Kechkar [1990].) Furthermore, it leads to a nonstandard assembly procedure. In Franca and Stenberg [1991] we, however, showed that with the assumption $S_h \subset V_h$ these additional jump terms above are unnecessary. This has a considerable practical significance since it allows the pressure to be eliminated at the element level by the penalty technique. Alternatively, one can use the augmented Lagrangian technique, see e.g. Brezzi and Fortin [1991]. Here we will briefly review the penalty technique. We replace the bilinear form with

$$B_h^\epsilon(\mathbf{w}, r; \mathbf{v}, q) = B(\mathbf{w}, r; \mathbf{v}, q) - \alpha \sum_{K \in \mathcal{C}_h} h_K^2 (-\Delta \mathbf{w} + \nabla r, -\Delta \mathbf{v} + \nabla q)_K - \epsilon (r, q),$$

and solve the problem

$$B_h^\epsilon(\mathbf{u}_h^\epsilon, p_h^\epsilon; \mathbf{v}, q) = F_h(\mathbf{v}, q) \qquad \forall (\mathbf{v}, q) \in \mathbf{V}_h \times P_h. \tag{4.3.16}$$

Now it is straightforward to show that if the original method is stable then the error estimates do not change if $0 \le \epsilon \le Ch^s$, $s = \min\{k + 1, l + 2\}$, i.e. we have

$$\|\mathbf{u} - \mathbf{u}_h^\epsilon\|_1 + \|p - p_h^\epsilon\|_0 \le C(h^k |\mathbf{u}|_{k+1} + h^{l+1} |p|_{l+1})$$

and also

$$\|\mathbf{u} - \mathbf{u}_h^\epsilon\|_0 \le C(h^{k+1} |\mathbf{u}|_{k+1} + h^{l+2} |p|_{l+1})$$

when (4.3.11) holds.

Let now $\{\phi_i\}_{i=1}^{N_u}$ and $\{\psi_i\}_{i=1}^{N_p}$ be the basis for V_h and P_h, respectively. The discretization (4.3.16) then leads to the following matrix equation

$$\begin{pmatrix} \mathbf{A} & \mathbf{B} \\ \mathbf{B}^T & \mathbf{C} \end{pmatrix} \begin{pmatrix} \mathbf{U} \\ \mathbf{P} \end{pmatrix} = \begin{pmatrix} \mathbf{F}_1 \\ \mathbf{F}_2 \end{pmatrix} \tag{4.3.17}$$

where $\mathbf{U}$ and $\mathbf{P}$ are the degrees of freedom for $\mathbf{u}_h^\epsilon$ and p_h^ϵ, respectively, and the matrices are given by

$$(\mathbf{A})_{ij} = (\nabla \phi_i, \nabla \phi_j) - \alpha \sum_{K \in \mathcal{C}_h} h_K^2 (\Delta \phi_i, \Delta \phi_j)_K,$$

$$(\mathbf{B})_{ij} = -(\text{div } \phi_j, \psi_i) + \alpha \sum_{K \in \mathcal{C}_h} h_K^2 (\Delta \phi_j, \nabla \psi_i)_K,$$

$$(\mathbf{C})_{ij} = -\epsilon (\psi_i, \psi_j) - \alpha \sum_{K \in \mathcal{C}_h} h_K^2 (\nabla \psi_i, \nabla \psi_j)_K,$$

$$(\mathbf{F}_1)_i = (\mathbf{f}, \phi_i) + \alpha \sum_{K \in \mathcal{C}_h} h_K^2 (\mathbf{f}, \Delta \phi_i)_K,$$

$$(\mathbf{F}_2)_i = -\alpha \sum_{K \in \mathcal{C}_h} h_K^2 (\mathbf{f}, \nabla \psi_i)_K.$$

Now, since $\epsilon > 0$ the matrix $\mathbf{C}$ is negative definite and it can be inverted on each element separately (we have assumed that the pressure approximation is discontinuous). By eliminating $\mathbf{P}$, we get the following system for $\mathbf{U}$ alone

$$\left(\mathbf{A} - \mathbf{B}\mathbf{C}^{-1}\mathbf{B}^T\right)\mathbf{U} = \mathbf{F}_1 - \mathbf{B}\mathbf{C}^{-1}\mathbf{F}_2.$$

Due to the assumption $0 < \alpha < C_I$ the matrix $\mathbf{A}$ is positive definite and since $\mathbf{C}$ is negative definite, the coefficient matrix above is positive definite.

We note here that for a continuous pressure or a discontinuous pressure with the "jump terms" of Hughes and Franca [1987] we of course obtain a discrete system of the form (4.3.17), but in those cases the matrix $\mathbf{C}$ cannot be eliminated on each element separately.

A problem with the stabilized formulation presented above is the choice of the parameter α. The most straightforward way to get some insight into the dependency of α is to perform a large number of calculations with different values of α for various problems. In this respect we refer to Hughes et al. [1986], Franca et al. [1988b], Karam Filho and Loula [1988], Pierre [1988] and Franca et al. [1990a], where the results of some tests of this kind are reported for various methods. Alternatively, one could try to directly get some reliable estimates for the constant C_I in (4.3.5) which is the upper limit for α. Some result in this direction can be found in Harari [to appear].

However, there are some stabilized formulations for which no upper bounds for the stability parameters are required. Of these we will first consider the following alternative.

Method II (Douglas and Wang [1989]). *Find* $\mathbf{u}_h \in \mathbf{V}_h$ *and* $p_h \in P_h$ *such that*

$$B_h(\mathbf{u}_h, p_h; \mathbf{v}, q) = F_h(\mathbf{v}, q) \qquad \forall (\mathbf{v}, q) \in \mathbf{V}_h \times P_h$$

with

$$B_h(\mathbf{w}, r; \mathbf{v}, q) = B(\mathbf{w}, r; \mathbf{v}, q) - \alpha \sum_{K \in \mathcal{C}_h} h_K^2 (-\Delta \mathbf{w} + \nabla r, \Delta \mathbf{v} + \nabla q)_K$$

and

$$F_h(\mathbf{v}, q) = F(\mathbf{v}, q) - \alpha \sum_{K \in \mathcal{C}_h} h_K^2 (\mathbf{f}, \Delta \mathbf{v} + \nabla q)_K. \qquad \square$$

We note that the only difference to the previous formulation is that the sign in front of $\Delta \mathbf{v}$ has been changed. This has, however, as consequence that the method is stable and optimally convergent for all positive values of α.

Theorem 4.3.3 *Assume that either* $\mathbf{S}_h \subset \mathbf{V}_h$ *or* $P_h \subset C^0(\Omega)$ *and that* $\alpha > 0$. *For Method II we then have the following error estimate*

$$\|\mathbf{u} - \mathbf{u}_h\|_1 + \|p - p_h\|_0 \leq C(h^k|\mathbf{u}|_{k+1} + h^{l+1}|p|_{l+1}).$$

Proof: The analysis differs from that of the earlier method only with respect to the verification of the stability. To this end we let $\gamma > 1$ be a parameter and estimate as follows

$$
\begin{aligned}
B_h(\mathbf{u}, p; \mathbf{u}, -p) &= \|\nabla\mathbf{u}\|_0^2 + \alpha \sum_{K \in \mathcal{C}_h} h_K^2 \| -\Delta\mathbf{u} + \nabla p\|_{0,K}^2 \\
&= \|\nabla\mathbf{u}\|_0^2 + \alpha \sum_{K \in \mathcal{C}_h} h_K^2 \|\Delta\mathbf{u}\|_{0,K}^2 \\
&\quad + 2\alpha \sum_{K \in \mathcal{C}_h} h_K^2(-\Delta\mathbf{u}, \nabla p)_K + \alpha \sum_{K \in \mathcal{C}_h} h_K^2 \|\nabla p\|_{0,K}^2 \\
&\geq \|\nabla\mathbf{u}\|_0^2 + \alpha(1-\gamma) \sum_{K \in \mathcal{C}_h} h_K^2 \|\Delta\mathbf{u}\|_{0,K}^2 \\
&\quad + \left(1 - \frac{1}{\gamma}\right) \alpha \sum_{K \in \mathcal{C}_h} h_K^2 \|\nabla p\|_{0,K}^2 \\
&\geq \left(1 + \alpha(1-\gamma)C_I^{-1}\right) \|\nabla\mathbf{u}\|_0^2 + \left(1 - \frac{1}{\gamma}\right) \alpha \sum_{K \in \mathcal{C}_h} h_K^2 \|\nabla p\|_{0,K}^2 \\
&\geq C_2\|\nabla\mathbf{u}\|_0^2 + C_3\alpha \sum_{K \in \mathcal{C}_h} h_K^2 \|\nabla p\|_{0,K}^2 \\
&\geq C(\|\nabla\mathbf{u}\|_0^2 + \sum_{K \in \mathcal{C}_h} h_K^2 \|\nabla p\|_{0,K}^2)
\end{aligned}
$$

when we choose $1 < \gamma < (1 + C_I \alpha^{-1})$. With the use of Lemma 4.3.1 the stability is now proved as in the proof of Theorem 4.3.1. The error estimate then follows as in Theorem 4.3.2.   □

Remark. In the case of a discontinuous pressure the original proposal of Douglas and Wang contained jump terms for the pressure. In Franca and Stenberg [1991] we showed that with the assumption $\mathbf{S}_h \subset \mathbf{V}_h$ these are unnecessary.   □

The theoretical results proved for the two methods, suggest that the method of Douglas and Wang might be more "robust," by which we mean that the accuracy of the finite element solution is less sensitive to the choice of α (except for linear velocities for which the methods coincide). The numerical results reported in Franca et al. [1990a] confirms this. On the other hand, the method of Douglas and Wang does not give a symmetric system of equations.

However, it is possible to get methods with symmetric discrete systems and without any upper bounds on the stability parameter. We rewrite the Stokes problem with $\nabla \mathbf{u}$ as a new unknown, "the augmented stress" σ:

$$
\begin{aligned}
\sigma - \nabla \mathbf{u} &= \mathbf{0} && \text{in } \Omega, \\
\operatorname{div} \mathbf{u} &= 0 && \text{in } \Omega, \\
-\mathbf{div}\,\sigma + \nabla p &= \mathbf{f} && \text{in } \Omega, \\
\mathbf{u} &= \mathbf{0} && \text{on } \Gamma.
\end{aligned}
$$

Here **div** denotes the vector divergence applied to matrix functions.

The velocity and the pressure we approximate as before with the spaces (4.3.1) and (4.3.2)–(4.3.3), respectively, and for the augmented stress we introduce the space

$$
\Sigma_h = \{\, \tau \in [L^2(\Omega)]^{N \times N} \mid \tau_{|K} \in [R_m(K)]^{N \times N} \ \forall K \in \mathcal{C}_h \}, \tag{4.3.18}
$$

with $m = k - 1$ for triangular and tetrahedral elements and $m = k$ for quadrilaterals and hexahedrons.

Remark. When $\varepsilon(\mathbf{u})$ is used in the formulation instead of $\nabla \mathbf{u}$ then σ is symmetric and Σ_h can be reduced to consist of symmetric matrices. $\qquad\square$

Let us define the method directly in penalty form.

Method III (Franca and Stenberg [1991]). *Find* $(\sigma_h, p_h, \mathbf{u}_h) \in \Sigma_h \times P_h \times \mathbf{V}_h$ *such that*

$$
B_h(\sigma_h, p_h, \mathbf{u}_h; \tau, q, \mathbf{v}) = F_h(\tau, q, \mathbf{v}) \quad \forall (\tau, q, \mathbf{v}) \in \Sigma_h \times P_h \times \mathbf{V}_h,
$$

with

$$
\begin{aligned}
B_h(\kappa, r, \mathbf{w}; \tau, q, \mathbf{v}) ={}& -(\kappa, \tau) + (\nabla \mathbf{w}, \tau) + (\kappa, \nabla \mathbf{v}) - (r, \operatorname{div} \mathbf{v}) - (\operatorname{div} \mathbf{w}, q) \\
& - \epsilon\,(r, q) - \alpha \sum_{K \in \mathcal{C}_h} h_K^2 (\mathbf{div}\,\kappa - \nabla r, \mathbf{div}\,\tau - \nabla q)_K
\end{aligned}
$$

and

$$
F_h(\tau, q, \mathbf{v}) = (\mathbf{f}, \mathbf{v}) + \alpha \sum_{K \in \mathcal{C}_h} h_K^2 (\mathbf{f}, \mathbf{div}\,\tau - \nabla q)_K. \qquad\square
$$

For the method we have the optimal estimate.

Theorem 4.3.4 *Assume that either* $\mathbf{S}_h \subset \mathbf{V}_h$ *or* $P_h \subset \mathcal{C}^0(\Omega)$ *and that* $\alpha > 0$. *For the approximation with Method III we then have the following error estimate*

$$
\|\sigma - \sigma_h\|_0 + \|p - p_h\|_0 + \|\mathbf{u} - \mathbf{u}_h\|_1 \le C(h^{m+1}|\sigma|_{m+1} + h^{l+1}|p|_{l+1} + h^k|\mathbf{u}|_{k+1}),
$$

for all ϵ in the range $0 \leq \epsilon \leq Ch^s$, $s = \min\{k+1, l+2\}$. $\square$

Proof: We will not give the proof in full detail. We first note that the method is consistent. Hence, in analogy with the earlier methods we have to verify the stability condition which now is

$$\sup_{\substack{(\tau,q,\mathbf{v})\in\Sigma_h\times P_h\times\mathbf{V}_h \\ (\tau,q,\mathbf{v})\neq(0,0,0)}} \frac{B_h(\boldsymbol{\sigma},p,\mathbf{u};\tau,q,\mathbf{v})}{\|\tau\|_0 + \|q\|_0 + \|\mathbf{v}\|_1} \geq C(\|\boldsymbol{\sigma}\|_0 + \|p\|_0 + \|\mathbf{u}\|_1) \quad \forall(\boldsymbol{\sigma},p,\mathbf{u}) \in \Sigma_h\times P_h\times\mathbf{V}_h.$$

To prove this we first use an inverse estimate similar to (4.3.5) and estimate as in the proof of Theorem 4.3.3 in order to get

$$
\begin{aligned}
B_h(\boldsymbol{\sigma},p,\mathbf{u};-\boldsymbol{\sigma},-p,\mathbf{u}) &= \|\boldsymbol{\sigma}\|_0^2 + \epsilon\|p\|_0^2 + \alpha\sum_{K\in\mathcal{C}_h} h_K^2\|\mathbf{div}\,\boldsymbol{\sigma} - \nabla p\|_{0,K}^2 \\
&\geq C_1(\|\boldsymbol{\sigma}\|_0^2 + \sum_{K\in\mathcal{C}_h} h_K^2\|\nabla p\|_{0,K}^2) + \epsilon\|p\|_0^2. \quad (4.3.19)
\end{aligned}
$$

The second step is to use Lemma 4.3.1 in the same way as earlier and conclude that there is a velocity $\mathbf{w} \in \mathbf{V}_h$, with $\|\mathbf{w}\|_1 \leq \|p\|_0$, such that

$$B_h(\boldsymbol{\sigma},p,\mathbf{u};\mathbf{0},0,-\mathbf{w}) \geq C_2\|p\|_0^2 - C_3(\|\boldsymbol{\sigma}\|_0^2 + \|\nabla\mathbf{u}\|_0^2 + \sum_{K\in\mathcal{C}_h} h_K^2\|\nabla p\|_{0,K}^2). \quad (4.3.20)$$

Next, the assumption that $m = k - 1$ for triangles (tetrahedrons) and $m = k$ for quadrilaterals (hexahedrons) implies that there is $\boldsymbol{\kappa} \in \Sigma_h$ such that

$$(\boldsymbol{\kappa},\nabla\mathbf{u}) = \|\nabla\mathbf{u}\|_0^2 \quad (4.3.21)$$

and

$$\|\boldsymbol{\kappa}\| \leq C\|\nabla\mathbf{u}\|_0. \quad (4.3.22)$$

This gives

$$B_h(\boldsymbol{\sigma},p,\mathbf{u};\boldsymbol{\kappa},0,\mathbf{0}) \geq C_4\|\nabla\mathbf{u}\|_0^2 - C_5(\|\boldsymbol{\sigma}\|_0^2 + \sum_{K\in\mathcal{C}_h} h_K^2\|\nabla p\|_{0,K}^2). \quad (4.3.23)$$

The stability estimate is now obtained from (4.3.19), (4.3.20) and (4.3.23) by taking

$$(\tau,q,\mathbf{v}) = (-\boldsymbol{\sigma} + \delta\boldsymbol{\kappa}, -p, \mathbf{u} - \delta^2\mathbf{w})$$

with δ positive and small enough. $\square$

Let us next discuss the implementation of the method. Let $\{\chi_i\}_{i=1}^{N_\sigma}$, $\{\phi_i\}_{i=1}^{N_u}$ and $\{\psi_i\}_{i=1}^{N_p}$ be the basis for Σ_h, $\mathbf{V}_h$ and P_h, respectively. The discretization then leads to the following matrix equation

$$\begin{pmatrix} \mathbf{A} & \mathbf{B} & \mathbf{C} \\ \mathbf{B}^T & 0 & \mathbf{D} \\ \mathbf{C}^T & \mathbf{D}^T & \mathbf{E} \end{pmatrix} \begin{pmatrix} \mathbf{S} \\ \mathbf{U} \\ \mathbf{P} \end{pmatrix} = \begin{pmatrix} \mathbf{F}_1 \\ \mathbf{F}_2 \\ \mathbf{F}_3 \end{pmatrix} \qquad (4.3.24)$$

where $\mathbf{S}$, $\mathbf{U}$ and $\mathbf{P}$ are the degrees of freedom for σ_h, $\mathbf{u}_h$ and p_h, respectively. The matrices are given by

$$
\begin{aligned}
(\mathbf{A})_{ij} &= -(\chi_i, \chi_j) - \alpha \sum_{K \in \mathcal{C}_h} h_K^2 (\operatorname{div} \chi_i, \operatorname{div} \chi_j)_K\,, \\
(\mathbf{B})_{ij} &= (\nabla \phi_j, \chi_i)\,, \\
(\mathbf{C})_{ij} &= \alpha \sum_{K \in \mathcal{C}_h} h_K^2 (\nabla \psi_j, \operatorname{div} \chi_i)_K\,, \\
(\mathbf{D})_{ij} &= -(\psi_j, \operatorname{div} \phi_i)\,, \\
(\mathbf{E})_{ij} &= -\epsilon\,(\psi_i, \psi_j) - \alpha \sum_{K \in \mathcal{C}_h} h_K^2 (\nabla \psi_i, \nabla \psi_j)_K
\end{aligned}
$$

and

$$
\begin{aligned}
(\mathbf{F}_1)_i &= \alpha \sum_{K \in \mathcal{C}_h} h_K^2 (\mathbf{f}, \operatorname{div} \chi_i)_K\,, \\
(\mathbf{F}_2)_i &= (\mathbf{f}, \phi_i)\,, \\
(\mathbf{F}_3)_i &= -\alpha \sum_{K \in \mathcal{C}_h} h_K^2 (\mathbf{f}, \nabla \psi_i)_K\,.
\end{aligned}
$$

Now, since the matrix $\mathbf{A}$ is negative definite and Σ_h consists of discontinuous functions, $\mathbf{S}$ can be condensed in the assembling phase of the calculations. It is also important to note that in the condensation on one element each row of the matrix unknown $\mathbf{S}$ can be eliminated separately and the matrices to invert are identical for all rows. Hence, this condensation is not as expensive as one might think at a first glance. This leads to the system

$$\begin{pmatrix} \widehat{\mathbf{A}} & \widehat{\mathbf{B}} \\ \widehat{\mathbf{B}}^T & \widehat{\mathbf{C}} \end{pmatrix} \begin{pmatrix} \mathbf{U} \\ \mathbf{P} \end{pmatrix} = \begin{pmatrix} \widehat{\mathbf{F}}_1 \\ \widehat{\mathbf{F}}_2 \end{pmatrix},$$

with

$$
\begin{aligned}
\widehat{\mathbf{A}} &= -\mathbf{B}^T \mathbf{A}^{-1} \mathbf{B}\,, \\
\widehat{\mathbf{B}} &= \mathbf{D} - \mathbf{B}^T \mathbf{A}^{-1} \mathbf{C}\,, \\
\widehat{\mathbf{C}} &= \mathbf{E} - \mathbf{C}^T \mathbf{A}^{-1} \mathbf{C}
\end{aligned}
$$

and

$$\hat{\mathbf{F}}_1 = \mathbf{F}_2 - \mathbf{B}^T \mathbf{A}^{-1} \mathbf{F}_1,$$
$$\hat{\mathbf{F}}_2 = \mathbf{F}_3 - \mathbf{C}^T \mathbf{A}^{-1} \mathbf{F}_1.$$

From (4.3.21)–(4.3.22) it follows that the matrix $\mathbf{B}$ has full rank and hence the negative definiteness of $\mathbf{A}$ implies that $\hat{\mathbf{A}}$ is positive definite. Next, we recall that we in (4.3.19) showed that the matrix

$$\begin{pmatrix} \mathbf{A} & \mathbf{C} \\ \mathbf{C}^T & \mathbf{E} \end{pmatrix}$$

is negative definite. Now, since $\mathbf{A}$ and $\mathbf{E}$ are both negative definite, this implies that $\hat{\mathbf{C}}$ is negative definite. Hence, we have a system of the same type as for Method II, cf. the equation (4.3.17). Again, if discontinuous pressures are used, we can eliminate $\mathbf{P}$ locally and we obtain a positive system for $\mathbf{U}$.

For quadrilateral and hexahedral elements in the above formulations we have used equal order interpolation for the augmented stress and the velocity. The reason is that we need a pair of spaces satisfying the stability condition (4.3.21)–(4.3.22). To get the optimal approximation properties it is sufficient to take $m = k - 1$. Since the degrees of freedom for the augmented stress are condensed it would be preferable to choose the space Σ_h as small as possible. Let us therefore close the paper by showing that it is possible to modify the above formulation with the same stabilization technique. Then any space for the augmented stress can be used, i.e. we let m be arbitrary and define

$$\Sigma_h = \{ \boldsymbol{\tau} \in [L^2(\Omega)]^{N \times N} \mid \boldsymbol{\tau}_{|K} \in [R_m(K)]^{N \times N} \quad \forall K \in \mathcal{C}_h \}, \tag{4.3.25}$$

or

$$\Sigma_h = \{ \boldsymbol{\tau} \in [\mathcal{C}^0(\Omega)]^{N \times N} \mid \boldsymbol{\tau}_{|K} \in [R_m(K)]^{N \times N} \quad \forall K \in \mathcal{C}_h \}. \tag{4.3.26}$$

Method IV (Franca and Stenberg [1991]). *Find* $(\boldsymbol{\sigma}_h, p_h, \mathbf{u}_h) \in \Sigma_h \times P_h \times V_h$ *such that*

$$B_h(\boldsymbol{\sigma}_h, p_h, \mathbf{u}_h; \boldsymbol{\tau}, q, \mathbf{v}) = F_h(\boldsymbol{\tau}, q, \mathbf{v}) \quad \forall (\boldsymbol{\tau}, q, \mathbf{v}) \in \Sigma_h \times P_h \times V_h,$$

with

$$\begin{aligned} B_h(\boldsymbol{\kappa}, r, \mathbf{w}; \boldsymbol{\tau}, q, \mathbf{v}) = {} & -(\boldsymbol{\kappa}, \boldsymbol{\tau}) + (\nabla \mathbf{w}, \boldsymbol{\tau}) + (\boldsymbol{\kappa}, \nabla \mathbf{v}) + \beta(\boldsymbol{\kappa} - \nabla \mathbf{w}, \boldsymbol{\tau} - \nabla \mathbf{v}) \\ & - (r, \operatorname{div} \mathbf{v}) - (\operatorname{div} \mathbf{w}, q) - \epsilon\,(r, q) \\ & - \alpha \sum_{K \in \mathcal{C}_h} h_K^2 (\operatorname{\mathbf{div}} \boldsymbol{\kappa} - \nabla r, \operatorname{\mathbf{div}} \boldsymbol{\tau} - \nabla q)_K. \qquad \square \end{aligned}$$

Theorem 4.3.5 *Assume that the finite element spaces satisfy either $\mathbf{S}_h \subset \mathbf{V}_h$ or $P_h \subset C^0(\Omega)$, and further that $0 < \beta < 1$ and $\alpha > 0$. For the approximation with Method IV we then have the following error estimate*

$$\|\boldsymbol{\sigma} - \boldsymbol{\sigma}_h\|_0 + \|p - p_h\|_0 + \|\mathbf{u} - \mathbf{u}_h\|_1 \le C(h^{m+1}|\boldsymbol{\sigma}|_{m+1} + h^{l+1}|p|_{l+1} + h^k|\mathbf{u}|_{k+1}),$$

for all ϵ in the range $0 \le \epsilon \le Ch^s$, $s = \min\{k+1, l+2, m+2\}$.

Proof: For the stability we note that

$$
\begin{aligned}
B_h(\boldsymbol{\sigma}, p, \mathbf{u}; -\boldsymbol{\sigma}, -p, \mathbf{u}) &= (1 - \beta)\|\boldsymbol{\sigma}\|_0^2 + \beta\|\nabla \mathbf{u}\|_0^2 + \epsilon\|p\|_0^2 \\
&\quad + \alpha \sum_{K \in \mathcal{C}_h} h_K^2 \|\mathbf{div}\, \boldsymbol{\sigma} - \nabla p\|_{0,K}^2 \\
&\ge C(\|\boldsymbol{\sigma}\|_0^2 + \|\nabla \mathbf{u}\|_0^2 + \sum_{K \in \mathcal{C}_h} h_K^2 \|\mathbf{div}\, \boldsymbol{\sigma} - \nabla p\|_{0,K}^2),
\end{aligned}
$$

when $0 < \beta < 1$. The rest of the stability and error analysis is as before. $\qquad\square$

Of course this formulation leads to a second stability parameter for which the optimal choice is not known. However, for this parameter we have the explicit upper bound 1.

The matrix equations arising from this formulation is as (4.3.24) with a positive definite matrix instead of the zero matrix in the middle.

Let us finally remark that this method seems to be useful for some models of viscoelastic fluids, see Marchal and Crochet [1987]. These models contain spatial derivatives of the augmented stress and hence a continuous approximation for that variable is desirable. In the Galerkin formulation (obtained from above by choosing $\alpha = 0$ and $\beta = 0$) this leads to expensive elements as can be seen from Marchal and Crochet [1987] and Fortin and Pierre [1989].

References

[1] Babuška, I. (1973). The finite element method with Lagrangian multipliers. *Numer. Math.*, **20**, 179–192.

[2] Brezzi, F. (1974). On the existence, uniqueness and approximation of saddle-point problems arising from Lagrange multipliers. *RAIRO Ser. Rouge* **8**, 129–151.

[3] Brezzi, F. and Pitkäranta, J. (1984). On the stabilization of finite element approximations of the Stokes problem. *Efficient Solutions of Elliptic Systems, Notes on Numerical Fluid Mechanics*, **10** (Ed. by W. Hackbusch), Vieweg, Wiesbaden, 11–19.

[4] Brezzi, F. and Douglas, J. (1988). Stabilized mixed methods for Stokes problem. *Numer. Math.* **53**, 225–236.

[5] Brezzi, F. and Fortin, M. (1991). *Mixed and Hybrid Finite Element Methods*, Springer-Verlag, Heidelberg.

[6] Brooks, A. N. and Hughes, T.J.R. (1982). Streamline upwind/Petrov-Galerkin formulations for convective dominated flows with particular emphasis on the incompressible Navier-Stokes equations. *Comput. Methods Appl. Mech. Engrg.* **32**, 199–259.

[7] Ciarlet, P. G. (1978). *The Finite Element Method for Elliptic Problems*, North-Holland, Amsterdam.

[8] Douglas, J. and Wang, J. (1989). An absolutely stabilized finite element method for the Stokes problem. *Math. Comp.* **52**, 495–508.

[9] Durán, R. and Notchetto, R. (1989). Pointwise accuracy of a stable Petrov-Galerkin approximation of the Stokes problem. *SIAM J. Num. Anal.* **26**, 1395–1406.

[10] Fortin, F. and Pierre, R. (1989). On the convergence of the mixed method of Crochet and Marchal for viscoelastic flows. *Comput. Methods Appl. Mech. Engrg.* **73**, 341–350.

[11] Franca, L. P. and Hughes, T. J. R. (1988a). Two classes of mixed finite element methods. *Comput. Methods Appl. Mech. Engrg.* **69**, 89–129.

[12] Franca, L.P., Hughes, T.J.R., Loula, A.F.D. and Miranda, I. (1988b). A new family of stable element for nearly incompressible elasticity based on a mixed Petrov-Galerkin finite element formulation. *Numer. Math.* **53**, 123–141.

[13] Franca, L. P., Frey, S. L. and Hughes, T. J. R. (1990a). Stabilized finite element methods: I. Application to the advective-diffusive model. *Report LNCC* **32/90**, Rio de Janeiro, Brazil. Also, *Comput. Methods Appl. Mech. Engrg.*, to appear.

[14] Franca, L. P., Karam Filho, J., Loula, A. F. D. and Stenberg, R. (1990b). A convergence analysis of a stabilized method for the Stokes flow. *Report LNCC* **19/90**, Rio de Janeiro, Brazil.

[15] Franca, L. P. and Stenberg, R. (1991). Error analysis of some Galerkin-least-squares methods for the elasticity equations. *SIAM J. Num. Anal.* To appear.

[16] Girault, V. and Raviart, P. A. (1986). *Finite Element Methods for Navier-Stokes Equations. Theory and Algorithms*, Springer-Verlag, Heidelberg.

[17] Harari, I. (to appear). Ph.D. Thesis, Division of Applied Mechanics, Stanford University, Stanford, California.

[18] Hughes, T. J. R. (1987). *The Finite Element Method. Linear Static and Dynamic Analysis*. Prentice-Hall.

[19] Hughes, T. J. R. and Brooks, A. N. (1979). A multidimensional upwind scheme with no crosswind diffusion. *Finite Element Methods for Convection Dominated Flows* (Ed. by T.J.R. Hughes), ASME, New York, 19–35.

[20] Hughes, T. J. R., Franca, L. P. and Balestra, M. (1986). A new finite element formulation for computational fluid dynamics: V. Circumventing the Babuška-Brezzi condition: A stable Petrov-Galerkin formulation Stokes problem accommodating equal-order interpolations. *Comput. Methods Appl. Mech. Engrg.* **59**, 85–99.

[21] Hughes, T. J. R. and Franca, L. P. (1987). A new finite element formulation for computational fluid dynamics: VII. The Stokes problem with various well-posed boundary conditions: symmetric formulations that converge for all velocity/pressure spaces. *Comput. Methods Appl. Mech. Engrg.* **65**, 85–96.

[22] Hughes, T. J. R., Franca, L. P. and Hulbert, G. M. (1989). A new finite element formulation for computational fluid dynamics: VIII. The Galerkin-least-squares method for advective-diffusive equations. *Comput. Methods Appl. Mech. Engrg.* **73**, 173–189.

[23] Johnson, C. and Nävert, U. (1981). An analysis of some finite element methods for advection-diffusion problems. *Analytical and Numerical Approaches to Asymptotic Problems in Analysis* (Ed. by O. Axelsson, L. S. Frank and A. Van Der Sluis), North-Holland, Amsterdam, 99–116.

[24] Johnson, C., Nävert, U. and Pitkäranta, J. (1984). Finite element methods for linear hyperbolic problem. *Comput. Methods Appl. Mech. Engrg.*, **45**, 285–312.

[25] Karam Filho, J. and Loula, A.F.D. (1988). New mixed Petrov-Galerkin finite element formulations for incompressible flow. *Proceedings of "II Encontro Nacional de Ciências Térmicas", ENCIT 88, Aguas de Lindóia, Brazil*, 172–175

[26] Marchal, J. M. and Crochet, M.J. (1987). A new mixed finite element for calculating viscoelastic flow. *J. of Non-Newtonian Fluid Mech.* **26**, 77–114.

[27] Pierre, R. (1988). Simple C^0-approximations for the computation of incompressible flows. *Comput. Methods Appl. Mech. Engrg.*, **68**, 205–228.

[28] Pierre, R. (1989). Regularization procedures of mixed finite element approximations for the Stokes problem. *Num. Meths. Part. Diff. Eqs.*, **3**, 241–258

[29] Silvester, D. J. and Kechkar, N. (1990). Stabilized bilinear-constant velocity-pressure finite elements for the conjugate gradient solution of the Stokes problem. *Comput. Methods Appl. Mech. Engrg.*, **79**, 71–86

[30] Stenberg, R. (1990). A technique for analysing finite element methods for viscous incompressible flow. *Int. J. Num. Meths. Fluids.* **11**, 935–948.

[31] Verfürth, R. (1984). Error estimates for a mixed finite element approximation of the Stokes problem. *RAIRO Anal. Numer.* **18**, 175–182.

5 Optimal Control and Optimization of Viscous, Incompressible Flows

Max D. Gunzburger, L. Steven Hou, and Thomas P. Svobodny

5.1 Introduction

The control of fluid motions for the purpose of achieving some desired objective is crucial to many technological applications. In the past, these control problems have been addressed either through expensive experimental processes or through the introduction of significant simplifications into the analyses used in the development of control mechanisms. Only recently have flow control problems been addressed, by scientists and mathematicians, in a systematic, rigorous manner. This interest is quickly expanding so that, at this time, flow control is becoming a very active and successful area of inquiry. For example, recent publications, e.g., [1]–[28], provide analyses of various aspects of flow control problems, and include one or more of the following components:

- the construction of mathematical models, invoking minimal assumptions about the physical phenomena;

- the analysis of the mathematical models to answer questions about the existence and regularity of solutions and to derive necessary conditions that optimal controls and states must satisfy;

- the construction and analysis of discretization methods for determining approximate solutions of the optimal control problems, and the rigorous derivation of error estimates; and

- the development of computer codes implementing discretization algorithms, both for the purpose of showing the efficacy of these methods, and also to solve problems of practical interest.

An optimal control or optimization problem is composed of two ingredients: a desired objective and control mechanisms that are used to (hopefully) achieve the desired objective. In a mathematical description of such problems, the desired objective is usually expressed in terms of the extremization of a functional depending on the state of the system, and possibly also on the control mechanisms. In many settings, including the ones considered here, there are also constraints that the candidate optimal states and controls are required to satisfy. We restrict our attention to incompressible viscous flows; the

109

problems we discuss, as well as many others, can also be posed for compressible and/or inviscid flows.

In Section 5.2, we discuss the constraints that the state and control variables are required to satisfy. Then, in Section 5.3, we provide a small sampling of flow control and optimization objectives of practical interest. In Section 5.4 we describe some of the possible control mechanisms that can be used to achieve the objectives described in Section 5.3. Then, in Section 5.5, we describe some specific flow control problems which serve to illustrate some of the possible combinations of constraints, objectives, and control mechanisms. In Section 5.6 we describe, in general terms, the issues that arise when one attempts to analyse and computationally simulate flow control and optimization problems and then we treat, in more detail, a particular problem, illustrating the resolution of some of these issues.

5.2 Constraints

We first discuss the constraints that the flow, i.e., the state, and the control variables are required to satisfy. These constraints take the form of a system of partial differential equations, along with initial and boundary conditions. Here we present the unsteady versions of these constraints; for stationary problems, one simply omits the initial conditions and all terms involving a time derivative in the differential equations.

5.2.1 Equations for Incompressible Flows

In flow control and optimization problems, candidate optimal controls and states are naturally required to describe physically realizable flows, i.e., they should satisfy the governing equations of fluid mechanics. The context here is incompressible flows, so that these governing equations are the continuity equation (or conservation of mass equation)

$$\operatorname{div} \mathbf{u} = 0 \quad \text{for } \mathbf{x} \in \Omega, t > 0 \tag{5.2.1}$$

and the Navier-Stokes equation (or momentum equation or equation of motion)

$$\rho \left(\frac{\partial \mathbf{u}}{\partial t} + \mathbf{u} \cdot \operatorname{grad} \mathbf{u} \right) - 2\mu \operatorname{div} \left(D(\mathbf{u}) \right) + \operatorname{grad} p = \rho \mathbf{f} \quad \text{for } \mathbf{x} \in \Omega, t > 0, \tag{5.2.2}$$

where the deformation tensor $D(\mathbf{u})$ is given by

$$D(\mathbf{u}) = \tfrac{1}{2} \left((\operatorname{grad} \mathbf{u}) + (\operatorname{grad} \mathbf{u})^T \right). \tag{5.2.3}$$

In (5.2.1)–(5.2.2), the velocity field $\mathbf{u}$ and the pressure field p are the state variables; Ω denotes the flow domain, ρ the constant density, $\mathbf{f}$ the body force per unit mass, and μ the constant coefficient of viscosity.

Equations (5.2.1) and (5.2.2) are derived based on the assumption, among others, that temperature variations do not affect the mechanical properties of the fluid and the flow, i.e., viscosity coefficients, velocities, pressures, forces, and moments. As a result, the energy equation normally uncouples from the continuity and momentum equations. In such cases, the velocity and pressure may be solved for without regard to the energy equation, and the latter need be considered only if one is also interested in the thermal properties of the flow. However, for some flow control and optimization problems, (5.2.1) and (5.2.2) are coupled to the energy equation through the objective functional and/or the control mechanisms; an example of this situation is discussed in Section 5.5.

Consistent with (5.2.1)–(5.2.2), we have the energy equation

$$\rho c_v \left(\frac{\partial T}{\partial t} + \mathbf{u} \cdot \operatorname{grad} T \right) - \Phi(\mathbf{u}) - \kappa \Delta T = Q \quad \text{for } \mathbf{x} \in \Omega, t > 0. \tag{5.2.4}$$

The dissipation function $\Phi(\mathbf{u})$ is given by

$$\Phi(\mathbf{u}) = 2\mu D(\mathbf{u}) : D(\mathbf{u}), \tag{5.2.5}$$

where the colon denotes the scalar product of the two tensors which surround it. In (5.2.4), Q denotes the heat source, κ the constant coefficient of thermal conductivity, and c_v the constant specific heat at constant volume; the temperature T is an additional state variable.

5.2.2   Initital and Boundary Conditions

In addition to the constraints provided by the above partial differential equations, the control and state variables are required to satisfy constraints that take the form of boundary and initial conditions. Corresponding to the system (5.2.1)–(5.2.3) for the velocity and pressure, one may specify a variety of boundary conditions; see [38] for a discussion of some of the possibilities. Here, we restrict our attention to two types of boundary conditions for (5.2.1)–(5.2.3), namely the specification of components of the velocity $\mathbf{u}$ and/or components of the stress $\mathbf{t}$. Thus, on a portion Γ_1 of the boundary Γ of the flow region Ω, one may require that

$$\mathbf{u} = \mathbf{u}_b(\mathbf{x}, t) \quad \text{for } \mathbf{x} \in \Gamma_1, t > 0, \tag{5.2.6}$$

and on another portion Γ_2 (we tacitly assume that Γ_1, Γ_2, and also Γ_3, which is introduced below, are pairwise disjoint and that $\Gamma = \cup_{i=1}^3 \Gamma_i$)

$$\mathbf{t} = -p\mathbf{n} + 2\mu D(\mathbf{u}) \cdot \mathbf{n} = \mathbf{t}_b(\mathbf{x}, t) \quad \text{for } \mathbf{x} \in \Gamma_2, t > 0, \tag{5.2.7}$$

where **n** denotes the unit outer normal vector to Ω. Both (5.2.6) and (5.2.7) involve the specification of vector-valued functions; it is also possible, and sometimes desirable, to specify some complementary components of (5.2.6) and (5.2.7). For example, an often used combination is the normal velocity and the tangential stress components, i.e., on a portion Γ_3 of the boundary Γ

$$\mathbf{u} \cdot \mathbf{n} = g_b(\mathbf{x}, t) \quad \text{and} \quad 2\mu\boldsymbol{\tau} \cdot D(\mathbf{u}) \cdot \mathbf{n} = \mathbf{h}_b(\mathbf{x}, t) \quad \text{for } \mathbf{x} \in \Gamma_3, t > 0, \quad (5.2.8)$$

where, at each point $\mathbf{x} \in \Gamma_3$, $\boldsymbol{\tau}(\mathbf{x})$ denotes a system of vectors spanning the tangent plane. Corresponding to (5.2.1)–(5.2.3), in the unsteady setting, one also has an initial condition for the velocity, i.e.,

$$\mathbf{u} = \mathbf{u}_0(\mathbf{x}) \quad \text{for } \mathbf{x} \in \Omega, t = 0. \quad (5.2.9)$$

If one has to also consider (5.2.4)–(5.2.5), either because one is interested in the thermal properties of the flow, or because one is forced to by a coupling induced by the objective functional and/or the control mechanisms, then boundary and initial conditions for the temperature are also required. The most common type of boundary conditions are the specification of the temperature or the heat flux. Thus, if Γ is disjointly divided into Γ_4 and Γ_5 (these do not in general have to coincide with any of Γ_i, i=1,2,3), we would have that

$$T = T_b(\mathbf{x}, t) \quad \text{for } \mathbf{x} \in \Gamma_4, t > 0, \quad (5.2.10)$$

and

$$- \kappa\mathbf{n} \cdot \operatorname{grad} T = q_b(\mathbf{x}, t) \quad \text{for } \mathbf{x} \in \Gamma_5, t > 0. \quad (5.2.11)$$

For unsteady problems, one also requires the initial condition

$$T = T_0(\mathbf{x}) \quad \text{for } \mathbf{x} \in \Omega, t = 0. \quad (5.2.12)$$

For flows in unbounded domains, e.g., exterior problems, there are additional requirements on the state variables which describe their asymptotic behavior at large distances from the origin.

5.2.3 *Flow Simulation* vs. *Flow Control* vs. *Flow Identification*

The partial differential equations and boundary and initial conditions presented above can be used in two distinct types of problems. First, for a *simulation* problem, we have that all the parameters appearing in the left-hand sides of (5.2.1)–(5.2.12) are specified, as are all the functions appearing in the right-hand sides of these equations, as are the flow domain Ω and the required subdivisions of its boundary Γ. Thus, the task at hand is, given these parameters, e.g., μ, κ, these functions, e.g., $\mathbf{f}, \mathbf{u}_b$, the flow domain Ω, and the various boundary segments Γ_i, determine the pressure p and velocity $\mathbf{u}$ from (5.2.1)–(5.2.3) and (5.2.6)–(5.2.9), and if also desired, the temperature T from (5.2.4)–(5.2.5) and (5.2.10)–(5.2.12). We again reiterate that if in a flow simulation study using the above mathematical model one is only interested in the mechanical properties of the flow, i.e., those that can be completely deduced from the pressure and velocity, then one only need consider (5.2.1)–(5.2.3) and (5.2.6)–(5.2.9).

If one or more of the left-hand side parameters, and/or if one or more of the right-hand side functions, and/or if the flow domain are *not* specified, then (5.2.1)–(5.2.12) becomes a *control* problem. In this case, (5.2.1)–(5.2.12) does not suffice to determine the state variables $\mathbf{u}$, p, and T, and the unknown parameters and/or functions and/or domain. For example, consider the system (5.2.1)–(5.2.3) and (5.2.6)–(5.2.9) for the velocity $\mathbf{u}$ and pressure p. Suppose that $\mathbf{u}_b$ is not known on Γ_1. Then, even if the left-hand side parameters, the other right-hand-side functions, the domain Ω, and the boundary segments $\Gamma_i, i = 1, 2, 3$, are all specified, we do not have enough information to determine the state $(\mathbf{u}, p)$ and the "control" $\mathbf{u}_b$. The additional information needed is deduced from a specified optimization objective; some such objectives are discussed in Section 5.3. How one deduces the needed additional information from an optimization objective is discussed within Section 5.6.

We again point out that some optimization objectives induce a coupling between the mechanical and thermal variables which is not present in (5.2.1)–(5.2.12); such a situation is discussed within Section 5.5.

What we call "control" or "optimization" problems also include what are known as *identification* problems. In both cases one is interested in determining controls, i.e., parameters and/or functions and/or domains, so that some desired objective is met. In a control problem, one assumes that, perhaps within some specified limits, one has the ability to vary the controls in any manner one desires. Furthermore, one can exercise this command in order to meet the objective, which itself is based on some favorable circumstance that one wishes the flow to realize. In an identification problem, one has no command over the "controls"; they are not at our disposal to vary as we see fit.

In order to illustrate the distinction between control and identification problems, consider the following two problems. First, one can try to minimize the drag on an airfoil by injecting or sucking fluid through orifices on the boundary. Since presumably one is able to adjust at will the way that fluid is injected or sucked through the orifices, this constitutes a control problem; one has command of the control, i.e., the injection or suction of fluid through the orifices, in order to achieve the objective, i.e., minimizing the drag on the airfoil. Then, the goal of studying this problem is to determine what control one has to use to meet the desired objective.

The second problem is to determine the coefficient of viscosity μ of a given fluid from experimental flow data. Thus, one has in hand some experimentally determined velocity fields for the fluid and then one tries to find a viscosity coefficient so that solutions of the appropriate partial differential equations and initial and boundary conditions match, in some sense, the experimental data. Since the given fluid has a definite, albeit unknown, viscosity coefficient, this is an identification problem; one has no command over the "control," i.e., the viscosity coefficient. However, the goal of studying this problem is again to determine what "control" meets the desired objective.

Clearly, the distinction between optimal control and identification problems is lost once they are reduced to a mathematical problem, i.e., once the mathematical modelling of the objective, the controls, and the constraints is completed.

5.3 Objectives

In this section we describe some of the objectives that could be goals of an optimization process in optimal control and optimization problems for viscous, incompressible flows.

5.3.1 Flow Matching

Perhaps the simplest objective that can be considered is to have, as well as possible, the flow field conform to a prescribed flow field. Here are some examples of such objectives.

Given a flow field at time $t = 0$, a final time t_f, and a velocity field $\mathbf{U}(\mathbf{x})$, we want to control the flow so that at time $t = t_f$, the velocity field of the flow is "close to" to $\mathbf{U}$ over a specified portion $\Omega_c \subset \Omega$ of the flow domain. ($\Omega_c = \Omega$ is permissible.) This, of course, is a controllability problem, and in general, one cannot expect to find controls such that the flow velocity at time $t = t_f$ exactly matches the given terminal flow $\mathbf{U}$. Instead, one must be content with minimizing a functional of the type

$$\mathcal{M}(\mathbf{u}) = \| \mathbf{u}(\mathbf{x}, t_f) - \mathbf{U}(\mathbf{x}) \|^\alpha,$$

where $\mathbf{u}(\mathbf{x}, t)$ denotes the flow velocity and the choice of norm and exponent is governed by both mathematical and physical considerations. For example, one choice that has been considered is

$$\mathcal{M}(\mathbf{u}) = \frac{1}{4} \int_{\Omega_c} |\mathbf{u}(\mathbf{x}, t_f) - \mathbf{U}(\mathbf{x})|^4 \, d\Omega.$$

One could also try to "match" a given velocity field $\mathbf{U}(\mathbf{x}, t)$ over the time interval $0 < t < t_f$ by minimizing

$$\mathcal{M}(\mathbf{u}) = \frac{1}{4} \int_0^{t_f} \int_{\Omega_c} |\mathbf{u}(\mathbf{x}, t) - \mathbf{U}(\mathbf{x}, t)|^4 \, d\Omega \, dt.$$

Similar type functionals, e.g.,

$$\mathcal{M}(\mathbf{u}) = \frac{1}{4} \int_{\Omega_c} |\mathbf{u}(\mathbf{x}) - \mathbf{U}(\mathbf{x})|^4 \, d\Omega,$$

may be used in flow matching problems for stationary flows wherein $\mathbf{u}$ is independent of t.

Another important flow matching problem involves attempting to have the flow field agree with a desired flow at portions, usually of the outflow type, of the boundary of the flow domain. Let Γ_o denote the outflow boundary, and let $\mathbf{U}_o$ be a prescribed function, defined for $\mathbf{x} \in \Gamma_o$. Then, for example, in a stationary flow situation, one would minimize

$$\mathcal{M}(\mathbf{u}) = \int_{\Gamma_o} |\mathbf{u}(\mathbf{x}) - \mathbf{U}_o(\mathbf{x})|^2 \, d\Gamma.$$

Instead of the velocity, it is sometimes desirable to have some other flow variable conform to a prescribed function. An example of such a situtation is provided below where the variable of interest is the temperature. Also included in this class of problems (but with different types of functionals) are maneuvering problems wherein the objective is to steer a submerged body along a desired path.

5.3.2 Viscous Drag Minimization

An important objective in many applications is the minimization of drag. Consider a body moving with constant velocity $\mathbf{V}$ in a fluid. The region Ω occupied by the fluid may be bounded externally by fixed walls (an *internal* flow), or its extent may be infinite in all directions (an *external* flow), or its extent may be infinite in some directions, and finite in others (a *channel* flow). For the sake of concreteness, we consider the first case. The results below also hold for the other cases provided the velocity decays sufficiently rapidly to zero at infinity.

Let Γ_b denote the surface of the obstacle and Γ_e the fixed walls so that $\Gamma = \Gamma_b \cap \Gamma_e$ denotes the boundary of the flow domain Ω. The force exerted by the fluid on the obstacle is given by

$$\mathbf{F} = -\int_{\Gamma_b} \mathbf{t} \, d\Gamma,$$

where $\mathbf{t}$ denotes the stress vector. Then, the component of the force on the obstacle in the direction opposite to that of the motion, i.e., the *drag*, is given by

$$\mathcal{D} = -\frac{\mathbf{V}}{|\mathbf{V}|} \cdot \mathbf{F} = \frac{\mathbf{V}}{|\mathbf{V}|} \cdot \int_{\Gamma_b} \mathbf{t} \, d\Gamma.$$

But, since $\mathbf{u} = \mathbf{V}$ on Γ_b,

$$\mathcal{D} = \frac{1}{|\mathbf{V}|} \int_{\Gamma_b} \mathbf{u} \cdot \mathbf{t} \, d\Gamma .$$

Now, $\mathbf{t} = \sigma \cdot \mathbf{n}$, where σ denotes the stress tensor; then,

$$\mathcal{D} = \frac{1}{|\mathbf{V}|} \int_{\Gamma_b} \mathbf{u} \cdot \sigma \cdot \mathbf{n} \, d\Gamma .$$

An application of the divergence theorem yields that

$$|\mathbf{V}|\mathcal{D} = \int_{\Omega} (\operatorname{grad} \mathbf{u}) : \sigma \, d\Omega + \int_{\Omega} \mathbf{u} \cdot \operatorname{div} \sigma \, d\Omega - \int_{\Gamma_e} \mathbf{u} \cdot \sigma \cdot \mathbf{n} \, d\Gamma ;$$

using (5.2.3), the symmetry of the stress tensor σ, and $\mathbf{u} = \mathbf{0}$ on Γ_e, one obtains

$$|\mathbf{V}|\mathcal{D} = \int_{\Omega} D : \sigma \, d\Omega + \int_{\Omega} \mathbf{u} \cdot \operatorname{div} \sigma \, d\Omega ,$$

where D denotes the deformation tensor. The stress tensor is given by $\sigma = -pI + 2\mu D(\mathbf{u})$, where I denotes the identity tensor. Then, $D(\mathbf{u}) : I = \operatorname{div} \mathbf{u}$ and, using the Navier-Stokes equations (5.2.2), $\operatorname{div} \sigma = \rho(\partial \mathbf{u}/\partial t + \mathbf{u} \cdot \operatorname{grad} \mathbf{u} - \mathbf{f})$; furthermore, for incompressible flows, $\operatorname{div} \mathbf{u} = 0$ in Ω. We then obtain

$$
\begin{aligned}
|\mathbf{V}|\mathcal{D} &= -\int_{\Omega} pD : I \, d\Omega + 2\mu \int_{\Omega} D(\mathbf{u}) : D(\mathbf{u}) \, d\Omega \\
&\quad + \rho \int_{\Omega} \mathbf{u} \cdot \left(\frac{\partial \mathbf{u}}{\partial t} + \mathbf{u} \cdot \operatorname{grad} \mathbf{u} - \mathbf{f} \right) d\Omega \\
&= -\int_{\Omega} p \operatorname{div} \mathbf{u} \, d\Omega + 2\mu \int_{\Omega} D(\mathbf{u}) : D(\mathbf{u}) \, d\Omega \\
&\quad + \rho \int_{\Omega} \mathbf{u} \cdot \left(\frac{\partial \mathbf{u}}{\partial t} + \mathbf{u} \cdot \operatorname{grad} \mathbf{u} - \mathbf{f} \right) d\Omega \\
&= 2\mu \int_{\Omega} D(\mathbf{u}) : D(\mathbf{u}) \, d\Omega - \rho \int_{\Omega} \mathbf{f} \cdot \mathbf{u} \, d\Omega \\
&\quad + \frac{\rho}{2} \int_{\Omega} \left(\frac{\partial}{\partial t} + \mathbf{u} \cdot \operatorname{grad} \right) (\mathbf{u} \cdot \mathbf{u}) \, d\Omega .
\end{aligned}
$$

But, since Ω is a material volume, it is well known [41] that

$$
\begin{aligned}
\rho \int_{\Omega} \left(\frac{\partial}{\partial t} + \mathbf{u} \cdot \operatorname{grad} \right) (\mathbf{u} \cdot \mathbf{u}) \, d\Omega &= \int_{\Omega} \rho \left(\frac{\partial}{\partial t} + \mathbf{u} \cdot \operatorname{grad} \right) (\mathbf{u} \cdot \mathbf{u}) \, d\Omega \\
&= \frac{d}{dt} \int_{\Omega} \rho(\mathbf{u} \cdot \mathbf{u}) \, d\Omega = \rho \frac{d}{dt} \int_{\Omega} (\mathbf{u} \cdot \mathbf{u}) \, d\Omega ,
\end{aligned}
$$

where we have used the fact that ρ is constant. Then,

$$\mathcal{D} = \frac{2\mu}{|\mathbf{V}|} \int_\Omega D(\mathbf{u}) : D(\mathbf{u})\, d\Omega - \frac{\rho}{|\mathbf{V}|} \int_\Omega \mathbf{f}\cdot\mathbf{u}\, d\Omega + \frac{\rho}{2|\mathbf{V}|} \frac{d}{dt}\left(\int_\Omega |\mathbf{u}|^2\, d\Omega\right).$$

Thus, for stationary flows, minimizing the drag on an obstacle is equivalent to minimizing a functional proportional to the integral of the dissipation function (5.2.5), along with a term resulting from the rate of work done by the body force, i.e.,

$$
\begin{aligned}
\mathcal{D}(\mathbf{u}) =& \frac{\mu}{2|\mathbf{V}|} \int_\Omega \left((\operatorname{grad}\mathbf{u}) + (\operatorname{grad}\mathbf{u})^T\right) : \left((\operatorname{grad}\mathbf{u}) + (\operatorname{grad}\mathbf{u})^T\right)\, d\Omega \\
& - \frac{\rho}{|\mathbf{V}|} \int_\Omega \mathbf{f}\cdot\mathbf{u}\, d\Omega \\
=& \frac{1}{|\mathbf{V}|} \int_\Omega \Phi\, d\Omega - \frac{\rho}{|\mathbf{V}|} \int_\Omega \mathbf{f}\cdot\mathbf{u}\, d\Omega.
\end{aligned}
\tag{5.3.1}
$$

This is the most convenient drag functional to use in connection with incompressible flows.

For unsteady flows, a more relevant functional is the total drag over the time interval of interest, say $0 < t < t_f$,

$$
\begin{aligned}
\int_0^{t_f} \mathcal{D}\, dt =& \frac{2\mu}{|\mathbf{V}|} \int_0^{t_f}\int_\Omega D(\mathbf{u}) : D(\mathbf{u})\, d\Omega - \frac{\rho}{|\mathbf{V}|} \int_0^{t_f}\int_\Omega \mathbf{f}\cdot\mathbf{u}\, d\Omega\, dt \\
& + \frac{\rho}{2|\mathbf{V}|} \int_\Omega |\mathbf{u}(\mathbf{x}, t_f)|^2\, d\Omega - \frac{\rho}{2|\mathbf{V}|} \int_\Omega |\mathbf{u}(\mathbf{x}, 0)|^2\, d\Omega \\
=& \frac{2\mu}{|\mathbf{V}|} \int_0^{t_f}\int_\Omega D(\mathbf{u}) : D(\mathbf{u})\, d\Omega - \frac{\rho}{|\mathbf{V}|} \int_0^{t_f}\int_\Omega \mathbf{f}\cdot\mathbf{u}\, d\Omega\, dt \\
& + \frac{\rho}{2|\mathbf{V}|} \int_\Omega |\mathbf{u}(\mathbf{x}, t_f)|^2\, d\Omega - \frac{\rho}{2|\mathbf{V}|} \int_\Omega |\mathbf{u}_0|^2\, d\Omega.
\end{aligned}
\tag{5.3.2}
$$

The last integral is merely a constant since the integrand involves the known initial condition (5.2.9). This functional expresses the fact that the total drag experienced by a body up to a given time t_f is due in part to dissipation due to friction, in part to kinetic energy changes in the flow, and in part on the rate at which work is done by the body forces.

Note that if we instead fix the coordinate system so that the obstacle is at rest, then it can be shown that (5.3.1) and (5.3.2) still give the drag on the obstacle in stationary flow and the time integral of the drag on the obstacle, respectively.

5.3.3 *Vorticity Minimization*

Let $\omega = \text{curl } \mathbf{u}$ denote the vorticity vector. In [1], the minimization of the functional

$$\mathcal{W}(\mathbf{u}) = \mu \int_\Omega |\omega|^2 \, d\Omega = \mu \int_\Omega |\text{curl } \mathbf{u}|^2 \, d\Omega \qquad (5.3.3)$$

is advanced as a means of minimizing "turbulence."

The functionals $\mathcal{W}$ and $\mathcal{D}$ are related. This follows from the relation, known as the Bobyleff-Forsythe formula [41], valid for incompressible flows,

$$\int_\Omega \Phi \, d\Omega = \mu \int_\Omega |\omega|^2 \, d\Omega + 2\mu \int_\Gamma \mathbf{n} \cdot \left(\frac{\partial \mathbf{u}}{\partial t} + \mathbf{u} \cdot \text{grad } \mathbf{u} \right) d\Gamma \, .$$

If $\mathbf{u} = \mathbf{0}$ on the boundary of the flow domain, or if $\mathbf{u}$ decays sufficiently rapidly at infinity, and if the flow is stationary, then, from (5.3.1), in the absence of body forces, we have that

$$\mathcal{W}(\mathbf{u}) = |\mathbf{V}| \mathcal{D}(\mathbf{u}) \, .$$

For unsteady flows, instead of (5.3.3), one would use

$$\int_0^{t_f} \mathcal{W}(\mathbf{u}) \, dt = \mu \int_0^{t_f} \int_\Omega |\text{curl } \mathbf{u}|^2 \, d\Omega \, dt$$

for some time of interest t_f.

5.3.4 *Avoiding Hot Spots*

In many applications it is desirable that temperatures and/or temperature gradients along flow boundaries, e.g., structural components, not be allowed to exceed certain specified values. In particular, one would like to avoid "hot spots" along bounding surfaces, i.e., places where temperature peaks occur, since often such phenomena lead to meltdown or to flexural failures. A candidate functional whose minimization would avoid such problems is

$$\mathcal{T}(T) = \int_0^{t_f} \int_{\Gamma_\mathrm{T}} |\text{grad}_s T|^2 \, d\Gamma \, dt \, ,$$

where T denotes the temperature, grad_s the surface gradient, and Γ_T the portion of the flow boundary along which one would like to avoid the above mentioned problems. For steady flows, one instead might minimize

$$\mathcal{T}(T) = \int_{\Gamma_\mathrm{T}} |\text{grad}_s T|^2 \, d\Gamma \, .$$

Alternately, one could minimize

$$\mathcal{T}(T) = \int_0^{t_f} \int_{\Gamma_T} |T(\mathbf{x}, t) - T_T(\mathbf{x}, t)|^2 \, d\Gamma \, dt$$

where $T_T(\mathbf{x}, t)$ is a given desired temperature distribution along Γ_T. For steady flows, one would instead minimize

$$\mathcal{T}(T) = \int_{\Gamma_T} |T(\mathbf{x}) - T_T(\mathbf{x})|^2 \, d\Gamma \,.$$

A simple choice is $T_T = $ constant. These are examples of "flow matching" problems wherein one tries to match something other than the velocity field.

5.4 Control Mechanisms

The objectives listed in Section 5.3 are to be achieved by controlling the flow. Here, we discuss some of the control mechanisms that can be used to achieve those goals. (Of course, in a particular application, it is possible to use more than one control mehanism.) We note that many other control mechanisms are possible, and here we provide only a sampling of the possibilities.

5.4.1 *Velocity Controls Along Portions Of The Boundary*

A very much used mechanism for control is to inject or suck fluid through orifices along bounding surfaces. (Such control mechanisms have long been used in experimental studies of boundary layer control and drag minimization.) Thus, if Γ_c denotes the portion of the boundary covered by the orifices, we would seek a control $\mathbf{g}$ such that one of the functionals of Section 5.3 is minimized, subject to the constraints imposed by appropriate flow equations and initial and boundary conditions, and such that

$$\mathbf{u} = \mathbf{g}(\mathbf{x}, t) \quad \text{for } \mathbf{x} \in \Gamma_c \,, t_0 < t < t_1 \,, \tag{5.4.1}$$

where the control acts over the time interval $t_0 < t < t_1$. (In the notation of Section 5.2, $\Gamma = (\cup_i^3 \Gamma_i) \cup \Gamma_c$, where this division of the boundary is into disjoint pieces.) For stationary problems, we instead have the time-independent control $\mathbf{g}(\mathbf{x})$ acting in the boundary condition

$$\mathbf{u} = \mathbf{g}(\mathbf{x}) \quad \text{for } \mathbf{x} \in \Gamma_c \,. \tag{5.4.2}$$

Note that since the density ρ is a constant, (5.4.1) and (5.4.2) may also be viewed as controls on the mass flow rate $\rho\mathbf{u}$.

5.4.2   Stress Type Controls Along Portions of the Boundary

Another useful control mechanism is boundary stresses, or more general, mixed stress-velocity controls. In this case control is effected by adjusting the function $\mathbf{h}$ in the boundary condition

$$\mathbf{t} + \alpha\mathbf{u} = -p\mathbf{n} + 2\mu D(\mathbf{u}) \cdot \mathbf{n} + \alpha\mathbf{u} = \mathbf{h}(\mathbf{x}, t) \quad \text{for } \mathbf{x} \in \Gamma_c, t_0 < t < t_1,$$

where $\alpha(\mathbf{x})$ is a given nonnegative function. For stationary flows, the control $\mathbf{h}$ would, of course, be independent of time.

5.4.3   Temperature and Heating/Cooling Controls

Another common control mechanism is to adjust the temperature, or even more often, the heat flux, along portions of the boundary of the flow domain in order to achieve one of the desired objectives. Within this class of controls we find "heating" and "cooling" controls. For example, one could seek a control q such that one of the functionals of Section 5.3 is minimized, subject to the appropriate constraints of Section 5.2, and also

$$- \kappa\mathbf{n} \cdot \operatorname{grad} T = q(x, t) \quad \text{for } \mathbf{x} \in \Gamma_c, t_0 < t < t_1, \tag{5.4.3}$$

where again Γ_c denotes the portion of the boundary along which one allows the control to act. If instead one wishes to control using the temperature, then one would seek a control $T_c(\mathbf{x}, t)$ such that one of the functionals of Section 5.3 is minimized, subject to the appropriate constraints of Section 5.2, and such that

$$T = T_c(\mathbf{x}, t) \quad \text{for } \mathbf{x} \in \Gamma_c, t_0 < t < t_1.$$

For stationary flows, these relations between the state and controls have their obvious counterparts.

5.4.4   Distributed Controls

One could try to effect control through the body force in the Navier-Stokes equation (5.2.2). Thus, one would seek a control $\mathbf{f}$, defined on the flow domain Ω or on a portion of Ω, such that one of the functionals of Section 5.3 is minimized and subject to the appropriate constraints of Section 5.2. Physically, one way to effect such control is through a magnetic field acting on an ionized fluid or on a liquid metal.

Another possible distributed control is a heat source in the energy equation (5.2.4). In this case, we would seek a control Q, defined on the flow domain Ω or on a portion of Ω, such that one of the functionals of Section 5.3 is minimized, subject to the constraints of Section 5.2. Physically, one way to effect such a control is through radiation mechanisms, or through a targeted laser beam.

5.4.5 Shape Control

The control mechanisms discussed so far are known as *value controls*; this refers to the fact that we try to effect control through the adjustment of the values of the data of the problem. Another class of controls are known collectively as *shape controls*; in this case control is effected by adjusting the shape of the flow domain. The shape of the flow domain may be changed in many ways. For example, one could use leading and/or trailing edge flaps, or movable walls, or rudders, or propeller pitch. A related problem is the *optimal design* problem. Here, we want to choose a flow domain, e.g., the exterior of an airfoil, so that some objective is achieved. Of course, choosing the flow domain is tantamount to choosing its boundary, i.e., in the example, the airfoil itself.

5.4.6 Limiting the Size of the Control

In practice, the size of the controls are limited by technological constraints. For example, if we use mass injection or suction through orifices as a control mechanism, then the rate at which we can inject fluid, i.e., the "size" of **g** in (5.4.1), is limited by the flow rates of the pumps that are at one's disposal. Likewise, if one wishes to control by heat addition at portions of the boundary, then the amount of heat we can add, i.e., the size of q in (5.4.3), is limited by the heaters one has available for use. However, if the size of the control (measured in an appropriate norm) is not *a priori* constrained to be within some specified bounds, optimal controls found as solutions of a mathematical control or optimization problem are usually unbounded, and therefore not physically realizable.

There are two ways to limit the size of the controls. The first method is to impose *a priori* limits on the size of the controls. For example, if the control mechanism (5.4.1) is used, then one seeks an optimal control **g** only among those functions that satisfy $\|\mathbf{g}\| \leq K$ for some *a priori* chosen constant K.

A second method is to penalize the functionals of Section 5.3 with some norm of the control. For example, instead of minimizing the functional $\mathcal{D}(\mathbf{u})$ given by (5.3.1), one minimizes

$$\mathcal{D}(\mathbf{u}) + \delta\|\mathbf{g}\| = \frac{1}{|\mathbf{V}|} \int_{\Omega} \Phi \, d\Omega + \delta\|\mathbf{g}\|. \tag{5.4.4}$$

The addition of the term involving the control **g** automatically limits the size of the control; there is now a tradeoff. In order to make the first term, i.e., the drag, small, one must make the second term big, i.e., use a large control, and conversely, small controls yield large drags. Clearly, when one minimizes (5.4.4), one is no longer purely minimizing the drag. However, if the norm for the control **g** and the scaling factor δ are chosen judiciously, then the minimization of (5.4.4) can still yield a great reduction in the drag, and at the same time effectively limit the size of the control.

The first method certainly seems preferable; however, it results in control and optimization problems that are more difficult to analyze. Thus, the great majority of the mathematical literature concerned with flow control deals with the second method.

5.5 Some Sample Flow Control and Optimization Problems

In this section, we put together constraints, objectives, and controls gathered from Sections 5.2–5.4 in order to formulate some specific flow control and optimization problems. Actually, just about any combination drawn from those three sections results in a physically interesting flow control problem, i.e., one which can be applied to a *bona fide* technological problem. In order to keep the exposition simple, the examples presented below are all in the context of stationary flows.

5.5.1 Drag Minimization Via Mass Injection and Suction

Consider steady, incompressible flow about an obstacle. Let Γ denote the surface of the obstacle, and Ω the domain exterior to the obstacle. Then, Γ denotes the boundary of Ω. Let Γ_c denote a portion of Γ which may be made up of one or more disjoint sections. Let Γ_0 be the remaining portion of Γ so that $\Gamma = \Gamma_c \cup \Gamma_0$ and $\Gamma_c \cap \Gamma_0 = \emptyset$. Below, Γ_c is seen to be that part of the boundary on which the control is allowed to act. Far away from the obstacle, the flow is uniform with velocity $\mathbf{V}$.

We are concerned with finding the velocity on Γ_c such that the drag on the obstacle is minimized. Thus, we seek a state, i.e., a velocity field $\mathbf{u}$ and a pressure field p, defined on Ω, and a control function $\mathbf{g}$, defined on Γ_c, such that

$$
\mathcal{J}_D(\mathbf{u}, \mathbf{g}) = \frac{\mu}{2} \int_\Omega \left((\operatorname{grad} \mathbf{u}) + (\operatorname{grad} \mathbf{u})^T \right) : \left((\operatorname{grad} \mathbf{u}) + (\operatorname{grad} \mathbf{u})^T \right) \, d\Omega
$$
$$
+ \frac{\mu}{2} \int_\Omega \left(|\operatorname{grad}_s \mathbf{g}|^2 + |\mathbf{g}|^2 \right) \, d\Omega
\tag{5.5.1}
$$

is minimized, subject to

$$
\operatorname{div} \mathbf{u} = 0 \quad \text{in } \Omega,
\tag{5.5.2}
$$

$$
\rho \left(\frac{\partial \mathbf{u}}{\partial t} + \mathbf{u} \cdot \operatorname{grad} \mathbf{u} \right) - 2\mu \operatorname{div} \left(D(\mathbf{u}) \right) + \operatorname{grad} p = \mathbf{0} \quad \text{in } \Omega,
\tag{5.5.3}
$$

$$
\mathbf{u} = 0 \quad \text{on } \Gamma_0,
\tag{5.5.4}
$$

and

$$
\mathbf{u} = \mathbf{g} \quad \text{on } \Gamma_c.
\tag{5.5.5}
$$

Since no *a priori* constraints are placed on the size of the control $\mathbf{g}$, we penalized the cost functional (5.5.1) with a norm of the control. The motivation for choosing the particular norm for **g** used in (5.5.1) is based on both practical and mathematical considerations. Comparing with (5.3.1) and (5.4.4), we see that the functional (5.5.1) is a penalized version of the drag functional. The problem of minimizing (5.5.1), subject to (5.5.2)–(5.5.5) is considered in detail in [13].

One can also pose a version of this problem wherein the control is sought within a finite dimensional set. First, let $\Gamma_c = \cup_{k=1}^{K} \Gamma_k$, where each Γ_k represents an orifice on the boundary of the obstacle. At each orifice, assume that the injected or sucked flow has a known profile, but an unknown magnitude, i.e.,

$$\mathbf{u} = \alpha_k \phi_k(\mathbf{x}) \quad \text{on } \Gamma_k, \, k = 1, \dots, K , \qquad (5.5.6)$$

where the functions ϕ_k are prescribed, but the constants α_k, $k = 1, \dots, K$, are to be determined. Further, suppose that α_k, $k = 1, \dots, K$, are constrained to satisfy

$$a_k \leq \alpha_k \leq b_k \quad \text{for } k = 1, \dots, K , \qquad (5.5.7)$$

where a_k and b_k, $k = 1, \dots, K$, are prescribed numbers such that $a_k < b_k$ for each k. The ability to specify both the functions ϕ_k and the bounds a_k and b_k is a realistic engineering assumption.

Now, the only controls at our disposal are the numbers α_k, $k = 1, \dots, K$; furthermore these numbers are constrained by (5.5.7). Clearly, the control set is a bounded subset of K-dimensional Euclidean space, and thus is finite dimensional. Since we are imposing the *a priori* bounds (5.5.7) on our controls, we may now directly minimize the drag functional, which is proportional to the functional

$$\mathcal{J}_d(\mathbf{u}) = \frac{\mu}{2} \int_\Omega \left((\operatorname{grad} \mathbf{u}) + (\operatorname{grad} \mathbf{u})^T \right) : \left((\operatorname{grad} \mathbf{u}) + (\operatorname{grad} \mathbf{u})^T \right) d\Omega . \qquad (5.5.8)$$

The optimization problem at hand is: given the functions ϕ_k, defined on Γ_k, and the numbers a_k and b_k, $k = 1, \dots, K$, minimize the functional (5.5.8) subject to (5.5.2)–(5.5.4) and (5.5.6)–(5.5.7). This problem is treated in detail in [11].

5.5.2 *Flow Matching Via Shape and Value Control*

The next problem arises in the design of wind tunnels. At a given position, say $x = x_0$, of the wind tunnel, one wants to have a desired flow. For example, one could try to match some flight condition at that position. In order to achieve this objective, one is willing to build the wind tunnel so that, within certain practical limits, the walls of the tunnel in the section upstream of x_0 have arbitrary shape. Also, one is willing, again within certain limits, to have an arbitrary flow at some position $x = x_i$ upstream of the position $x = x_0$ at which we wish to match the flow to a desired flow.

We restate the problem using the notation of Sections 5.2–5.4. Let Ω denote the flow domain, which has an inflow boundary Γ_i and an outflow boundary Γ_o. These two boundary segments are connected by walls which we denote by Γ_w. The geometry of the segments Γ_o and Γ_i are prescribed, but that of Γ_w is to be determined. Of course, we do know that Γ_w has to connect Γ_o and Γ_i. The objective is to have the flow on Γ_o match, as well as possible, a given flow, e.g., we want to minimize the functional

$$\mathcal{J}_m(\mathbf{u}) = \int_{\Gamma_o} |\mathbf{u} - \mathbf{U}_o|^2 \, d\Gamma, \qquad (5.5.9)$$

where $\mathbf{U}_o$ is a given function defined on Γ_o. In order to meet this objective, we are allowed to choose the shape of the walls Γ_w and the inflow, i.e., $\mathbf{u}$ on Γ_i. In addition the state and control variables are required to satisfy appropriate differential equations and boundary conditions.

The optimization problem in hand is the following. Given the inflow and outflow boundaries Γ_i and Γ_o, respectively. Given a function $\mathbf{U}_o$ defined on Γ_o. Then, find an inflow velocity $\mathbf{g}$ and the shape of the wall boundaries Γ_w such that (5.5.9) is minimized, and such that (5.5.2), (5.5.3),

$$\mathbf{u} = 0 \quad \text{on } \Gamma_w, \qquad (5.5.10)$$

$$\mathbf{u} = \mathbf{g} \quad \text{on } \Gamma_i, \qquad (5.5.11)$$

and

$$\mathbf{t} = -p\mathbf{n} + 2\mu D(\mathbf{u}) \cdot \mathbf{n} = \mathbf{0} \quad \text{on } \Gamma_o \qquad (5.5.12)$$

are satisfied. (The boundary condition (5.5.10) is simply the no-slip condition along a solid wall; (5.5.12) is a simple zero-stress outflow condition; other types of outflow conditions can easily be used instead.) In addition, limits on the "size" of the controls, i.e., of $\mathbf{g}$ and Γ_w, must be imposed.

A practical approximate solution of this problem would almost certainly involve reducing the constraint set to one with finite dimension. For example, the wall boundaries Γ_w could be assumed to be Bezier surfaces determined by the position of a finite set of points, and the inflow velocity $\mathbf{g}$ in (5.5.11) could be assumed to have some definite functional form containing a finite number of free constants.

5.5.3 Avoiding Hot Spots Via Heating and Cooling

As in the previous example, we have an inflow boundary Γ_i, an outflow boundary Γ_o, and solid walls Γ_w connecting the two. In this example, the geometry of all these boundary segments is prescribed, as are the inflow velocity $\mathbf{u}_i$ and temperature T_i. At the outflow, one can impose one's favorite outflow boundary conditions. On the walls, we have the no-slip boundary conditions for the velocity. Control is to be effected through heating and cooling along a portion $\Gamma_c \subset \Gamma_w$ of the wall boundary.

The objective is to have the temperature be as uniform as possible, i.e., to avoid "hot spots." To this end, one can minimize the functional

$$\mathcal{J}_T(T, q) = \alpha \int_{\Gamma_w} |T - T_w|^2 \, d\Gamma + \beta \int_{\Gamma} |q|^2 \, d\Gamma, \tag{5.5.13}$$

where q denotes the heat flux control to be applied on Γ_c, α and β are appropriately chosen scale factors, and T_w a given function defined on Γ_w. (Note that one would really prefer to minimize the $L^\infty(\Gamma_w)$-norm of $(T - T_w)$, but this leads to substantial analytical and computational difficulties.)

The constraints imposed on the state variables $\mathbf{u}, p$, and T and the control q are (5.5.1), (5.5.2),

$$\rho c_v \left(\frac{\partial T}{\partial t} + \mathbf{u} \cdot \operatorname{grad} T \right) - \Phi(\mathbf{u}) - \kappa \, \Delta T = 0 \quad \text{in } \Omega, \tag{5.5.14}$$

$$\mathbf{u} = \mathbf{u}_i \quad \text{on } \Gamma_i, \tag{5.5.15}$$

$$T = T_i \quad \text{on } \Gamma_i, \tag{5.5.16}$$

$$\mathbf{t} = -p\mathbf{n} + 2\mu D(\mathbf{u}) \cdot \mathbf{n} = \mathbf{0} \quad \text{on } \Gamma_o, \tag{5.5.17}$$

$$\mathbf{n} \cdot \operatorname{grad} T = 0 \quad \text{on } \Gamma_o, \tag{5.5.18}$$

$$\mathbf{u} = \mathbf{0} \quad \text{on } \Gamma_w, \tag{5.5.19}$$

$$\mathbf{n} \cdot \operatorname{grad} T = 0 \quad \text{on } \Gamma_w \backslash \Gamma_c, \tag{5.5.20}$$

and

$$\mathbf{n} \cdot \operatorname{grad} T = q \quad \text{on } \Gamma_c. \tag{5.5.21}$$

In summary, the optimization problem at hand is as follows. Given the functions $\mathbf{u}_i$ and T_i, defined on Γ_i, and given the function T_w, defined on Γ_w, find a control q, defined on Γ_c, and state variables $\mathbf{u}, p$ and T, defined in Ω, such that (5.5.13) is minimized and such that (5.5.1)–(5.5.2) and (5.5.14)–(5.5.21) are satisfied. This problem is treated in detail in [15].

The velocity and pressure fields $\mathbf{u}$ and p, respectively, can be determined from (5.5.1), (5.5.2), (5.5.15), (5.5.17), and (5.5.19), since these do not involve the temperature T and all the data in these equations are prescribed *a priori*; these equations uncouple from the thermal equations. After obtaining the velocity field $\mathbf{u}$, it then merely acts as a coefficient in the energy equation (5.5.14). Thus, our optimization problem separates into two steps as follows. First, given the function $\mathbf{u}_i$, defined on Γ_i, find a velocity field

$\mathbf{u}$ and a pressure field p, defined on Ω, that satisfy (5.5.1), (5.5.2), (5.5.15), (5.5.17), and
(5.5.19). Second, given this velocity field and the functions T_i and T_w, defined on Γ_i
and Γ_w, repectively, find a control q, defined on Γ_c, and a temperature field T, defined
in Ω, such that (5.5.13) is minimized and such that (5.5.14), (5.5.16), (5.5.18), (5.5.20),
and (5.5.21) are satisfied.

Instead of using the heat flux, control may be effected by the injection or suction of
fluid through the boundary. Thus, we replace the boundary condition (5.5.19) by

$$\mathbf{u} = \mathbf{0} \quad \text{on } \Gamma_w \backslash \Gamma_c \tag{5.5.22}$$

and

$$\mathbf{u} = \mathbf{g} \quad \text{on } \Gamma_c, \tag{5.5.23}$$

the boundary condition (5.5.21) by

$$T = T_c \quad \text{on } \Gamma_c, \tag{5.5.24}$$

and the functional (5.5.13) by

$$\mathcal{J}_T(T, \mathbf{g}, T_c) = \alpha \int_{\Gamma_w} |T - T_w|^2 \, d\Gamma + \beta \int_{\Gamma_c} \left(|\text{grad}_s T_c|^2 + |T_c|^2 \right) d\Gamma \\ + \gamma \int_{\Gamma_c} \left(|\text{grad}_s \mathbf{g}|^2 + |\mathbf{g}|^2 \right) d\Gamma. \tag{5.5.25}$$

The controls are the velocity $\mathbf{g}$ and the temperature T_c on Γ_c. Now, the optimization
problem is given as follows. Given the functions $\mathbf{u}_i$ and T_i, defined on Γ_i, and given the
function T_w, defined on Γ_w, find controls $\mathbf{g}$ and T_c, defined on Γ_c, and state variables
$\mathbf{u}, p$ and T, defined in Ω, such that (5.5.25) is minimized and such that (5.5.1)–(5.5.2),
(5.5.14)–(5.5.18), (5.5.20), and (5.5.22)–(5.5.24) are satisfied. Now the velocity and
pressure are coupled to the temperature; one cannot solve for the mechanical variables
separately from the temperature. There is an obvious coupling through the functional
(5.5.25). However, even if the velocity control $\mathbf{g}$ did not appear in (5.5.25), one could
not solve for the velocity and pressure independently of the temperature since the system
(5.5.1)–(5.5.2), (5.5.15), (5.5.17), and (5.5.22)–(5.5.23) contains the unknown velocity
control $\mathbf{g}$ which cannot be determined without consideration of the functional (5.5.25).

5.6 An Example of the Mathematical Analysis of a Flow Optimization Problem

The major battles that must be fought and won if one wishes to successfully study optimal
flow control or flow optimization problems are as follows.

1. Give a precise mathematical statement of the problem; in particular, one must specify the function classes in which one will seek optimal states and controls, as well as provide a description of the optimization objective and the constraints.

2. Show that optimal states and controls exist, and provide information about their regularity.

3. From the first-order necessary conditions holding at an extremum of the objective functional, and from the constraints, derive a system of equations, and perhaps inequalities, that optimal controls and states satisfy. This system we call the *optimality system*.

4. Devise schemes for the approximate solution of the optimality system.

5. Deduce error estimates for the approximate solutions.

6. Develop computer codes implementing the algorithms devised for approximating solutions of the optimality system.

Clearly, from a practical standpoint, the fourth and sixth of these battles are the most important to win. However, we have found that the chances of winning these two battles, and of knowing exactly what one has won, are greatly enhanced if one can win some or all of the other four battles.

In this section we describe in mathematical terms, using the context of a specific optimization problem, one approach to winning the first five of the above battles. By no means is the approach described here the only possible means of winning these battles. We merely use the particular analyses described here to illustrate how, in many instances, these battles can be won. Space limitations allow us to only provide sketches of the various proofs. Detailed proofs may be found in [13].

5.6.1 *Drag Minimization Via Boundary Velocity Control*

The plan of this section is as follows. First, we introduce the notation that will be used throughout this section. Then, we give a precise statement of the optimization problem and indicate how one can prove that an optimal solution exists, and how one can obtain regularity results for the optimal states and controls. Then, we prove the existence of Lagrange multipliers and then use the method of Lagrange multipliers to derive an optimality system. Finally, we consider finite element approximations and derive error estimates.

We consider the problem (5.5.1)–(5.5.5); however, for convenience, we amend the functional (5.5.1) slightly by nondimensionalizing all variables and by adding a term, i.e.,

$$\mathcal{J}_D(\mathbf{u}, p, \mathbf{g}) = \frac{1}{2Re} \int_\Omega |(\operatorname{grad} \mathbf{u}) + (\operatorname{grad} \mathbf{u})^T|^2 \, d\Omega - \int_\Omega \mathbf{f} \cdot \mathbf{u} \, d\Omega \\ - \int_\Omega p \operatorname{div} \mathbf{u} \, d\Omega + \frac{1}{2Re} \int_{\Gamma_c} (|\operatorname{grad}_s \mathbf{g}|^2 + |\mathbf{g}|^2) \, d\Gamma, \quad (5.6.1)$$

where $Re = \rho|\mathbf{V}|L/\mu$ denotes the Reynolds number, L being a typical length scale in the problem. Since the flow is incompressible, (5.5.1) and (5.6.1) are in fact identical. The optimization problem we study is to seek state pairs $(\mathbf{u}, p)$ and controls $\mathbf{g}$ such that $\mathcal{J}_D(\cdot, \cdot)$ is minimized, subject to the constraints (5.5.2)–(5.5.5).

In (5.5.2)–(5.5.5) and (5.6.1), Ω denotes a bounded domain in $\mathbb{R}^d, d = 2$ or 3 with a boundary Γ; Γ_u and Γ_c are portions of Γ such that $\overline{\Gamma}_u \cup \overline{\Gamma}_c = \overline{\Gamma}$ and $\Gamma_u \cap \Gamma_c = 0$. When finite element approximations are considered, we will assume that Ω is a convex polyhedral domain; otherwise, we will assume that either Ω is convex or Γ is of class $C^{1,1}$. Γ_c and Γ_u denote the portions of Γ where velocity controls are and are not applied, respectively.

The continuity equation (5.5.2) implies that the control $\mathbf{g}$ must satisfy

$$\int_{\Gamma_c} \mathbf{g} \cdot \mathbf{n} \, d\Gamma = 0;$$

also, if Γ_c has a boundary, we assume that

$$\mathbf{g} = \mathbf{0} \quad \text{on } \partial\Gamma_c, \quad (5.6.2)$$

where $\partial\Gamma_c$ denotes the boundary of Γ_c, the latter viewed as a subset of Γ. The relation (5.6.2) is imposed in order to ensure that solutions of our optimization problems are "sufficiently" regular.

5.6.2 Notation

Throughout, C will denote a positive constant whose meaning and value changes with context. Also, $H^s(\mathcal{D}), s \in \mathbb{R}$, denotes the standard Sobolev space of order s with respect to the set $\mathcal{D}$, where $\mathcal{D}$ is either the flow domain Ω, or its boundary Γ, or part of that boundary. Of course, $H^0(\mathcal{D}) = L^2(\mathcal{D})$. Corresponding Sobolev spaces of vector-valued functions will be denoted by $\mathbf{H}^s(\mathcal{D})$, e.g., $\mathbf{H}^1(\Omega) = [H^1(\Omega)]^d$. Dual spaces will be denoted by $(\cdot)^*$.

Of particular interest will be the space

$$\mathbf{H}^1(\Omega) = \{v_j \in L^2(\Omega) \mid \frac{\partial v_j}{\partial x_k} \in L^2(\Omega) \text{ for } j, k = 1, \ldots, d\}$$

and the subspaces

$$\mathbf{H}_0^1(\Omega) = \{\mathbf{v} \in \mathbf{H}^1(\Omega) \mid \mathbf{v} = \mathbf{0} \quad \text{on } \Gamma\}$$

and

$$L_0^2(\Omega) = \{q \in L^2(\Omega) \mid \int_\Omega q \, d\Omega = 0\}.$$

For functions defined on Γ_c we will use the subspaces

$$\mathbf{W}(\Gamma_c) = \begin{cases} \mathbf{H}_0^1(\Gamma_c) & \text{if } \Gamma_c \text{ has a boundary} \\ \mathbf{H}^1(\Gamma_c) & \text{otherwise} \end{cases}$$

and

$$\mathbf{W}_n(\Gamma_c) = \begin{cases} \mathbf{H}_n^1(\Gamma_c) \cap \mathbf{H}_0^1(\Gamma_c) & \text{if } \Gamma_c \text{ has a boundary} \\ \mathbf{H}_n^1(\Gamma_c) & \text{otherwise}, \end{cases}$$

where

$$\mathbf{H}_n^1(\Gamma_c) = \{\mathbf{g} \in \mathbf{H}^1(\Gamma_c) \mid \int_{\Gamma_c} \mathbf{g} \cdot \mathbf{n} \, d\Gamma = 0\},$$

and, whenever Γ_c has a boundary,

$$\mathbf{H}_0^1(\Gamma_c) = \{\mathbf{g} \in \mathbf{H}^1(\Gamma_c) \mid \mathbf{g} = \mathbf{0} \quad \text{on } \partial\Gamma_c\}.$$

Norms of functions belonging to $H^s(\Omega)$, $H^s(\Gamma)$ and $H^s(\Gamma_c)$ are denoted by $\|\cdot\|_s$, $\|\cdot\|_{s,\Gamma}$ and $\|\cdot\|_{s,\Gamma_c}$, respectively. Of particular interest are the $L^2(\Omega)$–norm $\|\cdot\|_0$ and the semi-norm

$$|v|_1^2 = \sum_{j=1}^d \left\| \frac{\partial v}{\partial x_j} \right\|_0^2$$

and norm

$$\|v\|_1^2 = |v|_1^2 + \|v\|_0^2$$

defined for functions belonging to $H^1(\Omega)$. Norms for spaces of vector valued functions will be denoted by the same notation as that used for their scalar counterparts.

The inner products in $L^2(\Omega)$ and $\mathbf{L}^2(\Omega)$ are both denoted by $(\cdot, \cdot)$, those in $L^2(\Gamma)$ and $\mathbf{L}^2(\Gamma)$ by $(\cdot, \cdot)_\Gamma$, and those in $L^2(\Gamma_c)$ and $\mathbf{L}^2(\Gamma_c)$ by $(\cdot, \cdot)_{\Gamma_c}$.

We will use the two bilinear forms

$$a(\mathbf{u}, \mathbf{v}) = \frac{1}{2} \int_\Omega \left((\operatorname{grad} \mathbf{u}) + (\operatorname{grad} \mathbf{u})^T \right) : \left((\operatorname{grad} \mathbf{v}) + (\operatorname{grad} \mathbf{v})^T \right) d\Omega$$

$$\forall \, \mathbf{u}, \mathbf{v} \in \mathbf{H}^1(\Omega)$$

and

$$b(\mathbf{v}, q) = -\int_\Omega q \operatorname{div} \mathbf{v} \, d\Omega \quad \forall \, \mathbf{v} \in \mathbf{H}^1(\Omega) \text{ and } \forall \, q \in L^2(\Omega)$$

and the trilinear form

$$c(\mathbf{u}, \mathbf{v}, \mathbf{w}) = \int_\Omega \mathbf{u} \cdot \text{grad} \, \mathbf{v} \cdot \mathbf{w} \, d\Omega \quad \forall \, \mathbf{u}, \mathbf{v}, \mathbf{w} \in \mathbf{H}^1(\Omega).$$

These forms are continuous in the sense that there exist constants c_a, c_b and $c_c > 0$ such that

$$|a(\mathbf{u}, \mathbf{v})| \leq c_a \|\mathbf{u}\|_1 \|\mathbf{v}\|_1 \quad \forall \, \mathbf{u}, \mathbf{v} \in \mathbf{H}^1(\Omega), \tag{5.6.3}$$

$$|b(\mathbf{v}, q)| \leq c_b \|\mathbf{v}\|_1 \|q\|_0 \quad \forall \, \mathbf{v} \in \mathbf{H}^1(\Omega) \text{ and } q \in L^2(\Omega) \tag{5.6.4}$$

and

$$|c(\mathbf{u}, \mathbf{v}, \mathbf{w})| \leq c_c \|\mathbf{u}\|_1 \|\mathbf{v}\|_1 \|\mathbf{w}\|_1 \quad \forall \, \mathbf{u}, \mathbf{v}, \mathbf{w} \in \mathbf{H}^1(\Omega). \tag{5.6.5}$$

Moreover, we have the coercivity properties

$$a(\mathbf{v}, \mathbf{v}) \geq C_a \|\mathbf{v}\|_1^2 \quad \forall \, \mathbf{v} \in \mathbf{H}_0^1(\Omega) \tag{5.6.6}$$

and

$$\sup_{\mathbf{0} \neq \mathbf{v} \in \mathbf{H}_0^1(\Omega)} \frac{b(\mathbf{v}, q)}{\|\mathbf{v}\|_1} \geq C_b \|q\|_0 \quad \forall \, q \in L_0^2(\Omega), \tag{5.6.7}$$

for some constants C_a and $C_b > 0$.

For details concerning the notation employed and/or for (5.6.3)–(5.6.7), one may consult [29], [37], [38] and [42].

5.6.3 The Optimization Problem and the Existence of Optimal Solutions

We begin by giving a precise statement of the optimization problem we consider. Let $\mathbf{g} \in \mathbf{W}_n(\Gamma_c)$ denote the boundary control and let $\mathbf{u} \in \mathbf{H}^1(\Omega)$ and $p \in L_0^2(\Omega)$ denote the state, i.e., the velocity and pressure fields, respectively. The state and control variables are constrained to satisfy the system (5.5.2)–(5.5.5), which we recast into the following particular weak form (see, e.g., [30], [37], [38], and [42]):

$$\tfrac{1}{Re} a(\mathbf{u}, \mathbf{v}) + c(\mathbf{u}, \mathbf{u}, \mathbf{v}) + b(\mathbf{v}, p) - (\mathbf{v}, \mathbf{t})_\Gamma = (\mathbf{f}, \mathbf{v}) \quad \forall \, \mathbf{v} \in \mathbf{H}^1(\Omega), \tag{5.6.8}$$

$$b(\mathbf{u}, q) = 0 \quad \forall \, q \in L_0^2(\Omega) \tag{5.6.9}$$

and

$$(\mathbf{u}, \mathbf{s})_\Gamma - (\mathbf{g}, \mathbf{s})_{\Gamma_c} = 0 \quad \forall \, \mathbf{s} \in \mathbf{H}^{-1/2}(\Gamma), \tag{5.6.10}$$

where $\mathbf{f} \in \mathbf{L}^2(\Omega)$ is a given function. One may show that, in a distributional sense,

$$\mathbf{t} = \left[-p\mathbf{n} + \tfrac{1}{Re} \left(\text{grad} \, \mathbf{u} + (\text{grad} \, \mathbf{u})^T \right) \cdot \mathbf{n} \right]_\Gamma,$$

i.e., **t** is the stress force on the boundary.

We remark on the use, in the weak formulation (5.6.8)–(5.6.10), of the Lagrange multiplier **t** to enforce the boundary condition on the velocity. In the first place, there are technical reasons for this choice, the most important one appearing in the proof of the error estimates for finite element approximations. We will remark on this point further below. From a practical point of view, the introduction of the Lagrange multiplier **t** does not introduce any new difficulties. It was shown in [39], in the context of finite element approximations of solutions of the Navier-Stokes equations, that one may in fact uncouple the computation of the multiplier **t** from that of the velocity and pressure fields. Indeed, one may devise schemes such that one may solve (a discretization) of (5.6.10) for the velocity on the boundary, and then solve for **u** and p from discretizations of (5.6.8)–(5.6.10) by using subsapces of $\mathbf{H}_0^1(\Omega)$ in a discretization of (5.6.10). Subsequently, one may compute an approximation to **t**, if one so desires. See [39] for details. Moreover, since **t** is the stress on the boundary, this method provides a systematic mechanism for computing this interesting variable.

The functional (5.6.1), using the notation introduced above, is given by

$$\mathcal{J}_D(\mathbf{u},p,\mathbf{g}) = \tfrac{1}{2Re}a(\mathbf{u},\mathbf{u}) + b(\mathbf{u},p) - (\mathbf{f},\mathbf{u}) + \tfrac{1}{2Re}\|\mathbf{g}\|_{1,\Gamma_c}^2 . \qquad (5.6.11)$$

(If Γ_c has a boundary we may replace the term $\tfrac{1}{2Re}\|\mathbf{g}\|_{1,\Gamma_c}^2$ by $\tfrac{1}{2Re}|\mathbf{g}|_{1,\Gamma_c}^2$.)

The *admissibility set* $\mathcal{U}_{ad}$ is defined by

$$\mathcal{U}_{ad} = \{(\mathbf{u},p,\mathbf{g}) \in \mathbf{H}^1(\Omega) \times L_0^2(\Omega) \times \mathbf{W}_n(\Gamma_c) : \mathcal{J}_D(\mathbf{u},p,\mathbf{g}) < \infty,$$
$$\text{and there exists a } \mathbf{t} \in \mathbf{H}^{-1/2}(\Gamma)$$
$$\text{such that } (5.6.8)–(5.6.10) \text{ are satisfied}\} .$$

Then, $(\hat{\mathbf{u}},\hat{p},\hat{\mathbf{g}}) \in \mathcal{U}_{ad}$ is called an *optimal solution* if there exists $\epsilon > 0$ such that

$$\mathcal{J}_D(\hat{\mathbf{u}},\hat{p},\hat{\mathbf{g}}) \leq \mathcal{J}_D(\mathbf{u},p,\mathbf{g}) \qquad (5.6.12)$$
$$\forall\, (\mathbf{u},p,\mathbf{g}) \in \mathcal{U}_{ad} \text{ satisfying} \|\mathbf{u}-\hat{\mathbf{u}}\|_1 + \|p-\hat{p}\|_0 + \|\mathbf{g}-\hat{\mathbf{g}}\|_{1,\Gamma_c} \leq \epsilon .$$

We first show that an optimal solution exists and prove a preliminary regularity result.

Theorem 5.6.1 *There exists an optimal solution* $(\hat{\mathbf{u}},\hat{p},\hat{\mathbf{g}}) \in \mathcal{U}_{ad}$. *Moreover, any optimal solution satisfies* $\hat{\mathbf{u}} \in \mathbf{H}^{3/2}(\Omega)$ *and* $\hat{p} \in H^{1/2}(\Omega) \cap L_0^2(\Omega)$ *and if* $\hat{\mathbf{t}} \in \mathbf{H}^{-1/2}(\Gamma)$ *is such that* $(\hat{\mathbf{u}},\hat{p},\hat{\mathbf{g}},\hat{\mathbf{t}})$ *is a solution of (5.6.8)–(5.6.10), then* $\hat{\mathbf{t}} \in \mathbf{L}^2(\Gamma)$.

Sketch of proof: First one shows that the admissibility set $\mathcal{U}_{ad}$ is not empty. This is done by setting $\mathbf{g} = \mathbf{0}$ in (5.6.8)–(5.6.10). Then, that system is merely an uncontrolled

Navier-Stokes problem for which, since $\mathbf{f} \in \mathbf{L}^2(\Omega)$, it is well known ([37] or [42]) that a norm-bounded solution $(\tilde{\mathbf{u}}, \tilde{p}, \tilde{\mathbf{t}})$ exists. Moreover, it is easily shown that $\mathcal{K}(\tilde{\mathbf{u}}, \tilde{p}, \mathbf{0}) \leq C(\|\tilde{\mathbf{u}}\|_1 + \|\tilde{p}\|_0 + \|\mathbf{f}\|_0)\|\tilde{\mathbf{u}}\|_1 < \infty$. Thus, $(\tilde{\mathbf{u}}, \tilde{p}, \mathbf{0}) \in \mathcal{U}_{ad}$.

Next one considers a minimizing sequence $\{\mathbf{u}^{(k)}, p^{(k)}, \mathbf{g}^{(k)}\}$ with elements belonging to $\mathcal{U}_{ad}$. Using the definition of $\mathcal{U}_{ad}$ and the functional (5.6.1), and regularity results about solutions of the Navier-Stokes equations, one easily shows that this sequence is uniformly bounded in $\mathbf{H}^1(\Omega) \times L_0^2(\Omega) \times \mathbf{W}_n(\Gamma_c)$. Then, one can extract a subsequence such that

$$
\begin{aligned}
\mathbf{g}^{(k)} &\rightharpoonup \hat{\mathbf{g}} \quad \text{in} \quad \mathbf{W}_n(\Gamma_c) \\
\mathbf{u}^{(k)} &\rightharpoonup \hat{\mathbf{u}} \quad \text{in} \quad \mathbf{H}^1(\Omega) \\
p^{(k)} &\rightharpoonup \hat{p} \quad \text{in} \quad L_0^2(\Omega) \\
\mathbf{t}^{(k)} &\rightharpoonup \hat{\mathbf{t}} \quad \text{in} \quad \mathbf{H}^{-1/2}(\Gamma) \\
\mathbf{u}^{(k)} &\rightarrow \hat{\mathbf{u}} \quad \text{in} \quad \mathbf{L}^2(\Omega) \\
\mathbf{u}^{(k)}|_\Gamma &\rightarrow \hat{\mathbf{u}}|_\Gamma \quad \text{in} \quad \mathbf{L}^2(\Gamma)
\end{aligned}
$$

for some $(\hat{\mathbf{u}}, \hat{p}, \hat{\mathbf{g}}, \hat{\mathbf{t}}) \in \mathbf{H}^1(\Omega) \times L_0^2(\Omega) \times \mathbf{W}_n(\Gamma_c) \times \mathbf{H}^{-1/2}(\Gamma)$. The last two convergence results above follow from the compact imbeddings $\mathbf{H}^1(\Omega) \subset \mathbf{L}^2(\Omega)$ and $\mathbf{H}^{1/2}(\Gamma) \subset \mathbf{L}^2(\Gamma)$.

The next step is to pass to the limit to show that $(\hat{\mathbf{u}}, \hat{p}, \hat{\mathbf{g}}, \hat{\mathbf{t}})$ satisfies (5.6.8)–(5.6.10). Then, one uses the weak lower semicontinuity of the functional (5.6.1) to conclude that the limit function $(\hat{\mathbf{u}}, \hat{p}, \hat{\mathbf{g}}, \hat{\mathbf{t}})$ is an optimal solution. Thus one can show that an optimal solution belonging to $\mathcal{U}_{ad}$ exists.

The regularity result easily follows by a standard argument, namely, moving the non-linear term to the right hand side and then using regularity results about the linear Stokes problem. $\square$

5.6.4 *The Existence of Lagrange Multipliers*

We wish to use the method of Lagrange multipliers to turn our constrained optimization problem into an unconstrained one. First, one should show that suitable Lagrange multipliers exist.

Let $B_1 = \mathbf{H}^1(\Omega) \times L_0^2(\Omega) \times \mathbf{W}_n(\Gamma_c) \times \mathbf{H}^{-1/2}(\Gamma)$ and $B_2 = (\mathbf{H}^1(\Omega))^* \times L_0^2(\Omega) \times \mathbf{H}^{1/2}(\Gamma)$ and let the nonlinear mapping $M : B_1 \rightarrow B_2$ denote the (generalized) constraint equations, i.e., $M(\mathbf{u}, p, \mathbf{g}, \mathbf{t}) = (\mathbf{f}, z, \mathbf{b})$ for $(\mathbf{u}, p, \mathbf{g}, \mathbf{t}) \in B_1$ and $(\mathbf{f}, z, \mathbf{b}) \in B_2$ if and only if

$$
\tfrac{1}{Re} a(\mathbf{u}, \mathbf{v}) + c(\mathbf{u}, \mathbf{u}, \mathbf{v}) + b(\mathbf{v}, p) - (\mathbf{v}, \mathbf{t})_\Gamma = (\mathbf{f}, \mathbf{v}) \quad \forall \, \mathbf{v} \in \mathbf{H}^1(\Omega),
$$

$$b(\mathbf{u}, q) = (z, q) \quad \forall\, q \in L_0^2(\Omega)$$

and

$$(\mathbf{u}, \mathbf{s})_\Gamma - (\mathbf{g}, \mathbf{s})_{\Gamma_c} = (\mathbf{b}, \mathbf{s})_\Gamma \quad \forall\, \mathbf{s} \in \mathbf{H}^{-1/2}(\Gamma).$$

Thus, the constraints (5.6.8)–(5.6.10) can be expressed as $M(\mathbf{u}, p, \mathbf{g}, \mathbf{t}) = (\mathbf{f}, 0, \mathbf{0})$.

Given $\mathbf{u} \in \mathbf{H}^1(\Omega)$, the operator $M'(\mathbf{u}) \in \mathcal{L}(B_1; B_2)$ may be defined as follows: $M'(\mathbf{u}) \cdot (\mathbf{w}, r, \mathbf{k}, \mathbf{y}) = (\bar{\mathbf{f}}, \bar{z}, \bar{\mathbf{b}})$ for $(\mathbf{w}, r, \mathbf{k}, \mathbf{y}) \in B_1$ and $(\bar{\mathbf{f}}, \bar{z}, \bar{\mathbf{b}}) \in B_2$ if and only if

$$\frac{1}{Re} a(\mathbf{w}, \mathbf{v}) + c(\mathbf{w}, \mathbf{u}, \mathbf{v}) + c(\mathbf{u}, \mathbf{w}, \mathbf{v}) + b(\mathbf{v}, r) - (\mathbf{v}, \mathbf{y})_\Gamma = (\bar{\mathbf{f}}, \mathbf{v}) \qquad (5.6.13)$$
$$\forall\, \mathbf{v} \in \mathbf{H}^1(\Omega),$$

$$b(\mathbf{w}, q) = (\bar{z}, q) \quad \forall\, q \in L^2(\Omega), \qquad (5.6.14)$$

and

$$(\mathbf{w}, \mathbf{s})_\Gamma - (\mathbf{k}, \mathbf{s})_{\Gamma_c} = (\bar{\mathbf{b}}, \mathbf{s})_\Gamma \quad \forall\, \mathbf{s} \in \mathbf{H}^{-1/2}(\Gamma). \qquad (5.6.15)$$

Lemma 5.6.2 *For* $\mathbf{u} \in \mathbf{H}^1(\Omega)$, *the operator* $M'(\mathbf{u})$ *from* B_1 *into* B_2 *has closed range.*

Sketch of proof: This result follows by showing that $M'(\mathbf{u})$ is a compact perturbation of a generalized Stokes operator $S \in \mathcal{L}(B_1; B_2)$, and then showing that the adjoint of S is a semi-Fredholm operator, i.e., has a closed range and a finite-dimensional kernel. Then it follows that S itself, and any compact perturbation of S, has closed range; see [40]. □

Lemma 5.6.3 *For* $\mathbf{u} \in \mathbf{H}^1(\Omega)$, *the operator* $M'(\mathbf{u})$ *from* B_1 *into* B_2 *is onto.*

Sketch of proof: Assume that $M'(\mathbf{u})$ is not onto. Then, the image of $M'(\mathbf{u})$ is strictly contained in B_2 and, by Lemma 5.6.2, is closed, so that there exists a nonzero $(\nu, \phi, \tau) \in (B_2)^* = \mathbf{H}^1(\Omega) \times L_0^2(\Omega) \times \mathbf{H}^{-1/2}(\Gamma)$ such that

$$< (\bar{\mathbf{f}}, \bar{z}, \bar{\mathbf{b}}), (\nu, \phi, \tau) >= 0 \quad \forall\, (\bar{\mathbf{f}}, \bar{z}, \bar{\mathbf{b}}) \text{ belonging to the range of } M'(\mathbf{u}),$$

where here $< \cdot, \cdot >$ denotes the duality pairing between B_2 and B_2^*. This equation implies that $(\nu, \phi, \tau) \in (B_2)^* = \mathbf{H}^1(\Omega) \times L_0^2(\Omega) \times \mathbf{H}^{-1/2}(\Gamma)$ satisfies

$$\frac{1}{Re} a(\mathbf{w}, \nu) + c(\mathbf{w}, \mathbf{u}, \nu) + c(\mathbf{u}, \mathbf{w}, \nu) + b(\mathbf{w}, \phi) + (\mathbf{w}, \tau)_\Gamma = 0 \qquad (5.6.16)$$
$$\forall\, \mathbf{w} \in \mathbf{H}^1(\Omega),$$

$$b(\nu, r) = 0 \quad \forall\, r \in L_0^2(\Omega), \qquad (5.6.17)$$

$$(\nu, \mathbf{y})_\Gamma = 0 \quad \forall\, \mathbf{y} \in \mathbf{H}^{-1/2}(\Gamma), \qquad (5.6.18)$$

and

$$(\mathbf{k}, \boldsymbol{\tau})_{\Gamma_c} = 0 \quad \forall \, \mathbf{k} \in \mathbf{W}_n(\Gamma_c). \tag{5.6.19}$$

The key now is to show that (5.6.17)–(5.6.19) implies the contradictory result $(\boldsymbol{\nu}, \phi, \boldsymbol{\tau}) = (\mathbf{0}, 0, \mathbf{0})$, so that then $M'(\mathbf{u})$ is indeed onto. In order to arrive at this contradictory result, one first shows that (5.6.17)–(5.6.19) may be extended to a domain Ω_e which is a smooth extension of Ω in such a way that $\boldsymbol{\nu} = \mathbf{0}$ on the boundary of Ω_e. In this case, (5.6.17)–(5.6.19), with the various forms redefined in terms of integrals over Ω_e, is merely an eigenvalue problem. Since the extension of Ω to Ω_e is arbitrary, then for a given value of Re, one can guarantee that there is some extended domain Ω_e such that the only solution of (5.6.17)–(5.6.19) is the trivial solution $(\boldsymbol{\nu}, \phi, \boldsymbol{\tau}) = (\mathbf{0}, 0, \mathbf{0})$ on Ω_e. Since $\Omega \subset \Omega_e$, this result also holds on Ω, which provides us with the desired contradiction. $\qquad\qquad\qquad\qquad\qquad\qquad\qquad\qquad\qquad\qquad\qquad\qquad\qquad\qquad\qquad \square$

For fixed $\mathbf{f} \in \mathbf{L}^2(\Omega)$ and given $\mathbf{u} \in \mathbf{H}^1(\Omega)$, $p \in L_0^2(\Omega)$, and $\mathbf{g} \in \mathbf{H}^1(\Gamma_c)$, we have that the operator $\mathcal{J}_D'(\mathbf{u}, p, \mathbf{g}) \in \mathcal{L}(B_1; \mathbb{R})$ may be defined as follows: $\mathcal{J}_D'(\mathbf{u}, p, \mathbf{g}) \cdot (\mathbf{w}, r, \mathbf{k}, \mathbf{y}) = \tilde{a}$ for $(\mathbf{w}, r, \mathbf{k}, \mathbf{y}) \in B_1$ and $\tilde{a} \in \mathbb{R}$ if and only if

$$\tfrac{1}{Re} a(\mathbf{w}, \mathbf{u}) + b(\mathbf{w}, p) + b(\mathbf{u}, r) - (\mathbf{f}, \mathbf{w}) + \tfrac{1}{Re}(\mathrm{grad}_s\, \mathbf{g}, \mathrm{grad}_s\, \mathbf{k})_{\Gamma_c} + \tfrac{1}{Re}(\mathbf{g}, \mathbf{k})_{\Gamma_c} = \tilde{a}.$$

Let $(\hat{\mathbf{u}}, \hat{p}, \hat{\mathbf{g}}) \in \mathbf{H}^1(\Omega) \times \mathbf{W}_n(\Gamma_c)$ denote an optimal solution in the sense of (5.6.13). Then, consider the nonlinear operator $N : B_1 \to \mathbb{R} \times B_2$ defined by

$$N(\mathbf{u}, p, \mathbf{g}, \mathbf{t}) = \begin{pmatrix} \mathcal{J}_D(\mathbf{u}, p, \mathbf{g}) - \mathcal{J}_D(\hat{\mathbf{u}}, \hat{p}, \hat{\mathbf{g}}) \\ M(\mathbf{u}, p, \mathbf{g}, \mathbf{t}) \end{pmatrix}.$$

Then, for $(\mathbf{u}, p, \mathbf{g}) \in \mathbf{H}^1(\Omega) \times L_0^2(\Omega) \times \mathbf{H}^{1/2}(\Gamma_c)$, the operator $N'(\mathbf{u}, p, \mathbf{g})$ from B_1 into $\mathbb{R} \times B_2$ may be defined as follows: $N'(\mathbf{u}, p, \mathbf{g}) \cdot (\mathbf{w}, r, \mathbf{k}, \mathbf{y}) = (\tilde{a}, \tilde{\mathbf{f}}, \tilde{z}, \tilde{\mathbf{b}})$ for $(\mathbf{w}, r, \mathbf{k}, \mathbf{y}) \in B_1$ and $(\tilde{a}, \tilde{\mathbf{f}}, \tilde{z}, \tilde{\mathbf{b}}) \in \mathbb{R} \times B_2$ if and only if

$$\tfrac{1}{Re} a(\mathbf{w}, \mathbf{u}) + b(\mathbf{w}, p) + b(\mathbf{u}, r) - (\mathbf{f}, \mathbf{w}) + \nu(\mathrm{grad}_s\, \mathbf{g}, \mathrm{grad}_s\, \mathbf{k})_{\Gamma_c} + \tfrac{1}{Re}(\mathbf{g}, \mathbf{k})_{\Gamma_c} = \tilde{a},$$

$$\tfrac{1}{Re} a(\mathbf{w}, \mathbf{v}) + c(\mathbf{w}, \mathbf{u}, \mathbf{v}) + c(\mathbf{u}, \mathbf{w}, \mathbf{v}) + b(\mathbf{v}, r) - (\mathbf{v}, \mathbf{y})_{\Gamma} = (\tilde{\mathbf{f}}, \mathbf{v}) \quad \forall \, \mathbf{v} \in \mathbf{H}^1(\Omega),$$

$$b(\mathbf{w}, q) = (\tilde{z}, q) \quad \forall \, q \in L_0^2(\Omega),$$

and

$$(\mathbf{w}, \mathbf{s})_{\Gamma} - (\mathbf{k}, \mathbf{s})_{\Gamma_c} = (\tilde{\mathbf{b}}, \mathbf{s})_{\Gamma} \quad \forall \, \mathbf{s} \in \mathbf{H}^{-1/2}(\Gamma).$$

Lemma 5.6.4 *For* $(\mathbf{u}, p, \mathbf{g}) \in \mathbf{H}^1(\Omega) \times L_0^2(\Omega) \times \mathbf{H}^{1/2}(\Gamma_c)$, *the operator* $N'(\mathbf{u}, p, \mathbf{g})$ *from* B_1 *into* $\mathbb{R} \times B_2$ *has closed range but is not onto.*

Sketch of proof: From Lemma 5.6.2, we have that $M'(\mathbf{u})$ has a closed range. Also, the continuity of the various bilinear and trilinear forms, i.e., (5.6.3)–(5.6.5), and of the inner products appearing in the definition of $M'(\mathbf{u})$, imply that this operator belongs to $\mathcal{L}(B_1, B_2)$ and therefore the kernel of $M'(\mathbf{u})$ is a closed subspace. Now, $\mathcal{J}'_D(\mathbf{u}, p, \mathbf{g})$ acting on the kernel of $M'(\mathbf{u})$ is either identically zero or onto $\mathbb{R}$, so that $\mathcal{J}'_D(\mathbf{u}, p, \mathbf{g})$ acting on the kernel of $M'(\mathbf{u})$ has a closed range, and therefore the operator $N'(\mathbf{u}, p, \mathbf{g})$ has a closed range in B_2.

The operator $N'(\mathbf{u}, p, \mathbf{g})$ is not onto because if it were, by the Implicit Function Theorem, we would have $(\tilde{\mathbf{u}}, \tilde{p}, \tilde{\mathbf{g}}) \in \mathcal{U}_{ad}$ such that $\mathcal{J}_D(\tilde{\mathbf{u}}, \tilde{p}, \tilde{\mathbf{g}}) < \mathcal{J}_D(\hat{\mathbf{u}}, \hat{p}, \hat{\mathbf{g}})$, contradicting the hypothesis that $(\hat{\mathbf{u}}, \hat{p}, \hat{\mathbf{g}})$ is an optimal solution. □

We are now prepared to show the existence of Lagrange multipliers.

Theorem 5.6.5 *Let* $(\hat{\mathbf{u}}, \hat{p}, \hat{\mathbf{g}}) \in \mathbf{H}^1(\Omega) \times L_0^2(\Omega) \times \mathbf{H}^1(\Gamma_c)$ *denote an optimal solution in the sense of (5.6.13). Then there exists a nonzero Lagrange multiplier* $(\hat{\boldsymbol{\nu}}, \hat{\phi}, \hat{\boldsymbol{\tau}}) \in \mathbf{H}^1(\Omega) \times L_0^2(\Omega) \times \mathbf{H}^{-1/2}(\Gamma)$ *satisfying the Euler equations*

$$\hat{\alpha}\mathcal{K}'(\hat{\mathbf{u}}, \hat{p}, \hat{\mathbf{g}}) \cdot (\mathbf{w}, r, \mathbf{k}, \mathbf{y}) + \; < (\hat{\boldsymbol{\nu}}, \hat{\phi}, \hat{\boldsymbol{\tau}}), M'(\hat{\mathbf{u}}) \cdot (\mathbf{w}, r, \mathbf{k}, \mathbf{y}) >= 0 \qquad (5.6.20)$$
$$\forall \; (\mathbf{w}, r, \mathbf{k}, \mathbf{y}) \in \mathbf{H}^1(\Omega) \times L_0^2(\Omega) \times \mathbf{W}_n(\Gamma_c) \times \mathbf{H}^{-1/2}(\Gamma),$$

where $< \cdot, \cdot >$ *denotes the duality pairing between* $\mathbf{H}^1(\Omega) \times L_0^2(\Omega) \times \mathbf{H}^{-1/2}(\Gamma)$ *and* $\mathbf{H}^1(\Omega))^* \times L_0^2(\Omega) \times \mathbf{H}^{1/2}(\Gamma)$.

Proof: From Lemma 5.6.4, we have that the range of $N'(\hat{\mathbf{u}}, \hat{p}, \hat{\mathbf{g}})$ is a closed, proper subspace of $\mathbb{R} \times B_2$. Then, the Hahn-Banach theorem implies that there exists a nonzero element of $(\hat{\alpha}, \hat{\boldsymbol{\nu}}, \hat{\phi}, \hat{\boldsymbol{\tau}}) \in \mathbb{R} \times (B_2)^* = \mathbb{R} \times \mathbf{H}^1(\Omega) \times L_0^2(\Omega) \times \mathbf{H}^{-1/2}(\Gamma)$ that annihilates the range of $N'(\hat{\mathbf{u}}, \hat{p}, \hat{\mathbf{g}})$, i.e., such that (5.6.20) is valid. Note that $\hat{\alpha} \neq 0$ since otherwise we would have, using Lemma 5.6.3, that $< (\tilde{\mathbf{f}}, \tilde{z}, \tilde{\mathbf{b}}), (\hat{\boldsymbol{\nu}}, \hat{\phi}, \hat{\boldsymbol{\tau}}) >= 0$ for all $(\tilde{\mathbf{f}}, \tilde{z}, \tilde{\mathbf{b}}) \in B_2$ so that $(\hat{\boldsymbol{\nu}}, \hat{\phi}, \hat{\boldsymbol{\tau}}) \equiv 0$, contradicting the fact that $(\hat{\alpha}, \hat{\boldsymbol{\nu}}, \hat{\phi}, \hat{\boldsymbol{\tau}}) \neq 0$. □

5.6.5 *The optimality system*

Using (5.6.14)–(5.6.15), setting (without loss of generality) $\hat{\alpha} = -1$, and dropping the $(\hat{\cdot})$ notation for optimal solutions, we may rewrite (5.6.20) in the form

$$\tfrac{1}{Re}a(\mathbf{w}, \boldsymbol{\nu}) + c(\mathbf{w}, \mathbf{u}, \boldsymbol{\nu}) + c(\mathbf{u}, \mathbf{w}, \boldsymbol{\nu}) + b(\mathbf{w}, \phi) + (\mathbf{w}, \boldsymbol{\tau})_\Gamma$$
$$= \tfrac{1}{Re}a(\mathbf{u}, \mathbf{w}) + b(\mathbf{w}, p) - (\mathbf{f}, \mathbf{w}) \quad \forall \; \mathbf{w} \in \mathbf{H}^1(\Omega), \qquad (5.6.21)$$

$$b(\boldsymbol{\nu}, r) = b(\mathbf{u}, r) = 0 \quad \forall \; r \in L_0^2(\Omega), \qquad (5.6.22)$$

$$(\boldsymbol{\nu}, \mathbf{y})_\Gamma = 0 \quad \forall\, \mathbf{y} \in \mathbf{H}^{-1/2}(\Gamma) \tag{5.6.23}$$

and

$$\tfrac{1}{Re}(\mathrm{grad}_s\, \mathbf{g}, \mathrm{grad}_s\, \mathbf{k})_{\Gamma_c} + \tfrac{1}{Re}(\mathbf{g}, \mathbf{k})_{\Gamma_c} = -(\mathbf{k}, \boldsymbol{\tau})_{\Gamma_c} \quad \forall\, \mathbf{k} \in \mathbf{W}_n(\Gamma_c), \tag{5.6.24}$$

where in (5.6.22) we have used (5.6.9).

Since for some $\mathbf{t} \in \mathbf{H}^{-1/2}(\Gamma)$ optimal solutions satisfy the constraints (5.6.8)–(5.6.10), we see necessary conditions for an optimum are that (5.6.8)–(5.6.10) and (5.6.21)–(5.6.24) are satisfied. This system of equations will be called the *optimality system*.

Using (5.6.8), we may replace (5.6.21) by

$$\tfrac{1}{Re}a(\mathbf{w}, \boldsymbol{\nu}) + c(\mathbf{w}, \mathbf{u}, \boldsymbol{\nu}) + c(\mathbf{u}, \mathbf{w}, \boldsymbol{\nu}) + b(\mathbf{w}, \phi) - (\mathbf{w}, \boldsymbol{\theta})_\Gamma = -c(\mathbf{u}, \mathbf{u}, \mathbf{w}) \tag{5.6.25}$$
$$\forall\, \mathbf{w} \in \mathbf{H}^1(\Omega),$$

where $\boldsymbol{\theta} = \mathbf{t} - \boldsymbol{\tau}$. The replacement of the right hand side of (5.6.21) by the right hand side of (5.6.26) facilitates the derivation of finite element error estimates.

Thus, the optimality system in terms of the variables $\mathbf{u}, p, \mathbf{t}, \mathbf{g}, \boldsymbol{\nu}, \phi$ and $\boldsymbol{\theta}$ is given by (5.6.8)–(5.6.10) and (5.6.22)–(5.6.26). Integrations by parts may be used to show that this system constitutes a weak formulation of the boundary value problem

$$-\frac{1}{Re}\mathrm{div}\left((\mathrm{grad}\,\mathbf{u}) + (\mathrm{grad}\,\mathbf{u})^T\right) + \mathbf{u}\cdot\mathrm{grad}\,\mathbf{u} + \mathrm{grad}\,p = \mathbf{f} \quad \text{in } \Omega, \tag{5.6.26}$$

$$\mathrm{div}\,\mathbf{u} = 0 \quad \text{in } \Omega, \tag{5.6.27}$$

$$\mathbf{u} = \begin{cases} \mathbf{g} & \text{on } \Gamma_c \\ \mathbf{0} & \text{on } \Gamma_u, \end{cases} \tag{5.6.28}$$

$$\tfrac{1}{Re}(-\Delta_s\mathbf{g} + \mathbf{g}) + \beta\mathbf{n} =$$
$$\phi\mathbf{n} - \tfrac{1}{Re}\left((\mathrm{grad}\,\boldsymbol{\nu}) + (\mathrm{grad}\,\boldsymbol{\nu})^T\right)\cdot\mathbf{n} \tag{5.6.29}$$
$$- (\mathbf{u}\cdot\mathbf{n})\boldsymbol{\nu} - p\mathbf{n} + \tfrac{1}{Re}\left((\mathrm{grad}\,\mathbf{u}) + (\mathrm{grad}\,\mathbf{u})^T\right)\cdot\mathbf{n} \quad \text{on } \Gamma_c,$$

$$\int_{\Gamma_c} \mathbf{g}\cdot\mathbf{n}\, d\Gamma = 0 \quad \text{and, if } \Gamma_c \text{ has a boundary, } \mathbf{g} = \mathbf{0} \text{ on } \partial\Gamma_c, \tag{5.6.30}$$

$$-\tfrac{1}{Re}\mathrm{div}\left((\mathrm{grad}\,\boldsymbol{\nu}) + (\mathrm{grad}\,\boldsymbol{\nu})^T\right) + \boldsymbol{\nu}\cdot(\mathrm{grad}\,\mathbf{u})^T \tag{5.6.31}$$
$$- \mathbf{u}\cdot\mathrm{grad}\,\boldsymbol{\nu} + \mathrm{grad}\,\phi = -\mathbf{u}\cdot\mathrm{grad}\,\mathbf{u} \quad \text{in } \Omega,$$

$$\mathrm{div}\,\boldsymbol{\nu} = 0 \quad \text{in } \Omega \tag{5.6.32}$$

and

$$\boldsymbol{\nu} = \mathbf{0} \quad \text{on } \Gamma. \tag{5.6.33}$$

Note that, in (5.6.30), Δ_s denotes the surface Laplacian and $\beta \in \mathbb{R}$ is an additional unknown constant that accounts for the single integral constraint of (5.6.30).

The optimality system (5.6.26)–(5.6.33) consists of the Navier-Stokes system (5.6.26)–(5.6.28), the system (5.6.31)-(5.6.33) whose left hand side is the adjoint of Navier-Stokes operator linearized about $\mathbf{u}$, and the surface Laplacian system (5.6.30)–(5.6.30).

Note that (5.6.24) may be expressed in the form

$$\tfrac{1}{Re}(\text{grad}_s\,\mathbf{g}, \text{grad}_s\,\mathbf{k})_{\Gamma_c} + \tfrac{1}{Re}(\mathbf{g}, \mathbf{k})_{\Gamma_c} + \beta \int_{\Gamma_c} \mathbf{k} \cdot \mathbf{n}\, d\Gamma = (\mathbf{k}, \boldsymbol{\theta} - \mathbf{t})_{\Gamma_c} \tag{5.6.34}$$

$$\forall\, \mathbf{k} \in \mathbf{W}(\Gamma_c)$$

and

$$\int_{\Gamma_c} \mathbf{g} \cdot \mathbf{n}\, d\Gamma = 0. \tag{5.6.35}$$

Although (5.6.24) and (5.6.34)–(5.6.35) are equivalent, the latter is more easily discretized. Also, note the relation between (5.6.30)–(5.6.30) and (5.6.34)–(5.6.35).

The use of the $\mathbf{H}^1(\Gamma_c)$-norm of $\mathbf{g}$ in the functional (5.6.1), or, equivalently, in (5.6.11), results in the appearance of the surface Laplacian in (5.6.30). The use of the more "natural" $\mathbf{H}^{1/2}(\Gamma_c)$-norm would have resulted in a much less attractive problem relating the control $\mathbf{g}$ to the variables $\mathbf{t}$ and $\boldsymbol{\theta}$. On the other hand, the use, in (5.6.1), of the weaker $\mathbf{L}^2(\Gamma_c)$-norm for $\mathbf{g}$ would not allow us to derive the regularity results of the following subsection. From a practical point of view, the use of the $\mathbf{H}^1(\Gamma_c)$-norm of $\mathbf{g}$ in the functional (5.6.1) results in less oscillatory optimal controls $\mathbf{g}$.

Our notion of an optimal solution is a local one; see (5.6.13). Moreover, there is no reason to believe that, in general, optimal solutions are unique. This is to be expected since the uncontrolled stationary Navier-Stokes equations are known to have multiple solutions for sufficiently large values of the Reynolds number. However, just as in the Navier-Stokes case ([37], [38], [42] or [43]), for sufficiently small values of the Reynolds number, i.e., for "small enough" data or "large enough" viscosity, one can guarantee that optimal solutions are unique.

5.6.6 *Regularity of Solutions of the Optimality System*

We now examine the regularity of solutions of the optimality system (5.6.8)–(5.6.10) and (5.6.22)–(5.6.26), or equivalently, (5.6.26)–(5.6.33). Note that if Γ_c has a boundary, we can only conclude that, for arbitrary $\epsilon > 0$, $\mathbf{u}|_{\Gamma} \in \mathbf{H}^{3/2-\epsilon}(\Gamma)$, and in this case we cannot obtain the following results. Thus, throughout this section, we assume that Γ_c does not have a boundary.

Theorem 5.6.6 *Suppose that* Γ_c *does not have a boundary* $\partial\Gamma_c$ *and that the given data satisfies* $\mathbf{f} \in \mathbf{L}^2(\Omega)$. *Suppose that* Ω *is of class* $C^{1,1}$. *Then, if* $(\mathbf{u}, p, \mathbf{g}, \boldsymbol{\nu}, \phi) \in \mathbf{H}^1(\Omega) \times L_0^2(\Omega) \times \mathbf{W}_n(\Gamma_c) \times \mathbf{H}^1(\Omega) \times L_0^2(\Omega)$ *denotes a solution of the optimality system (5.6.8)– (5.6.10) and (5.6.22)–(5.6.26), we have that* $(\mathbf{u}, p, \mathbf{g}, \boldsymbol{\nu}, \phi) \in \mathbf{H}^2(\Omega) \times H^1(\Omega) \times \mathbf{H}^{3/2}(\Gamma_c) \times \mathbf{H}^2(\Omega) \times H^1(\Omega)$. *If the boundary is sufficiently smooth, we also may conclude that* $\mathbf{g} \in \mathbf{H}^{5/2}(\Gamma_c)$.

Sketch of proof: The proof is fairly standard, making use of well-known regularity results for the Stokes and Poisson problems. From Theorem 1 we already know that any solution $\mathbf{u}$ of the optimality system satisfies $\mathbf{u} \in \mathbf{H}^{3/2}(\Omega)$. Then, moving the nonlinear terms in (5.6.31)–(5.6.33) to the right-hand-side and considering the resulting Stokes problem, on conludes that $\boldsymbol{\nu} \in \mathbf{H}^{3/2}(\Omega)$ as well. Then, from (5.6.30), it follows that $\mathbf{g} \in \mathbf{H}^2(\Gamma_c)$. Then we move the nonlinear terms in (5.6.26)–(5.6.28) to the right-hand-side; from the resulting Stokes problem one can show that $\mathbf{u} \in \mathbf{H}^2(\Omega)$ and $p \in H^1(\Omega)$. With this knowledge, we return to (5.6.31)–(5.6.33) and conclude that $\boldsymbol{\nu} \in \mathbf{H}^2(\Omega)$ and $\phi \in H^1(\Omega)$. Finally, for sufficiently smooth domains, we have that the data in (5.6.30) belongs to $\mathbf{H}^{1/2}(\Gamma_c)$ so that $\mathbf{g} \in \mathbf{H}^{5/2}(\Gamma_c)$.   □

The above result also holds for convex regions of $\mathbb{R}^2$, provided $\Gamma_c = \Gamma$. In general, we may show that if $\mathbf{f} \in \mathbf{H}^m(\Omega)$ and Ω is sufficiently smooth, then $(\mathbf{u}, p, \mathbf{g}, \boldsymbol{\nu}, \phi) \in \mathbf{H}^{m+2}(\Omega) \times H^{m+1}(\Omega) \times \mathbf{H}^{m+\frac{5}{2}}(\Gamma_c) \times \mathbf{H}^{m+2}(\Omega) \times H^{m+1}(\Omega)$. In particular, if $\mathbf{f}$ is of class $C^\infty(\overline{\Omega})$ and Ω is of class C^∞, then $\mathbf{u}, p, \mathbf{g}, \boldsymbol{\nu}$ and ϕ are all $C^\infty(\overline{\Omega})$ functions as well.

The hypotheses of Theorem 5.6.6 imply that $\mathbf{t} \in \mathbf{H}^{1/2}(\Gamma)$ and $\theta \in \mathbf{H}^{1/2}(\Gamma)$.

5.6.7   Finite element discretizations.

A finite element discretization of the optimality system (5.6.8)–(5.6.10) and (5.6.22)– (5.6.26) is defined as follows. First, one chooses families of finite dimensional subspaces $\mathbf{V}^h \subset \mathbf{H}^1(\Omega)$, $S^h \subset L^2(\Omega)$. These families are parametrized by the parameter h that tends to zero; commonly, this parameter is chosen to be some measure of the grid size in a subdivision of Ω into finite elements. We let $S_0^h = S^h \cap L_0^2(\Omega)$ and $\mathbf{V}_0^h = \mathbf{V}^h \cap \mathbf{H}_0^1(\Omega)$.

One may choose any pair of subspaces $\mathbf{V}^h$ and S^h that can be used for finding finite element approximations of solutions of the Navier-Stokes equations. Thus, concerning these subspaces, we make the following standard assumptions which are exactly those employed in well-known finite element methods for the Navier-Stokes equations. First, we have the approximation properties: there exist an integer k and a constant C, independent of $h, \mathbf{v}$ and q, such that

$$\inf_{\mathbf{v}^h \in \mathbf{V}^h} \|\mathbf{v} - \mathbf{v}^h\|_1 \leq Ch^m \|\mathbf{v}\|_{m+1} \quad \forall\, \mathbf{v} \in \mathbf{H}^{m+1}(\Omega), \; 1 \leq m \leq k, \tag{5.6.36}$$

and

$$\inf_{q^h \in S_0^h} \|q - q^h\|_0 \le Ch^m \|q\|_m \quad \forall\, q \in H^m(\Omega) \cap L_0^2(\Omega), \ 1 \le m \le k; \qquad (5.6.37)$$

next, we assume the *inf-sup condition*, or *Ladyzhenskaya-Babuska-Brezzi condition*: there exists a constant C, independent of h, such that

$$\inf_{0 \ne q^h \in S_0^h} \sup_{0 \ne v^h \in V^h} \frac{b(v^h, q^h)}{\|v^h\|_1 \|q^h\|_0} \ge C. \qquad (5.6.38)$$

This condition assures that finite element discretizations of the Navier-Stokes equations are stable. For thorough discussions of the approximation properties (5.6.36)–(5.6.37), see, e.g., [31] or [35], and for like discussions of the stability condition (5.6.38), see, e.g., [37] or [38]. The latter references may also be consulted for a catalogue of finite element subspaces that meet the requirements of (5.6.36)–(5.6.38).

Next, let $P^h = V^h|_\Gamma$, i.e., P^h consists of the restriction, to the boundary Γ, of functions belonging to P^h. For all choices of conforming finite element spaces V^h, e.g., Lagrange type finite element spaces, we then have that $P^h \subset H^{-1/2}(\Gamma)$. For the subspaces $P^h = V^h|_\Gamma$, we assume the approximation property: there exist an integer k and a constant C, independent of h and s, such that

$$\inf_{s^h \in P^h} \|s - s^h\|_{-1/2,\Gamma} \le Ch^m \|s\|_{m-\frac{1}{2}} \quad \forall\, s \in H^{m-\frac{1}{2}}(\Gamma), \ 1 \le m \le k, \qquad (5.6.39)$$

and the inverse assumption: there exists a constant C, independent of h and s^h such that

$$\|s^h\|_{s,\Gamma} \le Ch^{s-q} \|\nu\|_{q,\Gamma} \quad \forall\, s^h \in P^h, \ -1/2 \le q \le s \le 1/2. \qquad (5.6.40)$$

See [30] or [35] for details concerning (5.6.39) and (5.6.40).

Now, let $Q^h = V^h|_{\Gamma_c}$, i.e., Q^h consists of the restriction, to the boundary segment Γ_c, of functions belonging to V^h. Again, for all choices of conforming finite element spaces V^h we then have that $Q^h \subset H^1(\Gamma_c)$. Let $Q_0^h = Q^h \cap W(\Gamma_c)$. We assume the approximation property: there exist an integer k and a constant C, independent of h and k, such that

$$\inf_{k^h \in Q_0^h} \|k - k^h\|_{s,\Gamma_c} \le Ch^{m-s+\frac{1}{2}} \|k\|_{m+\frac{1}{2}}$$
$$\forall\, k \in W(\Gamma_c), \ 1 \le m \le k, \ 0 \le s \le 1. \qquad (5.6.41)$$

This property follows from (5.6.36), once one notes that the same type of polynomials are used in Q_0^h as are used in V^h.

Once the approximating subspaces have been chosen we seek $\mathbf{u}^h \in \mathbf{V}^h, p^h \in S_0^h, \mathbf{t}^h \in \mathbf{P}^h, \mathbf{g}^h \in \mathbf{Q}_0^h, \boldsymbol{\nu}^h \in \mathbf{V}^h, \phi^h \in S_0^h, \boldsymbol{\theta}^h \in \mathbf{P}^h$ and $\beta^h \in \mathbb{R}$ such that

$$\frac{1}{Re}a(\mathbf{u}^h, \mathbf{v}^h) + c(\mathbf{u}^h, \mathbf{u}^h, \mathbf{v}^h) + b(\mathbf{v}^h, p^h) - (\mathbf{v}^h, \mathbf{t}^h)_\Gamma = (\mathbf{f}, \mathbf{v}^h)$$
$$\forall \, \mathbf{v}^h \in \mathbf{V}^h, \tag{5.6.42}$$

$$b(\mathbf{u}^h, q^h) = 0 \quad \forall \, q^h \in S_0^h, \tag{5.6.43}$$

$$(\mathbf{u}^h, \mathbf{s}^h)_\Gamma - (\mathbf{g}^h, \mathbf{s}^h)_{\Gamma_c} = 0 \quad \forall \, \mathbf{s}^h \in \mathbf{P}^h, \tag{5.6.44}$$

$$\frac{1}{Re}(\text{grad}_s \, \mathbf{g}^h, \text{grad}_s \, \mathbf{k}^h)_{\Gamma_c} + \frac{1}{Re}(\mathbf{g}^h, \mathbf{k}^h)_{\Gamma_c} + \beta^h \int_{\Gamma_c} \mathbf{k}^h \cdot \mathbf{n} \, d\Gamma = (\boldsymbol{\theta}^h - \mathbf{t}^h, \mathbf{k}^h)_{\Gamma_c}$$
$$\forall \, \mathbf{k}^h \in \mathbf{Q}_0^h, \tag{5.6.45}$$

$$\int_{\Gamma_c} \mathbf{g}^h \cdot \mathbf{n} \, d\Gamma = 0, \tag{5.6.46}$$

$$\frac{1}{Re}a(\mathbf{w}^h, \boldsymbol{\nu}^h) + c(\mathbf{w}^h, \mathbf{u}^h, \boldsymbol{\nu}^h) + c(\mathbf{u}^h, \mathbf{w}^h, \boldsymbol{\nu}^h) + b(\mathbf{w}^h, \phi^h)$$
$$- (\mathbf{w}^h, \boldsymbol{\theta}^h)_\Gamma = -c(\mathbf{u}^h, \mathbf{u}^h, \mathbf{w}^h) \quad \forall \, \mathbf{w}^h \in \mathbf{V}^h, \tag{5.6.47}$$

$$b(\boldsymbol{\nu}^h, r^h) = 0 \quad \forall \, r^h \in S_0^h \tag{5.6.48}$$

and

$$(\boldsymbol{\nu}^h, \mathbf{y}^h) = 0 \quad \forall \, \mathbf{y}^h \in \mathbf{P}^h. \tag{5.6.49}$$

From a computational standpoint, this is a formidable system. Therefore, how one solves this system is a rather important question. We do not address this issue here.

The use of the $\mathbf{H}^1(\Gamma_c)$-norm of $\mathbf{g}$ in the functional (5.6.1), or, equivalently, in (5.6.11), results in the need to solve the surface problem (5.6.45)–(5.6.46). Had we used the $\mathbf{H}^{1/2}(\Gamma_c)$-norm instead, we would be faced with an undesirable computational problem involving the $\mathbf{H}^{1/2}(\Gamma_c)$-inner product. The avoidance of such a happenstance is the main motivation for using the $\mathbf{H}^1(\Gamma_c)$-norm of $\mathbf{g}$. In addition, the regularity brought to us through the use of the $\mathbf{H}^1(\Gamma_c)$-norm of $\mathbf{g}$ turns out to be an asset in deriving error estimates.

5.6.8 Quotation of Results Concerning the Approximation of a Class of Nonlinear Problems

The error estimates to be derived below make use of results of [34] and [36] (see also [37]) concerning the approximation of a class of nonlinear problems, and of [39] for the approximation of the Stokes equations with inhomogeneous velocity boundary conditions. Here, for the sake of completeness, we will state the relevant results, specialized to our needs.

The types of nonlinear problems considered in [34], [36], and [37] are of the type

$$F(\lambda, \psi) \equiv \psi + TG(\lambda, \psi) = 0 \qquad (5.6.50)$$

where $T \in \mathcal{L}(Y; X)$, G is a C^2 mapping from $\Lambda \times X$ into Y, X and Y are Banach spaces and Λ is a compact interval of $\mathbb{R}$. We say that $\{(\lambda, \psi(\lambda)) : \lambda \in \Lambda\}$ is a branch of solutions of (5.6.50) if $\lambda \to \psi(\lambda)$ is a continuous function from Λ into X such that $F(\lambda, \psi(\lambda)) = 0$. The branch is called a *regular branch* if we also have that $D_\psi F(\lambda, \psi(\lambda))$ is an isomorphism from X into X for all $\lambda \in \Lambda$. (Here, $D_\psi F(\cdot, \cdot)$ denotes the Frechet derivative of $F(\cdot, \cdot)$ with respect to the second argument.)

Approximations are defined by introducing a subspace $X^h \subset X$ and an approximating operator $T^h \in \mathcal{L}(Y; X^h)$. Then, we seek $\psi^h \in X^h$ such that

$$F^h(\lambda, \psi^h) \equiv \psi^h + T^h G(\lambda, \psi^h) = 0 \,. \qquad (5.6.51)$$

We will assume that there exists another Banach space Z, contained in Y, with continuous imbedding, such that

$$D_\psi G(\lambda, \psi) \in \mathcal{L}(X; Z) \quad \forall \lambda \in \Lambda \text{ and } \psi \in X \,. \qquad (5.6.52)$$

Concerning the operator T^h, we assume the approximation properties

$$\lim_{h \to 0} \|(T^h - T)y\|_X = 0 \quad \forall y \in Y \qquad (5.6.53)$$

and

$$\lim_{h \to 0} \|(T^h - T)\|_{\mathcal{L}(Z;X)} = 0 \,. \qquad (5.6.54)$$

Note that (5.6.52) and (5.6.54) imply that the operator $DG_\psi(\lambda, \psi) \in \mathcal{L}(X; X)$ is compact. Morevover, (5.6.54) follows from (5.6.53) whenever the imbedding $Z \subset Y$ is compact.

We can now state the first result of [34], [36], and [37] that will be used in the sequel. In the statement of the theorem, $D^2 G$ represents any and all second Frechet derivatives of G.

Theorem 5.6.7 *Let X and Y be Banach spaces and Λ a compact subset of $\mathbb{R}$. Assume that G is a C^2 mapping from $\Lambda \times X$ into Y and that $D^2 G$ is bounded on all bounded sets of $\Lambda \times X$. Assume that (5.6.52)–(5.6.54) hold and that $\{(\lambda, \psi(\lambda)); \lambda \in \Lambda\}$ is a branch of regular solutions of (5.6.50) . Then, there exists a neighborhood $\mathcal{O}$ of the origin in X and, for $h \leq h_0$ small enough, a unique C^2 function $\lambda \to \psi^h(\lambda) \in X^h$ such that $\{(\lambda, \psi^h(\lambda)); \lambda \in \Lambda\}$ is a branch of regular solutions of (5.6.51) and $\psi^h(\lambda) - \psi(\lambda) \in \mathcal{O}$ for all $\lambda \in \Lambda$. Moreover, there exists a constant $C > 0$, independent of h and λ, such that*

$$\|\psi^h(\lambda) - \psi(\lambda)\|_X \leq C\|(T^h - T)G(\lambda, \psi(\lambda))\|_X \quad \forall \lambda \in \Lambda \,. \qquad \square \qquad (5.6.55)$$

We now turn to the results of [39] concerning the approximation of the Stokes problem with the use of Lagrange multipliers to enforce velocity boundary conditions.

Theorem 5.6.8 *The Stokes problem: seek* $\tilde{\mathbf{u}} \in \mathbf{H}^1(\Omega), \tilde{p} \in L_0^2(\Omega)$ *and* $\tilde{\mathbf{t}} \in \mathbf{H}^{-1/2}(\Gamma)$ *such that*

$$a(\tilde{\mathbf{u}}, \mathbf{v}) + b(\mathbf{v}, \tilde{p}) - (\mathbf{v}, \tilde{\mathbf{t}})_\Gamma = (\tilde{\mathbf{f}}, \mathbf{v}) \quad \forall\, \mathbf{v} \in \mathbf{H}^1(\Omega),$$

$$b(\tilde{\mathbf{u}}, q) = 0 \quad \forall\, q \in L_0^2(\Omega)$$

and

$$(\tilde{\mathbf{u}}, \mathbf{s})_\Gamma = 0 \quad \forall\, \mathbf{s} \in \mathbf{H}^{-1/2}(\Gamma)$$

has a unique solution. Let (5.6.36)–(5.6.40) hold. Then, the discrete Stokes problem: seek $\tilde{\mathbf{u}}^h \in \mathbf{V}^h, \tilde{p}^h \in S_0^h$ *and* $\tilde{\mathbf{t}}^h \in \mathbf{P}^h$ *such that*

$$a(\tilde{\mathbf{u}}^h, \mathbf{v}^h) + b(\mathbf{v}^h, \tilde{p}^h) - (\mathbf{v}^h, \tilde{\mathbf{t}}^h)_\Gamma = (\tilde{\mathbf{f}}, \mathbf{v}^h) \quad \forall\, \mathbf{v}^h \in \mathbf{V}^h,$$

$$b(\tilde{\mathbf{u}}^h, q^h) = 0 \quad \forall\, q^h \in S_0^h$$

and

$$(\tilde{\mathbf{u}}^h, \mathbf{s}^h)_\Gamma = 0 \quad \forall\, \mathbf{s}^h \in \mathbf{P}^h$$

also has a unique solution. Moreover, as $h \to 0$,

$$\|\tilde{\mathbf{u}} - \tilde{\mathbf{u}}^h\|_1 + \|\tilde{p} - \tilde{p}^h\|_0 + \|\tilde{\mathbf{t}} - \tilde{\mathbf{t}}^h\|_{-1/2,\Gamma} \to 0$$

and, if $(\tilde{\mathbf{u}}, \tilde{p}, \tilde{\mathbf{t}}) \in \mathbf{H}^{m+1}(\Omega) \times H^m(\Omega) \times \mathbf{H}^{m-\frac{1}{2}}(\Gamma)$, *then there exists a constant* C, *independent of* h, *such that*

$$\|\tilde{\mathbf{u}} - \tilde{\mathbf{u}}^h\|_1 + \|\tilde{p} - \tilde{p}^h\|_0 + \|\tilde{\mathbf{t}} - \tilde{\mathbf{t}}^h\|_{-1/2,\Gamma} \le Ch^m (\|\tilde{\mathbf{u}}\|_{m+1} + \|\tilde{p}\|_m). \qquad \square$$

5.6.9 Error Estimates

We begin by recasting the optimality system (5.6.8), (5.6.9), (5.6.10), (5.6.22), (5.6.23), (5.6.26), (5.6.34) and (5.6.35) and its discretization (5.6.42)–(5.6.49) into a form that fits into the framework of (5.6.50)–(5.6.51). Let $\lambda = Re$. Let

$$
\begin{aligned}
X &= \mathbf{H}^1(\Omega) \times L_0^2(\Omega) \times \mathbf{H}^{-\frac{1}{2}}(\Gamma) \times \mathbf{W}(\Gamma_c) \times \mathbb{R} \times \mathbf{H}^1(\Omega) \times L_0^2(\Omega) \times \mathbf{H}^{-\frac{1}{2}}(\Gamma) \\
Y &= (\mathbf{H}^1(\Omega))^* \times \mathbf{H}^{\frac{1}{2}}(\Gamma) \times (\mathbf{H}^1(\Omega))^* \\
Z &= \mathbf{L}^{3/2}(\Omega) \times \mathbf{H}^1(\Gamma) \times \mathbf{L}^{3/2}(\Omega) \\
X^h &= \mathbf{V}^h \times S_0^h \times \mathbf{P}^h \times \mathbf{Q}_0^h \times \mathbb{R} \times \mathbf{V}^h \times S_0^h \times \mathbf{P}^h
\end{aligned}
$$

where $(\mathbf{H}^1(\Omega))^*$ denotes the dual space of $\mathbf{H}^1(\Omega)$. Note that $Z \subset Y$ with a compact imbedding.

Let the operator $T \in \mathcal{L}(Y; X)$ be defined in the following manner: $T(\zeta, \kappa, \eta) = (\tilde{\mathbf{u}}, \tilde{p}, \tilde{\mathbf{t}}, \tilde{\mathbf{g}}, \tilde{\beta}, \tilde{\nu}, \tilde{\phi}, \tilde{\theta})$ for $(\zeta, \kappa, \eta) \in Y$ and $(\tilde{\mathbf{u}}, \tilde{p}, \tilde{\mathbf{t}}, \tilde{\mathbf{g}}, \tilde{\beta}, \tilde{\nu}, \tilde{\phi}, \tilde{\theta}) \in X$ if and only if

$$a(\tilde{\mathbf{u}}, \mathbf{v}) + b(\mathbf{v}, \tilde{p}) - (\mathbf{v}, \tilde{\mathbf{t}})_\Gamma = (\zeta, \mathbf{v}) \quad \forall \mathbf{v} \in \mathbf{H}^1(\Omega), \tag{5.6.56}$$

$$b(\tilde{\mathbf{u}}, q) = 0 \quad \forall q \in L_0^2(\Omega), \tag{5.6.57}$$

$$(\tilde{\mathbf{u}}, \mathbf{s})_\Gamma = (\kappa, \mathbf{s})_\Gamma \quad \forall \mathbf{s} \in \mathbf{H}^{-1/2}(\Gamma), \tag{5.6.58}$$

$$(\mathrm{grad}_s \tilde{\mathbf{g}}, \mathrm{grad}_s \mathbf{k})_{\Gamma_c} + (\tilde{\mathbf{g}}, \mathbf{k})_{\Gamma_c} + \tilde{\beta} \int_{\Gamma_c} \mathbf{k} \cdot \mathbf{n} \, d\Gamma - (\mathbf{k}, \tilde{\theta} - \tilde{\mathbf{t}})_{\Gamma_c} = 0 \tag{5.6.59}$$
$$\forall \mathbf{k} \in \mathbf{W}(\Gamma_c),$$

$$\int_{\Gamma_c} \tilde{\mathbf{g}} \cdot \mathbf{n} \, d\Gamma = 0, \tag{5.6.60}$$

$$a(\mathbf{w}, \tilde{\nu}) + b(\mathbf{w}, \tilde{\phi}) - (\mathbf{w}, \tilde{\theta})_\Gamma = (\eta, \mathbf{w}) \quad \forall \mathbf{w} \in \mathbf{H}^1(\Omega), \tag{5.6.61}$$

$$b(\tilde{\nu}, r) = 0 \quad \forall r \in L_0^2(\Omega), \tag{5.6.62}$$

and

$$(\tilde{\nu}, \mathbf{y})_\Gamma = 0 \quad \forall \mathbf{y} \in \mathbf{H}^{-1/2}(\Gamma), \tag{5.6.63}$$

Note that this system is *weakly coupled*. First, one may separately solve the Stokes problems (5.6.56)–(5.6.58) for $\tilde{\mathbf{u}}, \tilde{p}$ and $\tilde{\mathbf{t}}$ and (5.6.61)–(5.6.63) for $\tilde{\nu}, \tilde{\phi}$ and $\tilde{\theta}$; then, one may solve the surface Laplacian problem (5.6.59)–(5.6.60) for $\tilde{\mathbf{g}}$ and $\tilde{\beta}$.

The need for introducing the Lagrange multiplier $\mathbf{t}$ in order to enforce the velocity boundary condition can now be made clear. Note the appearance of this multiplier in (5.6.59). Formally, we could eliminate $\tilde{\mathbf{t}}$ (and, similarly, $\tilde{\theta}$) by using the relation

$$\tilde{\mathbf{t}} = \left[-\tilde{p}\mathbf{n} + \nu(\mathrm{grad}\,\tilde{\mathbf{u}} + (\mathrm{grad}\,\tilde{\mathbf{u}})^T) \cdot \mathbf{n}\right]_\Gamma.$$

However, then we would not have that T is well defined from all of Y into X (as is required in Theorem 5.6.7) since the right hand side of the above expression does not make sense for general $\tilde{\mathbf{u}} \in \mathbf{H}^1(\Omega)$ and $p \in L_0^2(\Omega)$.

Analogously, the operator $T^h \in \mathcal{L}(Y; X^h)$ is defined as follows: $T^h(\zeta, \kappa, \eta) = (\tilde{\mathbf{u}}^h, \tilde{p}^h, \tilde{\mathbf{t}}^h, \tilde{\mathbf{g}}^h, \tilde{\beta}^h, \tilde{\nu}^h, \tilde{\phi}^h, \tilde{\theta}^h)$ for $(\zeta, \kappa, \eta) \in Y$ and $(\tilde{\mathbf{u}}^h, \tilde{p}^h, \tilde{\mathbf{t}}^h, \tilde{\mathbf{g}}^h, \tilde{\beta}^h, \tilde{\nu}^h, \tilde{\phi}^h, \tilde{\theta}^h) \in X^h$ if and only if

$$a(\tilde{\mathbf{u}}^h, \mathbf{v}^h) + b(\mathbf{v}^h, \tilde{p}^h) - (\mathbf{v}^h, \tilde{\mathbf{t}}^h)_\Gamma = (\zeta, \mathbf{v}^h) \quad \forall \mathbf{v}^h \in \mathbf{V}^h, \tag{5.6.64}$$

$$b(\tilde{\mathbf{u}}^h, q^h) = 0 \quad \forall \, q^h \in S_0^h, \tag{5.6.65}$$

$$(\tilde{\mathbf{u}}^h, \mathbf{s}^h)_\Gamma = (\kappa, \mathbf{s}^h)_\Gamma \quad \forall \, \mathbf{s}^h \in \mathbf{P}^h, \tag{5.6.66}$$

$$(\text{grad}_s \, \tilde{\mathbf{g}}^h, \text{grad}_s \, \mathbf{k}^h)_{\Gamma_c} + (\tilde{\mathbf{g}}^h, \mathbf{k}^h)_{\Gamma_c}$$
$$+\tilde{\beta}^h \int_{\Gamma_c} \mathbf{k}^h \cdot \mathbf{n} \, d\Gamma - (\mathbf{k}^h, \tilde{\boldsymbol{\theta}}^h - \tilde{\mathbf{t}}^h)_{\Gamma_c} = 0 \quad \forall \, \mathbf{k}^h \in \mathbf{Q}_0^h, \tag{5.6.67}$$

$$\int_{\Gamma_c} \tilde{\mathbf{g}}^h \cdot \mathbf{n} \, d\Gamma = 0, \tag{5.6.68}$$

$$a(\mathbf{w}^h, \tilde{\boldsymbol{\nu}}^h) + b(\mathbf{w}^h, \tilde{\phi}^h) - (\mathbf{w}^h, \tilde{\boldsymbol{\theta}}^h)_\Gamma = (\mathbf{w}^h, \boldsymbol{\eta})_\Gamma \quad \forall \, \mathbf{w}^h \in \mathbf{V}^h, \tag{5.6.69}$$

$$b(\tilde{\boldsymbol{\nu}}^h, r^h) = 0 \quad \forall \, r^h \in S_0^h \tag{5.6.70}$$

and

$$(\tilde{\boldsymbol{\nu}}^h, \mathbf{y}^h) = 0 \quad \forall \, \mathbf{y}^h \in \mathbf{P}^h. \tag{5.6.71}$$

The system (5.6.64)–(5.6.71) is weakly coupled in the same sense as the system (5.6.56)–(5.6.63).

Let Λ denote a compact subset of $\mathbb{R}_+$. Next, we define the *nonlinear* mapping $G : \Lambda \times X \to Y$ as follows: $G(\lambda, (\mathbf{u}, p, \mathbf{t}, \mathbf{g}, \beta, \boldsymbol{\nu}, \phi, \theta)) = (\zeta, \kappa, \eta)$ for $\lambda \in \Lambda$, $(\mathbf{u}, p, \mathbf{t}, \mathbf{g}, \beta, \boldsymbol{\nu}, \phi, \theta) \in X$ and $(\zeta, \kappa, \eta) \in Y$ if and only if

$$(\zeta, \mathbf{v}) = \lambda c(\mathbf{u}, \mathbf{u}, \mathbf{v}) - \lambda(\mathbf{f}, \mathbf{v}) \quad \forall \, \mathbf{v} \in \mathbf{H}^1(\Omega),$$

$$(\kappa, \mathbf{s})_\Gamma = -(\mathbf{g}, \mathbf{s})_{\Gamma_c} \quad \forall \, \mathbf{s} \in \mathbf{H}^{-1/2}(\Gamma)$$

and

$$(\boldsymbol{\eta}, \mathbf{w}) = \lambda c(\mathbf{w}, \mathbf{u}, \boldsymbol{\nu}) + \lambda c(\mathbf{u}, \mathbf{w}, \boldsymbol{\nu}) + \lambda c(\mathbf{u}, \mathbf{u}, \mathbf{w}) \quad \forall \, \mathbf{w} \in \mathbf{H}^1(\Omega).$$

It is easily seen that the optimality system (5.6.8), (5.6.9), (5.6.10), (5.6.22), (5.6.23), (5.6.26), (5.6.34) and (5.6.35) is equivalent to

$$(\mathbf{u}, \lambda p, \lambda \mathbf{t}, \mathbf{g}, \lambda \beta, \boldsymbol{\nu}, \lambda \phi, \lambda \theta) + TG(\lambda, (\mathbf{u}, \lambda p, \lambda \mathbf{t}, \mathbf{g}, \lambda \beta, \boldsymbol{\nu}, \lambda \phi, \lambda \theta)) = 0 \tag{5.6.72}$$

and that the discrete optimality system (5.6.42)–(5.6.49) is equivalent to

$$(\mathbf{u}^h, \lambda p^h, \lambda \mathbf{t}^h, \mathbf{g}^h, \lambda \beta^h, \boldsymbol{\nu}^h, \lambda \phi^h, \lambda \theta^h)$$
$$+TG(\lambda, (\mathbf{u}^h, \lambda p^h, \lambda \mathbf{t}^h, \mathbf{g}^h, \lambda \beta^h, \boldsymbol{\nu}^h, \lambda \phi^h, \lambda \theta^h)) = 0. \tag{5.6.73}$$

We have thus recast our continuous and discrete optimality problems into a form that enables us to apply Theorem 5.6.7.

We call a solution $\big(\mathbf{u}(\lambda), p(\lambda), \mathbf{t}(\lambda), \mathbf{g}(\lambda), \beta(\lambda), \boldsymbol{\nu}(\lambda), \phi(\lambda), \boldsymbol{\theta}(\lambda)\big)$ of the problem (5.6.8)–(5.6.10), (5.6.22)–(5.6.26) and (5.6.34)–(5.6.35), or equivalently, of (5.6.72), regular if the linear system

$$a(\check{\mathbf{u}}, \mathbf{v}) + \lambda c(\check{\mathbf{u}}, \mathbf{u}, \mathbf{v}) + \lambda c(\mathbf{u}, \check{\mathbf{u}}, \mathbf{v})+$$
$$\lambda b(\mathbf{v}, \check{p}) - \lambda(\mathbf{v}, \check{\mathbf{t}})_\Gamma = (\mathbf{f}_1, \mathbf{v}) \quad \forall \, \mathbf{v} \in \mathbf{H}^1(\Omega)\,,$$

$$b(\check{\mathbf{u}}, q) = 0 \quad \forall \, q \in L_0^2(\Omega)\,,$$

$$(\check{\mathbf{u}}, \mathbf{s})_\Gamma - (\check{\mathbf{g}}, \mathbf{s})_{\Gamma_c} = 0 \quad \forall \, \mathbf{s} \in \mathbf{H}^{-1/2}(\Gamma)\,,$$

$$\tfrac{1}{\lambda}(\operatorname{grad}_s \check{\mathbf{g}}, \operatorname{grad}_s \mathbf{k})_{\Gamma_c} + \tfrac{1}{\lambda}(\check{\mathbf{g}}, \mathbf{k})_{\Gamma_c}+$$
$$\check{\beta} \int_{\Gamma_c} \mathbf{k} \cdot \mathbf{n}\, d\Gamma - (\mathbf{k}, \check{\boldsymbol{\theta}} - \check{\mathbf{t}})_{\Gamma_c} = 0 \quad \forall \, \mathbf{k} \in \mathbf{W}(\Gamma_c)\,,$$

$$\int_{\Gamma_c} \check{\mathbf{g}} \cdot \mathbf{n}\, d\Gamma = 0\,,$$

$$a(\mathbf{w}, \check{\boldsymbol{\nu}}) + \lambda c(\mathbf{w}, \check{\mathbf{u}}, \boldsymbol{\nu}) + \lambda c(\mathbf{w}, \mathbf{u}, \check{\boldsymbol{\nu}}) + \lambda c(\mathbf{u}, \mathbf{w}, \check{\boldsymbol{\nu}})+$$
$$\lambda c(\check{\mathbf{u}}, \mathbf{w}, \boldsymbol{\nu}) + \lambda b(\mathbf{w}, \check{\phi}) - (\mathbf{w}, \check{\boldsymbol{\theta}})_\Gamma + \lambda c(\check{\mathbf{u}}, \mathbf{u}, \mathbf{w})$$
$$+\lambda c(\mathbf{u}, \check{\mathbf{u}}, \mathbf{w}) = (\mathbf{f}_2, \mathbf{v}) \quad \forall \, \mathbf{w} \in \mathbf{H}^1(\Omega)\,,$$

$$b(\check{\boldsymbol{\nu}}, r) = 0 \quad \forall \, r \in L_0^2(\Omega)\,,$$

and

$$(\check{\boldsymbol{\nu}}, \mathbf{y})_\Gamma = 0 \quad \forall \, \mathbf{y} \in \mathbf{H}^{-1/2}(\Gamma)$$

has a unique solution $(\check{\mathbf{u}}, \check{p}, \check{\mathbf{t}}, \check{\mathbf{g}}, \check{\beta}, \check{\boldsymbol{\nu}}, \check{\phi}, \check{\boldsymbol{\theta}}) \in X$ for every $\mathbf{f}_k \in (\mathbf{H}^1(\Omega))^*, k = 1, 2$. An analogous definition holds for regular solutions of the discrete optimality system (5.6.42)–(5.6.49), or equivalently, (5.6.73).

It can be shown, using techniques similar to those employed for the Navier-Stokes equations (see [43] and the references cited therein) that for almost all values of the Reynolds number that the optimality system (5.6.8), (5.6.9), (5.6.10), (5.6.22), (5.6.23), (5.6.26), (5.6.34) and (5.6.35), or equivalently, of (5.6.72), is regular, i.e., is locally unique. Thus, it is reasonable to assume that the optimality system has branches of regular solutions.

In order to apply the results of [34], [36], [37], and [39], we need to estimate the approximation properties of the operator T^h.

Proposition 5.6.9 *The problem (5.6.56)–(5.6.63) has a unique solution belonging to X. Assume that (5.6.36)–(5.6.41) hold. Then, the problem (5.6.64)–(5.6.71) has a unique solution belonging to X^h. Let $(\tilde{\mathbf{u}}, \tilde{p}, \tilde{\mathbf{t}}, \tilde{\mathbf{g}}, \tilde{\beta}, \tilde{\boldsymbol{\nu}}, \tilde{\phi}, \tilde{\boldsymbol{\theta}})$ and $(\tilde{\mathbf{u}}^h, \tilde{p}^h, \tilde{\mathbf{t}}^h, \tilde{\mathbf{g}}^h, \tilde{\beta}^h, \tilde{\boldsymbol{\nu}}^h, \tilde{\phi}^h, \tilde{\boldsymbol{\theta}}^h)$ denote the solutions of (5.6.56)–(5.6.63) and (5.6.64)–(5.6.71), respectively. Then, we also have that*

$$\|\tilde{\mathbf{u}} - \tilde{\mathbf{u}}^h\|_1 + \|\tilde{p} - \tilde{p}^h\|_0 + \|\tilde{\mathbf{t}} - \tilde{\mathbf{t}}^h\|_{-\frac{1}{2},\Gamma} + \|\tilde{\mathbf{g}} - \tilde{\mathbf{g}}^h\|_{1,\Gamma_c} + |\tilde{\beta} - \tilde{\beta}^h| +$$
$$\|\tilde{\boldsymbol{\nu}} - \tilde{\boldsymbol{\nu}}^h\|_1 + \|\tilde{\phi} - \tilde{\phi}^h\|_0 + \|\tilde{\boldsymbol{\theta}} - \tilde{\boldsymbol{\theta}}^h\|_{-\frac{1}{2},\Gamma} \to 0 \qquad (5.6.74)$$

as $h \to 0$. If, in addition, $(\tilde{\mathbf{u}}, \tilde{p}, \tilde{\mathbf{t}}, \tilde{\mathbf{g}}, \tilde{\beta}, \tilde{\boldsymbol{\nu}}, \tilde{\phi}, \tilde{\boldsymbol{\theta}}) \in \mathbf{H}^{m+1}(\Omega) \times H^m(\Omega) \cap L_0^2(\Omega) \times \mathbf{H}^{m-\frac{1}{2}}(\Gamma) \times \mathbf{H}^{m+1}(\Gamma_c) \times I\!\!R \times \mathbf{H}^{m+1}(\Omega) \times H^m \cap L_0^2(\Omega) \times \mathbf{H}^{m-\frac{1}{2}}(\Gamma)$, then

$$\|\tilde{\mathbf{u}} - \tilde{\mathbf{u}}^h\|_1 + \|\tilde{p} - \tilde{p}^h\|_0 + \|\tilde{\mathbf{t}} - \tilde{\mathbf{t}}^h\|_{-\frac{1}{2},\Gamma} + \|\tilde{\mathbf{g}} - \tilde{\mathbf{g}}^h\|_{1,\Gamma_c} + |\tilde{\beta} - \tilde{\beta}^h| +$$
$$\|\tilde{\boldsymbol{\nu}} - \tilde{\boldsymbol{\nu}}^h\|_1 + \|\tilde{\phi} - \tilde{\phi}^h\|_0 + \|\tilde{\boldsymbol{\theta}} - \tilde{\boldsymbol{\theta}}^h\|_{-\frac{1}{2},\Gamma} \qquad (5.6.75)$$
$$\leq Ch^m (\|\tilde{\mathbf{u}}\|_{m+1} + \|\tilde{p}\|_m + \|\tilde{\boldsymbol{\nu}}\|_{m+1} + \|\tilde{\phi}\|_m).$$

Sketch of proof: Consider the Stokes problems (5.6.56)-(5.6.58) and (5.6.61)–(5.6.63) and the discrete Stokes problems (5.6.64)–(5.6.66) and (5.6.69)–(5.6.71). Theorem 5.6.8 implies that each has a unique solution. Moreover, we have that (5.6.74) and (5.6.76) hold for the velocities, pressures and stresses, and their corresponding Lagrange multipliers.

Next, consider the problems (5.6.59)–(5.6.60) and (5.6.67)–(5.6.68) for β and the control $\mathbf{g}$ and their approximations. One may easily show that these problems fit into the framework of the Brezzi theory for mixed finite element methods (see [32] or [33]). Thus, using that theory, we may conclude that these problems both have unique solutions and that the estimates (5.6.74) and (5.6.76) for β and $\mathbf{g}$ hold. $\qquad \square$

Using this proposition and Theorem 5.6.7, we are led to the following result.

Theorem 5.6.10 *Assume that Λ is a compact interval of $I\!\!R_+$ and that there exists a branch $\{(\lambda, \psi(\lambda) = (\mathbf{u}, p, \mathbf{t}, \mathbf{g}, \beta, \boldsymbol{\nu}, \phi, \boldsymbol{\theta})) \in X : \lambda \in \Lambda\}$ of regular solutions of the optimality system (5.6.8)–(5.6.10), (5.6.22)-(5.6.23), (5.6.26) and (5.6.34)–(5.6.35). Assume that the finite element spaces $\mathbf{V}^h$, S^h, $\mathbf{P}^h$ and Q_0^h satisfy the conditions (5.6.36)–(5.6.41). Then, there exists a neighborhood $\mathcal{O}$ of the origin in X and, for $h \leq h_0$ small enough, a unique branch $\{(\lambda, \psi^h(\lambda) = (\mathbf{u}^h, p^h, \mathbf{t}^h, \mathbf{g}^h, \beta^h, \boldsymbol{\nu}^h, \phi^h, \boldsymbol{\theta}^h)) \in X^h) : \lambda \in \Lambda\}$ of solutions of the discrete optimality system (5.6.42)–(5.6.49) such that $\psi^h(\lambda) - \psi(\lambda) \in \mathcal{O}$ for all $\lambda \in \Lambda$. Moreover,*

$$\begin{aligned}\|\psi^h(\lambda) - \psi(\lambda)\|_X = \ & \|\mathbf{u} - \mathbf{u}^h\|_1 + \|p - p^h\|_0 + \|\mathbf{t} - \mathbf{t}^h\|_{-1/2,\Gamma} \\ & + \|\mathbf{g} - \mathbf{g}^h\|_{1,\Gamma_c} + |\beta - \beta^h| + \|\boldsymbol{\nu} - \boldsymbol{\nu}^h\|_1 \quad (5.6.76) \\ & + \|\phi - \phi^h\|_0 + \|\boldsymbol{\theta} - \boldsymbol{\theta}^h\|_{-1,2,\Gamma} \to 0 \end{aligned}$$

as $h \to 0$, *uniformly in* $\lambda \in \Lambda$.

If, in addition, the solution of the optimality system satisfies $(\mathbf{u}, p, \mathbf{t}, \mathbf{g}, \boldsymbol{\nu}, \phi, \boldsymbol{\theta}) \in$ $\mathbf{H}^{m+1}(\Omega) \times H^m(\Omega) \cap L_0^2(\Omega) \times \mathbf{H}^{m-\frac{1}{2}}(\Gamma) \times \mathbf{H}^{m+1}(\Gamma_c) \times \mathbf{H}^{m+1}(\Omega) \times H^m(\Omega) \cap L_0^2(\Omega) \times$ $\mathbf{H}^{m-\frac{1}{2}}(\Gamma)$ *for* $\lambda \in \Lambda$, *then there exists a constant* C, *independent of* h, *such that*

$$\|\mathbf{u} - \mathbf{u}^h\|_1 + \|p - p^h\|_0 + \|\mathbf{t} - \mathbf{t}^h\|_{-1/2,\Gamma} + \|\mathbf{g} - \mathbf{g}^h\|_{1,\Gamma_c} + |\beta - \beta^h| +$$
$$\|\boldsymbol{\nu} - \boldsymbol{\nu}^h\|_1 + \|\phi - \phi^h\|_0 + \|\boldsymbol{\theta} - \boldsymbol{\theta}^h\|_{-1,2,\Gamma} \tag{5.6.77}$$
$$\leq C h^m \big(\|\mathbf{u}(\lambda)\|_{m+1} + \|p(\lambda)\|_m + \|\boldsymbol{\nu}(\lambda)\|_{m+1} + \|\phi(\lambda)\|_m \big),$$

uniformly in $\lambda \in \Lambda$.

Sketch of proof: Clearly, G is a C^∞ polynomial map from $\mathbb{R}_+ \times X$ into Y. Therefore, using (5.6.3)–(5.6.5), $D^2 G(\mathbf{u}, \boldsymbol{\xi})$ is easily shown to be bounded on all bounded sets of X. Now, given $(\mathbf{u}, p, \mathbf{t}, \mathbf{g}, \beta, \boldsymbol{\nu}, \phi, \boldsymbol{\theta}) \in X$, a direct computation, using (5.6.3)–(5.6.5), yields that $D_\phi G(\lambda, (\mathbf{u}, p, \mathbf{t}, \mathbf{g}, \beta, \boldsymbol{\nu}, \phi, \boldsymbol{\theta})) \in \mathcal{L}(X; Y)$. On the other hand, since $(\mathbf{u}, p, \mathbf{t}, \mathbf{g}, \beta, \boldsymbol{\nu}, \phi, \boldsymbol{\theta}) \in X$ and $(\mathbf{v}, q, \mathbf{s}, \mathbf{k}, \alpha, \mathbf{w}, r, \mathbf{y}) \in X$, one can use the Sobolev imbedding theorem to show that $(\tilde{\boldsymbol{\zeta}}, \tilde{\boldsymbol{\kappa}}, \tilde{\boldsymbol{\eta}}) \in Z$ and therfore

$$D_\psi G(\lambda, (\mathbf{u}, p, \mathbf{t}, \mathbf{g}, \beta, \boldsymbol{\nu}, \phi, \boldsymbol{\theta})) \in \mathcal{L}(X; Z) \quad \text{for} \quad (\mathbf{u}, p, \mathbf{t}, \mathbf{g}, \beta, \boldsymbol{\nu}, \phi, \boldsymbol{\theta}) \in X.$$

Of course, Z is continuously imbedded into Y; moreover, the imbedding $Z \subset Y$ is compact.

Next, we turn to the approximation properties of the operator T^h. From Proposition 5.6.9, we have that (5.6.53) holds. Since the imbedding of Z into Y is compact, (5.6.54) follows from (5.6.73), and then (5.6.77) follows from (5.6.55). From Proposition 5.6.9 we also may conclude that there exists a constant C, independent of h, such that

$$\|(T - T^h) G(\lambda, \psi(\lambda))\|_X \leq C h^m \big(\|\mathbf{u}\|_{m+1} + \|p\|_m + \|\boldsymbol{\nu}\|_{m+1} + \|\phi\|_m \big).$$

Then (5.6.78) follows from (5.6.55). $\qquad\square$

The theories of [34], [36], [37], and [39] can also be used to derive an estimate for the error of $\mathbf{u}^h$ and $\boldsymbol{\nu}^h$ in the $\mathbf{L}^2(\Omega)$-norm and of $\mathbf{g}^h$ in the $\mathbf{H}^{1/2}(\Gamma_c)$-norm. We omit any reference to the proof.

Theorem 5.6.11 *Assume the hypotheses of Theorems 5.6.6 and 5.6.10. Then, there exists a constant* C, *independent of* h *such that*

$$\|\mathbf{u} - \mathbf{u}^h\|_0 + \|\boldsymbol{\nu} - \boldsymbol{\nu}^h\|_0 + \|\mathbf{g} - \mathbf{g}^h\|_{1/2,\Gamma_c} = O(h^{m+\frac{1}{2}}). \qquad\square$$

By other means, it can be shown that actually

$$\|\mathbf{u} - \mathbf{u}^h\|_0 + \|\boldsymbol{\nu} - \boldsymbol{\nu}^h\|_0 + \|\mathbf{g} - \mathbf{g}^h\|_{0,\Gamma_c} = O(h^{m+1}).$$

Note that in all cases the error in the approximation to the control is $1/2$-order higher than that obtainable from the error estimates for the velocity approximation and an application of trace theorems.

Acknowledgement

MDG was supported by AFOSR and ONR; LSH by the National Science and Engineering Council of Canada; and TPS by ONR.

References

References on the control and optimization of viscous incompressible flows.

[1] F. Abergel and R. Temam. (1990). On some control problems in fluid mechanics, *Theoret. Comput. Fluid Dynamics* **1**, 303–325.

[2] G. Arumugan and O. Pironneau. (1989). On the problems of riblets as a drag reduction device, *Optimal Cont. Appl. Meth.* **10**, 93–112.

[3] M. Bristeau, O. Pironneau, R. Glowinski, J. Periaux, P. Perrier, and G. Poirier. (1983). Application of optimal control and finite element methods to the calculation of transonic flows and incompressible viscous flows, *Nerical Methods in Applied Fluid Mechanics* (Ed. by B. Hunt), Academic, London, 203–312.

[4] C. Cuvelier. (1976). Optimal control of a system governed by the Navier-Stokes equations coupled with the heat equations, *New Developments in Differential equations* (Ed. by W. Eckhaus), North-Holland, Amsterdam, 81–98.

[5] C. Cuvelier. (1978). Resolution numérique d'un problème de controle optimal d'un couplage des équations de Navier-Stokes et celle de la chaleur, *Calcolo* **15**, 345–379.

[6] H. Fattorini and S. Sritharan. (To appear). Optimal control theory for viscous flow problems.

[7] A. Fursikov. (1982). Control problems and theorems concerning the unique solvability of a mixed boundary value problem for the three-dimensional Navier-Stokes and Euler equations, *Math. USSR Sbornik* **43** 251–273.

[8] A. Fursikov. (1983). On some control problems and results concerning the unique solvability of a mixed boundary value problem for the three-dimensional Navier-Stokes and Euler systems, *Soviet Math. Dokl.* **21**, 889–893.

[9] A. Fursikov. (1983). Properties of solutions of some extremal problems connected with the Navier-Stokes system, *Math. USSR Sbornik* **46**, 323–351.

[10] M. Gunzburger, L. Hou and T. Svobodny. (1989). Numerical approximation of an optimal control problem associated with the Navier-Stokes equations, *Appl. Math. Let.* **2**, 29–31.

[11] M. Gunzburger, L. Hou and T. Svobodny. (1991). Boundary velocity control of incompressible flow with an application to viscous drag reduction; *SIAM J. Control and Optimization*, **30**, 167–181.

[12] M. Gunzburger, L. Hou and T. Svobodny. (1991). Analysis and finite element approximation of optimal control problems for the stationary Navier-Stokes equations with distributed and Neumann controls; *Math. Comput.* **57**, 123–151.

[13] M. Gunzburger, L. Hou and T. Svobodny. (1991). Analysis and finite element approximation of optimal control problems for the stationary Navier-Stokes equations with Dirichlet controls; *Math. Mod. Numer. Anal.*, **25**, 711–748.

[14] M. Gunzburger, L. Hou and T. Svobodny. (To appear). Control of temperature distributions along boundaries of engine components; *Numerical Methods for Turbulent and Laminar Flows*.

[15] M. Gunzburger, L. Hou and T. Svobodny. (To appear). Heating and cooling control of temperature distributions along boundaries of flow domains, *J. Math. Sys. Estim. Control*.

[16] M. Gunzburger, L. Hou and T. Svobodny. (To appear). Optimal boundary control of nonsteady incompressible flow.

[17] M. Gunzburger, L. Hou and T. Svobodny. (1990). Optimal boundary control of nonsteady incompressible flow with an application to viscous drag reduction; *Proc. 29th Conference on Decision and Control*, IEEE, New York 377–378.

[18] L. Hou and T. Svobodny. (To appear). Optimization problems for the Navier-Stokes equations with regular boundary controls; *J. Math. Anal. Appl.*

[19] K. Ito and M. Desai. (1991). Optimal control of the Navier-Stokes equations, *Center for Applied Mathematical Sciences Report* **91-6**, U. Southern California, Los Angeles.

[20] L. Ji and J. Zhou. (1990). The boundary element method for boundary control of the linear Stokes flow, *Proc. 29th Conference on Decision and Control* IEEE, New York, 1192–1194.

[21] K.-T. Li and A.-X. Huang. (1984). Mathematical aspects of optimal control finite element method for Navier-Stokes problems, *J. Comput. Math.* **2**, 139–151.

[22] J. Lions. (1971). *Optimal Control of Systems Governed by Partial Differential Equations*, Springer, Berlin.

[23] J. Lions. (1985). *Control of Distributed Singular Systems*, Bordas, Paris.

[24] Y.-R. Ou. (To appear). Active control of forces around a rotating cylinder.

[25] Y.-R. Ou and J. Burns. (To appear). Optimal control of lift/drag ratio on a rotating cylinder.

[26] O. Pironneau. (1984). *Optimal Shape Design for Elliptic Systems*, Springer, New York.

[27] S. Sritharan. (1990). An optimal control problem in exterior hydrodynamics, *New Trends and Applications of Distributed Parameter Control Systems* (Ed. by G. Chen, E. Lee, L. Markus, and W. Littman), Marcel Dekker, 385–417.

[28] S. Sritharan. (To appear). Dynamic programming and the Navier-Stokes equations.

Other references.

[29] R. Adams. (1975). *Sobolev Spaces*, Academic, New York.

[30] I. Babuška. (1973). The finite element method with Lagrange multipliers, *Numer. Math.* **16**, 179–192.

[31] I. Babuška and A. Aziz. (1973). Survey lectures on the mathematical foundations of the finite element method, *The Mathematical Foundations of the Finite Element Method with Applications to Partial Differential Equations* (Ed. by A. Aziz), Academic, New York, 3–359.

[32] F. Brezzi. (1974). On the existence, uniqueness, and approximation of saddle-point problems arising form Lagrange multipliers, *RAIRO Anal. Numér.* **32**, 129–151.

[33] F. Brezzi. (1988). A survey of mixed finite element methods. *Finite Elements, Theory and Application* (Ed. by D. Dwoyer, M. Hussaini and R. Voigt), Springer, New York, 34–49.

[34] F. Brezzi, J. Rappaz and P.-A. Raviart. (1980). Finite-dimensional approximation of nonlinear problems. Part I: branches of nonsingular solutions, *Numer. Math.* **36**, 1–25.

[35] P. Ciarlet. (1978). *The Finite Element Method for Elliptic Problems*, North-Holland, Amsterdam.

[36] M. Crouzeix and J. Rappaz. (1989). *On Numerical Approximation in Bifurcation Theory*, Masson, Paris.

[37] V. Girault and P.-A. Raviart. (1986). *Finite Element Methods for Navier-Stokes Equations*, Springer, Berlin.

[38] M. Gunzburger. (1989). *Finite Element Methods for Incompressible Viscous Flows: A Guide to Theory, Practice and Algorithms*, Academic, Boston.

[39] M. Gunzburger and L. Hou. (To appear). Treating inhomogeneous essential boundary conditions in finite element methods, *SIAM J. Numer. Anal.*

[40] M. Schecter. (1971). *Principles of Functional Analysis*, Academic, New York.

[41] J. Serrin. (1959). Mathematical principles of classical fluid mechanics, *Handbüch der Physik* VIII/1 (ed. by S. Flügge and C. Truesdell), Springer, Berlin, 125–263.

[42] R. Temam. (1979). *Navier-Stokes Equations*, North-Holland, Amsterdam.

[43] R. Temam. (1983). *Navier-Stokes Equations and Nonlinear Functional Analysis*, SIAM, Philadelphia.

6 A Fully-Coupled Finite Element Algorithm, Using Direct and Iterative Solvers, for the Incompressible Navier-Stokes Equations

W. G. Habashi, M. F. Peeters, M. P. Robichaud and V-N. Nguyen

Abstract

This chapter presents a Finite Element solution method for the incompressible Navier-Stokes equations, in primitive variables form. To provide the necessary coupling between continuity and momentum, and enhance stability, a pressure dissipation in the form of a Laplacian is introduced into the continuity equation. The recasting of the problem variables in terms of pressure and an "auxiliary" velocity demonstrates how the effects of the pressure dissipation can be eliminated, while retaining its stabilizing properties. The method can also be interpreted as a Helmholtz decomposition of the velocity vector.

The governing equations are discretized by a Galerkin weighted residual method and, because of the modification to the continuity equation, equal interpolation for all the unknowns is permitted. Newton linearization is used and, at each iteration, the linear algebraic system is solved in a fully-coupled manner by direct or iterative solvers. For direct methods, a vector-parallel Gauss elimination method is developed that achieves execution rates exceeding 2.3 Gigaflops, i.e. over 86% of a Cray YMP-8 current peak performance. For iterative methods, preconditioned conjugate gradient-like methods are studied and good performances, competitive with direct solvers, are achieved. Convergence of such methods being sensitive to preconditioning, a hybrid dissipation method is proposed, with the preconditioner having an artificial dissipation that is gradually lowered, but frozen at a level higher than the dissipation introduced into the physical equations.

Convergence of the Newton-Galerkin algorithm is very rapid. Results are demonstrated for two-and three-dimensional incompressible flows.

6.1 Introduction

The simulation of complex aerodynamic fields by means of inviscid and viscous flow equations is rapidly becoming the preferred analysis and design tool in the aerospace

151

industry. There is no shortage of methods for discretizing the Navier-Stokes equations, with these methods differing in their discretization of the time or pseudo-time term, space terms, linearization and algebraic equations solution method. The predominant space discretization method in industrial practice are Finite Volume (FVM) Methods, with the Finite Element Method (FEM) often wrongly perceived as taxing on computer memory. During space discretization, methods also differ in applying the dissipation necessary to stabilize the numerical solution and two approaches are possible: centered schemes, with dissipation introduced through an explicit artificial viscosity or upwind schemes applied to the convective terms.

For time discretization, explicit and implicit approximations can be used. Explicit schemes trade the speed of convergence for simplicity by not requiring matrix inversion. They are easily vectorizable and parallelizable. To speed up convergence, various acceleration techniques are used such as local time stepping, residual averaging and multigrid methods. Large-scale problems are therefore more easily amenable to solution on today's computers, with a compromise between large solution times and manageable memory resources. Implicit schemes, on the other hand, allow much larger time steps at the cost of inverting some matrices at each step. These range from fully-coupled schemes, to ADI schemes, all the way to schemes that only require the solution of scalar tridiagonal matrices. It must be appreciated, however, that the hierarchy of simplifications in the solution of the coupled system of equations must be at the cost of additional iterations to obtain the same overall convergence of the nonlinear system.

The numerical solution of the incompressible Navier-Stokes equations presents some additional problems. First, the governing equations, namely continuity and momentum, are in terms of velocity and pressure, with none of the equations identifiable as governing the pressure. Various techniques have been developed to overcome this problem. Chorin [1] suggested adding the time derivative of pressure to the continuity equation, thereby identifying it as the pressure one. The added term links the equations and allows pressure to be updated from the continuity equation. The scheme is, however, unsuitable for the time-accurate prediction of unsteady flows. A different scheme has been suggested by Harlow and Welch [2]. Other known algorithms in this context are Patankar and Spalding's [3] SIMPLE and SIMPLER, in which a Poisson pressure or pressure correction equation is solved at every iteration to satisfy the conservation of mass. Other alternatives exist and have been applied in the finite difference and finite element contexts. Taylor and Hughes [4], for example, solve the equations simultaneously and, to provide stability, use unequal order elements for pressure and velocity. The consistency of representation of the variables by different degree polynomials in finite elements is known as the Babuska-Brezzi condition [5,6]. It has similarity to the necessity of staggered grids in finite difference solutions [3] to avoid odd-even decoupling of the pressure field, known as checkerboarding. Recently, a move has been initiated in finite volume methods away from staggered grids and towards the use of co-located solution methods [7].

It is suggested here to introduce a pressure Laplacian directly in the continuity equation, thereby interpreting it as a pressure Poisson equation. This error term, with a small coefficient proportional to the grid size, provides the necessary coupling between the equations and circumvents the requirements of the Babuska-Brezzi condition. An added feature of the present work is to show that, by defining an "auxiliary" velocity, the error due to the pressure dissipation term can be removed, at least for incompressible flows.

To speed up convergence, the iteration for the nonlinearity is carried out through a Newton linearization, followed by a direct or iterative solution of the discretized linear equations at each iteration.

The chapter is therefore divided into two sections. The first presents the numerical algorithm and its two- and three-dimensional results, while the second describes the direct and iterative solution methods developed for the large set of linear equations, resulting at each step of the Newton linearization, in three-dimensional problems.

6.2 Newton-Galerkin Finite Element Formulation of the Incompressible Navier-Stokes Equations

For simplicity the method will be described for two-dimensional flows. Extension to three dimensions is straightforward and results will be presented. The equations governing steady, two-dimensional, incompressible, viscous, laminar flow can be written as:

$$\frac{\partial}{\partial x}(u) + \frac{\partial}{\partial y}(v) = 0 \tag{6.2.1a}$$

$$\frac{\partial}{\partial x}(u^2 + p) + \frac{\partial}{\partial y}(uv) - \frac{1}{Re}\left(\frac{\partial^2 u}{\partial x^2} + \frac{\partial^2 u}{\partial y^2}\right) = 0 \tag{6.2.1b}$$

$$\frac{\partial}{\partial x}(uv) + \frac{\partial}{\partial y}(v^2 + p) - \frac{1}{Re}\left(\frac{\partial^2 v}{\partial x^2} + \frac{\partial^2 v}{\partial y^2}\right) = 0 \tag{6.2.1c}$$

We propose two approaches. In the first, hereafter referred to as Method 1, the continuity equation is simply augmented with a Laplacian on the right hand side to provide the necessary coupling to momentum and avoid pressure checkerboarding:

$$\frac{\partial}{\partial x}(u) + \frac{\partial}{\partial y}(v) = \lambda\left(\frac{\partial^2 p}{\partial x^2} + \frac{\partial^2 p}{\partial y^2}\right) \tag{6.2.2}$$

This pressure-Poisson formulation introduces a mass source term into the continuity equation, thus λ should be made as small as possible.

A second approach, thereafter referred to as Method 2, is also proposed to eliminate the error in mass continuity associated with the pressure dissipation. In this scheme the continuity equation is first written as:

$$\frac{\partial}{\partial x}(u^*) + \frac{\partial}{\partial y}(v^*) = \lambda \left(\frac{\partial^2 p}{\partial x^2} + \frac{\partial^2 p}{\partial y^2} \right) \tag{6.2.3}$$

with the auxiliary velocity vector, u^*, defined as:

$$u^* = u + \lambda \frac{\partial p}{\partial x} \qquad v^* = v + \lambda \frac{\partial p}{\partial y} \tag{6.2.4}$$

The continuity equation remains therefore exact in terms of the physical velocity, i.e. $\nabla \cdot \mathbf{u} = 0$. The momentum equations and the boundary conditions can be rewritten in terms of $\mathbf{u}^*$, the auxiliary velocity. This is straightforward for the convective terms but the viscous terms merit examination. Equation (6.2.4) is rewritten in vector form as:

$$\mathbf{u}^* = \mathbf{u} + \lambda \nabla p = \mathbf{u} + \hat{\mathbf{u}} \tag{6.2.5}$$

Since $\hat{\mathbf{u}}$ is the gradient of a scalar, $\hat{\mathbf{u}}$ is irrotational, hence:

$$\nabla \times \hat{\mathbf{u}} = 0 \tag{6.2.6}$$

The viscous term of the momentum equations can hence be rewritten as:

$$\begin{aligned} \frac{1}{Re} \nabla^2 \mathbf{u} &= \frac{1}{Re} \left[\nabla^2 \mathbf{u}^* - \nabla^2 \hat{\mathbf{u}} \right] \\ &= \frac{1}{Re} \left[\nabla^2 \mathbf{u}^* - \nabla(\nabla \cdot \hat{\mathbf{u}}) + \nabla \times (\nabla \times \hat{\mathbf{u}}) \right] \\ &= \frac{1}{Re} \left[\nabla^2 \mathbf{u}^* - \nabla(\nabla \cdot \mathbf{u}^*) \right] \end{aligned} \tag{6.2.7a}$$

or

$$\frac{1}{Re} \nabla^2 \mathbf{u} = -\frac{1}{Re} \nabla \times (\nabla \times \mathbf{u}^*) \tag{6.2.7b}$$

The viscous term is thus expressed exactly in terms of the auxiliary velocity $\mathbf{u}^*$. The use of Equation (6.2.7b) may, however, lead to some numerical problems due to poor conditioning of the resulting matrix. To avoid this, Equation (6.2.7b) is used with the second term neglected, introducing a small error, proportional to λ, in the viscous term. This error is much less important than the error in the continuity equation of Method 1, and decreases rapidly with Reynolds number. Method 2 therefore introduces into the continuity equation the stabilizing properties of an artificial viscosity, without the associated error, at the expense of a small error in the viscous terms. Both methods allow equal order of interpolation for pressure and velocity, with no restriction on the choice of elements as in [4,5,6].

Method 2 can also be interpreted, from equation (6.2.5), as a Helmholtz decomposition of the velocity vector into divergence and curl free parts. The velocity **u** represents the divergence free component of **u***, while the curl free component, **û**, is assumed proportional to the pressure. Thus:

$$\nabla \times \mathbf{u}^* = \nabla \times \mathbf{u} - \nabla \times \lambda \nabla p = \nabla \times \mathbf{u} \qquad (6.2.8)$$

i.e. the physical velocity vector, **u**, and the auxiliary one, **u***, have the same vorticity. In addition, the physical velocity satisfies the incompressibility condition:

$$\nabla \cdot \mathbf{u} = 0 \qquad (6.2.9)$$

One then has then the choice of solving the problem in terms of $(\mathbf{u}, p)$, $(\mathbf{u}^*, p)$ or $(\mathbf{u}, \mathbf{u}^*)$. Hafez and Soliman [8,9], for example, propose the $(\mathbf{u}, \mathbf{u}^*)$ system as an alternative to the velocity-vorticity formulation $(\mathbf{u}, \Omega)$ of [10,11] which requires special care in ensuring mass conservation, since the continuity equation is not imposed directly but only through its gradient. The $(\mathbf{u}, \mathbf{u}^*)$ system of equations of Hafez and Soliman would consist of solving:

$$\mathbf{u}^* = \mathbf{u} + \lambda \left[-\frac{D\mathbf{u}}{Dt} + \frac{1}{Re}\nabla^2\mathbf{u} \right] \qquad (6.2.10a)$$

$$\nabla^2\mathbf{u} = -\nabla \times (\nabla \times \mathbf{u}^*) \qquad (6.2.10b)$$

We prefer the $(\mathbf{u}^*, p)$ formulation because it allows an explicit and direct imposition of the continuity equation. Moreover, in three dimensions it consists of a four equation system versus six for the $(\mathbf{u}, \mathbf{u}^*)$ system, and is hence more amenable to a fully-coupled solution.

6.2.1 Finite Element Discretization

The discretization and linearization are similar for both methods and only the second formulation will be outlined in detail, again for two-dimensional flows. The finite element formulation starts by selecting element interpolation or shape functions for the vector of nodal unknowns $U = (u^*, v^*, p)$:

$$U = \sum_{j=1}^{4} N_j U_j \qquad (6.2.11)$$

All variables are interpolated by bilinear shape functions:

$$N_j = \frac{1}{4}(1 + \xi\xi_j)(1 + \eta\eta_j) \quad j = 1, \ldots, 4 \qquad (6.2.12)$$

expressed in terms of the normalized non-dimensional parent element coordinates ξ_j and η_j.

The Galerkin weighted residual form of the equations can be written as:

$$\iint_A W \left[\frac{\partial u^*}{\partial x} + \frac{\partial v^*}{\partial y} - \lambda \left(\frac{\partial^2 p}{\partial x^2} + \frac{\partial^2 p}{\partial y^2} \right) \right] dA = 0 \qquad (6.2.13a)$$

$$\iint_A W \left[\frac{\partial}{\partial x} \left[\left(u^* - \lambda \frac{\partial p}{\partial x} \right)^2 + p \right] + \frac{\partial}{\partial y} \left[\left(u^* - \lambda \frac{\partial p}{\partial x} \right) \left(v^* - \lambda \frac{\partial p}{\partial y} \right) \right] \right.$$
$$\left. - \frac{1}{Re} \left(\frac{\partial^2 u^*}{\partial x^2} + \frac{\partial^2 u^*}{\partial y^2} \right) \right] dA = 0 \qquad (6.2.13b)$$

$$\iint_A W \left[\frac{\partial}{\partial x} \left[\left(u^* - \lambda \frac{\partial p}{\partial x} \right) \left(v^* - \lambda \frac{\partial p}{\partial y} \right) \right] + \frac{\partial}{\partial y} \left[\left(v^* - \lambda \frac{\partial p}{\partial y} \right)^2 + p \right] \right.$$
$$\left. - \frac{1}{Re} \left(\frac{\partial^2 v^*}{\partial x^2} + \frac{\partial^2 v^*}{\partial y^2} \right) \right] dA = 0 \qquad (6.2.13c)$$

where W are weight functions, identical to the interpolation or shape functions, N.

After integration by parts, the weak-Galerkin form of the system, in terms of the auxiliary velocity, $\mathbf{u}^*$, is:

$$\iint_A \left[\left(u^* - \lambda \frac{\partial p}{\partial x} \right) \frac{\partial W}{\partial x} + \left(v^* - \lambda \frac{\partial p}{\partial y} \right) \frac{\partial W}{\partial y} \right] dA - \oint_S W (\mathbf{u} \cdot \mathbf{n}) dS = 0 \quad (6.2.14a)$$

$$\iint_A \left[\left\{ \left(u^* - \lambda \frac{\partial p}{\partial x} \right)^2 + p \right\} \frac{\partial W}{\partial x} + \left(u^* - \lambda \frac{\partial p}{\partial x} \right) \left(v^* - \lambda \frac{\partial p}{\partial y} \right) \frac{\partial W}{\partial y} \right.$$
$$\left. - \frac{1}{Re} \left(\frac{\partial u^*}{\partial x} \frac{\partial W}{\partial x} + \frac{\partial u^*}{\partial y} \frac{\partial W}{\partial y} \right) \right] dA \qquad (6.2.14b)$$
$$+ \oint_S \frac{1}{Re} W \left(\frac{\partial u^*}{\partial y} n_y - \frac{\partial v^*}{\partial y} n_x \right) dS - \oint_S W \left[(\mathbf{u} \cdot \mathbf{n}) u + p n_x \right] dS = 0$$

$$\iint_A \left[\left(u^* - \lambda \frac{\partial p}{\partial x} \right) \left(v^* - \lambda \frac{\partial p}{\partial y} \right) \frac{\partial W}{\partial x} + \left\{ \left(v^* - \lambda \frac{\partial p}{\partial y} \right)^2 + p \right\} \frac{\partial W}{\partial y} \right.$$

$$\left. - \frac{1}{Re} \left(\frac{\partial v^*}{\partial x} \frac{\partial W}{\partial x} + \frac{\partial v^*}{\partial y} \frac{\partial W}{\partial y} \right) \right] dA \qquad (6.2.14c)$$

$$+ \oint_S \frac{1}{Re} W \left(\frac{\partial v^*}{\partial x} n_x - \frac{\partial u^*}{\partial x} n_y \right) dS - \oint_S W \left[(\mathbf{u} \cdot \mathbf{n}) v + p n_y \right] dS = 0$$

6.2.2 Newton Linearization

After substituting the shape and weight functions into equations (6.2.14a–c), Newton's method can be introduced by setting:

$$U^{n+1} = U^n + \Delta U \qquad (6.2.15)$$

for the vector of nodal unknowns $U = (u^*, v^*, p)$. Upon neglecting second order terms, the continuity and momentum equations yield, respectively:

Continuity:

$$\left[K_{ij}^{pp} \right] \{ \Delta p \} + \left[K_{ij}^{pu} \right] \{ \Delta u^* \} + \left[K_{ij}^{pv} \right] \{ \Delta v^* \} = - \{ R_i^p \} \qquad (6.2.16a)$$

where the element contributions to the matrices are:

$$\left[k_{ij}^{pp} \right] = -\lambda \iint_A \left(\frac{\partial W_i}{\partial x} \frac{\partial N_j}{\partial x} + \frac{\partial W_i}{\partial y} \frac{\partial N_j}{\partial y} \right) dA$$

$$\left[k_{ij}^{pu} \right] = \iint_A \frac{\partial W_i}{\partial x} N_j dA$$

$$\left[k_{ij}^{pv} \right] = \iint_A \frac{\partial W_i}{\partial y} N_j dA$$

and where the contribution to the residual is:

$$\{ r_i^p \} = \iint_A \left[\left(u^* - \lambda \frac{\partial p}{\partial x} \right) \frac{\partial W_i}{\partial x} + \left(v^* - \lambda \frac{\partial p}{\partial y} \right) \frac{\partial W_i}{\partial y} \right] dA - \oint_S W_i (\mathbf{u} \cdot \mathbf{n}) dS$$

X-momentum:

$$\left[K_{ij}^{up}\right]\{\Delta p\} + \left[K_{ij}^{uu}\right]\{\Delta u^*\} + \left[K_{ij}^{uv}\right]\{\Delta v^*\} = -\{R_i^u\} \qquad (6.2.16b)$$

where the element contributions to the matrices are:

$$\left[k_{ij}^{up}\right] = \iint\limits_{A}\left[\frac{\partial W_i}{\partial x}\left(-2\lambda u_j^*\frac{\partial W_j}{\partial x} + 2\lambda^2\frac{\partial p_j}{\partial x}\frac{\partial W_i}{\partial x} + N_j\right)\right.$$

$$\left. + \frac{\partial W_i}{\partial y}\left(-\lambda u_j^*\frac{\partial W_j}{\partial y} - \lambda u_j^*\frac{\partial W_j}{\partial x} + \lambda^2\frac{\partial p_j}{\partial x}\frac{\partial W_j}{\partial y} + \lambda^2\frac{\partial p_j}{\partial y}\frac{\partial W_j}{\partial x}\right)\right]dA$$

$$\left[k_{ij}^{uu}\right] = \iint\limits_{A}\left[\frac{\partial W_i}{\partial x}\left(2u_j^* - 2\lambda\frac{\partial p_j}{\partial x}\right)N_j + \frac{\partial W_i}{\partial y}\left(v_j^* - \lambda\frac{\partial p_j}{\partial y}\right)N_j\right.$$

$$\left. - \frac{1}{Re}\left(\frac{\partial W_i}{\partial x}\frac{\partial N_j}{\partial x} + \frac{\partial W_i}{\partial y}\frac{\partial N_j}{\partial y}\right)\right]dA$$

$$\left[k_{ij}^{uv}\right] = \iint\limits_{A}\frac{\partial W_i}{\partial y}\left(u_j^* - \lambda\frac{\partial p_j}{\partial x}\right)N_j dA$$

and where the contribution to the residual is:

$$\{r_i^U\} = \iint\limits_{A}\left\{\left[\left(u^* - \lambda\frac{\partial p}{\partial x}\right)^2 + p\right]\frac{\partial W_i}{\partial x} + \left(u^* - \lambda\frac{\partial p}{\partial x}\right)\left(v^* - \lambda\frac{\partial p}{\partial y}\right)\frac{\partial W_i}{\partial y}\right.$$

$$\left. - \frac{1}{Re}\left(\frac{\partial u^*}{\partial x}\frac{\partial W_i}{\partial x} + \frac{\partial u^*}{\partial y}\frac{\partial W_i}{\partial y}\right)\right\}dA$$

$$- \oint\limits_{S}W_i\left[(\mathbf{u}\cdot\mathbf{n})u + pn_x\right]dS + \oint\limits_{S}\frac{1}{Re}W_i\frac{\partial u^*}{\partial n}dS$$

Y-momentum:

$$\left[K_{ij}^{up}\right]\{\Delta p\} + \left[K_{ij}^{uu}\right]\{\Delta u^*\} + \left[K_{ij}^{uv}\right]\{\Delta v^*\} = -\{R_i^u\} \qquad (6.2.16c)$$

where the element contributions to the matrices are:

$$[k_{ij}^{vp}] = \iint_A \left[\frac{\partial W_i}{\partial x} \left(-2\lambda u_j^* \frac{\partial W_j}{\partial x} + 2\lambda^2 \frac{\partial p_j}{\partial x} \frac{\partial W_j}{\partial x} + N_j \right) \right.$$

$$\left. + \frac{\partial W_i}{\partial y} \left(-\lambda u_j^* \frac{\partial W_j}{\partial y} - \lambda u_j^* \frac{\partial W_j}{\partial x} + \lambda^2 \frac{\partial p_j}{\partial x} \frac{\partial W_j}{\partial y} + \lambda^2 \frac{\partial p_j}{\partial y} \frac{\partial W_j}{\partial x} \right) \right] dA$$

$$[k_{ij}^{vu}] = \iint_A \frac{\partial W_i}{\partial y} \left(v_j^* - \lambda \frac{\partial p_j}{\partial y} \right) N_j dA$$

$$[k_{ij}^{vv}] = \iint_A \left[\frac{\partial W_i}{\partial x} \left(u_j^* - \lambda \frac{\partial p_j}{\partial x} \right) N_j + \frac{\partial W_i}{\partial y} \left(2v_j^* - 2\lambda \frac{\partial p_j}{\partial y} \right) N_j \right.$$

$$\left. - = \frac{1}{Re} \left(\frac{\partial W_i}{\partial x} \frac{\partial N_j}{\partial x} + \frac{\partial W_i}{\partial y} \frac{\partial N_j}{\partial y} \right) \right] dA$$

and where the contribution to the residual is:

$$\{r_i^v\} = \iint_A \left\{ \left(u^* - \lambda \frac{\partial p}{\partial x} \right) \left(v^* - \lambda \frac{\partial p}{\partial y} \right) \frac{\partial W_i}{\partial x} \right.$$

$$+ \left[\left(v^* - \lambda \frac{\partial p}{\partial y} \right)^2 + p \right] - \frac{1}{Re} \left(\frac{\partial v^*}{\partial x} \frac{\partial W_i}{\partial x} + \frac{\partial v^*}{\partial y} \frac{\partial W_i}{\partial y} \right) \right\} dA$$

$$- \oint_S W_i \left[(\mathbf{u} \cdot \mathbf{n}) v + p n_y \right] dS + \oint_S \frac{1}{Re} W_i \frac{\partial v^*}{\partial n} dS$$

All integrals are evaluated numerically by a (4×4) Gauss-Legendre quadrature, after expressing them in local coordinates, ξ, η. Typically:

$$[k_{ij}] = \int_{-1}^{1} \int_{-1}^{1} k_{ij} \left[x(\xi, \eta), y(\xi, \eta) \right] |J| d\xi d\eta$$

whᵉⱼe J is the Jacobian of the local transformation at each of the four Gaussian points of an element.

6.2.3 Boundary Conditions

The boundary conditions are detailed separately for both Methods 1 and 2.

Method 1

For the continuity equation the boundary conditions are:

At inlet: the contour integral of equation (6.2.14a) is calculated using the specified inlet velocity, **u**,

At exit: the pressure is specified as a Dirichlet boundary condition,

On walls: the contour integral of equation (6.2.14a) drops out naturally because of the no-penetration condition at the wall.

For the momentum equations the boundary conditions are:

At inlet: u, v profiles are imposed as Dirichlet boundary conditions.

At exit: $u_n = v_n = 0$, i.e. the streamlines are parallel; the first contour integral in equations (6.2.14b) and (6.2.14c) drops out and the second one is calculated at the exit,

On walls: $u = 0$, $v = 0$, a Dirichlet boundary condition.

Method 2

For the continuity equation, the boundary conditions are identical to Method 1.

For the momentum equations, when using the Helmholtz decomposition of Method 2, there is no longer an explicit appearance of **u**. Thus, equation (6.2.5) must be used for boundary conditions where **u*** must be specified as follows:

At inlet: u^*, v^* inlet profiles are imposed as Dirichlet boundary conditions derived from equation (6.2.4), i.e. using the imposed inlet velocity profile, and the pressure gradient from the previous iteration. The inlet values of u^* and v^* thus change from iteration to iteration. The values for u and v, however, remain constant at the inlet,

At exit: $u_n^* = v_n^* = 0$, i.e. the streamlines are parallel. This means that the first contour integral in equations (6.2.14b) and (6.2.14c) drops out and the second one is calculated at the exit,

On walls: $u^* = \lambda p_x$, $v^* = \lambda p_y$, a Dirichlet boundary condition using the pressure obtained from the previous iteration. Similar to the inlet conditions, the values of u^* and v^* on walls change from iteration to iteration. The values for u and v, however, remain constant ($= 0$).

6.2.4   *Solution to the Equations*

After FEM discretization and Newton linearization, the following representative delta form of the equations can be obtained for two-dimensional flows in terms of the cell-vertex unknowns of pressure and velocity components at the 4 vertices of a bilinear finite element:

$$\sum_{e=1}^{E}\left[\sum_{j=1}^{4}\left\{\left[k_{ij}^{p}\right]_{p}\Delta p_{j}+\left[k_{ij}^{u}\right]_{p}\Delta u_{j}+\left[k_{ij}^{v}\right]_{p}\Delta v_{j}\right\}\right] = -(R_i)_p$$

$$\sum_{e=1}^{E}\left[\sum_{j=1}^{4}\left\{\left[k_{ij}^{p}\right]_{u}\Delta p_{j}+\left[k_{ij}^{u}\right]_{u}\Delta u_{j}+\left[k_{ij}^{v}\right]_{u}\Delta v_{j}\right\}\right] = -(R_i)_{u}. \quad (6.2.17a)$$

$$\sum_{e=1}^{E}\left[\sum_{j=1}^{4}\left\{\left[k_{ij}^{p}\right]_{v}\Delta p_{j}+\left[k_{ij}^{u}\right]_{v}\Delta u_{j}+\left[k_{ij}^{v}\right]_{v}\Delta v_{j}\right\}\right] = -(R_i)_{v}.$$

which can be written in the matrix form:

$$\begin{bmatrix} [K^{\mathbf{u}}]_{\mathbf{u}} & [K^{p}]_{\mathbf{u}} \\ [K^{\mathbf{u}}]_{p} & [K^{p}]_{p} \end{bmatrix} \left\{ \begin{array}{c} \Delta\mathbf{u} \\ \Delta p \end{array} \right\} = -\left\{ \begin{array}{c} R_{\mathbf{u}} \\ R_{p} \end{array} \right\} \qquad (6.2.17b)$$

This system can be solved in a fully-coupled manner, using direct or iterative solvers and section 6.3 of this chapter addresses the development of such solvers.

6.2.5   Results

Let us define Method 1 as being the solution of equations (6.2.1b, 6.2.1c and 6.2.2) and Method 2 as the solution of (6.2.1b, 6.2.1c and 6.2.3).

Results are first presented for a sudden expansion geometry, at a Reynolds number (Re) of 100, using both methods to assess the effect of the pressure dissipation on the continuity equation. Tests are carried out using Method 1 to determine the effect of grid size on the required pressure dissipation parameter λ. The flow is calculated on a fine grid (42×19 elements), using the lowest value of λ (= 0.005, for this case) that gives smooth pressure contours and the results are shown in Figure 6.1. The grid is then coarsened and the lowest value for λ that suppresses wiggles is heuristically determined on each grid. These tests indicate that λ should be proportional to Δ^3; Δ being the mesh size. Tests are then again run on the finest grid (42×19 elements) using both approaches with a high λ (= 0.1). The resulting pressure contours are compared in Figures 6.2 and 6.3 to those of Method 1, with low λ, shown in Figure 6.1. It can be noted that, while both contours are smooth, Method 1 has weaker pressure gradients, reflecting the error in the continuity equation introduced by λ. Moreover, the pressure contours from Method 2 at the highest λ (= 0.1), Figure 6.2, compare well with those calculated by Method 1 at the lowest λ (= 0.005) in Figure 6.1, demonstrating that method 2 is virtually independent of the pressure dissipation coefficient λ, even at this

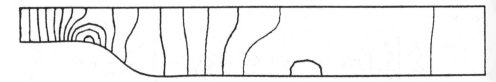

Figure 6.1: Newton-Galerkin Incompressible Navier-Stokes algorithm, fine grid, low pressure dissipation coefficient, $\lambda = 0.005$, Method 1.

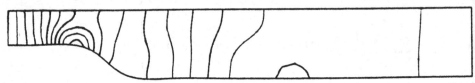

Figure 6.2: Newton-Galerkin Incompressible Navier-Stokes algorithm, fine grid, high pressure dissipation coefficient, $\lambda = 0.1$, Method 2.

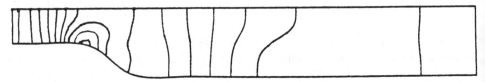

Figure 6.3: Newton-Galerkin Incompressible Navier-Stokes algorithm, fine grid, high pressure dissipation coefficient, $\lambda = 0.1$, Method 1.

relatively low Reynolds number. Figure 6.4 shows the convergence history of method 2 for the sudden expansion geometry, demonstrating its rapid convergence to machine accuracy. Quadratic convergence is, however, not achieved for method 2 because of the lagging of the boundary conditions implementation in the Newton scheme.

The second test case selected is the driven cavity problem at a Reynolds number of 400, discretized on a (50×50) grid. The centerline velocity profile is plotted in Figure 6.5 and the streamlines are shown in Figure 6.6. Both compare very well with the stream function-vorticity results of [12] and [13].

The third problem considered is flow in a converging-diverging channel, calculated by Method 1. The minimum channel width is 60% of the channel inlet. The grid for this problem consists of 1150 elements covering only half of the channel, with a symmetry boundary condition at the centerline. Incompressible streamlines for $Re = 100$ are shown in Figure 6.7. The length of the separation zone is 0.98 times the inlet channel width, which compares well with the value of 1.00 obtained in [12] by the stream function-vorticity method.

The method is demonstrated for three-dimensional flow in a gas turbine pipe diffuser,

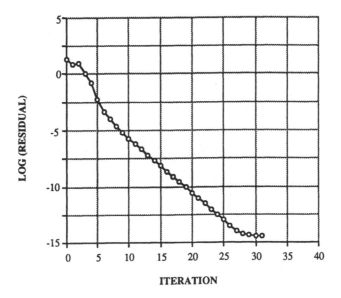

Figure 6.4: Newton-Galerkin incompressible Navier-Stokes algorithm, convergence history, Method 2.

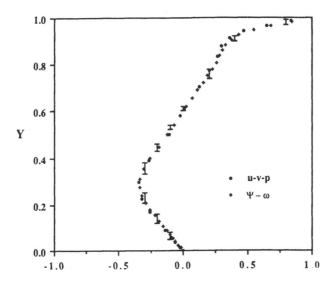

Figure 6.5: Incompressible viscous flow in a cavity at $Re = 400$, (50×50) grid, centerline velocity profile.

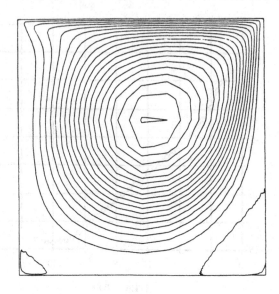

Figure 6.6: Incompressible viscous flow in a cavity at $Re = 400$, (50×50) grid, streamlines.

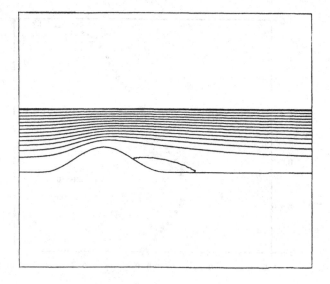

Figure 6.7: Converging-diverging channel $Re = 100$, 1150 elements, streamlines.

at $Re = 100$, calculated by Method 1. This component is used to diffuse the flow downstream of a centrifugal compressor, before entering the combustor. The grid used is shown in Figure 6.8a and the quadratic convergence of Method 1 is demonstrated in Figure 6.8b, while secondary velocities in the middle of the bend are shown in Figure 6.8c.

6.3 Solution of the Fully-Coupled System of Equations, For Three-Dimensional Flows, By Direct and Iterative Methods

The coupled system of equations resulting from two-dimensional formulations, includes the pressure and two velocity components and is amenable to a direct solver on a large class of computers and workstations. For example, two-dimensional flows, with three variables $(\mathbf{u}, p)$ per cell-vertex, discretized on an $(N \times M)$ structured grid would require the solution of 3NM equations with a bandwidth of $3N$. The number of operations for such a direct solution is estimated as $27MN^3$. For the equivalent three-dimensional problem on an $(L \times N \times M)$ grid, the number of equations increases to $4LMN$ and the bandwidth to $4NL$. The number of operations increases to $64MN^3L^3$, or $2.37L^3$ times the two-dimensional solution. The practicality of using coupled methods of solutions is therefore not clear cut for three-dimensional flow situations and this section presents the development of solvers for direct and iterative solution of large systems, on vector and parallel computers.

6.3.1 Gauss Elimination on Vector-Parallel Supercomputers

Storaasli, Nguyen and Agarwal [14], using the computing power of a Cray YMP with 8 processors were able to achieve impressive execution rates for the direct solution of a symmetric set of 54,870 equations for the structural analysis of the Space Shuttle Solid Rocket Booster. Similar ideas are used here, but with important refinements:

- the Navier-Stokes system matrix is non-symmetric, i.e. the solver is developed for a general, variable bandwidth matrix,
- larger sets of equations, with a larger bandwidth, are solved,
- no special language, other than Fortran, is used,
- the parallel-vector strategy is highly optimized.

In the following, the underlying ideas of the vector-parallel Gauss elimination are discussed.

6.3.2 The Vector-Parallel Gauss Elimination

The matrix is stored in a continuous vector containing the entries row-by-row, in a skyline mode. Two pointers need to be defined: one for the start of each row and another for

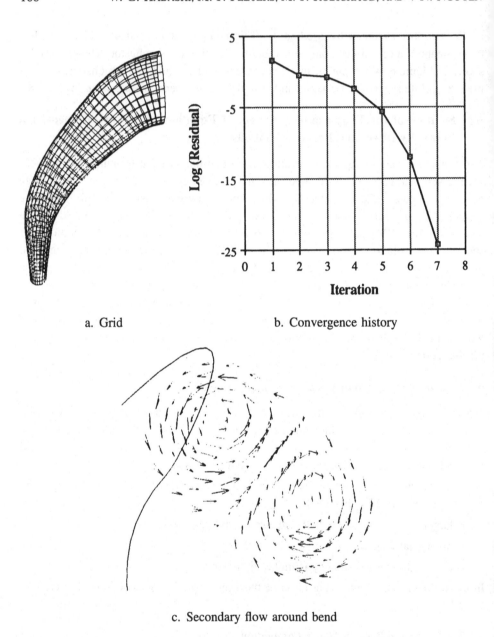

a. Grid   b. Convergence history

c. Secondary flow around bend

Figure 6.8: Incompressible viscous flow in a gas turbine pipe diffuser, $Re = 100$.

the farthest row above it affecting its elimination.

The classical Gauss elimination procedure starts the elimination from the top of the matrix. The elimination-row is first divided by its diagonal, and has multiples of it subtracted from all the following rows, to eliminate the column corresponding to its diagonal. In the approach of Storaasli et al [14], the procedure is inverted, with a row selected to be operated on and all previous rows affecting it being used to eliminate the corresponding columns of that row. This is obviously more amenable to parallel computing since at the row being eliminated synchronization is needed only with a number of preceding rows equal to the number of processors. This is much less than the continuous synchronization that would be required by the classical Gauss elimination. In addition, the fact that many rows are available to be used for elimination of the selected row allows loop unrolling, which will be described later.

The decomposition proceeds by assigning an equation to a processor. The elimination is performed in parallel on the processors, using all previously completed factorized rows. As soon as a processor has completed the factorization of a row, it operates on the next unfactorized one. The vectorization is carried out on the row operations using loop unrolling of various levels. The vector length is controlled by the bandwidth and the stride is 1 since all vector components are contiguous.

6.3.3   Dynamic Assignment of Equations to Processors

Since each processor must be initiated, taking some finite time to come on stream, speed can be gained by dynamically assigning equations to be operated on to the available processors. This is illustrated in Figure 6.9 for a 3-processor case.

Assuming that equation 1 has been assigned to the first processor, it would be divided by its diagonal for the elimination of subsequent rows. In the figure, it is shown that rows 2 and 3 are successively assigned to processor 1, as long as processors 2 and 3 are not fully active. When processor 2 is operational the next equation to be eliminated is assigned to it, in this case equation 4; processor 1 will then host equation 5. When processor 3 is activated it is shown that it hosts equation 8, and so on. In addition, the figure illustrates that in the second access of processor 3 no elimination is required, since no rows affect the current one, and processor 3 immediately operates on the next available equation.

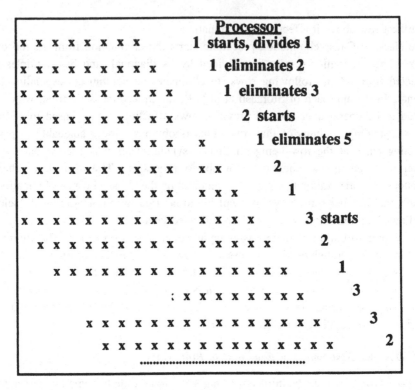

Figure 6.9: Dynamic assignment of equations to processors.

6.3.4 Dynamic Loop Unrolling

Loop unrolling is a technique to minimize the fetching and storing of data to and from memory in a compute intensive application. It consists of explicitly writing out portions of a DO-loop to minimize the number of times data is stored back to memory. As an example of a level-3 loop unrolling, consider the following:

```
      DO 100 I =1, M
      DO 100 J =1, N
      A(J) = A(J) + B(I)*C(J,I)
100   CONTINUE
```

An unrolling of the above DO-loop can be written as:

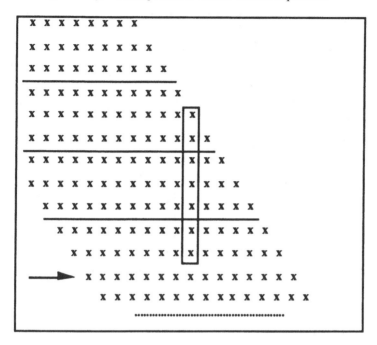

Figure 6.10: Static block unrolling.

```
DO 100 I =1, M, 3
DO 100 J =1, N
A(J) = A(J) + B(I)*C(J,I) + B(I+1)*C(J,I+1) +
       B(I+2)*C(J,I+2)
100    CONTINUE
```

Memory access is costly, and substantial savings are achieved by the unrolled form since A(J) remains in the vector register without being repeatedly stored. For the particular type of application on Cray computers, the optimal level of unrolling is found to be 7. Static and dynamic loop unrolling are illustrated in Figures 6.10–6.11 for a level-3 loop unrolling.

Storaasli et al [14] use level-9 loop unrolling and divide the matrix into blocks of 9 rows each. It can be seen for the level-3 example of Figure 6.10, that when a particular row is being eliminated (shown by an arrow), normally two special blocks, each having less than 3 rows will occur. This means that special loop unrolling statements must be written for both blocks, with lower level of unrolling and hence less efficiency.

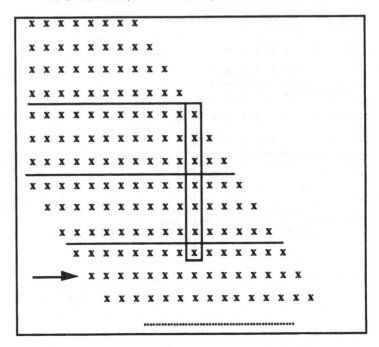

Figure 6.11: Dynamic block unrolling.

By dynamically sizing blocks to start at the first row affecting the elimination, it is clear from Figure 6.11 that only one special short block can occur. Over the large number of operations involved in the matrix decomposition this can translate into a sizeable saving.

6.3.5 Dynamic Elimination

In a static elimination procedure, a block is only processed if all information within that block is ready, i.e. the whole block has already been operated on, as indicated for $(L-1)$ blocks of Figure 6.12. This implies a wait state where a processor is idle if the entire block operations are not yet complete, as shown in the same figure for block (L).

On the other hand, in a dynamic elimination, instead of spending CPU cycles in an idle state, the processor, in this case processor 1, is allowed to start operating on the completed portion of the equation block (L), even if the whole block is not yet complete. While this partial-block operation affects the level of loop unrolling, the penalty is less than allowing a processor to remain idle. At the termination of a partial-block operation, it is not unusual for the previously uncompleted rows to have been operated on and the row being eliminated can then be completely processed.

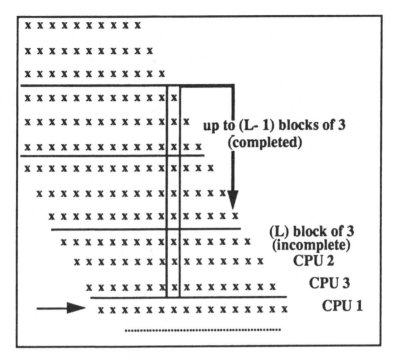

Figure 6.12: Static and dynamic elimination.

6.3.6 Direct Solver: Results

The proposed Gauss elimination has been tested on two types of Cray supercomputers: the Supernet XMP-2/8 in Montreal, with two processors and 8 Megawords of memory and two Cray Research YMP-8 computers, both with 8 processors and having 32 and 128 Megawords of memory, respectively. The equation sets solved are from the discretization of the Navier-Stokes equations for a three-dimensional incompressible flow in the gas turbine pipe diffuser of Figure 6.8.

All results pertain to the application of the Gauss elimination algorithm to the solution of the matrix at a single Newton step. Test cases size information is included in Table 6.1, with the number of equations ranging from about 6,800 to over 100,000 and the bandwidth from 476 to 1,269.

The results in Tables 6.2 to 6.4 illustrate the performance achieved on the three Cray computers. It should be noted, however, that the results of Table 6.3 are based on the first production YMP which has a slower clock cycle than the one used in Table 6.4. All YMP results are obtained in dedicated mode while the XMP results are timeshared.

	# Elems	# Eqns	Max Band	Storage Mwords
Test1	1,785	6,804	476	5.5
Test2	2,208	8,496	492	7.5
Test3	3,168	12,192	688	14.7
Test4	3,762	14,630	964	25.1
Test5	21,936	50,470	1,269	116
Test6	26,078	100,334	620	110
Test7	7,500	24,021	2,258	91.7

Table 6.1: 3-D Navier-Stokes Gauss solver, test case information.

	1-CPU Mflops	2-CPU Mflops	Speed 1 to 2
Test1	170	337	99.1%

Table 6.2: 3-D Navier-Stokes Gauss solver on CRAY XMP-2/8.

On the XMP, speeds of 170 MFlops are attained on a single processor, i.e. of the order of 76% of the peak rate of this machine. Parallelization on two-processors indicate an efficiency of 99%.

For larger problems to be solved in core, and to test parallelization on a larger number of processors, YMP class computers were used. Table 6.3 presents the results obtained on 8-processors with 32 Megawords of memory and using multitasking. The largest test case attempted had around 15,000 equations with a maximum bandwidth of 1,000, requiring 25 MWords of storage. Execution speeds are 30% higher on the YMP than the XMP and results indicate that the speedup from 1 to 2 processors is similarly nearly 99% efficient. With 8 processors the efficiency is of the order of 96%. Results also indicate that the larger the problem, the better the parallel efficiency.

The largest test cases attempted were on a 128 Megawords YMP and are shown in Table 6.4. Autotasking has been used for these cases. Test case 5 has a large number of unknowns and a moderate bandwidth, while test case 6 has half the unknowns but double the bandwidth. The ratio between the two speeds also works out exactly to be proportional to NB^2, where N is the number of equations and B the bandwidth. An even more impressive execution rate of 2.307 Gigaflops has been obtained in test case 7 selected for its large bandwidth, which improves vectorization by increasing the vector length.

	1-CPU Mflops	2-CPU Mflops	4-CPU Mflops	8-CPU Gflops	Speed 1 to 8
Test1	224	443	869	1.577	88%
Test2	227	449	884	1.635	90%
Test3	238	469	933	1.796	94%
Test4	247	490	973	1.899	96%

Table 6.3: 3-D Navier-Stokes Gauss solver on CRAY YMP-8/32.

	8-CPU Gflops	Elapsed secs
Test5	2.276	59
Test6	1.980	31
Test7	2.307	77

Table 6.4: 3-D Navier-Stokes Gauss solver on CRAY YMP-8/128.

It can be concluded that the execution rates demonstrated here, coupled with the rapid convergence of the Newton-Galerkin algorithm, make it currently possible to solve large-scale incompressible Navier-Stokes problems, of about 100,000 unknowns, to machine accuracy, in less than 5 minutes on a Cray YMP-8.

6.3.7 *An Iterative Solver: Preconditioned Conjugate Gradient-like Algorithms*

Direct methods such as Gauss elimination require $O(NEQ^{2.33})$ operations for the factorization step, and $O(NEQ^{1.67})$ operations for the substitution step, while storage requirements are proportional to $O(NEQ^{1.67})$ on an $N \times N \times N$ mesh, where NEQ is the number of equations. Iterative methods for the Newton correction, on the other hand, offer the advantage of $O(NEQ)$ storage and, under certain conditions, preconditioned conjugate-gradient methods can produce a machine accurate solution in $O(NEQ^{1.17})$ operations [15]. These estimates must, however, be tempered by the sensitivity of iterative methods to matrix conditioning and by the difficulty of vectorization for many preconditioning schemes, although some progress is being achieved in adapting iterative methods to the capabilities of new architecture computers [16].

The choice of iterative methods for the systems arising from the linearization of the Navier-Stokes equations is limited by the non-symmetry and non-positive definiteness of the matrix. Classical Conjugate Gradient (CG) methods, highly efficient for symmetric problems, become inapplicable. One must use variants, based either on minimization of the residuals, such as the Generalized Minimum Residual (GMRES) method [17] or on extensions of the bi-conjugate gradient, such as the Conjugate Gradient Squared (CGS) method [18] or CGSTAB [19].

6.3.8 Preconditioning For Non-Symmetric CG-like Methods

Preconditioning is essential to improving the robustness of iterative methods and is achieved here through an incomplete factorization applied to the non-symmetric tangent matrix arising from the Newton linearization. Given a matrix $[K]$, one computes $[S] = [L][U]$, an approximation of the factorization of $[K]$, and transforms the system,

$$[K]\{\Delta x\} = -\{R\} \qquad (6.3.1)$$

into the preconditioned one,

$$[L]^{-1}[K][U]^{-1}\{\Delta z\} = -[L]^{-1}\{R\} \qquad (6.3.2)$$

where

$$\{\Delta z\} = [U]\{\Delta x\} \qquad (6.3.3)$$

This might be labeled a centered preconditioning, different from the left preconditioning,

$$[S]^{-1}[K]\{\Delta x\} = -[S]^{-1}\{R\} \qquad (6.3.4)$$

generally employed in the implementation of the GMRES algorithm. It can be verified that (6.3.2) implicitly sets on the finite dimensional spaces new scalar products associated with $[L]^T[L]$ and $[U]^T[U]$, which are positive definite, while a left preconditioning brings in a scalar product defined by $[S]^{-1}$ which lacks this property when used with iterative methods such as CGS.

6.3.9 Time-Marching and Hybrid Dissipation

Iterative methods, even with the above preconditioning, are not generally robust enough for the ill-conditioned matrices arising from the 3-D steady incompressible Navier-Stokes equations at high Reynolds number. A time-marching procedure has been introduced to improve the conditioning of the matrix $[K]$ in such cases through the addition of a mass matrix, $[M]/\Delta t$, on the diagonal:

$$\begin{bmatrix} [K^u]_u + [M]/\Delta t & [K^p]_u \\ [K^u]_p & [K^p]_p + [M]/(RT\Delta t) \end{bmatrix} \begin{Bmatrix} \Delta u \\ \Delta p \end{Bmatrix} = -\begin{Bmatrix} R_u \\ R_p \end{Bmatrix} \qquad (6.3.5)$$

This technique, although efficient, stops being an effective strategy when it dictates the use of a very small time step to obtain convergence of the iterative solver. In such cases the convergence of the outer nonlinear, or Newton, iteration towards steady-state would be very slow. For such problems, the robustness of Preconditioned Conjugate Gradient (PCG)-like methods can be enhanced by adding streamwise diffusion to both sides of the iterative equation, in the form:

$$[K(\lambda, \mu_{art})] \left\{ \begin{array}{c} \Delta \vec{V} \\ \Delta p \end{array} \right\} = - \left\{ \begin{array}{c} R_{\vec{V}}(\mu_{art}) \\ R_p(\lambda) \end{array} \right\} \qquad (6.3.6)$$

This leads to an algorithm where the iteration matrix $[K]$ is computed with progressively lower values of the parameters λ and μ_{art}, referred to as λ^{LHS}, μ_{art}^{LHS}, but higher than those in the residual, denoted by λ^{RHS}, μ_{art}^{RHS}. The residual is hence computed with the smallest possible values of these parameters for which the outer Newton iteration converges. This hybrid artificial diffusion algorithm can be described as follows:

Hybrid artificial diffusion algorithm

1. Set: $\mu_{art}^{RHS} = \mu_{art}^{LHS}$, $\lambda^{RHS} = \lambda^{LHS}$

2. $\vec{V}$, p being given, compute $\|R_V, R_p\|_0$

   Newton Iteration:

3. Solve $\Delta \vec{V}$ and Δp_i with PCGS at each Newton iteration,

$$[K(\lambda^{LHS}, \mu_{art}^{LHS})] \left\{ \begin{array}{c} \Delta \vec{V} \\ \Delta p \end{array} \right\} = - \left\{ \begin{array}{c} R_{\vec{V}}(\mu_{art}^{RHS}) \\ R_p(\lambda^{RHS}) \end{array} \right\} \qquad (6.3.7)$$

4. Update $\vec{V}$ and p:

$$\left\{ \begin{array}{c} \vec{V}_{i+1} \\ p_{i+1} \end{array} \right\} = \left\{ \begin{array}{c} \vec{V}_i \\ p_i \end{array} \right\} + \left\{ \begin{array}{c} \Delta \vec{V}_i \\ \Delta p_i \end{array} \right\} \qquad (6.3.8)$$

   till $\|R_V, R_p\|_{i+1}/\|R_V, R_p\|_0 < 10^{-m}$, repeat from 3.

5. Lower λ^{RHS}, μ_{art}^{RHS} and repeat from 2, if necessary. ($m = 2$ for the first cycle and $m = 4$ for a second and subsequent cycles)

Normally two such cycles are sufficient.

6.3.10 *Iterative Solvers: Results*

Numerical results are presented for the gas turbine pipe diffuser of Figure 6.8. The solutions are obtained with the Hybrid Artificial diffusion method described above, with $\lambda^{RHS} = \lambda^{LHS} = 0.05$ and $\mu_{art}^{RHS} = \mu_{art}^{LHS} = 0.0$ for the three grids of Table 6.5.

	# Planes	Nodes per plane	# Eqns
Grid 1	25	135	11,521
Grid 2	40	216	30,889
Grid 3	60	368	81,185

Table 6.5: 3-D Navier-Stokes iterative solver, test case information.

Time step size	# Newton iterations	Average CGS # iterations per Newton step	CPU hours (SGI 4D/310)
0.5	97	7.1	1.70
1.0	51	11.9	1.21
5.0	14	26.3	0.65
∞	6	28.3	0.35

Table 6.6: Convergence properties of PCGS as function of time step size.

Three iterative methods, CGS, CGSTAB and GMRES(k), preconditioned with incomplete factorization, have been studied for the effects of time step, Reynolds number and grid size. CGS and CGSTAB are found to have moderate storage requirements (6 vectors) compared to GMRES(k), for which it is necessary to store the k previous search directions, with the value of k generally around 15.

The effect of time step on convergence properties is presented in Table 6.6 for Grid 1, at $Re = 100$, using the PCGS method. It can be seen that increasing the time step from 0.5 to 5.0 reduces the number of outer Newton iterations by a factor of nearly 7, while the number of inner PCGS iterations, per Newton step, increases only by a factor of 4. This shows that the highest time step for which PCGS converges will lead to the smallest computational effort. The table also shows that the steady-state formulation, when possible, requires much fewer Newton steps and leads to quadratic convergence.

Figure 6.13 contrasts the storage requirements of a skyline-mode direct method with those of incomplete decomposition PCG iterative methods. The storage varies as $O(NEQ^{1.03})$ for the iterative methods, compared to $O(NEQ^{1.52})$ for direct solvers. For the largest test case here, the iterative method requires 30 times less memory, a ratio that dramatically increases with grid size, to reach 150 for 1,000,000 equations.

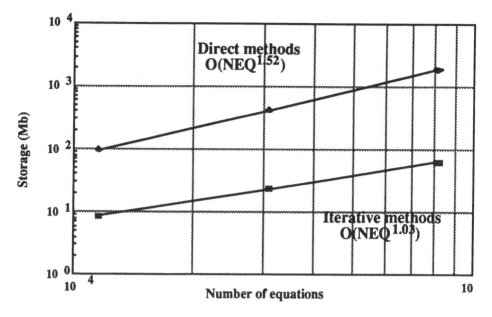

Figure 6.13: Memory requirements of iterative and direct methods.

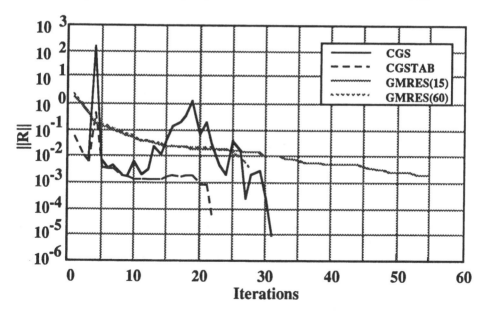

Figure 6.14: Convergence of iterative methods, $Re = 100$.

To assess the convergence properties of the iterative methods, tests were performed for laminar flow, at different Reynolds numbers, using a steady-state formulation and no streamwise diffusion. Solutions were obtained from $Re = 100$ up to $Re = 675$, using a continuation scheme. Convergence could not be achieved, without diffusion, for a steady-state formulation, at higher Reynolds numbers. GMRES(k) was used with 15 descent directions, a commonly used value, and with 60 descent directions, a very large number aimed at testing GMRES behavior with little or no restart. Convergence history for a given Newton step is presented for $Re = 100$ in Figure 6.14 and for $Re = 675$ in Figure 6.15.

It can be seen that the GMRES convergence is monotonic, while the CGS and CGSTAB convergence is faster but non-monotonic. This behavior is amplified at the higher Reynolds number, but with CGSTAB a little more stable than CGS.

Figure 6.16 shows the effect of Reynolds number on the average number of inner iterations per Newton step. It clearly shows that CGS and GMRES, with a large number of descent directions (60), are the more robust of the methods. It must be emphasized, however, that there is a substantial difference in the memory requirements of these two methods that is function of the size of the preconditioner. In this case, GMRES(60) requires 25% more memory.

Figures 6.17 and 6.18 show the effect of grid size on the computational requirements of iterative methods for $Re = 100$ and 675. It can be seen that computational time varies as $O(NEQ^{1.3})$ for all methods at $Re = 100$, and as $O(NEQ^{1.4})$ for CGS and GMRES(60) at $Re = 675$. This compares favorably with $O(NEQ^{1.17})$ for the CG method on elliptic problems.

Although supercomputers are much faster than workstations, the overall solution times of iterative methods on workstations have been shown to be quite comparable to direct methods on supercomputers. Considering that the requirements of direct methods increase much more rapidly puts such interactive methods in a very strong position for the solution of large-scale Navier-Stokes problems.

6.4 Conclusion

This chapter presents a pressure dissipation scheme for solving the incompressible Navier-Stokes equations, with little, if no artificial viscosity. The method is simple and, more importantly, is more in line with the general spirit of finite element techniques. The method is justified both in a heuristic way and formally as a Helmholtz decomposition. The method allows equal interpolation of all variables and removes cumbersome restrictions on the choice of elements.

A fast converging Newton linearization is demonstrated. The solution of the resulting

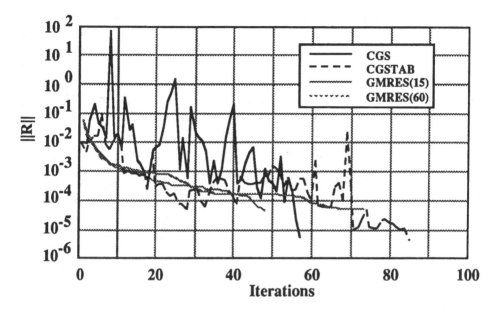

Figure 6.15: Convergence of iterative methods, $Re = 675$.

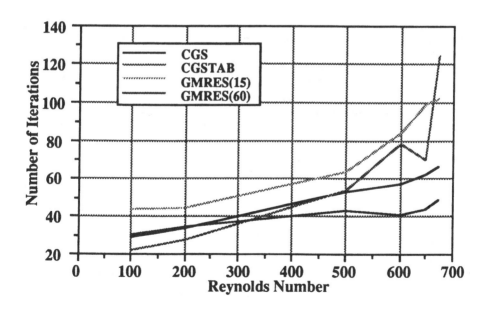

Figure 6.16: Number of inner iterations versus Reynolds number.

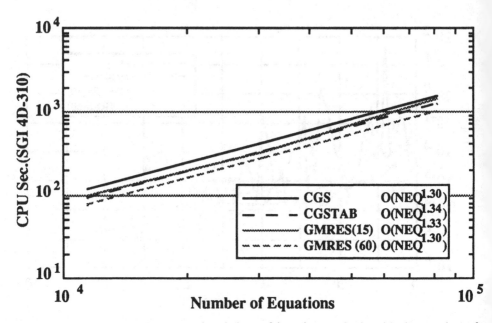

Figure 6.17: Variation of computational time of iterative methods with the number of equations, at $Re = 100$.

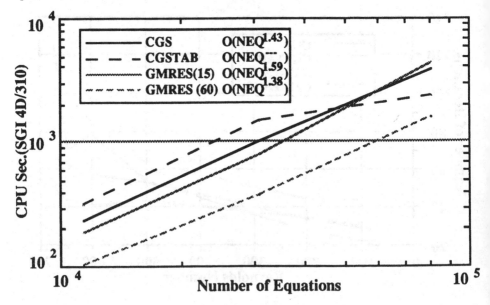

Figure 6.18: Variation of computational time of iterative methods with the number of equations, at $Re = 675$.

fully-coupled set of linear equations removes time step size restrictions of pseudo-time marching techniques. The availability of supercomputers like the Cray YMP makes the use of large-scale direct solvers for the solution of CFD problems possible and it has been clearly shown that Gaussian elimination procedures vectorize and parallelize well on such computers. Execution rates of over 2.3 Gigaflops can be attained and, to overcome memory limitations, out-of-core versions of the method can be developed to handle larger CFD problems.

The application of preconditioned conjugate gradient-like iterative methods to the solution of the equations has also been successfully demonstrated. The incomplete decomposition reduces the storage requirements dramatically. The introduction of time dependent terms in the equations improves the matrix conditioning. When this is coupled with a hybrid artificial diffusion method, i.e. higher in the iteration matrix than in the physical matrix, a robust scheme can be obtained, that is competitive with direct solvers.

It is concluded that the continuing evolution of hardware, at the supercomputer and workstation levels, will make it possible to increasingly use the clearly advantageous properties of fully-coupled solution strategies for fluid dynamics problems.

References

[1] Chorin, A. J., "A Numerical Method for Solving Incompressible Viscous Flow Problems," *Journal of Computational Physics*, Vol. 2, p. 12, 1967.

[2] Harlow, F. H. and Welch, J. E., "Numerical Calculation of Time-Dependent Viscous Incompressible Flow of Fluid with Free Surface," *Physics of Fluids*, Vol. 8, p. 2182, 1965.

[3] Patankar, S. V. and Spalding, D. B., "A Calculation Procedure for Heat, Mass and Momentum Transfer in Three-Dimensional Parabolic Flows," *International Journal of Heat and Mass Transfer*, Vol. 15, pp. 1787–1806, 1972.

[4] Taylor, C. and Hughes, T. G., *Finite Element Programming of the Navier-Stokes Equations*, Pineridge Press, Swansea, 1981.

[5] Babuska, I., "Error Bounds for Finite Element Method," *Numerical Mathematics*, Vol. 16, pp. 322–333, 1971.

[6] Brezzi, F., "On the Existence, Uniqueness and Approximation of Saddle Point Problems Arising from Lagrange Multipliers," *R.A.I.R.O.*, Série Rouge, R2, pp. 123–151, 1974.

[7] Rhie, C. M., "A Pressure Based Navier-Stokes Solution Using the Multi-Grid Method," *AIAA Paper 86-0207*, 1986.

[8] Hafez, M. M. and Soliman, M., "A Velocity Decomposition Method for Viscous Incompressible Flow Calculations: Part I," *Proceedings of the Seventh International Conference on Finite Element Methods in Flow Problems*, University of Alabama in Huntsville Press, pp. 805–809, 1989.

[9] Hafez, M. M. and Soliman, M., "A Velocity Decomposition Method for Viscous Incompressible Flow Calculations: Part II," *Proceedings of the AIAA 9th Computational Fluid Dynamics Conference*, Buffalo, N.Y., pp. 359–369, 1989.

[10] Fasel, H., "Investigation of the Stability of Boundary Layers by a Finite-Difference Model of the Navier-Stokes Equations," *Journal of Fluid Mechanics*, Vol. 78, pp. 355–383, 1976.

[11] Guevremont, G., Habashi, W. G., Hafez, M. M., and Peeters, M. F., "A Velocity-Vorticity Finite Element Formulation of the Compressible Navier-Stokes Equations," *Proceedings of the 4th International Conference on Computational Engineering Science*, Atlanta, Springer-Verlag, pp. 51.x.1–51.x.4, April 1988.

[12] Peeters, M. F., Habashi, W. G., and Dueck, E. G., "Finite Element Stream Function- Vorticity Solutions of the Incompressible Navier-Stokes Equations," *International Journal for Numerical Methods in Fluids*, Vol. 7, No. 1, pp. 17–27, January 1987.

[13] Olson, M. D., "Comparison Problem No.1: Recirculating Flow in a Square Cavity," Report No. 22, University of British Columbia, Vancouver, B.C., 1979.

[14] Storaasli, O. O., Nguyen, D. T., and Agarwal, T. K., "Parallel-Vector Solution of Large-Scale Structural Analysis Problems on Supercomputers," *AIAA Journal*, Vol. 28, No. 7, pp. 1211-1216, 1990.

[15] Axelsson, O. A. and Barker, V. A., *Finite Element Solution of Boundary Value Problems*, Academic Press, 1984.

[16] Van der Vorst, H. A., "The Performance of FORTRAN Implementations for Preconditioned Conjugate Gradients on Vector Computers," *Parallel Computing*, Vol. 3, pp. 49–58, 1986.

[17] Saad, Y. and Schultz, M. H., "GMRES: A Generalized Minimum Residual Algorithm for Solving Nonsymmetric Linear Systems," *SIAM Journal of Scientific and Statistical Computing*, Vol. 17, pp. 856–869, 1986.

[18] Sonneveld, P., "CGS: A Fast Lanczos-type Solver for Nonsymmetric Linear Systems," *SIAM Journal of Scientific and Statistical Computing*, Vol. 20, pp. 36–52, 1989.

[19] Van der Vorst, H. A. and Sonneveld, P., "CGSTAB: A More Smoothly Converging Variant of CGS," Technical Report 90-50, Delft University of Technology, Delft, May 1990.

7 Numerical Solution of the Incompressible Navier-Stokes Equations in Primitive Variables on Unstaggered Grids

M. Hafez and M. Soliman

Abstract

A numerical algorithm for enforcing the conservation of mass in incompressible flow simulation is discussed and the details of the implementations in terms of standard finite volumes and finite elements are given. It is also demonstrated that both segregated and coupled iterative techniques are applicable and numerical results of test cases for two and three dimensional cavity flows are presented.

7.1 Introduction

In many applications, numerical simulations of three dimensional incompressible flows are needed. The problem of satisfying exactly the continuity equation, for these flows, is well known [1],[2].

It is conceivable to update the velocity field using the momentum equations but it is not clear how to update the pressure to conserve mass, since no pressure term appears in the continuity equation. Various methods have been introduced to tackle this problem including the penalty method, the artificial compressibility method, the artificial viscosity methods, the projection and the pressure correction methods. (The discussion, here, is limited to methods based on primitive variables. The vector potential and/or velocity vorticity formulations are not covered, see for example the work of Osswald, Ghia & Ghia [19] which still requires staggered orthogonal grids to enforce mass conservation.)

In the primitive variable methods, either staggered grids are used or the continuity equation is modified. For example in the penalty method of Temam [3], a small term proportional to the pressure is added to the continuity equation, while in Chorin's artificial compressibility method [4], the continuity equation is modified by an artificial time dependent term proportional to the time derivative of the pressure. Other methods are based on adding second or fourth order space derivatives of the pressure [5],[6]. On the other hand, in the pressure correction methods of Harlow and Welch [7] and Patanker and Spalding [8], mass conservation is forced at each iteration via the solution of a Poisson's equation for the pressure, and staggered grids are used to implement these methods, since on a regular grid an odd-even decoupling of the pressure field is possible. To avoid such a difficulty, Chorin [9] in his projection method, constructed special finite difference operators for the boundary nodes in the case of simple geometries (see also Abdallah [10]).

Recently, there have been many attempts [11]–[15] to extend the above methods to complex geometries via body fitted curvilinear coordinate systems usually with staggered arrangements of velocities and pressure in a number of ways. The governing equations are solved either in terms of Cartesian, covariant or contravariant velocity components. Again, if collocated schemes are employed, the continuity equation is modified [16].

Alternative formulations based on finite volumes, which are applicable on unstructured grids, have been introduced and applied by many authors (see for example Dwyer et al [17]). A novel technique in terms of complementary control volumes, is examined rigorously by Nicolaides [18].

Currently, at NASA Ames, there is a group of scientists, working with D. Kwak, on the solution of incompressible Navier-Stokes equations. They have produced four major codes to simulate three dimensional flows in generalized coordinates. The INS3D code is developed based on the pseudocompressibility approach for steady problems and approximate factorization. To obtain time-accurate solutions via the pseudocompressibility formulation, it is necessary to satisfy the continuity equation at each time step by subiteration in pseudo time (see also Merkle [20]). In order to use a large time step in the pseudo time iteration, an upwind differencing scheme, based on flux-difference splitting, is used combined with an implicit line relaxation scheme in the INS3D-UP code.

A fast code, based on a finite volume scheme with an LU-SGS implicit algorithm has been also developed under the name INS3D-LU. Finally the INS3D-FS code is a generalized flow solver based on a fractional step method (see also Kim and Moin [21]) and a finite volume approach on a staggered grid. The continuity equation is solved in terms of the contravariant components of the velocity and hence, mass is conserved exactly in a discrete sense.

More details can be found in a recent VKI lecture by Kwak [20], where many impressive results are presented. It is still desirable, however, to solve the equations in the physical space using standard finite volumes and/or finite elements techniques.

In the framework of finite elements, the use of mixed interpolations (i.e., different shape functions for pressure and velocity) is the analogue of the use of staggered grids in finite differences. There is, however, a well developed mathematical theory for Stokes flows, where the shape functions must satisfy the celebrated Babuska-Brezzi inf-sup condition (see references [23],[24],[25]). Few elements can be shown to be stable but, in general, it is not straight forward to check this condition, particularly for three dimensional problems.

In the present effort, such a condition is circumvented and equal order interpolations of velocity and pressure are used by modifying the continuity equation. Similar attempts should be mentioned, see for example references [26]–[32]. Also methods based on least squares formulations [33],[34] belong to this category. Some modifications of the momentum equations, particularly for high Reynolds number flows, are proposed as well.

Following this introduction, the problem formulations are described together with various discretization procedures and solution techniques. Finally numerical results and some concluding remarks are discussed.

An Appendix entitled "The Incompressible Limit of Compressible Flows" is also included where a unified approach of compressible and incompressible flows is presented. Based on this approach, a code has been developed to simulate compressible as well as low Mach number flows including the special case of $M = 0.0$.

7.2 Problem Formulation

The conservation of mass and the incompressible Navier-Stokes equations are given by:

$$\nabla \cdot \bar{q} = 0 \tag{7.2.1}$$

$$\bar{q}_t + \nabla \cdot (\bar{q}\,\bar{q}) = -\nabla p + \frac{1}{R_e} \nabla^2 \bar{q} \tag{7.2.2}$$

where R_e is the Reynolds number. Other forms [25] of the nonlinear convection terms are also used. For example the divergence form can be replaced by $(\bar{q} \cdot \nabla)\bar{q}$ (since $\nabla \cdot \bar{q} = 0$). An average of the two expressions yields a skew-symmetric form

$$\frac{1}{2} \left(\nabla \cdot (\bar{q}\,\bar{q}) + (\bar{q} \cdot \nabla)\bar{q} \right) = (\bar{q} \cdot \nabla)\bar{q} + \frac{1}{2}(\nabla \cdot \bar{q})\bar{q} \tag{7.2.3}$$

which conserves kinetic energy when discretized.

Similarly, the viscous terms can be written in other forms. Using vector identities,

$$\nabla^2 \bar{q} = \nabla(\nabla \cdot \bar{q}) - \nabla \times \nabla \times \bar{q}. \tag{7.2.4}$$

The first term in the right-hand side of equation (7.2.4) vanishes and the second term can be rewritten in terms of the vorticity vector.

Dirichlet boundary conditions, $\bar{q} = 0$, are imposed at solid surfaces. Also, the level of the pressure can be adjusted with an arbitrary constant.

Various methods have been developed to solve the above problem. In the penalty method, the continuity equation is modified to read

$$\varepsilon p + \nabla \cdot \bar{q} = 0 \qquad \varepsilon > 0, \varepsilon \to 0. \tag{7.2.5}$$

Upon eliminating the pressure terms from the Navier-Stokes equations one obtains

$$\bar{q}_t + (\bar{q} \cdot \nabla)\bar{q} - \frac{\nabla(\nabla \cdot q)}{\varepsilon} - \frac{1}{R_e}\nabla^2 \bar{q} = -\frac{1}{2}(\nabla \cdot \bar{q})\bar{q}. \qquad (7.2.6)$$

The term on the right-hand side of equation (7.2.6) has been introduced for stability consideration. The quality of the solution depends on the choice of ε, since there is a mass source in the continuity equation proportional to εp.

Chorin's artificial compressibility method improves the situation for steady flow calculations by balancing this term using values of the pressure at the previous iteration level in an artificial time process, hence the continuity equation becomes

$$\varepsilon p^n + \nabla \cdot \bar{q} = \varepsilon p^{n-1}. \qquad (7.2.7)$$

Now, at convergence, mass is conserved.

For truly unsteady flows, equation (7.2.7) can be modified as follows

$$\varepsilon \left(\frac{\partial p}{\partial t}\right)^n + \nabla \cdot \bar{q} = \varepsilon \left(\frac{\partial p}{\partial t}\right)^{n-1} \qquad (7.2.8)$$

hence subiterations are needed at each time step.

Other methods to balance the pressure term in equation (7.2.5) rely on the neighboring values leading to space derivatives terms which are dissipative. For example, artificial viscosity in terms of second order space derivatives of the pressure can be used to modify the continuity equation, i.e.,

$$\nabla \cdot \bar{q} = \varepsilon \nabla^2 p. \qquad (7.2.9)$$

Fourth order dissipation terms can be viewed as balancing the Laplacian of the pressure term in the right-hand side of equation (7.2.9).

In the present method the Laplacian of the pressure in equation (7.2.9) is balanced using the divergence of the momentum equations.

From equation (7.2.2), one obtains

$$\nabla^2 p = -\nabla \cdot \nabla \cdot (\bar{q}\,\bar{q}) = \nabla \cdot \bar{g} \qquad (7.2.10)$$

where $g = -\nabla \cdot (\bar{q}\,\bar{q})$, hence equation (7.2.9) reads

$$\nabla \cdot \bar{q} = \varepsilon(\nabla^2 p - \nabla \cdot \bar{g}). \qquad (7.2.11)$$

Notice on the discrete level the right-hand side of equation (7.2.11) does not necessarily vanish.

Unlike the standard pressure correction methods, here the velocity and the pressure variables are not staggered and equation (7.2.11) is solved in a coupled manner with equation (7.2.2) and also ε is not related to the time step Δt. In fact for steady problems, the time dependent term $\bar{q}_t$ can be dropped, the momentum and the modified continuity equations are linearized, and the resulting equations are solved via a direct Gaussian elimination procedure for banded matrices.

For unsteady problems, no artificial time dependent terms for the pressure is needed if the modified equation (7.2.11) is solved simultaneously with equation (7.2.2) using a direct solver at each time step. Application of direct solvers requires, however, large memories and it is not feasible for large scale computations, particularly, for three dimensional problems.

At any rate, the above methods can be combined together to give the following modified form of the continuity equation:

$$\varepsilon_1 \left(\frac{\partial p}{\partial t}\right)^n + \nabla \cdot \bar{q} = \varepsilon_1 \left(\frac{\partial p}{\partial t}\right)^{n-1} + \varepsilon_2 (\nabla^2 p - \nabla \cdot \bar{g}). \tag{7.2.12}$$

Equation (7.2.12) and the momentum equation (7.2.2) can be written in the conservation form

$$\overline{W}_t + \nabla \cdot (\overline{F}_1 + \overline{F}_2) = \overline{S} \tag{7.2.13}$$

where $\overline{W}$ is a vector which consists of the pressure and the velocity components and $\nabla \cdot \overline{F}_1$ represents the first order derivative terms while $\nabla \cdot \overline{F}_2$ represents the modifications of the continuity equation as well as the viscous terms.

Standard finite volumes and finite elements techniques are applicable to discretize equation (7.2.13). Moreover, one can solve the resulting algebraic equations in a segregated or a coupled manner. The latter is more efficient in general.

7.3 Modifications of the Momentum Equations

Momentum equations can be also modified to enhance the stability or the accuracy of the calculations. Using the vector identity given by equation (7.2.4), the following expression:

$$(\nabla^2 \bar{q} + \nabla \times \bar{\omega})$$

must vanish for smooth flows. On the discrete level, however this may not be the case, and hence a proposed form of the modified momentum equations reads

$$\bar{q}_t + \nabla \cdot (\bar{q}\,\bar{q}) = -\nabla p + \frac{1}{R_e} \nabla^2 \bar{q} + \varepsilon (\nabla^2 \bar{q} + \nabla \times \bar{\omega}). \tag{7.3.1}$$

The term $\varepsilon \nabla^2 \bar{q}$ looks like artificial viscosity except it is balanced by the $\nabla \times \bar{\omega}$ term. The vector $\bar{\omega}$ can be calculated using the vorticity transport equation. This choice, however, is expensive particularly for three dimensional problems. A proper approximation of $\bar{\omega}$ using its definition ($\bar{\omega} = \nabla \times \bar{q}$) may be sufficient to obtain the modifications proposed in equation (7.3.1). The calculation of $\bar{\omega}$ and the term $\nabla \times \bar{\omega}$ can be lagged. Thus, using a deferred correction procedure, this term appears only in the residuals evaluated from the previous time step or iteration. Further investigation is needed to study this problem.

7.4   Remarks about the Present Method

The present modifications of both the continuity and the momentum equations can be obtained via a partial least squares procedure. For example Poisson's equation for the pressure (7.2.10), can be viewed as the result of the minimization of the integral of the squares of the momentum equations with respect to the pressure, assuming the continuity equation is satisfied. No viscous or time dependent terms appear in equation (7.2.10) because of the mass conservation assumption.

On the other hand, the modification of the momentum equations can be viewed as the result of the minimization (with respect to the velocity) of the integral of the squares of the kinematics equations (the continuity and the vorticity definition) where the vorticity is assumed to satisfy the momentum balance.

In this framework, the parameter ε in equation (7.2.11) or equation (7.3.1) is related to a Lagrange multiplier.

A full least squares procedure, applied to the continuity and the momentum equations coupled together results in a symmetric problem at the expense of increasing the complexity of the formulation.

The present modifications can be also viewed as higher order viscosity terms. Using a centered finite difference scheme on a Cartesian grid, the discrete form of the right-hand side of equation (7.2.11), can be shown to be equivalent to fourth order derivatives of the pressure in the two directions. A similar interpretation holds for the modifications of the momentum equations. In this regard, a finite element discretization is preferred for it's stabilization effects (see the theoretical discussions of Brezzi [29] and Hughes [30]).

An alternative approach is to add directly a biharmonic operator for the pressure and the velocity components to the continuity and the momentum equations respectively. To implement such a modification, the biharmonic operator is written as the Laplacian squared. For example, the continuity equation is modified by $\varepsilon \nabla^2 r$ where $r = \nabla^2 p$.

7.5 Discretization Techniques

To have the capability of modelling complex geometries with unstructured grids, applications of finite volumes or finite element techniques are required.

Using Gauss' theorem, equations (7.2.11) and (7.2.2) are derived from the integral formulation

$$\iint \bar{q} \cdot \overline{dA} = \varepsilon \iint (\nabla p - \bar{g}) \cdot \overline{dA} \qquad (7.5.1)$$

$$\frac{\partial}{\partial t} \iiint \bar{q}\, dV + \iint \bar{q}\bar{q} \cdot \overline{dA} = - \iint p\, \overline{dA} + \iint \tau \cdot \overline{dA} \qquad (7.5.2)$$

where τ is the stress tensor.

For convenience, the discretization process is demonstrated for two dimensional flows, where the domain can be divided into triangles or quadrilaterals usually made up from triangles.

In the first case, a control volume can be constructed around each node by joining the centers of the triangles surrounding it, thus constructing a polygon whose edges bisect the triangle edges. The generalization to three dimensions for tetrahedral leads to polyhedral where the tetrahedral edges are perpendicular to the polyhedral faces. Also the polyhedral edges are perpendicular to the tetrahedral faces and pass through the centers of these faces (see reference [18]).

Assuming linear interpolation in each element for both pressure and velocity components and applying equations (7.5.1) and (7.5.2) yields a discrete system of equations for the unknowns at all nodes. The values of the velocity components at the nodes on the solid surface are set to zero but the pressure is unknown at this nodes. Also, in evaluating the right-hand side of equation (7.5.1), the contour integral coinciding with the boundary is neglected.

The simplest example of control volumes using quadrilaterals is two sets of squares of the same size, where the corners of one of the sets are positioned at the centers of the other and vice-versa. The control volumes for the general case can be constructed by joining the centers of the four elements surrounding each node. The derivatives of the variables at the centers of the elements are calculated in terms of the values of the variables at the four corners assuming linear approximations in both directions. Thus, more accurate representation (compared to the triangular case) of the viscous terms are possible.

For finite element discretizations, the standard Galerkin method with equal interpolations for velocity and pressure is applied to equations (7.2.11) and (7.2.2). For triangles, linear shape functions are used. The four node bilinear elements are more accurate particularly for stretched grids. In both cases, it is clear that the modifications of the continuity equation will not identically vanish. Multiplying the equations by the shape function and integrating by parts the second order terms, yields a system of algebraic equations to be solved for the unknowns at the nodes. Again, the contour integral on the solid surfaces associated with the modification term is neglected.

Finally, for a regular Cartesian grid, it is well known that the above techniques reduce to central differences. For example the Laplacian on triangles is presented by a four point stencil while using bilinear elements gives a nine point scheme.

7.6 Solution Algorithms

One of the most efficient algorithms in CFD is approximate factorization. Unfortunately, it is not applicable for unstructured grids. Fully coupled implicit techniques are used in finite elements. Frontal solvers using out of core memory have been developed and are recommended for two dimensional problems. On the other hand, fully explicit methods are fully vectorizable and have been successfully applied for compressible flows. For incompressible flows, finite volumes or finite elements can be used for space discretization of equations (7.2.12) and (7.2.2) leading to a large system of ordinary differential equations in time, and standard Runge Kutta methods may be applied to solve such systems.

Between the two extremes, fully implicit and fully explicit, lies the block relaxation algorithm, where the nodal equations are divided into blocks. Each block is solved implicitly using a direct solver assuming the unknowns in the other blocks are known as in Gauss Seidel iteration (black and white ordering of the blocks may be adopted).

Also, a segregated approach can be used where the pressure is updated from the modified continuity equation via the solution of a Poisson's equation, then the velocity field is updated from the momentum equation. A fully converged solution of the Poisson's equation is not necessary at each time step but it helps the overall convergence at the expense of extra calculations.

7.7 Numerical Results

A standard test problem is the calculation of the driven cavity flows. For regular grids, the above discretization procedures are greatly simplified. In particular, finite volumes techniques result in a central difference scheme which can be easily implemented in both two and three dimensions. In the following, the modifications of the continuity equation have been tested and some numerical results are presented.

In Figure 7.1, the convergence history of the residuals of the segregated approach for the two dimensional case at $R_e = 100$, is plotted. At each iteration a Poisson's equation is solved for the pressure. The convergence history of the residuals of the coupled

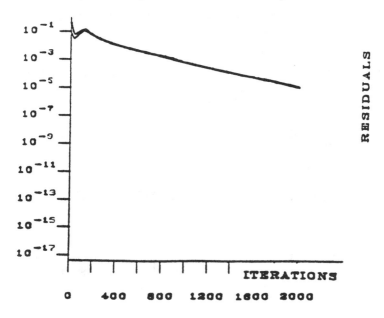

Figure 7.1: Convergence history of the residuals of the segregated approach (2-D).

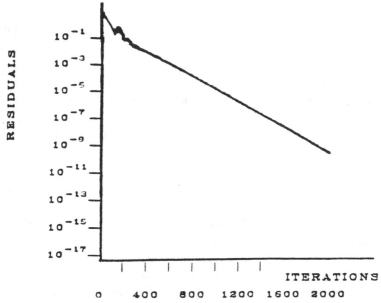

Figure 7.2: Convergence history of the residuals of the coupled approach (block relaxation) (2-D).

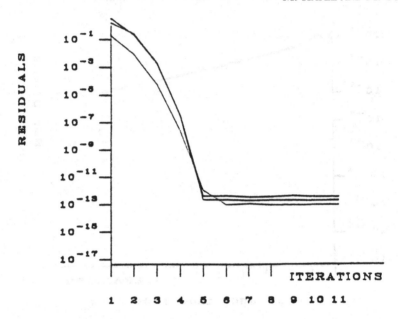

Figure 7.3: Convergence history of the residuals – direct solver (2-D).

approach using block relaxation is shown in Figure 7.2. The pressure and the velocity along a line are solved simultaneously, and no solution of the Poisson's equation by itself is required. It is noticed that overrelaxation of the pressure correction accelerates the convergence. The value of the overrelaxation parameter used in this calculation is 1.7. The convergence history of the residuals using Newton's method and a direct solver at each step is shown in Figure 7.3 for comparison. The convergence is quadratic, however a large memory is required. Currently, an effort [37] is focused on the use of the out of core memory. Preliminary results are very encouraging. Calculations on grids of 256×256 points are feasible.

The vorticity and the pressure contours for $R_e = 100$ and 400 are shown in Figures 7.4 and 7.5. In reference [1], it is suggested to use a smooth function for the tangential velocity component at the top to avoid singularities at the corners. The results of the calculations based on this boundary condition are shown in Figure 7.6.

In Figure 7.7, vorticity contours for $R_e = 30,000$ are plotted based on the converged results obtained via analytical continuation in Reynolds number, using out of core memory and a direct solver for the steady stream function/vorticity equations (see ref. [37]). In this calculation, a second order centered scheme is used with no upwinding or artificial viscosity and with the fine grid, the details of the solution are resolved. (The

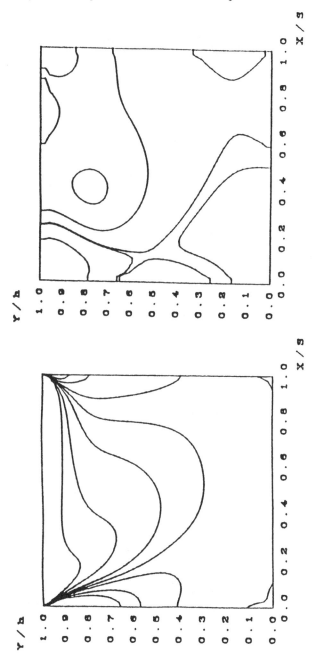

Figure 7.4: Vorticity and pressure contours (71 × 71), $R_e = 100$.

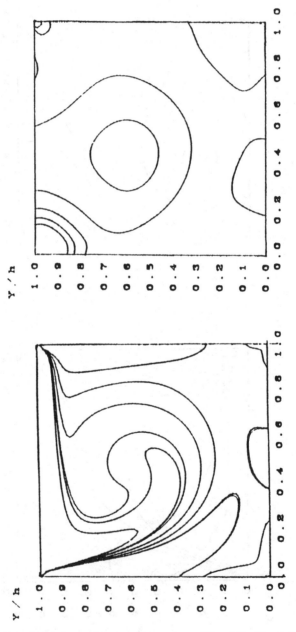

Figure 7.5: Vorticity and pressure contours (71 × 71), $R_e = 400$.

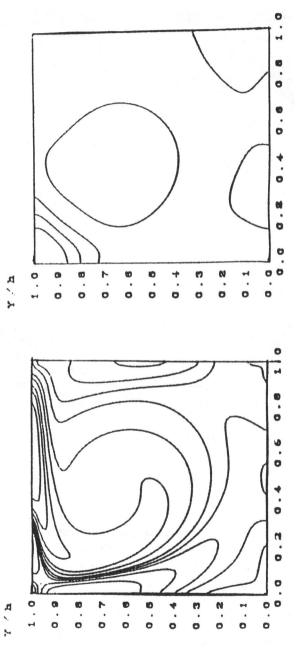

Figure 7.6: Vorticity and pressure contours (71×71), $R_e = 400$ (Peyert B. C.).

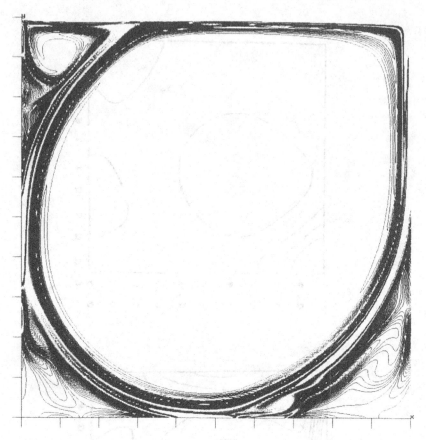

Figure 7.7: Vorticity contours for $R_e = 3 \times 10^4$ (256 × 256).

corresponding calculation based on the present primitive variable method is underway.) Obviously, such a solution is not stable to small perturbations, and can not be obtained by time marching or relaxation techniques.

The results of the three dimensional driven cavity problem are presented in Figures 7.8–7.10. Figure 7.8 shows the convergence history of the residuals of the coupled approach based on block relaxation. The vorticity contours in the midplanes for $R_e = 100$ and 400 are shown in Figures 7.9 and 7.10 respectively.

7.8 Concluding Remarks

Two and three dimensional solutions of the incompressible Navier-Stokes equations in primitive variables on unstaggered grids have been presented. Based on a simple mod-

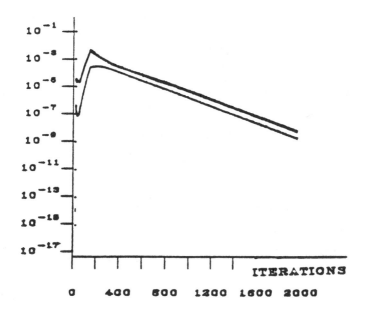

Figure 7.8: Convergence history of the residuals of the coupled approach.

ification of the continuity equation, it is shown that standard finite volumes and finite elements techniques are applicable. Several solution algorithms are discussed and compared for the present test problem. The block relaxation algorithm is promising. The present method is currently applied to a problem of a complex geometry using unstructured grids.

Appendix 7.A: The Incompressible Limit of Compressible Flows

Usually compressible and incompressible flows are simulated by different codes. This is because the performance of most of the compressible flow codes degenerates at low Mach numbers and indeed can not handle the special case of $M = 0.0$. In reference [36], a new formulation was introduced to tackle this problem. First, the density is not used as one of the variables and it is eliminated in terms of the pressure and the temperature via the perfect gas equation of state. This step by itself, will not solve the problem simply because the equation of state is not valid for incompressible (constant density) flows. The formulation is completed using the perturbations of the pressure and temperature relative to reference values.

If the pressure is normalized by $\rho_\infty U_\infty^2$, where ρ_∞ and U_∞^2 are the reference values for the density and velocity, the incompressible limit can not be defined since

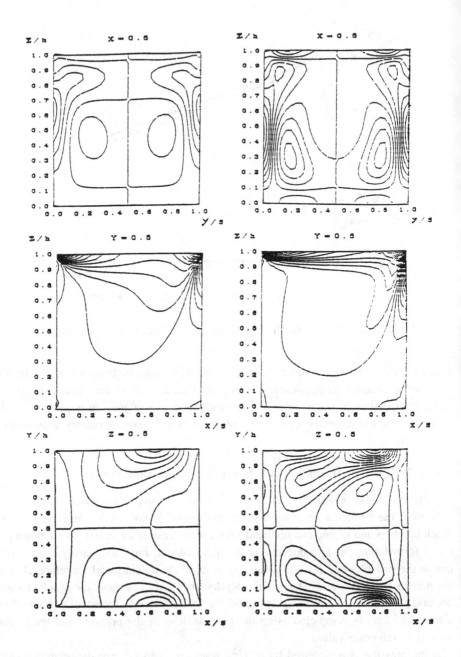

Figure 7.9: Vorticity contours, $R_e = 100$ Figure 7.10: Vorticity contours, $R_e = 400$
$(32 \times 32 \times 32)$. $(32 \times 32 \times 32)$.

$$\tilde{p} = \frac{p}{\rho_\infty U_\infty^2} = \frac{1}{\gamma M_\infty^2} \frac{p}{p_\infty} . \tag{7.A.1}$$

As the Mach number M_∞ approaches zero, $\tilde{p}_\infty$ becomes unbounded. A new variable p^* is introduced, where

$$p^* = \tilde{p} - \tilde{p}_\infty \tag{7.A.2}$$

and $\tilde{p}_\infty = \frac{1}{\gamma M_\infty^2}$.

Similarly, the temperature is usually normalized by U_∞^2/c_p,

$$\tilde{T} = \frac{c_p T}{U_\infty^2} = \frac{1}{(\gamma - 1)M_\infty^2} \frac{T}{T_\infty} . \tag{7.A.3}$$

Again $\tilde{T}$ is not defined as $M_\infty \to 0$.

A new variable T^* is introduced where

$$T^* = \tilde{T} - \tilde{T}_\infty \tag{7.A.4}$$

and $\tilde{T}_\infty = \frac{1}{(\gamma-1)M_\infty^2}$.

The equation of state in terms of the normalized pressure and temperature reads

$$\tilde{p} = \tilde{\rho}\tilde{T}\frac{\gamma}{\gamma - 1} \tag{7.A.5}$$

where $\tilde{\rho} = \rho/\rho_\infty$.

Hence, the density in terms of the new variables p^* and T^* is given by

$$\tilde{\rho} = \frac{\gamma M_\infty^2 p^* + 1}{(\gamma - 1)M_\infty^2 T^* + 1} . \tag{7.A.6}$$

It is clear from (7.A.6) that $\tilde{\rho}$ approaches 1.0 as $M_\infty^2 \to 0$.

A single code was developed, based on the new variables and using staggered grids to simulate compressible as well as incompressible flows. For unstaggered grids, the continuity equation can be modified as discussed in the main text.

The continuity equation for compressible flows is

$$\tilde{\rho}_t + \nabla \cdot (\tilde{\rho}\bar{q}) = 0.0 . \tag{7.A.7}$$

Equation (7.A.7) will be modified to read

$$\tilde{\rho}_t + \nabla \cdot (\tilde{\rho}\bar{q}) = \varepsilon(\nabla^2 p^* - \nabla \cdot \bar{g}) \tag{7.A.8}$$

where $\tilde{\rho}$ is given by (7.A.6) and consequently

$$\tilde{\rho}_t = \frac{\gamma M_\infty^2 \tilde{\rho}}{\gamma M_\infty^2 p^* + 1} p_t^* - \frac{(\gamma - 1)M_\infty^2 \tilde{\rho}}{(\gamma - 1)M_\infty^2 T^* + 1} T_t^* . \tag{7.A.9}$$

The modified mass, momentum and energy equations can be solved in terms of p^*, T^* and $\bar{q}$ and for all speeds on unstaggered grids using standard discretization and iterative techniques. The present unified approach avoids also some problems of simulating compressible flows with low Mach number regions.

In passing, it is noticed that generally in the extensions of the pressure correction methods [35] to simulate compressible flows, an isentropic relation between the pressure and the density corrections is assumed and subiterations are needed to simulate general nonisentropic flows. Such an assumption is not needed in the present method.

References

[1] Peyret, R. & Taylor, T. (1984), "Computational Methods for fluid Flow," Springer Verlag.

[2] Anderson, D., Tannehill, J. & Pletcher, R. (1984), "Comp. Fluid Mech. and Heat Transfer," McGraw Hill.

[3] Temam, R. (1977), "Navier-Stokes Equations," North Holland.

[4] Chorin, A. (1967), "A Numerical Method for Solving Incompressible Viscous Flow Problems," J.C.P., vol. 2, pp. 12–26.

[5] Steger, J. & Kutler, P. (1977), "Implicit Finite Difference Procedure for the Computation of Vortex Wakes," AIAA J., vol. 15, pp. 581–590.

[6] Kwak, D., Chang, C., Shanks, S. & Chakrawarthy, S. (1984), "An Incompressible Navier-Stokes Flow Solver in Three Dimensional Curvilinear Coordinate System using Primitive variables," AIAA-84- 0253.

[7] Harlow, F. and Welsh, J. (1965), "Numerical Calculation of Time-Dependent Viscous Incompressible Flow with Free Surface," Phy. Fluids, vol. 80, pp. 2182–2189.

[8] Patanker, S. & Spalding, D. (1972), "A Calculation Procedure for Heat, Mass and Momentum Transfer in Three-Dimensional Parabolic Flows," Int. J. Heat and Mass Transfer, vol. 15, pp. 1787–18060.

[9] Chorin, A. (1968), "Numerical Solution of Navier-Stokes Equations," Math. Comp., vol. 22, pp. 745–762.

[10] Abdallah, S. (1987), "Numerical Solutions for the Incompressible Navier-Stokes Equations on a Non-staggered Grid," J.C.P., vol. 70, pp. 193–202.

[11] Rhie, C. & Chow, W. (1983), "Numerical Study of the Turbulent Flow Past an Airfoil with Trailing Edge Separation," AIAA J., vol. 21, pp. 1523–1532.

[12] Karki, K. & Patanker, S. (1988), "A Pressure Based Calculation Procedure for Viscous Flows at All Speeds in Arbitrary Configurations," AIAA Paper 88-0058.

[13] Peric, M., Kessler, R. & Scheuerer, G. (1988), "Comparison of Finite Volume Numerical Methods with Staggered and Collocated Grids," Computers and Fluids, vol. 16, pp. 389–403.

[14] Shyy, W., Tonh, S. & Correa, S. (1985), "Numerical Recirculating Flow Calculations Using a Body-fitted Coordinate System," Num. Heat Transfer, vol. 8, pp. 99–113.

[15] Joshi, D. & Vanka, S. (1990), "A Multigrid Calculation Procedure for Internal Flows in Complex Geometries," AIAA Paper 90-0442.

[16] Mansour, M. & Hamed, A. (1990), "Implicit Solution of the Incompressible Navier-Stokes Equations on a Non-staggered Grid," J.C.P. , vol. 86, pp. 147–167.

[17] Dwyer, H., Soliman, M. & Hafez, M. (1986), "Time Accurate Solutions of the Navier-Stokes equations for Reacting Flows," Proceedings of the 10th Int. Conference on Num. Methods in Fluid Dynamics, Beijing, China, Springer Verlag.

[18] Nicolaides, R. (1978), "Flow Discretization by Complementary Volume Techniques," AIAA Paper 89.

[19] Osswald, G., Ghia, K. & Ghia, U. (1987), "A Direct Algorithm for Solution of Three-Dimensional Unsteady Navier-Stokes Equations," AIAA Paper 87-1139.

[20] Merkle, C. & Athavale, M. (1987), "Time Accurate Unsteady Incompressible Flow Algorithms Based on Artificial Compressibility," AIAA Paper 87-1137.

[21] Kim, J. & Moin, P. (1985), "Application of a Fractional Step Method to Incompressible Navier-Stokes Equations," J.C.P, vol 59, pp. 308-323.

[22] Kwak, D. (1989) "Computation of Viscous Incompressible Flows," VKI Lecture Series.

[23] Gunzberger, M. (1989), "Finite Element Methods for Viscous Incompressible Flows," Academic Press.

[24] Pironneau, O. (1989), "Finite Element Methods for Fluids," J.Wiley.

[25] Gresho, P. (1991), "Incompressible Fluid Dynamics: Some Fundamental Formulation Issues," in Annual Review of Fluid Mech.

[26] Comini, G. & del Giudice, S. (1982), "Finite Element Solution of the Incompressible Navier-Stokes Equations," Num. Heat Transfer, vol. 5, pp. 463–478.

[27] Kawahara, M. & Ohmiya, K. (1985), "Finite Element Analysis of Density Flow Using the Velocity Correction Method," Int. J. Numer. Methods Fluids, vol. 5, pp. 981–993.

[28] Pierre, R. (1988), "Simple C0 Approximations for the Computations of Incompressible Flows," Comp. Methods Appl. Mech. Eng., vol. 68, pp. 205–227.

[29] Brezzi, F. & Douglas, J. (1989), "Stabilized mixed methods for the Stokes problem," Numer. Math. vol. 2, pp. 225–235.

[30] Hughes, T. & Franca, L.P. (1987), "A New Finite Element Formulation for Computational Fluid Dynamics: VII. The Stokes Problem with Various Well-Posed Boundary Conditions: Symmetric Formulations that Converge for all Velocity/Pressure Spaces," Comp. Methods in Appl. Mech. and Eng., vol. 65, pp. 85–96.

[31] Zienkiewicz, O., Szmlter, J. & Peraire, J. (1990), "Compressible and Incompressible Flow; An Algorithm for All Seasons," Comp. Methods Appl. Mech. Eng., vol. 78, pp. 105–121.

[32] Peeters, M. & Habashi, W. (1989), "Finite Element Solutions of the Three Dimensional Navier-Stokes Equations," in Finite Element Analysis in Fluids, by Chang, T. and Karr, G. (Eds).

[33] Glowinski, R. (1985), "Numerical Methods for Nonlinear Variational Problems," Springer Verlag.

[34] Jiang, B. & Povinelli, L. (1989), "Least-squares Finite Element Method for Fluid Dynamics," NASA TM 102352-ICOMP-89-23.

[35] Harlow, F. & Amsden, A. (1971), "Numerical Fluid Dynamics Calculation Method for All Flow Speeds," J.C.P., vol. 8, pp. 197–213.

[36] Hafez, M. & Ahmed, J. (1989), "Vortex Breakdown Simulation, Part III: Compressibility Effects," in Fourth Symposium on Numerical and Physical Aspects of Aerodynamics Flows, Long Beach, California.

[37] Reuther,J., McKillop, A. & Hafez, M. (in prep), "Solutions of Incompressible Steady/Unsteady Navier Stokes Equations Using Direct Solver and Out of Core Memory."

8 Spectral Element and Lattice Gas Methods for Incompressible Fluid Dynamics

George Em Karniadakis, Steven A. Orszag,

Einar M. Rønquist, and Anthony T. Patera

8.1 Introduction

The rapid development and introduction of new supercomputer systems over the last decade has opened new opportunities for numerical studies of incompressible fluid flows. A new awareness has also developed that the emerging hardware technologies influence both the nature and implementation of effective algorithms to solve these problems. Specific software implementation issues concern such varied questions as code dependencies, locality, round-off errors, storage access and capacity, input-output, workstation interfaces, cache utilization, ... and are affected significantly by such computer hardware characteristics as the graininess of parallel systems, vector length, instruction conflicts, instruction set design, network access, distribution and paging of memory, to mention but a few. One significant result of this complex environment has been the stimulation of new ideas to make optimal use of the new supercomputer architectures and to achieve both high accuracy and high computational efficiency in the fluid simulations.

In this paper, we shall review some novel methods that are especially well suited for various aspects of incompressible fluid flow simulation studies. The key dynamical feature of these flows is the absence of shock waves, so many of the results to be stated for incompressible flows should also carry over to shock-free flows at moderate Mach numbers. Even within the context of incompressible flows, there is a wealth of dynamical phenomena to be studied, including laminar flows, transition to turbulence, turbulence, free surface flows, heat transfer, particle transport, fluid-structural interactions, and multiphase flows, among other phenomena. In the present paper it is simply not possible to do justice to this broad range of problems; we shall concentrate on recent developments in the numerical methods themselves, leaving to other works the full discussion of applications. We concentrate on two methods here, namely, the spectral element (or p-type finite element) method and the lattice gas (or cellular automaton) method. Most of the paper discusses new algorithmic developments for spectral element methods. These techniques are particularly attractive in that they suggest a natural mapping of the problem onto medium-grained parallel architectures, that other computer implementation issues are

203

handled straight forwardly, and that they give high-order accurate results that are well suited for a broad range of demanding scientific and engineering applications. These methods extend naturally to solve the variety of fluid flow phenomena discussed above.

Numerical methods that solve incompressible flow problems have undergone a natural evolution in the last several decades. Early work was almost exclusively based on finite-difference methods. Then, in the 1960s, finite-element methods came to the fore. Spectral methods underwent significant development through the 1970s and 80s and, today, it remains true that most work on the direct numerical simulation of turbulence and much work on the direct numerical simulation of transition is based on these latter methods. In the 1990s, we can expect another revolution in the way that numerical techniques will be used to solve these flow problems. The revolution is simply that the best features of all the previous methods will be combined into effective, efficient, and accurate flow solvers; in other words, the emerging numerical techniques will be *hybrid methods* which gather the most effective ideas of methods developed over the last two or three decades.

Indeed, the ultimate step in the development of increasingly high-order numerical methods was the maturation of spectral methods which are, at least formally, infinite-order accurate. However, these methods which require the use of series (based on orthogonal polynomial expansions) are global in character so they are quite unsuitable for complex geometry problems. On the other hand, the development of weighted-residual methods led naturally to finite-element numerical techniques which handle complex geometries in a natural, efficient way. It remained until the early 1980s for techniques to be developed that combined the ideas of spectral methods with those of finite element methods and allow the general formulation of so-called spectral element techniques. These latter methods enjoy the geometric flexibility of finite element methods and the accuracy and convergence properties of spectral methods.

In the later half of the 80s, a new class of numerical approaches to solving the Navier-Stokes equations was developed. These new algorithms are based on discrete lattice models of interacting "particles," whose continuum description is governed the continuum fluid flow equations. Perhaps the most interesting of these methods is the cellular automaton model of Frisch, Hasslacher, and Pomeau (hereafter called FHP) [20]. In the FHP model, two dimensional fluid motion is represented using a triangular grid which is sparsely populated by discrete particles that move with unit velocity. In Section 8.6, some details of these methods are described.

This paper is organized as follows. In Section 8.2, recent developments on splitting methods (fractional-step techniques) are discussed. Splitting methods are particularly important in the treatment of boundary conditions on the pressure, outflow boundary conditions, and for the achievement of high-order accuracy in time-integration of the flow equations. In Section 8.3, we review the general formulation of spectral element schemes. Here the method is explored for elliptic equations, then for the linearized Stokes equations, and finally for the Navier-Stokes equations. In this section, the method is explained for so-called conforming spectral elements. In Section 8.4, we survey numerical methods to solve elliptic problems, like those posed by solution of the Poisson equation for the pressure and implicit Helmholtz equations encountered in implicit time-stepping solutions of the unsteady Stokes equations. Both direct and iterative solvers are discussed in some depth in this section and their computational complexity is analyzed. In Section 8.5, we discuss hybrid spectral schemes. In particular, we describe recent work on extension of spectral element methods to non-conforming elements (sometimes called mortar elements or unstructured elements). In addition, in this section, we discuss iterative relaxation methods to allow the solution of problems in "patched" domains to be accomplished. In Section 8.6, lattice gas methods simulating the Navier-Stokes equations are discussed. Here, we concentrate on the so-called lattice Boltzmann approach which involves the description of a fluid in terms of a probability distribution function for lattice particles. Finally, in Section 8.7, some brief conclusions and future trends are indicated.

8.2   High-Order (Time-Accurate) Fractional Step Methods

The choice of temporal discretization for partial differential equations is typically made in two steps. First, the temporal treatment of each individual spatial operator is considered, mainly in order to decide between an explicit or implicit approach. An important factor in this evaluation is the available solver technology for implicit operator treatment, i.e., for inversion of the discrete spatial operator. In addition, accuracy requirements and stability restrictions must be considered. Second, in the case of multiple spatial operators, a global solution strategy must be designed with the overall objective to minimize the computational cost. This second step often leads to the conclusion that an operator splitting approach [58], in which each of the discrete spatial operators is treated separately, is most cost-effective. In the case of the unsteady Navier-Stokes equations, some possibilities are splitting the implicitly treated viscous operator from the implicitly treated pressure/divergence constraint [12], [52],[46], [33], [41], or using splitting to relax the stiffness typically associated with an explicit treatment of the nonlinear convection operator and an implicit treatment of the Stokes operator [18], [49], [5], [26], [33], [41].

Some recent advances in extending the operator splitting methods to high-order temporal accuracy is described in [41] in a general context and to the time-dependent Navier-Stokes equations in particular. A more efficient approach has been suggested recently in [35], [54] where high-order pressure boundary conditions are developed which combined with the construction of mixed explicit/implicit stiffly stable schemes led to arbitrarily high-order time accurate discretizations with no additional overhead. This approach will be the focus of the remainder of this section.

8.2.1 The Classical Splitting Scheme

We consider here Newtonian, incompressible flows with constant properties, which are governed by the Navier-Stokes equations written in the form,

$$\frac{\partial \mathbf{v}}{\partial t} = -\nabla p + \nu \mathbf{L}(\mathbf{v}) + \mathbf{N}(\mathbf{v}) \quad \text{in } \Omega \tag{8.2.1a}$$

subject to the incompressibility constraint

$$Q \equiv \nabla \cdot \mathbf{v} = 0 \quad \text{in } \Omega \tag{8.2.1b}$$

where $\mathbf{v} \, (= u\hat{x} + v\hat{y} + w\hat{z})$ is the velocity vector, p is the static pressure, and ν is the kinematic viscosity. Here $\mathbf{L}$ and $\mathbf{N}$ represent the linear and nonlinear operators respectively and are defined as,

$$\mathbf{L}(\mathbf{v}) \equiv \nabla^2 \mathbf{v} = \nabla(\nabla \cdot \mathbf{v}) - \nabla \times (\nabla \times \mathbf{v}) \tag{8.2.1c}$$

$$\mathbf{N}(\mathbf{v}) \equiv -\frac{1}{2}[\mathbf{v} \cdot \nabla \mathbf{v} + \nabla(\mathbf{v} \cdot \mathbf{v})]. \tag{8.2.1d}$$

The nonlinear terms are written here in a skew-symmetric form following [50] in order to minimize aliasing effects. To proceed, we would like to integrate (8.2.1a) using high-order time-stepping schemes. Such schemes are routinely used for the numerical solution of ordinary differential equations; however their use in the present context leads to two issues: (1) the derivation of mixed explicit-implicit schemes and appropriate treatment of the pressure term, and (2) the investigation of stability and accuracy properties of such mixed schemes.

In this section we set up the framework for the mixed explicit-implicit Adams-Bashforth/Adams-Moulton family for time-stepping discretizations. Later, in Section 8.2.2 we will consider more general multistep families.

Upon integration of (8.2.1a) over one time step Δt we obtain,

$$\mathbf{v}^{n+1} - \mathbf{v}^n = -\int_{t_n}^{t_{n+1}} \nabla p \, dt + \nu \int_{t_n}^{t_{n+1}} \mathbf{L}(\mathbf{v}) dt + \int_{t_n}^{t_{n+1}} \mathbf{N}(\mathbf{v}) dt \tag{8.2.2a}$$

where the superscript index n refers to time level $t_n \equiv n\Delta t$. We then rewrite the pressure term as follows

$$\int_{t_n}^{t_{n+1}} \nabla p \, dt = \Delta t \nabla \bar{p}^{n+1} \tag{8.2.2b}$$

so that $\bar{p}^{n+1}$ is the scalar field that ensures that the final velocity field is incompressible at the end of time level $(n+1)$. It has been common practice (for efficiency reasons) to approximate the nonlinear terms via an explicit scheme, for example, a J_e-order scheme from the Adams-Bashforth family as follows,

$$\int_{t_n}^{t_{n+1}} \mathbf{N}(\mathbf{v})dt = \Delta t \sum_{q=0}^{J_e-1} \beta_q \mathbf{N}(\mathbf{v}^{n-q}) \tag{8.2.2c}$$

where β_q are appropriately chosen weights [22]. The linear terms are approximated via implicit schemes for stability reasons; such an approximation, using for example, a scheme of order J_i from the Adams-Moulton family gives,

$$\int_{t_n}^{t_{n+1}} \mathbf{L}(\mathbf{v})dt = \Delta t \sum_{q=0}^{J_i-1} \gamma_q \mathbf{L}(\mathbf{v}^{n+1-q}) \tag{8.2.2d}$$

where γ_q are appropriately chosen weights for the implicit scheme [22]. The above system of equations (8.2.1) along with the incompressibility constraint form a strongly coupled system.

The solution to this semi-discrete system of equations can be obtained by further splitting equation (8.2.2a) into three substeps as follows,

$$\frac{\hat{\mathbf{v}} - \mathbf{v}^n}{\Delta t} = \sum_{q=0}^{J_e-1} \beta_q \mathbf{N}(\mathbf{v}^{n-q}) \qquad \text{in } \Omega \tag{8.2.3a}$$

$$\frac{\hat{\hat{\mathbf{v}}} - \hat{\mathbf{v}}}{\Delta t} = -\nabla \bar{p}^{n+1} \qquad \text{in } \Omega \tag{8.2.3b}$$

$$\frac{\mathbf{v}^{n+1} - \hat{\hat{\mathbf{v}}}}{\Delta t} = \nu \sum_{q=0}^{J_i-1} \gamma_q \mathbf{L}(\mathbf{v}^{n+1-q}) \qquad \text{in } \Omega \tag{8.2.3c}$$

with Dirichlet boundary conditions $\vec{v}_0$

$$\mathbf{v}^{n+1} = \vec{v}_0 \qquad \text{on } \partial\Omega. \tag{8.2.3d}$$

Here $\hat{\mathbf{v}}$, $\hat{\hat{\mathbf{v}}}$ are intermediate velocity fields defined in (8.2.3b–c). The classical splitting method proceeds by introducing two further assumptions: first that the field $\hat{\hat{\mathbf{v}}}$ satisfies the incompressibility constraint, and thus

$$\nabla \cdot \hat{\hat{\mathbf{v}}} = 0 \qquad \text{in } \Omega; \tag{8.2.4a}$$

and second that the same field $\hat{\mathbf{v}}$ satisfies also the prescribed Dirichlet condition in the direction $\mathbf{n}$ normal to the boundary.

$$\hat{\mathbf{v}} \cdot \mathbf{n} = \vec{\mathbf{v}}_0 \cdot \mathbf{n} . \tag{8.2.4b}$$

Incorporating these assumptions into equation (8.2.3b) we arrive at a separately solvable elliptic equation for the pressure with Neumann boundary conditions in the form,

$$\nabla^2 \bar{p}^{n+1} = \nabla \cdot \left(\frac{\hat{\mathbf{v}}}{\Delta t}\right) \quad \text{in } \Omega \tag{8.2.4c}$$

$$\frac{\partial \bar{p}^{n+1}}{\partial n} = -\frac{\vec{\mathbf{v}}_0 \cdot \mathbf{n} - \hat{\mathbf{v}} \cdot \mathbf{n}}{\Delta t} \quad \text{on } \partial\Omega . \tag{8.2.4d}$$

The final field $\mathbf{v}^{n+1}$ is then obtained by solving the Helmholtz equation (8.2.3c–d) with the field $\hat{\mathbf{v}}$ acting as a forcing term.

The above splitting approach although very efficient in practice, produces solutions that often suffer from large splitting errors which may lead to erroneous results [42]. The reason for that is primarily the imposition of the incorrect boundary condition (8.2.4b), which is inconsistent with the continuous equations (8.2.1a–b). To illustrate this, consider, for example, flow inside a impermeable-wall box. The boundary condition (8.2.4d) reduces to $\frac{\partial \bar{p}^{n+1}}{\partial n} = 0$; the correct boundary condition can be obtained from the semi-discrete equation (8.2.2a)

$$\frac{\partial \bar{p}^{n+1}}{\partial \mathbf{n}} = \mathbf{n} \cdot \left[\sum_{q=0}^{J_e-1} \beta_q \mathbf{N}(\mathbf{v}^{n-q}) + \nu \sum_{q=0}^{J_i-1} \gamma_q \mathbf{L}(\mathbf{v}^{n+1-q})\right] \quad \text{on } \Omega . \tag{8.2.4e}$$

The right hand side of this equation is not zero and is independent of the discretization parameter Δt, so that the errors induced by (8.2.4d) may be of $\mathcal{O}(1)$. On the other hand, imposition of the boundary condition (8.2.4e) involves terms at time level $(n+1)$ and leads to a coupled system.

The exact form of the pressure equation as derived in [46] is,

$$\nabla^2 \bar{p}^{n+1} = \nabla \cdot \left[\left(\frac{\hat{\mathbf{v}}}{\Delta t}\right) + \nu \sum_{q=0}^{J_i-1} \gamma_q \mathbf{L}(\mathbf{v}^{n+1-q})\right] \quad \text{in } \Omega \tag{8.2.4f}$$

which follows from the requirement that an elliptic equation for the divergence be homogeneous (see also Section 8.2.2). In practice, however, equation (8.2.4c) is sufficiently accurate [46].

8.2.2   High-Order Splitting Scheme

High-Order Pressure Boundary Condition   The obvious alternative therefore is to approximate the linear terms on the boundary via an explicit type scheme of order J_e, so that equation (8.2.4e) be replaced by,

$$\frac{\partial \bar{p}^{n+1}}{\partial n} = \mathbf{n} \cdot [\sum_{q=0}^{J_e-1} \beta_q \mathbf{N}(\mathbf{v}^{n-q}) + \nu \sum_{q=0}^{J_e-1} \beta_q \mathbf{L}(\mathbf{v}^{n-q})] \qquad \text{on } \partial\Omega. \qquad (8.2.5a)$$

A similar approach was suggested in [46], where a first order Euler-forward scheme ($J_e = 1$) was employed for the linear terms; it was shown in [46] that such a boundary condition leads to instabilities. However, extending the ideas in [46], a stable scheme can be constructed by rewriting the viscous linear terms in terms of a solenoidal part which we approximate by an explicit scheme and an irrotational part ∇Q which is approximated by an implicit scheme of appropriate order, as follows

$$\frac{\partial \bar{p}^{n+1}}{\partial n} = \mathbf{n} \cdot [\sum_{q=0}^{J_e-1} \beta_q \mathbf{N}(\mathbf{v}^{n-q}) + \nu \sum_{q=0}^{J_i-1} \gamma_q \nabla Q^{n+1-q} + \nu \sum_{q=0}^{J_e-1} \beta_q (-\nabla \times (\nabla \times \mathbf{v})^{n-q})]. \quad (8.2.5b)$$

Note that in this latter equation we drop the term $\gamma_0 \nabla Q^{n+1}$ (since we require that $Q^{n+1} = 0$ in order to honor the incompressibility constraint). To demonstrate how the form (8.2.5b) of the pressure boundary condition prevents propagation and accumulation of time-differencing errors, we consider a first-order scheme ($J_e = J_i = 1$); here for simplicity we drop the non-linear terms, and define $\omega_s \equiv \mathbf{n} \cdot \nabla \times (\nabla \times \mathbf{v})$. The exact equation (8.2.4e) at time level $(n + 1)$ therefore reads

$$\frac{\partial \bar{p}^{n+1}}{\partial n} = \nu (\frac{\partial Q^{n+1}}{\partial n} - \omega_s^{n+1}). \qquad (8.2.6a)$$

If we expand ω^{n+1} around time level n and solve for the normal derivative of the divergence we obtain,

$$\frac{\partial Q^{n+1}}{\partial n} = \frac{1}{\nu} \frac{\partial \bar{p}^{n+1}}{\partial n} + \omega_s^n + \Delta t \frac{\partial \omega_s^n}{\partial t} + \cdots \qquad (8.2.6b)$$

Finally, incorporating the pressure boundary condition (8.2.5b) we obtain,

$$\frac{\partial Q^{n+1}}{\partial n} \propto \Delta t \frac{\partial \omega_s^n}{\partial t} \qquad (8.2.6c)$$

If instead we write the viscous terms as the Laplacian operator on $\mathbf{v}$, we have

$$\frac{\partial Q^{n+1}}{\partial n} = \frac{1}{\nu} \frac{\partial \bar{p}^{n+1}}{\partial n} - \mathbf{n} \cdot \nabla^2 \mathbf{v}^n + \frac{\partial Q^n}{\partial n} + \Delta t \frac{\partial \omega_s^n}{\partial t} + \cdots \qquad (8.2.6d)$$

and due to the pressure boundary condition we obtain,

$$\frac{\partial Q^{n+1}}{\partial n} \propto \frac{\partial Q^n}{\partial n} + \Delta t \frac{\partial \omega_s^n}{\partial t} . \tag{8.2.6e}$$

This latter equation indicates a possible algebraic instability arising from modes corresponding to unit amplification.

A. Pressure Compatibility Condition   Solution of the Poisson equation (8.2.4c) for the pressure $\bar{p}^{n+1}$ with Neumann boundary conditions (8.2.5b) requires that a compatibility condition hold. In particular, the equation for the pressure along with the boundary condition are (here we drop the superscripts),

$$\nabla^2 \bar{p} = \nabla \cdot (\frac{\hat{\mathbf{v}}}{\Delta t}) \tag{8.2.7a}$$

$$\frac{\partial \bar{p}}{\partial n} = \sum_{q=0}^{J_e-1} \beta_q (-\nu \omega_s + \mathbf{n} \cdot \mathbf{N})^{n-q} . \tag{8.2.7b}$$

It is required that,

$$
\begin{aligned}
\int_\Omega \nabla^2 \bar{p}\, dv &= \int_{\partial\Omega} \frac{\partial \bar{p}}{\partial n} dS \\
&= \int_{\partial\Omega} \sum_{q=0}^{J_e-1} \beta_q (-\nu \omega_s + \mathbf{n} \cdot \mathbf{N})^{n-q} dS \\
&= \sum_{q=0}^{J_e-1} \beta_q \int_{\partial\Omega} (-\nu \omega_s + \mathbf{n} \cdot \mathbf{N})^{n-q} dS \\
&= \sum_{q=0}^{J_e-1} \beta_q \int_\Omega \nabla \cdot (-\nu \omega_\mathbf{s} + \mathbf{N})^{n-q} dv \\
&= \sum_{q=0}^{J_e-1} \beta_q \int_\Omega \nabla \cdot (\mathbf{N})^{n-q} dv
\end{aligned}
\tag{8.2.8}
$$

The last simplification is due to the earlier definition of ω_s. The solvability condition therefore requires that,

$$\sum_{q=0}^{J_e-1} \beta_q \int_\Omega \nabla \cdot (\mathbf{N})^{n-q} dx = \int_\Omega \nabla \cdot (\frac{\hat{\mathbf{v}}}{\Delta t}) dx \tag{8.2.9}$$

This equation holds by the definition of $\hat{\mathbf{v}}$ in equation (8.2.3a); here for simplicity we assumed zero-Dirichlet boundary conditions for the velocity $\mathbf{v}^n$.

B. Divergence-Boundary Layer Analysis   An estimate of the error incurred in the velocity field due to divergence errors can be obtained by applying a simplified boundary layer analysis for the divergence $Q \equiv Q^{n+1} = \nabla \cdot \mathbf{v}^{n+1}$. More specifically, taking the divergence of equation (8.2.2a) and using the substitutions (8.2.2b–d) we obtain,

$$\frac{Q}{\Delta t} - \gamma_0 \nu \nabla^2 Q = \frac{Q^n}{\Delta t} + \nabla \cdot (\nu \sum_{q=1}^{J_i-1} \gamma_q \mathbf{L}(\mathbf{v}^{n+1-q})) + \nabla \cdot (\sum_{q=0}^{J_i-1} \beta_q \mathbf{N}(\mathbf{v}^{n-q})) - \nabla^2 \bar{p}^{n+1}.$$

(8.2.10a)

Since the objective is to obtain $Q = 0$ at the time level $(n + 1)$ we set the right hand side to zero to obtain (8.2.4f) and the divergence equation,

$$Q - \gamma_0 \nu \Delta t \nabla^2 Q = 0.$$

(8.2.10b)

It is clear, therefore, that there exists a (numerical) boundary layer of thickness $\ell = \sqrt{\gamma_0 \nu \Delta t}$, so that $Q = Q_w e^{-s/\ell}$, and thus the boundary divergence is $Q_w = -\ell(\frac{\partial Q}{\partial n})_w$. (Here s is a general coordinate normal to the boundary). Similar order of magnitude analysis gives $Q_w = O(\frac{\partial v}{\partial n})$, and thus

$$v \propto Q_w \ell \propto (\frac{\partial Q}{\partial n})_w \gamma_0 \nu \Delta t.$$

(8.2.10c)

This relation demonstrates that the time-differencing error of the velocity field is one order smaller in Δt than the corresponding error in the boundary divergence. This result agrees with the classical splitting scheme of first-order corresponding to the inviscid-type boundary condition (8.2.4b). In particular, the order $O(1)$ errors in $\frac{\partial Q}{\partial n}$ result in first order $O(\Delta t)$ errors in the velocity field. Similarly, a first-order time treatment of the pressure boundary condition should be expected to produce second-order results in the velocity field.

The above argument demonstrates clearly that the time accuracy of the global solution is directly dependent on the boundary values of the divergence, (i.e. $\frac{\partial Q}{\partial n}$), and therefore (see equation (8.2.6a)) on the treatment of pressure boundary conditions.

C. Numerical Results   In this section we employ Adams explicit and implicit schemes up to third-order to implement the formulation described in Section 8.2. The accuracy of the overall scheme can be characterized by the set of parameters (J_e, J_p, J_i) denoting the individual accuracy of the explicit integration (nonlinear terms), the pressure boundary condition, and the implicit integration (viscous terms), respectively.

Example 1: Stokes Channel Flow   As a first test problem we consider a time-dependent Stokes flow problem between parallel plates which has been studied extensively before analytically and numerically [46],[16]. This problem, although linear, embodies all the essential features of the incompressible Navier-Stokes equations and serves as a model to assess the effect of the treatment of the pressure boundary condition on the overall time-accuracy of the scheme. The choice of a compatible initial conditions is very important for some numerical algorithms ([16]) in order to obtain a unique pressure solution; a compatible two-dimensional initial field is given by

$$u = [k \cos \lambda \sinh(ky) + \lambda \cosh k \sin(\lambda y)] \sin(kx) \qquad (8.2.11a)$$

$$v = [-\cos \lambda \cosh(ky) + \cosh k \cos(\lambda y)] \cos(kx) \qquad (8.2.11b)$$

where $\lambda = (-\frac{\sigma}{\nu} - k^2)^{1/2}$, and k is the streamwise (x-direction) wave number. We would like to point out, however, that in the method proposed here the initial inconsistency can be eliminated in one time step with the use of equation (8.2.4f), which is of elliptic character [46]. Here, we consider the case $\nu = 1$ and $k = 1$ which corresponds to streamwise periodicity length $L_x = 2\pi$. All eigenvalues σ for this system are real and negative so the system is stable. We consider the least negative (most unstable) symmetric mode ($\sigma = -9.317739$), which is the dominant mode resolved by the direct numerical simulation.

In the following tests the accuracy of the various schemes is examined by computing a decay-rate $\tilde{\sigma}$ which is defined as (following [16]),

$$\tilde{\sigma} = -\frac{1}{\nu T} \ln \frac{v(y = 0, t + T)}{v(y = 0, t)} \qquad (8.2.12)$$

where the time period T is taken to be $T = 0.3$; with the above parameters the energy of the initial field has been reduced by almost five orders of magnitude after the period T. A very high-resolution spectral element mesh was employed to eliminate any residual spatial discretization errors. We first investigate the effect of the pressure boundary condition for constant time step $\Delta t = 0.01$; in Table 8.1 we summarize the results of several simulations corresponding to different combinations (J_p, J_i). The classical splitting scheme corresponds to $\frac{\partial p}{\partial n} = 0$; $(J_p = 0)$ (first row); in the second and third rows results are presented for $(J_p = 1, 3)$ respectively; in the second and third column the reported results correspond to Euler-backward and Crank-Nicolson integration schemes for the linear terms ($J_i = 1, 2$) respectively. It is seen that although the latter effects the accuracy of $\tilde{\sigma}$, it is actually the pressure treatment that dictates the accuracy with the smallest error (10^{-4}) to occur for the highest-order pressure boundary condition ($J_e = 3$).

To examine in more detail the order of accuracy we carried out simulations for various time steps Δt for two different integration schemes: first with ($J_p = 1, J_i = 2$) corresponding to Euler-forward for the pressure boundary condition and Crank-Nicolson

	Int. order	
$\frac{\partial p}{\partial n}$	EB	CN
$J_p = 0$	-8.78424	-9.185184
$J_p = 1$	-8.916906	-9.327403
$J_p = 3$	-8.905312	-9.314893

Table 8.1: Effect of pressure boundary condition $\Delta t = 10^{-2}$.

θ	0 (CN)	0.1	0.2	0.5	0.5 (EB)	Uzawa
$\tilde{\sigma}$	-9.314897	-9.229633	-9.146342	-8.984164	-8.905312	-8.905312

Table 8.2: θ-scheme accuracy, $\Delta t = 10^{-2}$.

for the viscous terms (EF/CN); and secondly with $(J_p = 3, J_i = 2)$ corresponding to third-order Adams-Bashforth, Crank-Nicolson respectively (AB3/CN). We also obtained, for reference, results of a similar spectral element simulation that is based on the Uzawa (non-splitting) first order scheme [50]. In Figure 8.1 we plot the error in the decay-rate $\tilde{\sigma}$ versus time step Δt for these three schemes. It is seen that indeed the (AB3/CN) scheme obtains second-order accuracy consistent with the aforementioned analysis. It is also verified that the (EF/CN) scheme obtains second order accuracy; this result is also in complete agreement with the analysis reported in [46] where it was shown that a first-order pressure boundary condition of the form employing the tangential velocity (equation (8.2.5b)) can result in second-order overall accuracy; it is also consistent with our previously described order of magnitude analysis (Section 8.2.2 (B)). Finally, the zero order pressure boundary condition leads to identical results as the Uzawa scheme, i.e. first order.

To examine the stability of the above schemes and in particular the effect of the explicit treatment of pressure boundary condition we carried out a number of tests with relatively large $(\Delta t = \mathcal{O}(1))$ time steps. After systematic experimentation we concluded that schemes that incorporate the Euler-backwards integration for the linear terms are stable irrespective of the order of the boundary condition. However, use of the Crank-Nicolson scheme may lead to instabilities for large time-steps; here for example this instability first appears at $\Delta t = 0.1$. Such an instability has also been realized by others [16] and is referred to as short-wave instability. To overcome this we replaced the Crank-Nicolson rule with the θ-scheme that introduces damping and is more stable [11]; indeed the resulted overall scheme eliminates the short-wave instability. However, its accuracy

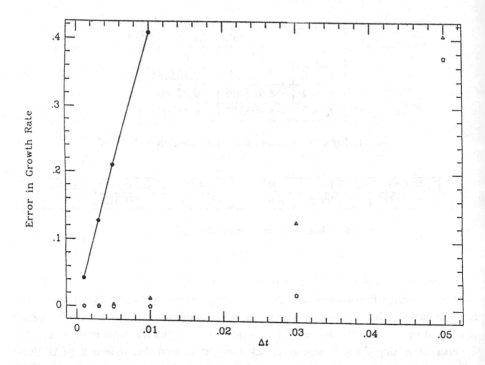

Figure 8.1: Error in decay-rate versus time-step for the channel Stokes flow. Solid line: Uzawa scheme. △: Euler-Forward/Crank-Nicolson scheme. o: Adams-Bashforth 3rd/Crank-Nicolson scheme.

although formally of second-order it degrades relatively fast as $\theta \to 0.5$ (approaching the Euler-backwards limit); this is verified in Table 8.2, where we compare the predicted decay-rate $\tilde{\sigma}$ using various values of the parameter θ.

Example 2: Grooved Channel Flow   To further investigate numerically the stability of the scheme we solved an inflow/outflow Stokes problem in the grooved channel geometry (Figure 8.2). The results from this experiment verified our previous conclusion, i.e. the instability present at relatively large Δt is due to the Crank-Nicolson rule and can be suppressed using a θ-scheme. This is illustrated graphically in Figure 8.3, where the time-history of the streamwise velocity component at a fixed point is plotted for various values of θ; it is seen that a minimum amount of damping is required to eliminate the aforementioned instability. In order to also examine the effect of the new formulation on the incompressibility of the simulated field we compute the streamwise velocity profile

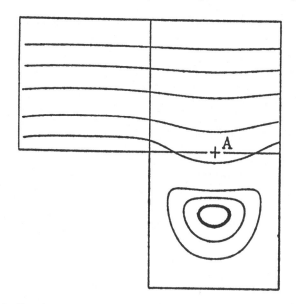

Figure 8.2: Steady-state streamlines of Stokes flow in a grooved channel.

at the outflow using three different formulations: (A) the Uzawa formulation in which the incompressibility constraint is satisfied exactly, (B) the classical splitting scheme with zero-order pressure boundary condition, and (C) the new formulation. After 50 time steps when a steady state is established (starting from zero initial conditions) there is a significant error (15 percent) in the mass exiting the domain for simulation (B), whereas there is only a 0.2 percent error corresponding to formulation (C). An efficient formulation for outflow boundary conditions addressing the problems associated with outflow boundaries in unbounded domains is developed and tested in [55].

The implementation of the above formulation to higher than second-order time-accurate schemes requires the incorporation of a higher-order integration rule for the linear terms. To achieve, for example, third-order accuracy a third-order Adams-Bashforth scheme for the pressure BC should be followed by a third-order Adams-Moulton scheme for the viscous corrections. The latter is only conditionally stable, however, with a relatively small region of stability that diminishes for higher wave numbers; this illustrated graphically in Figure 8.4 where we plot the stability diagram of a mixed Adams-Bashforth/Adams-Moulton scheme. Numerical experimentation for the same range of parameters as in the previous tests verified that indeed a mixed AB3/AM3 scheme is inappropriate for all practical reasons; results are summarized in Table 8.3. In the following section we will introduce a new class of integration schemes that can readily be

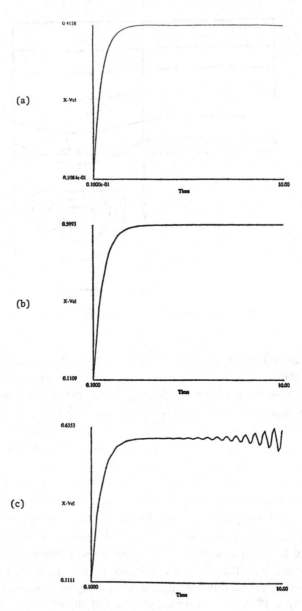

Figure 8.3: Velocity versus time at point (A) (see Figure 8.2 for different integrations schemes: (a) $\theta = 0$; $\Delta t = 10^{-2}$, (b) $\theta = 0.1$; $\Delta t = 10^{-1}$, (c) $\theta = 0.05$; $\Delta t = 10^{-1}$.

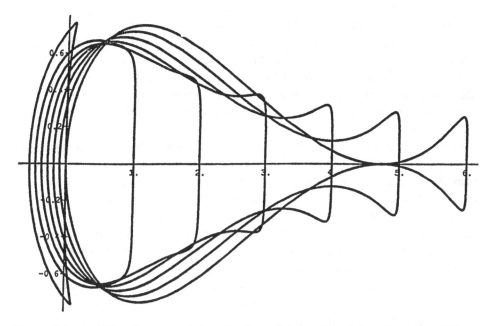

Figure 8.4: Stability diagram of the mixed explicit-implicit Adams-Bashforth/Adam-Moulton of third-order. The curves correspond to different values of the damping coefficient μ ($\mu = 0$ corresponds to the explicit scheme).

Δt	$\partial p/\partial n$								
	$J_p = 0$			$J_p = 1$			$J_p = 2$		
	EB	CN	AM3	EB	CN	AM3	EB	CN	AM3
0.005	0.4087		0.4087				0.40339		0.4073
0.01		0.4234	*		0.4158	*		0.4158	*
0.1					0.5992	*	0.5991	*	*

* = unstable

Table 8.3: Stokes flow in a groove. Values of U at a fixed point.

used to construct arbitrarily high-order stable schemes according to the formulation of Section 8.2.2.

Stiffly Stable Schemes

A. Stability Properties   To extend the formulation described in Section 8.2 to higher-order algorithms we need to include high-order implicit schemes for the integration of the linear term. In order to avoid severe constraints A-stable methods should be used; however, according to Dahlquist (1963) theorem, multistep methods that are A-stable cannot have order greater than two. An alternative approach can be followed by adopting stiffly stable methods commonly used in chemical kinetic studies. According to Gear [22] a method is stiffly stable if it is accurate for all components around the origin in the stability diagram and absolutely stable away from the origin in the left imaginary plane [22].

Stiffly stable methods have not been studied thoroughly especially in the context of solving the Navier-Stokes equations which have both diffusive and convective contributions. As mentioned earlier, efficiency dictates a mixed explicit-implicit discretization for the convective and diffusive terms respectively. The stability properties of such mixed schemes are best analyzed by following an approach similar to the one given by Gear [22] in studying implicit stiffly-stable schemes. Our contribution is in extending that approach to investigate the effect of an implicit viscous term on the stability properties of an explicit scheme. Geometrically the viscous contribution modifies the marginal stability curve in the stability diagram; this curve is defined here as the locus in the complex plane of all points corresponding to amplification factor of one. The model equation for our investigation is given by,

$$\frac{\partial u}{\partial t} = \mu \frac{\partial^2 u}{\partial x^2} + U \frac{\partial u}{\partial x} \qquad (8.2.13a)$$

where U is a constant convective velocity ($U = 1$), and μ is a damping constant. If we consider periodicity conditions we can obtain the modal equation (for mode k and eigenfunction u^*),

$$\frac{\partial u^*}{\partial t} = -\mu k^2 u^* + iku^* ; \qquad (8.2.13b)$$

for $\mu = 0$ we recover the marginal stability curve of an explicit scheme. To construct a family of curves we simply assign different values to the constant μk^2.

Following this approach we analyzed several widely used mixed schemes combining different order schemes from the Adams family. We also developed and analyzed mixed pairs consisting of the stiffly-stable schemes for the explicit part and appropriate explicit companion schemes [53]. For example, in Figure 8.5 we plot the stability diagram of a third-order stiffly-stable mixed scheme: we see that this scheme is similar to schemes of the Adams family, where a small amount of dissipation stabilizes the convection terms; however with the important difference that the stability region of these combined schemes is significantly broader as seen by comparing with Figure 8.4.

B. Splitting Method   Unlike the case considered in Section 8.2.1 where integration over one time step of equation (8.2.1a) resulted naturally in an Adams-Moulton time-stepping

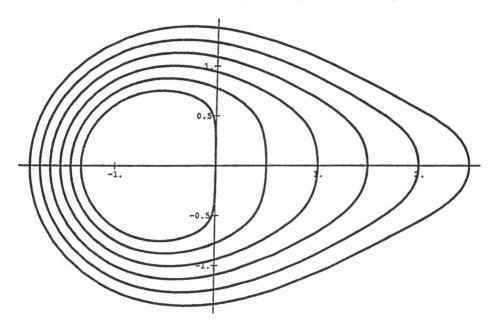

Figure 8.5: Stability diagram of the mixed explicit (second-order) – implicit (third-order) stiffly-stable scheme ($\mu = 0$ corresponds to explicit scheme).

scheme, here we shall consider stiffly-stable type schemes in order to enhance stability. Stiffly-stable methods for ordinary differential equations are based on backwards-differentiation. For a general multistep method we approximate, for example, the time derivative in the equation

$$\frac{\partial u}{\partial t} = f \quad as \quad \frac{1}{\Delta t}\left(\gamma_0 u^{n+1} - \sum_{i=0}^{J-1} \alpha_i u^{n-i}\right) \qquad (8.2.14a)$$

where for consistency we require $\gamma_0 = \sum_{i=0}^{J-1} \alpha_i$. It follows that (8.2.14a) can be rewritten as

$$\frac{1}{\Delta t}\sum_{i=0}^{J-1} \alpha_i (u^{n+1} - u^{n-i}). \qquad (8.2.14b)$$

We also make use of the following exact relations

$$\frac{1}{\Delta t}\sum_{i=0}^{J-1} \alpha_i (u^{n+1} - u^{n-i}) = \frac{1}{\Delta t}\sum_{i=0}^{J-1} \alpha_i \int_{(n-i)\Delta t}^{(n+1)\Delta t} \frac{\partial u}{\partial t} dt \qquad (8.2.14c)$$

Coef.	1st order	2nd order	3rd order
γ_0	1	3/2	11/6
α_0	1	2	3
α_1	0	-1/2	-3/2
α_2	0	0	1/3
β_0	1	2	5/2
β_1	0	-1	-2
β_2	0	0	1/2

Table 8.4: Stiffly-stable schemes coefficients.

$$= \frac{1}{\Delta t} \sum_{i=0}^{J-1} \alpha_i \int_{(n-i)\Delta t}^{(n+1)\Delta t} f \, dt. \qquad (8.2.14d)$$

We can now split the integrand on the right hand side f into several parts as $f = f_1 + f_2 + f_3$ and evaluate each part independently explicitly or implicitly. Following this approach for the Navier-Stokes equations therefore we obtain,

$$\frac{\gamma_0 \mathbf{v}^{n+1} - \sum_{q=0}^{J_i-1} \alpha_q \mathbf{v}^{n-q}}{\Delta t} = -\nabla \bar{p}^{n+1} + \sum_{q=0}^{J_e-1} \beta_q \mathbf{N}(\mathbf{v}^{n-q}) + \nu \mathbf{L}(\mathbf{v}^{n+1}) \qquad (8.2.15)$$

where the coefficients γ_0, α_q are the standard coefficients of (implicit) stiffly stable schemes corresponding to order J_i. The coefficients β_q for the explicit contributions are different than the ones defined in equation (8.2.2c), and can be readily computed by the method of undetermined coefficient and employing Taylor series expansions. In Table 8.4, we summarize the values of coefficients $\gamma_0, \alpha_q, \beta_q$ for schemes up to third-order; the first-order scheme corresponds to Euler-forward/backward integration rule.

To proceed with the splitting method we follow the three-substeps (as in Section 8.2.1) that satisfy equation (8.2.15), i.e.

$$\frac{\hat{\mathbf{v}} - \sum_{q=0}^{J_i-1} \alpha_q \mathbf{v}^{n-q}}{\Delta t} = \sum_{q=0}^{J_e-1} \beta_q \mathbf{N}(\mathbf{v}^{n-q}) \qquad \text{in } \Omega \qquad (8.2.16a)$$

$$\frac{\hat{\hat{\mathbf{v}}} - \hat{\mathbf{v}}}{\Delta t} = -\nabla \bar{p}^{n+1} \qquad \text{in } \Omega \qquad (8.2.16b)$$

$$\frac{\gamma_0 \mathbf{v}^{n+1} - \hat{\hat{\mathbf{v}}}}{\Delta t} = \nu \nabla^2 \mathbf{v}^{n+1} \qquad \text{in } \Omega. \qquad (8.2.16c)$$

The boundary condition for the pressure is again given by equation (8.2.5b), where the coefficients β_q are the modified coefficients given in Table 8.4.

C. Normal Mode Analysis The methods introduced in the previous sections can be analyzed for both stability and accuracy using a normal mode analysis applied to a general geometry domain for a Stokes problem.

Continuous Problem Taking the divergence of equation (8.2.1a) and dropping the non-linear terms we obtain an equation for the pressure given by,

$$\nabla^2 p = 0 \qquad (8.2.17a)$$

while taking the $\nabla \times \nabla \times$ of (8.2.1a) and using the incompressibility constraint (equation 8.2.1b) we obtain,

$$(\frac{\partial}{\partial t} - \nu\nabla^2)\nabla^2\mathbf{v} = 0 . \qquad (8.2.17b)$$

To proceed we assume the existence of normal mode expansions in the form

$$\mathbf{v}(\mathbf{x}, t) = \sum_{i=1}^{\infty} e^{\sigma_i t}\mathbf{v}_i(\mathbf{x}) \qquad (8.2.17c)$$

$$p(\mathbf{x}, t) = \sum_{i=1}^{\infty} e^{\sigma_i t}p_i(\mathbf{x}) \qquad (8.2.17d)$$

where $\mathbf{v}_i(\mathbf{x})$ are the velocity normal modes known to have decay rates σ_i with non-positive real parts, and $p_i(\mathbf{x})$ are the pressure modes. Substituting above in equation (8.2.17b) we obtain the equations that the modes satisfy,

$$\nabla^2 p_i = 0 \qquad (8.2.17e)$$
$$(\frac{\sigma_i}{\nu} - \nabla^2)\nabla^2\mathbf{v}_i = 0 \qquad (8.2.17f)$$

It follows from the above relations that the modes $p_i(\mathbf{x})$ are harmonic and satisfy a maximum principle theorem, and that the $\mathbf{v}_i$ have a harmonic part and an oscillatory part since $Re[\sigma_i] \leq 0$.

Semi-Discrete Problem Using an implicit integration multistep scheme we can write the Stokes equation in a semi-discrete form as follows,

$$\frac{\gamma_0 \mathbf{v}^{n+1} - \sum_{q=0}^{J_i-1} \alpha_q \mathbf{v}^{n-q}}{\Delta t} = -\nabla \bar{p}^{n+1} + \nu \sum_{q=0}^{J_i-1} \beta_q \nabla^2(\mathbf{v}^{n+1-q}) \qquad (8.2.18a)$$

Assuming for the normal modes that $(v_i^n, p_i^n) = \kappa^n(\tilde{v}_i, \tilde{p}_i)$ we obtain,

$$(\frac{\gamma_0 \kappa - P(\kappa)}{\Delta t})\tilde{v}_i = -\kappa \nabla \tilde{p} + \nu R(\kappa)\nabla^2 \tilde{v}_i \qquad (8.2.18b)$$

where we define the linear operators P, R as follows

$$P(\kappa) = \sum_{q=0}^{J_i-1} \alpha_q \kappa^{-q} \quad \text{and} \quad R(\kappa) = \sum_{q=0}^{J_i-1} \beta_q \kappa^{1-q}. \qquad (8.2.18c)$$

Following a similar approach as before for the continuous problem we now obtain,

$$\nabla^2 p_i = 0 \qquad (8.2.18d)$$
$$(\frac{\sigma_i}{\nu} - \nabla^2)\nabla^2 \tilde{v}_i = 0 \qquad (8.2.18e)$$

where $\sigma_i = \frac{\gamma_0 - P(\kappa_i)}{\Delta t R(\kappa_i)}$. This expression is general and is valid for any multistep method; for example for a Crank-Nicolson scheme, $\gamma_0 = 1, P = 1$, and $R = \frac{\kappa+1}{2}$, and for a second-order stiffly stable scheme, $\gamma_0 = 3/2, P = 2 - 1/2\kappa$, and $R = \kappa$.

We find therefore by comparing with the results of the continuous case that the same modes and same σ_i are appropriate for the time discretized problem. However, negative σ_i implies that κ is less than unity for any stable time-stepping scheme.

Splitting Formulation   Uncoupling of the governing equations (8.2.1) can be obtained by introducing a non-divergent intermediate velocity projection v^*. Assuming again a modal decomposition with amplification factor $\tilde{\kappa}$ we get,

$$\tilde{v}^* - P(\tilde{\kappa})\tilde{v} = -\nabla \tilde{p}\tilde{\kappa}\Delta t \qquad (8.2.19a)$$
$$\nabla \cdot v^* = 0 \qquad (8.2.19b)$$
$$\gamma_0 \tilde{\kappa} v - v^* = \nu \Delta t R(\tilde{\kappa})\nabla^2 v. \qquad (8.2.19c)$$

Elimination of v followed by an operation with $\nabla \times \nabla \times$ and using (8.2.19b) gives,

$$[\frac{\gamma_0 \tilde{\kappa} - P(\tilde{\kappa})}{\nu \Delta t R(\tilde{\kappa})} - \nabla^2]v^* = (\frac{\tilde{\sigma}}{\nu} - \nabla^2)\nabla^2 v^* = 0 \qquad (8.2.20)$$

where here we define a decay-rate $\tilde{\sigma}$ equal to the first term in the parenthesis above as before. The final velocity v satisfies a different equation obtained from (8.2.19c) of the form,

$$(\frac{\tilde{\sigma}}{\nu} - \nabla^2)\nabla^2 [\frac{\gamma_0 \tilde{\kappa}}{\nu \Delta t R(\tilde{\kappa})} - \nabla^2]v = 0. \qquad (8.2.21)$$

We see therefore that $\mathbf{v}^*$ satisfies an equation similar to that of the continuous problem and thus has two non-divergent modes (the pressure and the time-stepping mode corresponding to the Laplacian and time-dependent operator in equation (8.2.18d–e) respectively). However, the final velocity $\mathbf{v}$ has an extra mode producing a numerical boundary layer of thickness $\ell \propto (\nu\Delta t)^{1/2}$ (last operator in equation (8.2.21)). In the following, we analyze the effect of this splitting error on the accuracy and stability of the overall scheme.

D. Two-Dimensional Example with One Periodic Direction Here we re-examine the flow problem described in Section 8.2.2.C (example 1) and analyzed also in [46]. For the x-direction being periodic we can write the equation for the modes corresponding to wave number k as follows,

$$\mathbf{v}(x,y) = (\tilde{u}(y), \tilde{v}(y))e^{ikx} \tag{8.2.22}$$

From [46] we get the (symmetric) eigenvalue equation for a non-split formulation as,

$$k \tanh k = -\mu \tan \mu \quad \text{where} \quad \mu^2 = -k^2 - \frac{\sigma}{\nu} \tag{8.2.23}$$

For the current problem the general operators appearing in (8.2.21) can be rewritten as

$$\nabla^2 = D^2 - k^2; \quad \nabla^2 - \frac{\tilde{\sigma}}{\nu} = D^2 + \tilde{\mu}^2; \quad \nabla^2 - \frac{\gamma_0\tilde{\kappa}}{\nu\Delta t\tilde{R}} = D^2 - \lambda^2 \tag{8.2.24a}$$

where we have used above the following definitions,

$$\tilde{\mu}^2 = -k^2 - \frac{\tilde{\sigma}}{\nu} \quad ; \quad \lambda^2 = k^2 + \frac{\gamma_0\tilde{\kappa}}{\nu\Delta t\tilde{R}} \tag{8.2.24b}$$

$$\tilde{P} = P(\tilde{\kappa}) \quad ; \quad \tilde{R} = R(\tilde{\kappa}) \tag{8.2.24c}$$

$$D = \frac{d}{dy} \, . \tag{8.2.24d}$$

The general form of the solution $\tilde{v}, \tilde{v}^*$ that is symmetric about $y = 0$ is

$$\tilde{v}^* = \gamma_0 A^* \cosh ky + \gamma_0 B^* \cos \tilde{\mu}y \tag{8.2.25a}$$

$$\tilde{v} = \tilde{A} \cosh ky + \tilde{B} \cos \tilde{\mu}y + \tilde{C} \cosh \lambda y \, . \tag{8.2.25b}$$

We now consider each of the modes (k, μ, λ) separately; the first two are non-divergent as was mentioned earlier. Considering, for example, first the mode k as

$$\tilde{v}^*{}_k = \gamma_0 A^* \cosh ky; \quad \tilde{v}_k = \tilde{A} \cosh ky \tag{8.2.26a}$$

and substituting in (8.2.19c) we obtain

$$\tilde{A} = \frac{A^\star}{\tilde{\kappa}}. \tag{8.2.26b}$$

The corresponding velocity in the x-direction is obtained from the divergence-free condition

$$\tilde{u}_k = -\frac{D\tilde{v}_k}{ik} = i \sinh ky \frac{A^\star}{\tilde{\kappa}}. \tag{8.2.26c}$$

Similarly, we obtain for the second mode the relations

$$\tilde{B} = \frac{\gamma_0 B^\star}{\tilde{P}} \quad and \quad \tilde{u}_\mu = -\frac{i\tilde{\mu}}{k}\tilde{B}\sin\tilde{\mu}y. \tag{8.2.27}$$

Finally, for the non-divergent third mode for which $v^\star_\lambda = 0$ we obtain again from (8.2.19a)

$$-\tilde{P}\mathbf{v}_\lambda = -\tilde{\kappa}\nabla p \Delta t \tag{8.2.28a}$$

and therefore the field $\mathbf{v}_\lambda$ is irrotational, i.e. $\nabla \times \mathbf{v}_\lambda = 0$; this implies that the following relation holds

$$D\tilde{u}_\lambda = ik\tilde{v}_\lambda. \tag{8.2.28b}$$

To summarize, we can now express the modes $(\tilde{u}, \tilde{v})$ as follows (setting $\tilde{C} = C^\star$),

$$\tilde{v} = \frac{A^\star}{\tilde{\kappa}}\cosh ky + \frac{\gamma_0 B^\star}{\tilde{P}}\cos\tilde{\mu}y + C^\star\cosh\lambda y \tag{8.2.29a}$$

$$\tilde{u} = \frac{iA^\star}{\tilde{\kappa}}\sinh ky - \frac{i\gamma_0\tilde{\mu}B^\star}{k\tilde{P}}\sin\tilde{\mu}y + iC^\star\frac{k}{\lambda}\sinh\lambda y \tag{8.2.29b}$$

At this point we can find $C^\star$ in terms of $A^\star$ by employing the boundary condition at $y = 1$ ($\tilde{u} = \tilde{v} = 0$) i.e.,

$$C^\star = -\frac{A^\star\cosh k(\tilde{\mu}\tan\tilde{\mu} + k\tanh k)}{\tilde{\kappa}\cosh\lambda(\tilde{\mu}\tan\tilde{\mu} + \frac{k^2}{\lambda}\tanh\lambda)} \tag{8.2.30}$$

Using the exact eigenvalue relation (equation (8.2.23)) and the definition equation (8.2.24b) we also obtain

$$\tilde{\mu}\tan\tilde{\mu} - \mu\tan\mu = \mathcal{O}(\frac{\Delta\sigma}{\nu}) \tag{8.2.31}$$

where we define $\Delta\sigma = \tilde{\sigma} - \sigma$. Substituting this last equation in (8.2.30) and normalizing appropriately with $\cosh\lambda$ we see that the amplitude $C^\star$ of the boundary layer error term is proportional to $\frac{\Delta\sigma}{\nu}$, i.e.,

$$C^\star \propto \frac{\Delta\sigma}{\nu} \tag{8.2.32}$$

This last equation suggests that the error in growth rate which characterizes the time-accuracy of the scheme is directly proportional to the amplitude of the divergent mode.

The boundary condition for the pressure can be found from (8.2.19) applied at the boundary. This is an exact equation and has the form,

$$\tilde{\kappa}\nabla p = \nu\Delta t + R(\tilde{\kappa})\nabla^2\mathbf{v} \qquad (8.2.33)$$

where we assumed here $\mathbf{v} = 0$ at the boundary. However, this relation results in a coupled system since $\nabla^2\mathbf{v}^{n+1}$ is not known at the pressure step, so that an explicit treatment should be sought. For example, two first-order relations that can be used are,

$$\tilde{\kappa}\nabla p \;=\; \nu\nabla^2\tilde{\mathbf{v}} \qquad (8.2.34a)$$

$$\tilde{\kappa}\nabla p \;=\; -\nu\nabla\times\nabla\times\tilde{\mathbf{v}} \qquad (8.2.34b)$$

where in the latter equation we also incorporated the incompressibility constraint, i.e. $\nabla\cdot\mathbf{v} = 0$. It follows therefore from (8.2.19a) with $\mathbf{v}(y = 1) = 0$, that

$$v^\star + \nu\Delta t D^2\tilde{v} \;=\; 0 \qquad (8.2.35a)$$

$$v^\star - i\nu k\Delta t Du \;=\; 0. \qquad (8.2.35b)$$

Substitution of $v^\star, \tilde{u}, \tilde{v}$ in (8.2.35a–b) gives

$$A^\star(\gamma_0 + \frac{\nu\Delta t k^2}{\kappa})\cosh k + B^\star(\gamma_0 - \frac{\nu\Delta t\tilde{\mu}^2}{\tilde{P}}\cos\tilde{\mu}) + C^\star\nu\lambda^2\Delta t\cosh\lambda = 0 \quad (8.2.36a)$$

$$A^\star(\gamma_0 + \frac{\nu\Delta t k^2}{\kappa})\cosh k + B^\star(\gamma_0 - \frac{\nu\Delta t\tilde{\mu}^2}{\tilde{P}}\cos\tilde{\mu} + C^\star\nu k^2\Delta t\cosh\lambda = 0 \quad (8.2.36b)$$

coupled with the boundary conditions $\tilde{u} = \tilde{v} = 0$; to satisfy the boundary conditions and after substitution from (8.2.36a) the following determinant vanishes:

$$\begin{vmatrix} \gamma_0\tilde{\kappa} + \nu\Delta t^2 k^2 & \tilde{P} - \tilde{\mu}^2\nu\Delta t & \lambda^2\nu\Delta t \\ 1 & 1 & 1 \\ k\tanh k & -\tilde{\mu}\tan\tilde{\mu} & k^2\frac{\tanh\lambda}{\lambda} \end{vmatrix} = 0 \qquad (8.2.37)$$

The determinant for the case (8.2.36b) has a similar form with the term $\lambda^2\nu\Delta t$ above replaced by the term $k^2\nu\Delta t$.

We solve the above determinental equation for the particular case of a second order stiffly stable-scheme. The eigenvalue $\tilde{\kappa}$ corresponding to (8.2.36a) agrees with the analytical expansion for κ of the non-split scheme up to first order terms, which implies a reduction of the accuracy order of the overall scheme to order one, despite the second order time-stepping scheme employed. However, the eigenvalue due to (8.2.36b) agrees with that of κ to second order as also found by our boundary layer analysis. The expansion for the amplification factor $\tilde{\kappa}$ of the splitting scheme is,

$$
\begin{align}
\tilde{\kappa} &= \kappa + \Delta\kappa \tag{8.2.38a}\\
&= (1 + \sigma\Delta t + \sigma^2\Delta t^2/2 + \sigma^3\Delta t^3/3 + \mathcal{O}(\Delta t^4)) \tag{8.2.38b}\\
&+ (\kappa_1\Delta t^2 + \kappa_2\Delta t^{5/2} + \kappa_3\Delta t^3 + \kappa_4\Delta t^{7/2} + \mathcal{O}(\Delta t^4)) \tag{8.2.38c}
\end{align}
$$

where we find that $\kappa_1 = \kappa_2 = 0$ and that the next two coefficients κ_3, κ_4 are given by

$$
\kappa_3 = \frac{4\mu^2\nu\sigma^2 \sin 2\mu}{\sigma + 3\sin 2\mu} \tag{8.2.39a}
$$

$$
\kappa_4 = \frac{8\sqrt{2}}{3\sqrt{3}} \frac{\mu\nu^{3/2}k^2\sigma^2 \cos^2 \mu}{2\mu + \sin 2\mu}. \tag{8.2.39b}
$$

We also note that $\Delta\sigma$ which determines the boundary layer amplitude is given by

$$
\Delta\sigma = \kappa_3\Delta t^2 + \kappa_4\Delta t^{5/2} + \mathcal{O}(\Delta t^3) \tag{8.2.40}
$$

It is seen therefore that the decay-rate $\tilde{\sigma}$ computed using the stiffly-stable splitting scheme is accurate to second order in Δt if the rotational boundary condition (equation (8.2.5b)) is employed for the pressure. We also see using (8.2.32–8.2.40) that the error in the numerical boundary layer is of second order.

Numerical Results

Example 1: Stokes Channel Flow   In this section we investigate numerically the stability and accuracy properties of the schemes described in Section 8.2.2. As a first test we consider the Stokes flow problem in a smooth channel also examined in Section 8.2.2. Here, we demonstrate that the new schemes are not only stable but they also retain their formal accuracy. To this end, we carry out simulations for several values of the time step Δt for the third-order scheme ($J_i = 3, J_p = 3$), and compute the decay-rate $\tilde{\sigma}$. The set of parameters as well as the initial field remains the same as in Section 8.2.2. The results of these simulations are plotted in Figure 8.6 as a function of Δt^3; the straight line proves that the formal third-order accuracy of the scheme is indeed retained. In Figure 8.7 we also plot the divergence across the channel at a fixed position x for the two types of the pressure BC; we see that the rotational form eliminates almost completely any residual divergence errors. A detailed study on the efficient removal of boundary-divergence errors is given elsewhere [54].

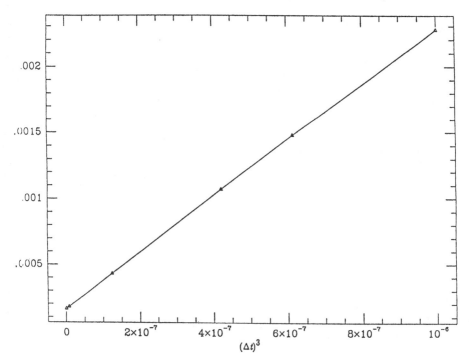

Figure 8.6: Error in decay-rate versus $(\Delta t)^3$ for the channel Stokes flow.

Example 2: Wannier Flow As a second test problem we consider a two-dimensional Stokes flow past a circular cylinder placed next to a moving wall. The available exact solution due to Wannier (1950) [56] for this complex-geometry flow allows for reliable evaluation of the time-differencing error. This problem and its variants have been recently used for code verification purposes [6], [36], [43]. The exact solution and the particular parameters we use in the present test are given by,

$$
\begin{aligned}
u &= -\frac{2(A + Fy)}{K_1}[(s + y) + \frac{K_1}{K_2}(s - y)] \\
&\quad - F \ln(\frac{K_1}{K_2}) - \frac{B}{K_1}[(s + 2y) - \frac{2y(s + y)^2}{K_2}] - D \qquad (8.2.41a) \\
v &= \frac{2x}{K_1 K_2}(A + Fy)(K_2 - K_1) - \frac{2Bxy(s + y)}{K_1^2} - \frac{2Cxy(s - y)}{K_2^2} \qquad (8.2.41b)
\end{aligned}
$$

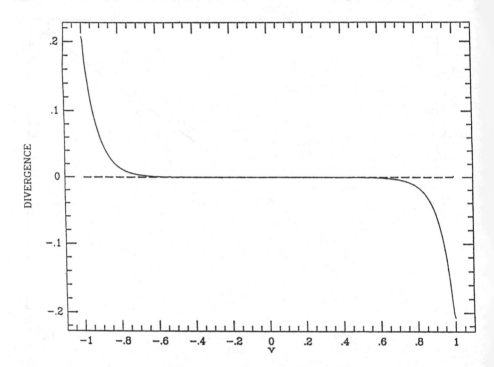

Figure 8.7: Profile of divergence of the velocity field of the channel Stokes flow for two different types of pressure boundary conditions. Solid line: $\frac{\partial p}{\partial n} = \nu \sum_{q=0}^{2} \beta_q \nabla^2 v^{n-q} \cdot \hat{n}$. Dashed line: $\frac{\partial p}{\partial n} = -\nu \sum_{q=0}^{2} \beta_q \nabla \times \omega \cdot \hat{n}$.

where we define:

$$A = -\frac{Ud}{\ln(\Gamma)}; \quad B = \frac{2(d+s)U}{\ln(\Gamma)}; \quad C = \frac{2(d-s)U}{\ln(\Gamma)}; \quad D = -U; \quad F = \frac{U}{\ln(\Gamma)} \tag{8.2.42a}$$

$$K_1 = x^2 + (s+y)^2; \quad K_2 = x^2 + (s-y)^2; \quad s^2 = d^2 - R^2; \quad \Gamma = \frac{d+s}{d-s}. \tag{8.2.42b}$$

Here, we denote by $R = 0.25$ the cylinder radius, by $d = 0.5$ the distance of the cylinder center from the wall, and by $U = 1$ the velocity of the moving wall. Dirichlet BC were used at the boundaries of the truncated domain computed from the exact solution. The computed steady-state solution is shown in Figure 8.8 in the form of streamline patterns, and is indistinguishable from the exact solution. In Figure 8.9 we plot the convergence history to the final steady-state presented as the L_2-error at $\Delta t = 10^{-2}$ versus time for four different schemes: (1) the classical splitting scheme using the inviscid-type

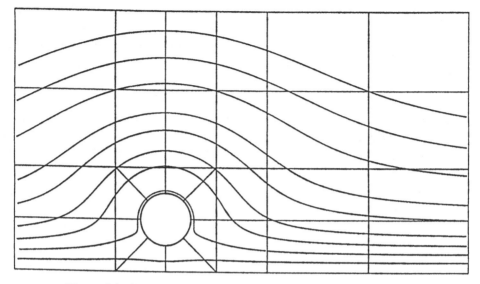

Figure 8.8: Steady-state streamline pattern of the Wannier flow.

boundary condition (8.2.4b); and the new stiffly stable schemes corresponding to (A) first-order; (B) second-order, and (C) third-order, respectively. The superiority of the new pressure boundary condition is reflected in the difference between errors induced by (1) and curve (A), whereas the high-order time-accuracy gain is realized by comparing the curves (A), (B), and (C). Numerical experiments suggest that for parameters ($J_i =$ 2; $J_e = 2$) the time-accuracy is second-order.

Example 3: Kovasznay Flow   Finally, we test the stiffly stable schemes in the context of Navier-Stokes computations. To this end, we consider the laminar flow behind a two-dimensional grid, the exact solution of which was given by Kovasznay (1947) [37]. The solution is given as a function of the Reynolds number R in the form

$$u \;=\; 1 - e^{\lambda x}\cos(2\pi y) \tag{8.2.43a}$$

$$v \;=\; \frac{\lambda}{2\pi}e^{\lambda x}\sin(2\pi y) \tag{8.2.43b}$$

where $\lambda = R^2/2 - (R^2/4 + 4\pi^2)^{1/2}$. The inflow/outflow boundary conditions are also defined by the above relations. The computed steady-state streamline pattern is plotted in Figure 8.10 at $R = 40$. A pair of bound eddies occur just downstream of the inflow, whereas the streamlines become parallel and equidistant at infinity. The computed solution is indistinguishable from the exact solution of Kovasznay [37]. In Figure 8.11

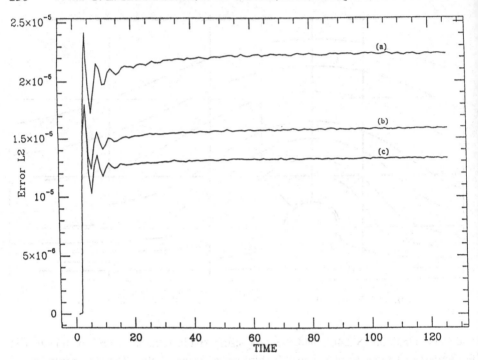

Figure 8.9: Error (L_2) versus time for the Wannier flow. (a): first-order stiffly-stable scheme. (b): second-order stiffly-stable scheme. (c) third-order stiffly-stable scheme. The corresponding error using the classical splitting scheme is three orders of magnitude larger.

we plot the L_2 error of the solution versus time for different orders of integration and compare the stiffly stable schemes with the classical splitting scheme (curve A). We see again that a large error is incurred due to inviscid-type boundary condition (equation 8.2.4b), and that the order of accuracy is increased with the order of integration. Numerical tests suggest that the stiffly-stable schemes are superior as regards *accuracy* to the schemes based on the Adams-family integration rules. As regards *stability*, the new schemes are also more stable than any mixed schemes of the Adams-family in agreement with the stability diagrams corresponding to the linear problem. Numerical experimentation indicates that for the current problem there is at least an order of magnitude gain in stability in comparing the second order schemes, while the stability of the third-order stiffly stable scheme is dictated by the explicit part and an approximate CFL value of one; the corresponding third-order Adams-Bashforth/Adams-Moulton (AB3/AM3) scheme is unstable for all values of time step Δt.

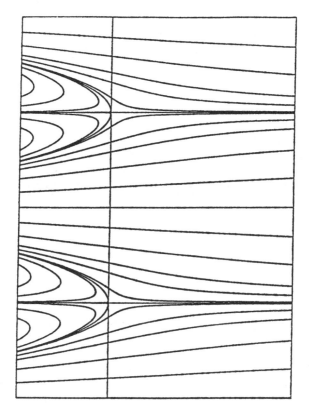

Figure 8.10: Steady-state streamline patterns for the Kovasznay flow at Reynolds number $R = 40$.

8.3 High-Order Spatial Discretization Schemes

Having discussed the semi-discrete formulation and appropriate time-stepping schemes in Section 8.2, we review here high-order spatial discretization schemes based on spectral element concepts.

8.3.1 Elliptic Problems

In this section we consider solution to the linear self-adjoint elliptic Poisson's equation in d space dimensions: Find u defined over $\Omega \in \mathcal{R}^d$ such that

$$-\nabla^2 u = f \quad \text{in } \Omega \tag{8.3.1a}$$

$$u = 0 \quad \text{on } \partial\Omega \tag{8.3.1b}$$

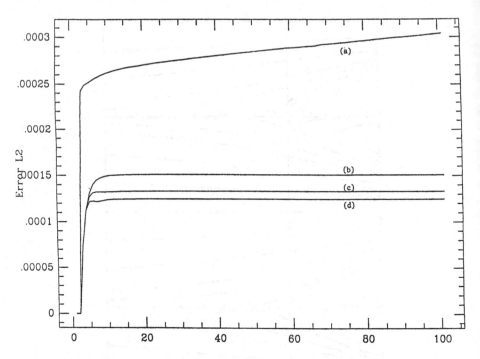

Figure 8.11: Error (L_2) versus time for the Kovasznay flow. (a): classical splitting. (b): first-order mixed stiffly-stable. (c): second-order mixed stiffly-stable. (d): third-order mixed stiffly-stable scheme.

The equivalent variational formulation of (8.3.1) is: Find $u \in H_0^1(\Omega)$ such that

$$(\nabla u, \nabla v) = (f, v) \qquad \forall v \in H_0^1(\Omega), \tag{8.3.2a}$$

where the inner product $(\cdot, \cdot)$ is defined as

$$\forall \phi, \psi \in L^2(\Omega) \qquad (\phi, \psi) = \int_\Omega \phi \psi \, d\Omega, \tag{8.3.2b}$$

Here $L^2(\Omega)$ is the space of all functions which are square integrable over Ω, while $H_0^1(\Omega)$ is the space of all functions which are square integrable, whose derivatives are also square integrable over Ω, and which satisfy homogeneous boundary conditions.

We shall consider here numerical approximations to the Poisson problem based on the variational form (8.3.2a): Find $u_h \in X_h$ such that

$$(\nabla u, \nabla v)_h = (f, v)_h \qquad \forall v \in X_h, \tag{8.3.3}$$

where for each value of the parameter h, X_h is a finite-dimensional space that approaches X as the discretization parameter h goes to zero. In (8.3.3) $(\cdot, \cdot)_h$ denotes evaluation of the continuous inner product $(\cdot, \cdot)$ by numerical quadrature.

Choosing an appropriate discrete space X_h with associated basis, we arrive at a set of algebraic equations given in matrix form as

$$\underline{A}\,\underline{u} = \underline{B}\,\underline{f} \qquad (8.3.4)$$

where $\underline{A}$ is the discrete Laplace operator, $\underline{B}$ is the mass matrix, and underscore refers to basis coefficients. In (8.3.4) we assume that the homogeneous boundary conditions are imposed by eliminating appropriate rows and columns.

Spectral Element Discretization   The general framework outlined above is common to several numerical techniques, e.g., the h-type ([51], [13], [23]), p-type and h-p-type finite element method ([28]), the spectral Galerkin technique ([25], [11]), the spectral element method ([48], [34], [39], [50]), and the mortar element method ([7], [2], [9]). The main differences between the various methods are the choice of approximation space X_h, the choice of basis for X_h, and the type of quadrature for the inner-products $(\cdot, \cdot)_h$.

We shall consider here the *conforming* spectral element method, in which case X_h is a *subspace* of X. The domain Ω is broken up into K disjoint subdomains,

$$\overline{\Omega} = \cup_{k=1}^{K} \overline{\Omega}^k, \qquad (8.3.5)$$

and for reasons of efficiency, the (possibly deformed) subdomains are taken to be quadrilaterals in $\mathcal{R}^2$, and hexahedra (bricks) in $\mathcal{R}^3$. Within each element (or subdomain) the solution is approximated as high-order polynomials in each spatial direction, and the approximation space X_h can be expressed as

$$X_h = H_0^1(\Omega) \cap \mathcal{P}_{N,K}(\Omega) \qquad (8.3.6a)$$

where

$$\mathcal{P}_{n,K}(\Omega) = \left\{ \Phi \in \mathcal{L}^2(\Omega); \Phi_{|\Omega_k} \in \mathcal{P}_n(\Omega_k), k = 1, \ldots, K \right\}, \qquad (8.3.6b)$$

and $\mathcal{P}_n(\Omega_k)$ denotes the space of all polynomials of degree less than or equal to n in each spatial direction. The spatial discretization parameter, h, is thus characterized by two numbers, the number of elements, K, and the polynomial degree within each element, N.

A detailed discussion of the error estimates for the spectral element discretization is given in [39]. We shall here only state the main result in the case of rectilinear elements:

$$\| u - u_h \|_1 \leq C \{ N^{1-\sigma} \| u \|_\sigma + N^{-\rho} \| f \|_\rho \} \qquad (8.3.7)$$

for $u \in H_0^\sigma(\Omega)$ and $f \in H^\rho(\Omega)$, and where C is a constant. In practice, this means that the discrete solution u_h converges spectrally fast to the exact solution u for K fixed, $N \to \infty$, with exponential convergence obtaining for *locally* analytic data and solution. This rapid convergence rate derives from the good approximation properties of the polynomial space X_h, and the accuracy associated with the Gauss quadrature and interpolation. The fact that the spectral element method exploits *all* the regularity of the data and the solution implies that less degrees-of-freedom are required to obtain a fixed accuracy compared to a low-order ($N = 1$) finite element method ([51], [13]). More importantly, if efficient iterative solvers are used, see Section 8.4.2, the computational cost (operation count) Z required to achieve a discrete solution u_h to a prescribed accuracy ϵ is typically less for a high-order method than a low-order method, as long as the error required, ϵ, is sufficiently small [50].

In order to arrive at (8.3.4) we also need to choose an appropriate basis for X_h in (8.3.3). The choice of basis does not effect the error estimates, however, it greatly effects the conditioning and sparsity of the resulting set of algebraic equations, and it is critical for the efficiency of parallel iterative solution procedures. Within each element Ω^k we express an element $w_h \in X_h$ in terms of high-order Lagrangian interpolants through the tensor-product Gauss-Lobatto points. For two-dimensional problems ($d = 2$) we can write

$$w_h(x,y) \mid_{\Omega^k} = \sum_{p=0}^{N} \sum_{q=0}^{N} w_{pq}^k h_p(r) h_q(s)$$
$$\mathbf{x} \in \Omega^k \to (r,s) \in]-1,1[^2, \tag{8.3.8}$$

where r and s are the local coordinates corresponding to translations of x and y, respectively, the $h_p(z)$ are the one-dimensional N-th order Lagrangian interpolants through the Gauss-Lobatto Legendre points ξ_p ($h_p \in \mathcal{P}_N(]-1,1[), h_p(\xi_q) = \delta_{pq}$), and w_{pq}^k is the value of w_h at the local node (ξ_p, ξ_q) in element Ω^k. For a function $w_h \in \mathcal{P}_{N,K}$ the representation (8.3.8) is sufficient; however, a function $w_h \in X_h$ must also honor the C^0 continuity requirement across elemental boundaries and satisfy the homogeneous boundary conditions. The tensor-product form (8.3.8) is essential for obtaining efficient matrix-vector products [45] which are computationally the most expensive part of an iterative solver, see Section 8.4.2.

The inner-products in (8.3.3) are evaluated using Gauss-Lobatto numerical quadrature [15]. Expressing the discrete solution $u_h \in X_h$, the testfunctions $v \in X_h$, and the data $f \in \mathcal{P}_{N,K}$ in terms of the basis (8.3.8), and choosing v to be non-zero at only one global collocation point, we arrive at a set of algebraic equations of the form (8.3.4). Note that for Legendre spectral element discretizations the quadrature points are the same as the collocation points, resulting in a *diagonal* mass matrix $\underline{B}$. This fact will prove useful as regards solution algorithms and time-stepping procedures, (see Sections 8.2 and 8.4).

In the case of deformed geometry, an isoparametric mapping is used, that is, the geometry is also approximated as piecewise high-order (N) polynomials. Although no general error bound is currently available, the numerical evidence to date indicates that spectral accuracy is achieved in relatively deformed elements ([44], [50]). As an example, we consider the Poisson problem (8.3.1) in the domain $\Omega =]0,1[\times]0,1+\frac{1}{4}\sin\pi y[$ with $f = 0$ and $u = \sin x \cdot e^{-y}$. In Figure 8.12 we plot the L^∞-error as a function of the total number of degrees-of-freedom in one spatial direction, $N_t = 2N+1$, demonstrating exponential convergence to the analytic solution [50].

In summary, this section has introduced the key ingredients of the spectral element discretization: variational forms; the piecewise high-order polynomial approximation space X_h, characterized by a discretization pair (K, N); tensor-product spaces, quadratures and bases; and convergence to the exact solution for K fixed and $N \to \infty$.

8.3.2 Steady Stokes Equations

In this section we consider the steady Stokes problem in d space dimensions: Find a velocity $\mathbf{v}$ and a pressure p in a domain $\Omega \in \mathcal{R}^d$ such that

$$-\mu\nabla^2\mathbf{v} + \nabla p = \mathbf{f} \quad \text{in } \Omega, \tag{8.3.9a}$$

$$-\nabla \cdot \mathbf{v} = 0 \quad \text{in } \Omega, \tag{8.3.9b}$$

subject to homogeneous Dirichlet velocity boundary conditions on the domain boundary $\partial\Omega$,

$$\mathbf{v} = \mathbf{0} \quad \text{on } \partial\Omega. \tag{8.3.9c}$$

Here $\mathbf{f}$ is the prescribed force and μ is the viscosity. The solution to the Stokes problem (8.3.9) is of interest not only in its own right, but also in that it constitutes the major building block in many Navier-Stokes solvers.

The equivalent variational formulation of (8.3.9) is: Find $(\mathbf{v}, p)$ in $X \times M$ such that

$$\mu(\nabla\mathbf{v}, \nabla\mathbf{w}) - (p, \nabla \cdot \mathbf{w}) = (\mathbf{f}, \mathbf{w}) \quad \forall\mathbf{w} \in X, \tag{8.3.10a}$$

$$-(\nabla \cdot \mathbf{v}, q) = 0 \quad \forall q \in M, \tag{8.3.10b}$$

where the proper spaces for $\mathbf{v}$ and p such that (8.3.10) is well posed are [10], [23].

$$X = [H_0^1(\Omega)]^d \tag{8.3.11a}$$

$$M = L_0^2(\Omega) = L^2(\Omega) \cap \{\phi \in L^2(\Omega); \int_\Omega \phi \, d\Omega = 0\}. \tag{8.3.11b}$$

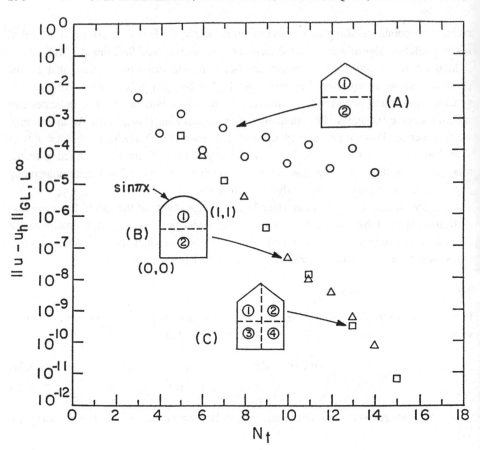

Figure 8.12: Error versus number of degrees of freedom (in the x-direction). Exponential convergence is achieved for appropriate spectral element discretization.

Here $L_0^2(\Omega)$ is the space of all functions which are square integrable over Ω with zero average, while $H_0^1(\Omega)$ is the space of all functions which are square integrable, whose derivatives are also square integrable over Ω, and which satisfy the homogeneous boundary conditions (8.3.9c).

We shall consider here numerical approximations to the Stokes problem based on the variational form (8.3.10): Find $(\mathbf{v}_h, p_h) \in (X_h, M_h)$ such that

$$\mu(\nabla \mathbf{v}_h, \nabla \mathbf{w})_h - ((p_h, \nabla \cdot \mathbf{w}))_h \;=\; (\mathbf{f}, \mathbf{w})_h \qquad \forall \mathbf{w} \in X_h, \qquad (8.3.12a)$$

$$-((\nabla \cdot \mathbf{v}_h, q))_h \;=\; 0 \qquad \forall q \in M_h, \qquad (8.3.12b)$$

where for each value of the parameter h, $X_h \subset X$ and $M_h \subset M$ are compatible subspaces of X and M [10], [3], [23], [8], that approach X and M as the discretization parameter h goes to zero. In (8.3.12) $(\cdot, \cdot)_h$ and $((\cdot, \cdot))_h$ denote evaluation of the continuous inner product $(\cdot, \cdot)$ by numerical quadrature (note however that the $(\cdot, \cdot)_h$ and $((\cdot, \cdot))_h$ may be different).

Choosing appropriate (compatible) discrete spaces X_h and M_h with associated bases, we arrive at a set of algebraic equations given in matrix form as

$$\underline{A}\,\underline{v}_i - \underline{D}_i^T\,\underline{p} \ = \ \underline{B}\,\underline{f}_i, \qquad i = 1, \ldots, d, \tag{8.3.13a}$$

$$-\underline{D}_i\,\underline{v}_i \ = \ 0, \tag{8.3.13b}$$

where $\underline{A}$ is the discrete Laplace operator ($\mu = 1$), $\underline{B}$ is the mass matrix, $\mathbf{D} = (\underline{D}_1, \ldots, \underline{D}_d)$ is the discrete gradient operator, and underscore refers to basis coefficients. In (8.3.13) we assume that the homogeneous boundary conditions are imposed by eliminating appropriate rows and columns. Note that in the limit as the discretization parameter $h \Rightarrow 0$, $(X_h, M_h) \Rightarrow (X, M)$, and (8.3.12) applies even for the continuous case.

Spectral Element Discretization   Again, we shall here consider the conforming spectral element method [48], [39], [50], in which the domain Ω is broken up into K disjoint subdomains,

$$\overline{\Omega} = \cup_{k=1}^{K} \overline{\Omega}^k. \tag{8.3.14}$$

The approximation spaces can be expressed as

$$X_h \ = \ X \cap \mathcal{P}_{N,K}(\Omega) \tag{8.3.15a}$$

$$M_h \ = \ M \cap \mathcal{P}_{N-2,K}(\Omega), \tag{8.3.15b}$$

where $\mathcal{P}_{n,K}(\Omega)$ is defined as before. We refer to [40], and [8] for a justification of the choice of discrete spaces. By choosing different polynomial degrees for the velocity and pressure, the main conclusions from the theoretical analysis can be summarized as: (i) the discrete solution is unique, that is, there exist no spurious pressure modes, and (ii) spectral convergence is obtained as the polynomial degree, N, is increased for fixed number of elements, K.

In order to arrive at (8.3.13) we also need to choose appropriate bases for X_h and M_h in (8.3.12). Within each element Ω^k we express the velocity and pressure in terms of high-order Lagrangian interpolants through the tensor-product Gauss-Lobatto and Gauss points, respectively.

The inner-products in (8.3.12) are evaluated using Gauss numerical quadrature [15], Gauss Legendre for $((\cdot, \cdot))_h$ and Gauss-Lobatto Legendre for $(\cdot, \cdot)_h$. Choosing appropriate test functions we arrive at a set of algebraic equations of the form (8.3.13).

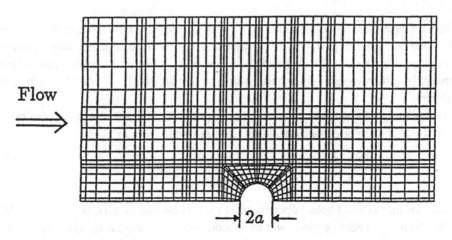

Figure 8.13: The plot shows a particular spectral element discretization ($K = 26$, $N = 6$) when solving steady Stokes flow past a sphere (axisymmetric). The length of the domain is 30 (axial direction) and the height is 15 (radial direction). The radius of the sphere is $a = \sqrt{2}$.

As a numerical example we consider parallel Stokes flow past a sphere [50]. This is an axisymmetric problem, and the spectral element mesh is shown in Figure 8.13. The exact solution for the drag is given as $F = 6\pi\mu Ua$, [57] where U is the (axial) free-stream velocity far away from the sphere, and a is the radius of the sphere. The computed drag is compared to the exact solution, and in Figure 8.14 we plot the relative error as a function of the polynomial degree, N, keeping the number of elements, K, fixed. Due to the smooth nature of the solution and the geometry, exponential convergence is obtained.

8.3.3 Steady Navier-Stokes Equations

Finally, we consider the steady Navier-Stokes equations in d space dimensions: Find a velocity $\mathbf{v}$ and a pressure p in a domain $\Omega \in \mathcal{R}^d$ such that

$$-\mu\nabla^2\mathbf{v} + \mathbf{v} \cdot \nabla\mathbf{v} + \nabla p = \mathbf{f} \quad \text{in } \Omega, \tag{8.3.16a}$$

$$-\nabla \cdot \mathbf{v} = 0 \quad \text{in } \Omega, \tag{8.3.16b}$$

subject to homogeneous Dirichlet velocity boundary conditions on the domain boundary $\partial\Omega$,

$$\mathbf{v} = \mathbf{0} \quad \text{on } \partial\Omega. \tag{8.3.16c}$$

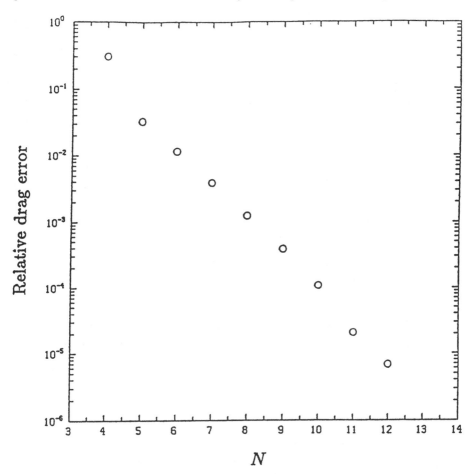

Figure 8.14: A plot of the relative error in the drag as a function of the polynomial degree N (for fixed $K = 26$) for Stokes flow past a sphere. Exponential convergence is obtained as N is increased.

Following the same procedure as for the steady Stokes problem, we arrive at the discrete problem: Find $(\mathbf{v}_h, p_h) \in (X_h, M_h)$ such that

$$
\begin{aligned}
\mu(\nabla \mathbf{v}_h, \nabla \mathbf{w})_h + (\mathbf{v} \cdot \nabla \mathbf{v}, \mathbf{w})_h - ((p_h, \nabla \cdot \mathbf{w}))_h = (\mathbf{f}, \mathbf{w})_h \qquad \forall \mathbf{w} \in X_h, \\
-((\nabla \cdot \mathbf{v}_h, q))_h = 0 \qquad \forall q \in M_h.
\end{aligned}
\tag{8.3.17}
$$

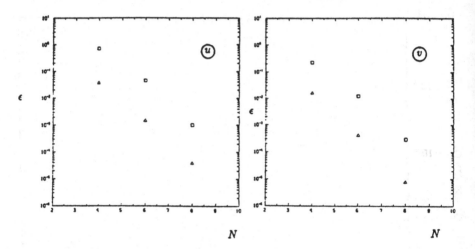

$$N \qquad\qquad N$$

Figure 8.15: Convergence of the L_2 ($\triangle$) and H^1 ($\square$) error norms for the Kovasznay solution in u and v velocity components, $K = 8$.

For conforming spectral element discretizations, the discrete spaces X_h and M_h are given in (8.3.15), and the final set of algebraic equations can be written in matrix form as

$$\underline{A}\,\underline{v}_i + \underline{C}(\mathbf{v})\,\underline{v}_i - \underline{D}_i^T\,\underline{p} \;=\; \underline{B}\,\underline{f}_i, \qquad i = 1,\ldots,d, \qquad (8.3.18a)$$

$$-\underline{D}_i\,\underline{v}_i \;=\; 0, \qquad\qquad\qquad\qquad (8.3.18b)$$

where $\underline{C}(\mathbf{v})$ is the discrete nonlinear (quadratic) convection operator.

We now illustrate the spatial convergence rate associated with the spectral element discretization of the steady Navier-Stokes equations for the Kovasznay flow described in Section 8.2.2. We solve this problem numerically in the case of $Re = 40$, $\lambda = \frac{1}{2}Re - \sqrt{\frac{1}{4}Re^2 + 4\pi^2}$, by integrating the time-dependent Navier-Stokes equations to steady state (see Section 8.2), imposing the exact solution on the domain boundary. Figure 8.15 shows the L_2 and H^1-error in the numerical solution at steady state as a function of the polynomial degree, N, inside each of the $K = 8$ spectral elements; spectral convergence is clearly achieved for this smooth solution.

8.4   Solvers

8.4.1   Elliptic Solvers

High order methods, such as the one presented above, are of little practical importance unless efficient algorithmic schemes are employed to solve the global system equations of the form (8.3.4). There have been numerous approaches proposed for solution of these equations broadly categorized as iterative or direct algorithms. The iterative approach is the only viable one for dynamically-deformed geometries or non-constant property flows; the reason is that in a direct approach extremely large matrices need to be inverted and stored at every time step which leads to enormous CPU and storage requirements. However, for fixed geometry and constant property flows, the direct approach is the best alternative as it is very general and robust; besides, the system matrices need only be inverted once, stored, and retrieved during the time-stepping. In this section we briefly discuss three different methods to solve the set of algebraic equations (8.3.4) resulting from spectral element spatial discretization of the Poisson equation. The elliptic solver is the single most important solver as regards the solution of the time-dependent Navier-Stokes equations since each time step typically involves one or more elliptic (Poisson or Helmholtz) solves. We shall here consider one direct method (static condensation), and two iterative methods (preconditioned conjugate gradient iteration and intra-element multigrid).

Direct Solvers We first present a direct approach appropriate for high-order finite element methods, which can be very effectively implemented in multi-processor computing environments. Let us re-write equation (47) in the form,

$$\underline{A}\underline{\phi} = \underline{g} \tag{8.4.1}$$

referring to the global spectral element discrete system. We then apply the *static condensation algorithm*, grouping the nodes and corresponding degrees-of-freedom into those lying on boundaries of spectral elements, $[^b\phi^k]$, and those located in the interior of elements $[^i\phi^k]$. The advantage of using static condensation in the spectral element method is obvious, since the majority of nodes are in the interior of elements (dense subconstructs), with no coupling between adjacent elemental interior nodes; this latter feature suggests that the major computational work (associated mostly with solution of elliptic equations) can be done in *parallel*.

The decomposed equations for one element can be written in matrix form as,

$$\left[\begin{array}{cc} [a^k] & [b^k]^T \\ [b^k] & [c^k] \end{array} \right] \left[\begin{array}{c} {}^b\phi^k \\ {}^i\phi^k \end{array} \right] = \left[\begin{array}{c} {}^bg^k \\ {}^ig^k \end{array} \right] \tag{8.4.2}$$

Solving separately for boundary nodes first, we obtain

$$\sum_{k}{}' ([a^k] - [b^k]^T[c^k]^{-1}[b^k])[^b\phi^k] = \sum_{k}{}' [^bg^k] - [b^k]^T[c^k]^{-1}[^ig^k] \tag{8.4.3}$$

where $\sum'$ denotes direct stiffness summation over adjoint elements. The equations for interior nodes can be handled separately for each element after the elemental boundary unknowns have been obtained:

$$[c^k][{}^i\phi^k] = [{}^ig^k] - [b^k][{}^b\phi^k] \tag{8.4.4}$$

The inversion of the global boundary system matrix in (8.4.3) (with relatively small rank) can be performed using a parallel LDL^T decomposition only once at a preprocessing stage, before the time-stepping begins. Thereafter, at each time step only the required forward- and back-solves and matrix multiplies are carried out. For a two-dimensional problem the amount of computational work as operation count per time step is approximately $\mathcal{O}(K^{3/2}N^2)$ for the system (8.4.3), while for the interior nodes the work is $\mathcal{O}(K^2N^4)$, where the domain is discretized with K elements with elemental resolution $N \times N$. More specifically, to solve the *two-dimensional* Poisson equation on a *serial* computer using a static condensation scheme the computational cost is given by

$$S_D = s_1 K^{3/2} N^2 + s_2 K N^4 + s_3 K N \tag{8.4.5a}$$

where $s_i, i = 1, 2, 3$ are constants for the contributions from the solution of the boundary unknowns, the interior unknowns, and the direct stiffness summation, respectively.

This solution scheme is particularly suited for *parallel* implementation as equation (8.4.4) that contributes the most to the overall cost S_D can be solved entirely in parallel without any communication overhead. The corresponding parallel cost for a P-headed processor is $\mathcal{O}(K/P)N^4$. Let us assume also that a parallel LDL^T algorithm is employed for the boundary unknowns (equation (8.4.3)). The overall parallel cost therefore is

$$Z_D = z_1(K^{3/2}/P)N^2 + z_2(K/P)N^4 + z_3(K/P)N + C_D \tag{8.4.5b}$$

where the last term includes all communication cost incurred in solving (64) in parallel as well as the cost associated with the direct stiffness summation (i.e. $C_D = C_{D_1} + C_{D_2}$). For simplicity the assumption is made here that a direct link exists between processors which operate on elements with common edges. If we now denote by $\Delta(m)$ the time-per-word to send m words across a direct link and δ the basic clock size for calculation for a given architecture, then $C_{D_2} = \sigma(N)N$, [19], where we define $\sigma(m) = \Delta/\delta$. The other component of communication cost depends on the specific configuration of the computer employed for the computation. As an example of a parallel static condensation algorithm the one-dimensional Burger's equation was implemented on a Intel iPSC/860-32 hypercube using 4 and 8 spectral elements mapped to a subcube of 4 and 8 processors respectively [29]. The results of this computation are listed in Table 8.5; we see that the

Time Steps	Serial Version (time (sec))	Parallel Version (time (sec))	Parallel Efficiency
1000	16.99111	5.37763	78.99
2000	33.26173	9.51513	87.39
3000	49.70778	13.55956	91.64
4000	66.13617	17.57693	94.06
5000	82.46719	21.52940	95.78
10000	164.18371	43.54075	94.27
10000	164.18713	43.32031	94.75

Table 8.5: Spectral element simulation of Burger's equation on an Intel/i860-32 hyper-cube, using $K = 4$ element with $N = 45$ collocation points per element. The CPU times for the one processor and for a subcube with four processors are shown in columns 2 and 3, respectively. The last column is the parallel efficiency.

parallel efficiency can be greatly increased reaching values of 94 percent for high values of the resolution parameter N.

Our primary interest is in the solution of the time-dependent Navier-Stokes equations at high Reynolds numbers. This would of course require not only an efficient way of solving the system equations as we described above, but also an intelligent way of managing the large volume of data for problems of industrial complexity. Typically the amount of data involved is much larger than will fit in the local memory of multi-processors, and thus it is necessary to maintain the data on external storage devices. As the computation proceeds, selective loading of small pieces of data into central memory is also performed. In addition, depending on the demands of the problem, the data can be packed and stored in 32-bit words without effecting the accuracy of the computation.

Iterative Solvers   The natural choice of solution algorithms in a parallel environment is an iterative procedure given that such techniques can be both highly local and concurrent. Other considerations such as memory requirements for large three-dimensional problems, variable timestep methods, adaptive mesh refinement, and treatment of time-dependent domains $\Omega(t)$, also suggest the use of iterative solvers. In order to evaluate the efficiency of an iterative approach we must consider two issues: (1) the operation count (or number of clock cycles) per iteration, Z^e; (2) the number of iterations, N_ϵ^A, required to achieve convergence to $\mathcal{O}(\epsilon)$. The total computational cost to invert the matrix $\underline{A}$ in (47) is then $Z = N_\epsilon^A Z^e$. In what follows we restrict ourselves to iterative solvers that require only residual evaluation, diagonal preconditioning, and inner-product summation.

Starting with the first issue, the work per iteration is essentially the work to evaluate matrix vector products such as those that appear in (8.3.4). Due to the tensor-product spaces, tensor-product quadratures, and tensor-product bases described in Section 8.3.1, these products can be evaluated efficiently using sum-factorization techniques in $\mathcal{O}(KN^{d+1})$ operations in $\mathcal{R}^d$ [45]. Thus, the number of clock cycles required per iteration on a *single* processor scales as $Z_1^e \sim KN^{d+1}$, an operation count which also applies to general-geometry isoparametric spectral element discretizations of non-separable equations [50].

The second issue is concerned with the convergence rate of an iterative approach; the goal is that the number of iterations N_ϵ^A be independent of the discretization parameter $h = (K, N)$. To invert the symmetric positive definite matrix $\underline{A}$ in (8.3.4) we consider two different iterative procedures. The first is a conjugate gradient iteration [24] with $\underline{P} = diag(\underline{A})$ as a diagonal preconditioner. Since this procedure has an order-dependent convergence rate with $N_\epsilon^A \sim K_1 N$ (here K_1 is the number of elements in *one* spatial direction, [39], the performance will deteriorate significantly for large problems.

The convergence rate can be improved by the use of an intra-element spectral element multigrid technique [50]. The key ingredients are: variational forms allowing for consistent derivation of restriction and prolongation operators; hierarchical spaces defined by varying the polynomial degree, N, within K *fixed* elements; and highly parallelizable $\underline{P}$- preconditioned Jacobi iteration as a smoother. This procedure results in an order-independent convergence rate in $\mathcal{R}^1$ ($N_\epsilon^A \sim \mathcal{O}(1)$), with only a weak N-dependence in in $\mathcal{R}^2$ and $\mathcal{R}^3$ ($N_\epsilon^A \sim N^{1/2}$).

As an illustration we consider the two-dimensional Laplace equation on a rectilinear domain $\Omega =]-0.5, 0.5[\times]-0.25, 0.25[$, with imposed Dirichlet boundary conditions on the domain boundary $\partial\Omega$ such that the solution is given as $u = 0.5 + x$. The initial guess for the solution is taken to be zero. The fact that the solution can be exactly represented as a first-order polynomial does not affect the multigrid convergence rate, however, it allows us to monitor the error $\| u - u_h \|$ which will be entirely due to incomplete iteration, with no contribution from discretization errors. In Figure 8.16 we plot the convergence histories corresponding to a spatial discretization of $K = 50$ and $K = 8$ equal spectral elements, each of order $N = 8$. Since most of the computational complexity is associated with performing matrix-vector products $\underline{A}\,\underline{u}$ on the fine grid, the savings using a multigrid approach is clearly seen. The convergence rate is independent of the the number of elements. For more general problems the current multigrid algorithm works well as long as the aspect ratios for the subdomains are reasonably small.

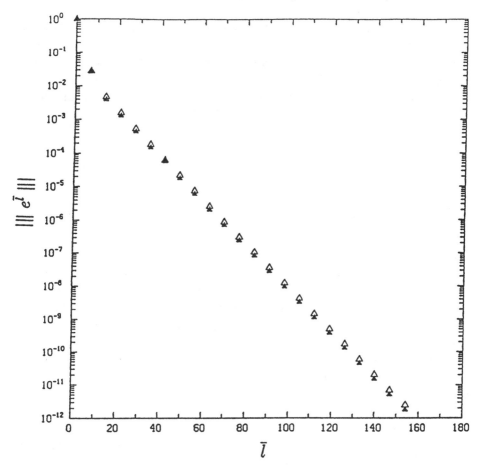

Figure 8.16: A plot of the iteration error $|||\underline{e}^l|||$ as a function of $\bar{l}$, the number of matrix-vector products $\underline{Au}$ on the fine grid, when solving a two-dimensional Poisson equation with data $f = 0$ and solution $u = 0.5 + x$. The plot compares convergence histories for the ($J = 2$, $m = 5$) spectral element multigrid with (non-optimal) Chebyshev acceleration for two different values of K, $K = 8$ (solid triangles) and $K = 50$ (open triangles), with fixed $N_J = 8$.

8.4.2   Steady Stokes Solvers

In this section we consider the solution of the algebraic system of equations (8.3.13) resulting from the spectral element discretization of the Stokes problem (8.3.9). It should be noted that the algorithms presented here are equally appropriate for other types of variational discretizations, and are in fact extensions of the classical Uzawa algorithm used in finite element analysis [23].

Our approach to solving (8.3.13) is a global iterative procedure in which the original saddle problem is decoupled (by block Gaussian elimination) into two positive (semi-) definite symmetric forms, one for the velocity and one for the pressure. Thus, the discrete saddle problem (8.3.13) can be replaced with the discrete equivalent statement

$$\underline{A}\,\underline{v}_i \;-\; \underline{D}_i^T \underline{p} = \underline{B}\,\underline{f}_i, \quad i = 1, \ldots, d, \tag{8.4.6a}$$

$$\underline{S}\,\underline{p} \;=\; -\underline{D}_i \underline{A}^{-1} \underline{B}\,\underline{f}_i, \tag{8.4.6b}$$

where the discrete pressure operator,

$$\underline{S} = \underline{D}_i \underline{A}^{-1} \underline{D}_i^T, \tag{8.4.6c}$$

is a positive semi-definite symmetric matrix.

We now make several comments regarding the system (8.4.6). First, we note that the equation set (8.4.6) does not correspond to a re-discretization of the continuous problem, that is, (8.4.6) is equivalent to (8.3.13). This implies that the theoretical error estimates derived for (8.2.19)–(8.2.20) directly apply here (in the case of spectral element discretizations we refer to [39] and [50]). Second, since the system matrices $\underline{S}$ and $\underline{A}$ are symmetric positive (semi-) definite, standard elliptic solvers like conjugate gradient iteration or multigrid techniques can readily be applied. The system (8.4.6) is solved by first solving (8.4.6b) for the pressure $\underline{p}$ and then solving (8.4.6a) for the velocity $\underline{v}_i$, $i = 1, \ldots, d$ with $\underline{p}$ known. Third, the pressure operator $\underline{S}$ is completely full due to the embedded inverse $\underline{A}^{-1}$ and thus clearly necessitates an *iterative* approach.

Heuristically, we expect the *continuous* pressure operator, s, to be close to the identity operator, I, and therefore to be well-conditioned. To see this we formally apply the Uzawa decoupling procedure to the continuous equations (8.3.9) and neglect boundary conditions,

$$s \sim \nabla \cdot (\nabla^2)^{-1} \nabla \sim I. \tag{8.4.7}$$

In the discrete case we do *not* expect $\underline{S}$ to be close to the identity matrix, $\underline{I}$, but rather the variational equivalent of the identity operator, the mass matrix $\underline{\tilde{B}}$. Hence, we expect

$$\underline{\tilde{B}}^{-1} \underline{S} \sim \underline{I}, \tag{8.4.8}$$

suggesting that we can invert $\underline{S}$ efficiently by conjugate gradient iteration, using the mass matrix $\underline{\tilde{B}}$ as a preconditioner. Note here the importance of the proper choice of bases and numerical quadratures in order to define a matrix $\underline{\tilde{B}}$ that is easy to invert, that is, in order for $\underline{\tilde{B}}$ to be diagonal.

We see that for general discretizations, each matrix-vector product evaluation $\underline{S}\underline{x}$ requires d standard elliptic Laplacian solves in $\mathcal{R}^d$. In order for this approach to be efficient for large multi-dimensional problems, the discrete Laplace operator $\underline{A}$ must be inverted by a fast solver, such as a good preconditioned conjugate gradient solver [39] or an intra-element multigrid solver [50]. In summary, the pressure is computed from (8.4.6b) by effecting a *nested* inner/outer iteration procedure.

If the condition number of the matrix $\tilde{\underline{B}}^{-1}\underline{S}$ is order unity, we see that the above algorithm requires only order d elliptic solves, and hence represents an ideal decoupling of the Stokes problem. Analytical and numerical results [38] indicate that for a given spatial discretization $h = (K, N)$, the condition number of $\underline{S}$ with respect to $\tilde{\underline{B}}$ scales like $N^{1/2}$ as $K, N \rightarrow \infty$. This result is related to the inf-sup parameter [10] associated with the approximation spaces X_h and M_h, and reflects the fact that this parameter is weakly resolution dependent. The result also tells us that the number of outer iterations in the pressure solver scales like $N^{1/4}$, independent of K. For a more in depth analysis of the steady Stokes solver, we refer the interested reader to [38].

8.5 Hybrid Spectral Element Schemes

Over the last two decades, a large number of numerical techniques have been proposed for the solution of the incompressible Navier-Stokes equations. Although the differences among these discretization techniques might have initially been very clear, there has been an increasing trend (especially this past decade) towards construction of hybrid algorithms with components that exhibit different properties but typically share a common root. A typical example of such a confluence of numerical algorithms is the spectral element method presented in the previous sections, which is based on two weighted-residual techniques: finite element and spectral methods. The combination of spectral-like accuracy with the flexibility in handling complex geometries have made the method quite succesfull in a number of applications in fluid dynamics, including flows in the transitional and turbulent regimes [33].

There exists however a large number of complex geometry flows where a straightforward application of the method either is not possible or it may lead to prohibitively expensive calculations. Examples include high Reynolds number turbulent flows, flows in unbounded domains, or flows in arbitrarily complex geometries. An example of the latter case is flow over randomly roughened surfaces; there are currently no numerical algorithms to handle such domains with random boundary. This limitation in current methodology has given rise to the development of new formulations that provide nonconforming discretizations or couple spectral element with low-order finite element or finite difference discretizations. Nonconforming formulations naturally lead to optimal discretizations and adaptive re-meshing procedures. Incorporation of local low-order techniques, on the other hand, can provide an efficient way of local "refinement patches" or handling of geometries with extreme disparity in length scales.

In this section, we review some recent work related to development of hybrid (spectral element based) schemes. First, we present a variational approach in formulating a non-conforming spectral element scheme and introduce the concept of "mortar elements." This formulation leads naturally to spectral element/finite element hybrid domains. We then present a general iterative relaxation procedure that provides the framework for similar spectral element/finite element coupling as well as spectral element/finite difference coupling.

8.5.1 Variational NonConforming Spectral Elements

Although the spectral element method described in Sections 8.3–8.4 offers a great deal of flexibility over classical mono-domain spectral techniques, the methods remain somewhat more difficult to implement in complex domains than low-order finite element techniques. First, the intrinsically smaller size of the latter as compared to the former leads to simpler meshing. Second, the availability of triangles and tetrahedra for low-order discretizations yields further flexibility and allows for more systematic meshing procedures. The p-type finite element method [4] does allow for triangular and tetrahedral elements, however these elements do not admit tensor product sum factorization, and thus can be relatively inefficient, in particular in three space dimensions.

A promising approach to maintaining the spectral element tensor product forms while simultaneously improving geometric flexibility is the use of non-conforming spectral element methods. One such approach, the mortar element method [7] allows not only for nonconforming spectral element approximations, but also for spectral element/finite element discretizations.

In the (R^2) mortar element discretization [2] the approximation space consists of two parts; the usual $P_N(\Omega^k)$ space in the interior of each spectral element; and a mortar space consisting of high order polynomials on segments the union of which comprises the sum of all elemental edges and vertices. From the mortar space the trace of elemental functions is uniquely determined as follows: on element edges the function is the L^2 projection of the mortar values; on element vertices the function is equal to the mortar value.

The important feature of the mortar element method is that it preserves locality, unlike Lagrange-multiplier techniques [17], as well as local elemental structure. The technique is optimal as regard discretization error, from which it follows that for analytic functions the discretization error goes to zero exponentially fast in the H^1-norm. Numerical examples are presented in the next section.

The ability to accept non-conforming spectral element approximations is a key ingredient in recently developed sliding-mesh, arbitrary-Lagrangian-Eulerian spectral element techniques for solution of the Navier-Stokes equations in time dependent geometry. In these techniques, independent meshes are associated with sliding or rotating subdomains, thereby allowing for treatment of problems involving relative motions which cannot be transformed away by an appropriate change of reference frame. Examples of impeller-in-chamber mixing calculations and rotor-stator start-up flows are described in [1].

8.5.2   Iterative Patching Procedure

Here we consider techniques for solving a general second-order elliptic partial differential equation where the global domain Ω is subdivided into a number of smaller, non-overlapping domains Ω_i. The emphasis is on generality in the discretizations within each subdomain and the ability, in the context of parallel computers, to update each subdomain simultaneously. Although for "conforming" discretizations (i.e. the same discrete representation of the solution in each subdomain) direct methods are still possible, in the case of fundamentally different discretizations we are forced to consider iterative procedures. One such method is that proposed in [21] and referred to here as "Zanolli" patching. The method consists of solving a sequence of alternating Dirichlet/Neumann problems, maintaining C^1 continuity and relaxing interface values to achieve C^0 continuity to within some pre-defined tolerance. As shown in [21], this procedure results in very fast convergence for the case of spectral collocation within subdomains. Here, it will be shown to perform equally well for mixed discretizations (spectral/finite difference). Also, because the original procedure is inherently serial, modifications to allow for parallel execution of the algorithm will be examined.

Sequential Algorithm   Consider the solution of the Helmholtz equation in one dimension, given by:

$$
\begin{aligned}
\phi_{xx} - \mu^2\phi &= f(x) \\
\phi(a) &= 0 \\
\phi(b) &= 0
\end{aligned}
\tag{8.5.1}
$$

The global domain $\Omega(a, b)$ is now subdivided into two domains, $\Omega_1(a, \delta)$ and $\Omega_2(\delta, b)$ where δ is the location of the interface or "patch." The Zanolli patching procedure is applied as follows: we look for a sequence of functions $\phi_1^n \in \Omega_1$ and $\phi_2^n \in \Omega_2$ which satisfy the following:

$$
\left\{
\begin{aligned}
\phi_{1,xx}^n - \mu^2\phi_1^n &= f \quad \text{in } \Omega_1 \\
\phi_1^n(a) &= 0 \\
\phi_1^n(\delta) &= \lambda^n
\end{aligned}
\right.
\tag{8.5.2}
$$

$$\begin{cases} \phi_{2,xx}^n - \mu^2 \phi_2^n = f & \text{in } \Omega_2 \\ \phi_2^n(b) = 0 \\ \phi_{2,x}^n(\delta) = \phi_{1,x}^n(\delta) \end{cases} \qquad (8.5.3)$$

where n denotes the iteration, λ^1 is a given real number, and subsequent λ^n's are computed as:

$$\lambda^{n+1} = \theta \cdot \phi_2^n(\delta) + (1 - \theta) \cdot \lambda^n. \qquad (8.5.4)$$

In this context, θ is a *relaxation parameter* which under certain conditions guarantees that the procedure (8.5.2–8.5.3) will always converge. The extension to two-dimensional problems is straightforward, with λ^n being replaced by $\lambda^n(s)$ (s denotes a local coordinate system along the patch) and the Neumann condition (8.5.3) replaced by an equivalent normal flux balance.

One of the important results of [21] is a theoretical prediction of optimal θ's and a method for choosing θ *dynamically* so as to accelerate convergence. Defining error functions

$$\begin{aligned} e_1^n &= \phi_1^n - \phi_1^{n-1} \\ e_2^n &= \phi_2^n - \phi_2^{n-1} \\ z^n(\theta) &= \theta \cdot \phi_2^n + (1 - \theta) \cdot \phi_1^n \qquad \text{on } \Gamma \end{aligned} \qquad (8.5.5)$$

where Γ is the line separating the two patched regions. The unique real number θ which minimizes $\|z^n(\theta) - z^{n-1}(\theta)\|^2$ is given by:

$$\theta^n = \frac{(e_1^n, e_1^n - e_2^n)}{\|e_1^n - e_2^n\|^2} \qquad (8.5.6)$$

where $(\cdot, \cdot)$ is the normal inner product in $\mathcal{L}^2$ and $\|\cdot\|$ is the associated norm.

Several examples of the performance of this algorithm for spectral collocation are given in [21], and for 1-D spectral-finite difference discretizations in [30]. In Figure 8.17 we compare the H^1 convergence of the variational conforming formulation with the Zanolli patching for a Poisson equation $-\nabla^2 u = f$ with $f = 8\pi^2 \sin 2\pi x \sin 2\pi y$ on the domain $]0, 1[\times] - 1, 3[$ decomposed into three elements as shown in the Figure; exponential convergence is obtained in both cases. Here th the Zanolli procedure corresponding to slightly better convergence. To demonstrate the effectiveness of the Zanolli algorithm for two-dimensional problems with non-conforming discretizations we solve the Helmholtz equation in a complex domain (see Figure 8.18). The global domain is subdivided into two approximately equally sized subdomains and discretized using spectral elements in Ω_1 and finite differences in Ω_2. In Figure 8.19 we show convergence of the solution at the interface for several values of θ, including the case where θ is updated dynamically. The performance seen here is typical, namely 5–10 iterations for convergence independent of the complexity of the the solution.

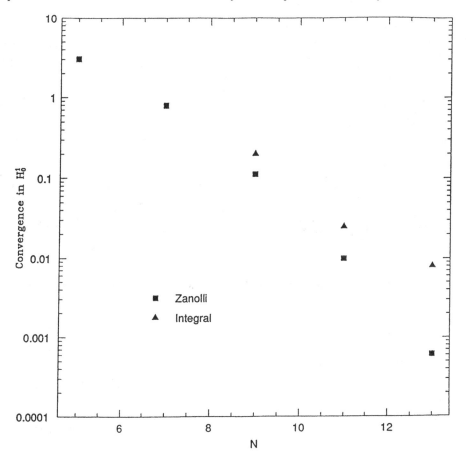

Figure 8.17: Error versus number of resolution per element for the Zanolli patching (squares) and the vertical patching (triangles).

Parallel Algorithm   One of the drawbacks to the Zanolli procedure is that it is a "serial algorithm." Because of the coupling (through the derivative term) between (8.5.2–8.5.3) the solution on each subdomain must be computed in sequence, limiting the application of this procedure to sequential processing. A simple modification to (8.5.2–8.5.3) which allows the computations to proceed *in parallel* is:

1. Set
$$
\begin{cases}
\phi_1^n(\delta) &= \theta \cdot \phi_2^{n-1}(\delta) + (1 - \theta) \cdot \phi_1^{n-1}(\delta) \\[2mm]
\phi_{2,x}^n(\delta) &= \phi_{1,x}^{n-1}(\delta)
\end{cases}
\tag{8.5.7a}
$$

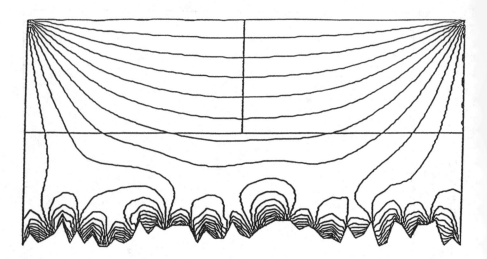

Figure 8.18: Solution contours of a Helmholtz equation in a domain with a random boundary.

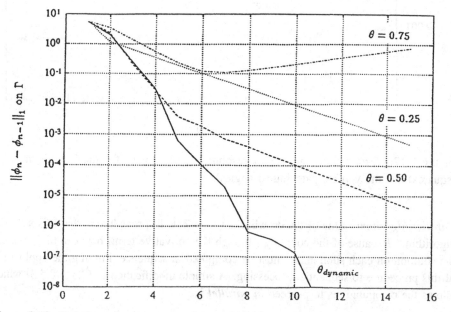

Figure 8.19: Relative convergence at the patching interface as a function of the relaxed parameter θ.

$$2.\ \text{Solve} \quad \begin{cases} \phi_{1,xx}^n - \mu^2 \phi_1^n = f & \text{in } \Omega_1 \\[2mm] \phi_{2,xx}^n - \mu^2 \phi_2^n = f & \text{in } \Omega_2 \end{cases} \qquad (8.5.7b)$$

Note that (1) denotes a communication and (2) a calculation step. While altering the convergence properties of the original scheme, this allows for each subdomain to be updated simultaneously. On a medium- to fine-grained machine, it also allows each Ω_i to be distributed over an appropriate collection of processors such that domains with differing workloads may be updated in the same amount of "real" time.

Convergence of the procedure (8.5.7) is analyzed in [30] and only summarized here. Figure 8.20 shows convergence rates for a typical two-dimensional configuration for various fixed values of θ, and Figure 8.21 shows the corresponding rates for the parallel procedure. Note that in the unrelaxed case ($\theta = 1$) the parallel procedure produces two uncoupled solutions which converge at half the rate of the serial version. However, a value of θ exists which yields acceptable convergence at better than half the serial rate. This result suggests that the amount of work represented by individual subdomains should be significant for the parallel procedure to be efficient. For efficiency measurements on typical two-dimensional cases the reader is referred to [30], while several flow simulations are reported in [31].

8.6   Discrete Lattice Gas Methods

Lattice gas (or cellular automata (CA)) models offer a novel and potentially usable alternative to numerical solution of the Navier-Stokes equations for the simulation of hydrodynamics. These new methods have stimulated much thinking about the fundamental statistical mechanics of many-particle systems. The basic idea of CA methods is to represent the fluid as an ensemble of interacting low-order bit computers situated at regularly spaced lattice sites. In the Frisch-Hasslacher-Pomeau (FHP) [20] model of two-dimensional hydrodynamics, the underlying lattice is a close-packed equilateral triangular lattice with sites at triangle vertices; each site has a seven-bit state with the first six bits specifying the presence or not of a particle traveling at angle $\mathbf{e}_j = j\, 60^\circ\ (0 \le j < 6)$ along the legs of the triangular lattice and the last bit specifying the existence or not of a particle at rest at the lattice site. Each particle (except a rest particle) moves one lattice distance in one fundamental time interval. After the particles propagate they then interact according to certain collision rules [20].

There are two main difficulties with these CA methods. First, their work requirements increase rapidly with Reynolds number. In fact, the work requirements for CA calculations increase slightly more rapidly than for corresponding calculations of the Navier-Stokes equations [47]. Also, the methods represent fluid dynamics only in the

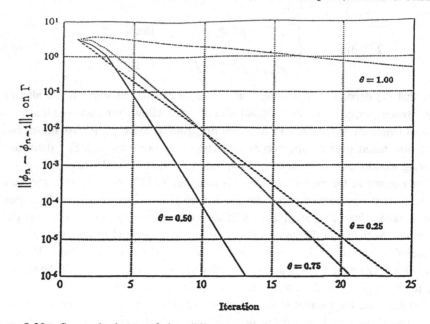

Figure 8.20: Spectral element-finite difference convergence rates for the sequential Zanolli patching algorithm.

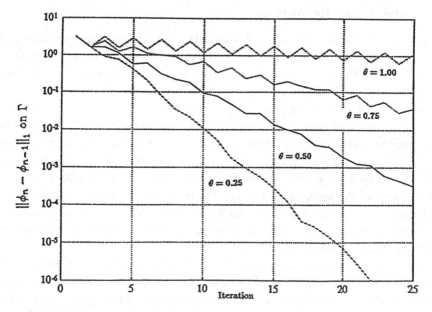

Figure 8.21: Spectral element-finite difference convergence for the modified (parallel) Zanolli algorithm.

limit of zero Mach number. For finite Mach numbers there is a finite error made in the representation of incompressible flow. Finally, CA methods are statistical in nature and suffer from statistical fluctuations. The velocity field is computed as the average velocity over a large number N of CA sites, so that there is an error of order $1/\sqrt{N}$ in the evaluation of this velocity field. These difficulties taken together suggest that CAs are most useful in the low-Reynolds number regions of the flow. For high Reynolds number flows the above reasoning suggests that CAs may be most useful in the wall regions where the local Reynolds number is small. Indeed, CAs offer the tremendous advantage over conventional methods of solution of the Navier-Stokes equations in that they may represent arbitrarily complex geometries, even random geometries, in a fairly straightforward way.

A very attractive way to avoid the difficulties with noisiness of CA systems is to use a lattice Boltzmann approach [59], [32]. This approach, which we will denote the LB method, seems to be more efficient than Monte Carlo CA methods for moderate to low Reynolds numbers. The idea is to integrate a kinetic equation for the CA system; here the kinetic equation is for average particle distribution functions along each of the discrete allowed particle velocities at each lattice site. For a two-dimensional CA system there are seven distribution functions (corresponding to the seven bits) at each lattice site. These functions are smooth, non-random functions governed by nonlinear partial differential equations which are integrated in space-time to obtain the flow description. Velocities are determined as averages over a number of lattice sites of the LB system. The method extends easily to three dimensions in which the lattice is a 24-bit or 25-bit projection of a four-dimensional FCHC lattice onto three-dimensional space.

8.6.1 Comparison of LB and Spectral Element Methods

Some results for solution of discrete LB systems in flow past a cylinder are given in Figures 8.22 and 8.23, [32]. In Figure 8.22 we show streamlines, and in Figure 8.23 we show a comparison of the drag on the cylinder with experimental data. These results support the mathematical analysis showing the reduction of the discrete LB system to Navier-Stokes dynamics as the lattice spacing gets smaller.

Several tests have been performed recently by Zanetti (private communications, Princeton University) to validate the use of Lattice Boltzmann equation method instead of the Navier Stokes equation in complicated geometries; comparisons were made with results obtained using spectral element methods. The LBE technique appears to be able to reproduce, within a reasonable accuracy, results obtained by spectral elements calculations. The computational cost is comparable (within a factor two) to the computational cost of a regular-grid explicit finite-difference Navier Stokes solver. However, the LBE has the

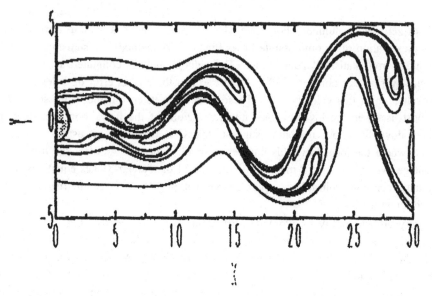

Figure 8.22: Instantaneous streamlines of vortex shedding from cylinder at Reynolds number $R = 52.8$ (Higuera and Succi, 1989).

advantage, with respect to standard finite differences methods of being more microscopic in nature and thus it can be easily modified to simulate a class of problems such as flows of immiscible fluids [27] which are very difficult to handle by more standard techniques. Another important advantage of the method, again directly related to its semi-microscopic nature, is its algorithmical simplicity; the code used to run the simulations described below is less than 200 lines of Fortran and yet it can easily handle complicated boundary conditions.

The specific method used by Zanetti (private communications) is a slightly modified LBE method in which the method of Higuera and Succi [32] is made Galilean invariant by using a different choice of equilibrium distribution function. The resulting numerical scheme has some of the flavor of a finite volume technique. It is however more microscopic in nature because it still involves the integration of a Boltzmann equation, albeit on a very limited phase space with only a finite number of possible particle velocities. This new model successfully passed a series of consistency tests: tests of rotational symmetry, tests of the stress tensor, tests of Galilean invariance, sound and shear waves.

Here we discuss results obtained by Zanetti for flow in moderately complicated geometries and compare the resulting LB solution with those obtained by spectral elements. We consider the simulation of a sheared flow in a channel with a square obstacle. We

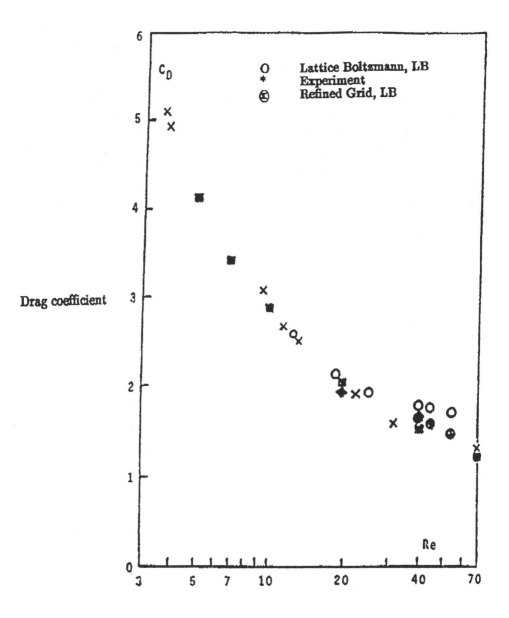

Figure 8.23: Cylinder drag coefficient as a function of Reynolds number. Experimental data is by Tritton, 1959 (Higuera and Succi, 1988).

choose this flow because it has rather simple boundary conditions: periodic in the flow direction and no-slip on the bottom wall and on the obstacle. The flow is driven by a moving top wall. Moreover, in the range of Reynolds number considered here the flow is steady.

The simulation domain was mapped onto a parallelogram of 128×128 lattice sites, with a top wall velocity of $U = 0.042$, and the reduced density, d, of the LBE fluid was chosen to be $d = 0.3$; the kinematic viscosity was adjusted so that the macroscopic Reynolds number defined using U, the width of the channel and the kinematic viscosity of the fluid was $Re = 110$. The same simulation was repeated using higher resolution, i.e., 192×192 lattice sites, with little discrepancy, less than 2% between the two runs. The magnitude of this discrepancy is consistent with that expected by naive consideration on the relative thickness of Knudsen layers [14].

The results of the LBE simulations were rescaled appropriately for the comparison with the spectral element results. It is worth mentioning that there are no free parameters in this conversion process, except, of course for an arbitrary constant added to the pressure. In the reference solution computed using the spectral element code, the flow domain was subdivided into twenty eight ($K = 28$) elements and, to test for convergence, the same problem was run with three different spectral resolutions, respectively 5×5, 7×7 and 9×9 modes per element.

We will now compare the spectral element results (solid line) to the LBE results (+). This is done at three stations. In Figure 8.24 we display pressure contours, while in Figure 8.25 we compare profiles of pressure, v_x, and v_y as a function of y downstream with respect to the obstacle at $x = 0.625$. The results in Figure 8.26 represent the same three quantities along a vertical line upstream of the obstacle, while these in Figure 8.27 show how the two sets of results compare along a horizontal line at $y = 0.135$. The LBE solution resolves the reference solution rather well. The discrepancies between the two solution methods seem to be particularly prominent, of the order of ten per cent, where the local velocity is small, typically less than one or two per cent of the wall velocity $V_0 = 1$. This small discrepancy between the two methods is probably physical in origin, i.e. due to weak compressibility effects. In fact, the amplitude of compressibility corrections to the local density scales as the square of Mach number, and the latter varies widely between different regions of the flow. Thus, we may expect that the system will try to compensate for these spurious corrections by rearranging the flow in subtle ways. This effect is very interesting and deserves further investigation in the future.

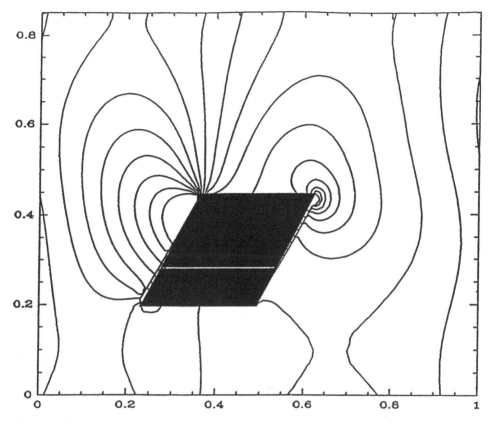

Figure 8.24: Pressure contour plot. The shaded area is the obstacle. The top wall is moving rightwards with unit velocity; the bottom wall and the obstacle are still. The Reynolds number is $R = 110$.

8.7 Discussion

Future developments in the computational fluid dynamics of incompressible flows will be governed by developments in computer architectures as well as by developments of new algorithms for these problems. For the next decade or so, the major development in computer hardware seems to be the maturation of parallel computers. As we have seen in this paper, spectral element methods are an attractive choice for several of the most promising parallel architectures now being considered. These methods are now developing into effective computational tools for sophisticated engineering problems and a wide variety of applications are expected in the near future.

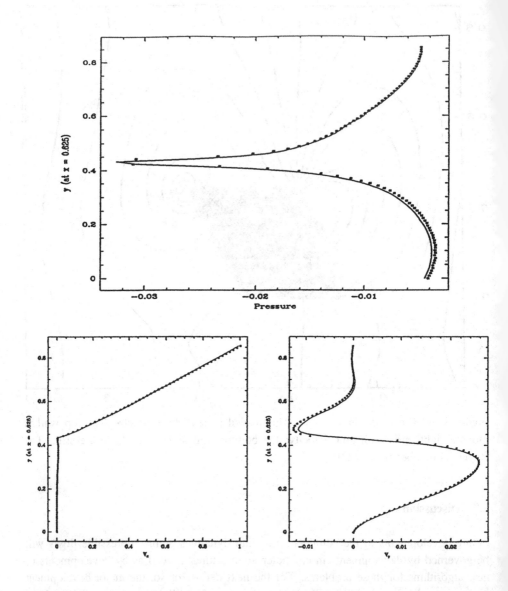

Figure 8.25: Pressure, v_x, and v_y as a function of y at $x = 2/3$. The LBE data is indicated with $+$.

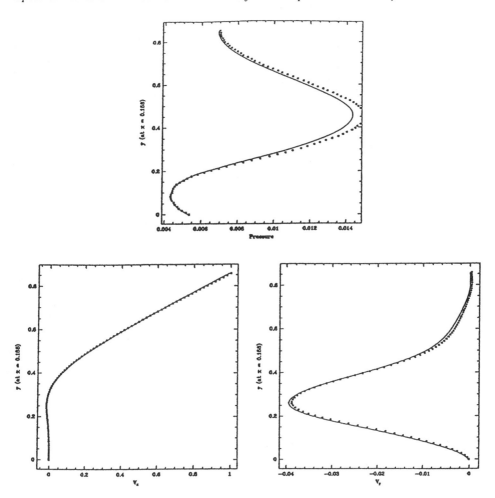

Figure 8.26: Pressure, v_x, and v_y as a function of y at $x = 0.185$. The LBE data is indicated with +.

Most work to date in computational fluid dynamics has focused on the use of structured, non-moving grids. However, recent developments suggest that these restrictions will soon be effectively removed and much wider classes of difficult complex geometry problems addressed. The development of hybrid schemes, such as combined spectral element and low-order difference or lattice gas methods, is likely to be useful for problems with very complex, time dependent geometries. In particular, unstructured meshes, such as those encountered in non-conforming spectral-element schemes, will likely play an increasing

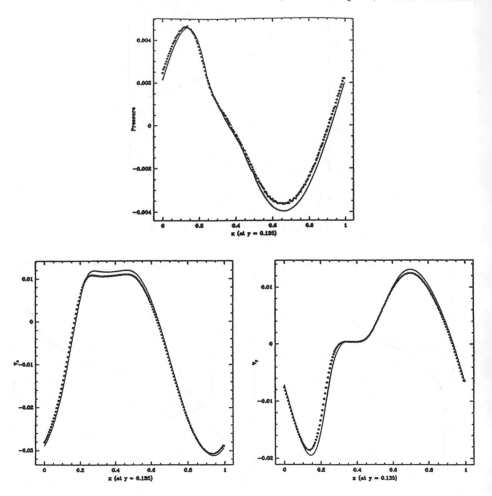

Figure 8.27: Pressure, v_x, and v_y as a function of y at $x = 0.135$. The LBE data is indicated with $+$.

role in computations. The ability to solve problems in time-varying geometries involves similar algorithmic breakthroughs. At the present time, prototypical problems involving non-conforming spectral-element schemes have been solved and engineering applications are under way.

Another current trend is to develop advanced methods for large-eddy simulation in which the effects of small-scale turbulence are modeled and large-scale effects are computed explicitly. This development involves the inclusion of the effects of highly varying spatial transport coefficients in the models. New algorithmic developments to allow effective implicit solutions of such problems with rapidly varying transport properties have now been made and their implementation for engineering applications is underway. Another development over the past few years which we will see increasing application is the derivation of effective boundary conditions for difficult flow problems. Examples include the techniques for the imposition of high-order accurate boundary conditions for the pressure in incompressible flows and methods to impose effective outflow boundary conditions and radiation boundary conditions for spatially unbounded problems. These developments are particularly important because they allow the simulation of flows in unbounded domains within relatively small bounded geometries.

In the decade of the 1990's, we fully expect that the solution of typical incompressible flow problems, whether they be high Reynolds number or complex geometry or multiphase, will be attainable using "black box" computer programs running on "user-friendly" super workstations available to practicing engineers and scientists. This development will enable working engineers and scientists to gain new insights into the physics and engineering of fluids and, hopefully, enable great advances in these fields.

Acknowledgements

We would like to acknowledge the use of unpublished results of our colleagues R. Henderson, A. Tomboulides and G. Zanetti at Princeton. We should also like to thank Prof. Yvon Maday from University of Paris 6 for many helpful discussions. One of us (GEK) would like to acknowledge financial support by AFOSR Grant number 90-0261, by ONR Contract N00014-90-1312, and by NSF Grants CTS-8906911, CTS-8906432 and CTS-8914422. Another one (SAO) would like to acknowledge support by DARPA under ONR Contract N00014-86-K-0759, ONR under Contract N00014-82-C-0451, AFOSR under Grant 90-0124 and NSF Grant OCE-9010851. Another one (EMR) would like to acknowledge support by NASA under Contracts NAS1-19102 and NAS3-26132. Finally, (ATP) would like to acknowledge support by NASA and by DARPA under ONR Contract N00014-89-J-1610 and ONR Contract N00014-88-K-0188.

References

[1] Anagnostou G. *Nonconforming sliding spectral element methods for the unsteady incompressible Navier-Stokes equations*. PhD thesis, Massachusetts Institute of Technology, 1991.

[2] Anagnostou G., Maday Y., Mavriplis C., and Patera A. T. On the mortar element method; generalizations and implementation. In *Third Int. Conf. on Domain Decomposition Methods, ed. R. Glowinski, SIAM*, 1990.

[3] Babuska I. Error bounds for the finite element method. *Num. Math.*, 16:322, 1971.

[4] Babuska I. and Dorr M. Error estimates for the combined h-p version of the finite element method. *Num. Math.*, 25:257, 1981.

[5] Benqué J. P., Ibler B., Keramsi A., and Labadie G. A new finite element method coupled with a temperature equation. In *Proc. Fourth Int. Symp. on Finite Element in Flow Problems, ed. T. Kawai, North Holland*, 1982.

[6] Beris A. N., Armstrong R. C., and Brown R. A. Finite element calculation of viscoelastic flow in a journal bearing: small eccentricities. *J. Non-Newtonian Fluid Mechanics*, 16:141, 1984.

[7] Bernardi C., Debit N., and Maday Y. Coupling spectral and finite element methods: First results. *Math. Comp.*, 54:21, 1990.

[8] Bernardi C., Maday Y., and Metivet B. Spectral approximation of the periodic-nonperiodic Navier-Stokes equations. *Num. Math.*, 51:655, 1987.

[9] Bernardi C., Maday Y., and Patera A. T. A new nonconforming approach to domain decomposition; the mortar element method. Technical Report R 89027, Publications du Laboratoire D'Analyse Numerique, 1990.

[10] Brezzi F. On the existence, uniqueness and approximation of saddle-point problems arising from lagrange multipliers. *Rairo Anal. Numer.*, 8 R2:129, 1974.

[11] Canuto C., Hussaini M., Quarteroni A., and Zang T. *Spectral Methods in Fluid Dynamics*. Springer-Verlag, 1987.

[12] Chorin A. J. Numerical solution of incompressible flow problems. In *Studies in Numerical Analysis 2, ed. J. M. Ortega and W. C. Rheinboldt, SIAM*, 1970.

[13] Ciarlet P. *The finite element method for elliptic problems*. North-Holland, 1978.

[14] Cornubert R., d'Humières D., and Levermore D.. A Knudsen layer theory for lattice gases. *Physica D*, 47, 1991.

[15] Davis P. J. and Rabinowitz P. *Methods of Numerical Integration*. Academic Press, 1985.

[16] Deville M. O., Kleiser L., and Montigny-Rannou F. Pressure and time treatment for Chebyshev spectral solution of a Stokes problem. *Int. J. Num. Meth. in Fluids*, 4:1149, 1984.

[17] Dorr M. Domain decomposition via Lagrange multipliers. *Num. Math.*

[18] Ewing R. E. and Russel T. F. Multistep Galerkin methods along characteristics for convection-diffusion problems. In *Advances in Computer Methods for Partial Differential Equations, ed. R. Vichnevetsky and Stepleman R. S., IMACS*, 1981.

[19] Fischer P. and Patera A. T. Parallel spectral element solution of the Stokes problem. *J. Comput. Phys.*, to appear, 1991.

[20] Frisch U., Hasslacher B., and Pomeau Y. Lattice Gas Automaton for the Navier-Stokes equation. *Phys. Rev. Lett.*, 56:1505, 1986.

[21] Funaro D., Quarteroni A., and Zanolli P. An iterative procedure with interface relaxation for domain decomposition methods. Technical report, Instituto di Analisi Numerica del Consiglio Nazionale delle Ricerche, Pavia, Italy, 1985.

[22] Gear C. W. *Numerical Initial Value Problems in Ordinary Differential Equations*. Prentice-Hall, 1973.

[23] Girault V. and Raviart P. A. *Finite element approximations of the Navier-Stokes equations*. Springer, 1986.

[24] Golub G. H. and Van Loan C. F. *Matrix Computations*. Johns Hopkins University Press, 1983.

[25] Gottlieb D. and Orszag S. A. *Numerical Analysis of Spectral Methods: Theory and Applications*. SIAM, Philadelphia, 1977.

[26] Gresho P. M., Chan S. T., Lee R. L., and Upson C. D. A modified finite element method for solving the time-dependent, incompressible Navier-Stokes equations. Part 1: Theory. *Int. J. Num. Meth. in Fluids*, 4:557, 1984.

[27] Gunstensen A. K., Rothman D. H., Zaleski S., and Zanetti G. Lattice Boltzmann model of immiscible fluids. *Phys. Rev. A*, 43(8):4320, 1991.

[28] Guo B. and Babuska I. The h-p version of the finite element method Part 1: The basic approximation results. *Comp. Mech.*, 1:21, 1986.

[29] Henderson R. D. *in preparation*. PhD thesis, Princeton University.

[30] Henderson R. D. and Karniadakis G. E. A hybrid spectral element-finite-difference method for parallel computers. In *Proc. Unstructured scientific computations on scalable multi-processors, ICASE, ed. R. Voight, MIT Press*, 1990.

[31] Henderson R. D. and Karniadakis G. E. Hybrid spectral element methods for flows over rough walls. In *Proc. Fifth Conf. on Domain Decomposition Methods, SIAM*, 1991.

[32] Higuera F. and Succi S. Simulating the flow past a cylinder with the lattice Boltzmann equation. *Europhysics Letters*, 8:517, 1989.

[33] Karniadakis G. E. Spectral element simulations of laminar and turbulent flows in complex geometries. *Appl. Num. Math.*, 6:85, 1989.

[34] Karniadakis G. E., Bullister E. T., and Patera A. T. A spectral element method for solution of two- and three-dimensional time dependent Navier-Stokes equations. In *Finite Element Methods for Nonlinear Problems, Springer-Verlag*, page 803, 1985.

[35] Karniadakis G. E., Israeli M., and Orszag S. A. High-order splitting methods for incompressible Navier-Stokes equations. *J. Comput. Phys.*, to appear, 1991.

[36] Korczak K. Z. and Patera A. T. An isoparametric spectral element method for solution of the Navier-Stokes equations in complex geometry. *J. Comput. Phys.*, 62:361, 1986.

[37] Kovasznay L. I. G. Laminar flow behind a two-dimensional grid. In *Proc. Cambridge Phil. Society*, page 44, 1948.

[38] Maday Y., Meiron D., Patera A. T., and Rønquist E. M. Analysis of iterative methods for the steady and unsteady stokes problem: Application to spectral element discretizations. *J. Comput. Phys.*, submitted.

[39] Maday Y. and Patera A. T. Spectral element methods for the Navier-Stokes equations. In *State of the Art Surveys in Computational Mechanics, eds. A. K. Noor and J. T. Oden, ASME*, 1989.

[40] Maday Y., Patera A. T., and Rønquist E. M. A well-posed optimal spectral element approximation for the stokes problem. *SIAM J. Numer. Anal.*, to appear, 1991.

[41] Maday Y., Patera A. T., and Rønquist E. M. An operator-integration-factor splitting method for time-dependent problems: application to incompressible fluid flow. *J. Sc. Comp.*, to appear, 1991.

[42] Marcus P. S. Simulation of Taylor-Couette flow. Part 1. Numerical methods and comparison with experiment. *J. Fluid Mech.*, 146:45, 1984.

[43] Maslanik M. M., Sani R. L., and Gresho P. M. An isoaparametric finite element Stokes flow test problem. *preprint*, 1989.

[44] Métivet B. *Resolution des equations de Navier-Stokes par metodes spectrales*. PhD thesis, Université de Pierre et Marie Curie, 1987.

[45] Orszag S. A. Spectral methods for problems in complex geometry. *J. Comput. Phys.*, 37:70, 1980.

[46] Orszag S. A., Israeli M., and Deville M. O. Boundary conditions for incompressible flows. *J. Sc. Comp.*, 1(1):75, 1986.

[47] Orszag S. A. and Yakhot V. Reynolds number scaling of Cellular Automaton hydrodynamics. *Phys. Rev. Lett.*, 56:1691, 1986.

[48] Patera A. T. A spectral element method for Fluid Dynamics; Laminar flow in a channel expansion. *J. Comput. Phys.*, 54:468, 1984.

[49] Pironneau O. On the transport-diffusion algorithm and its application to the Navier-Stokes equations. *Num. Math.*, 38:309, 1982.

[50] Rønquist E. M. *Optimal spectral element methods for the unsteady three-dimensional incompressible Navier-Stokes equations.* PhD thesis, Massachusetts Institute of Technology, 1988.

[51] Strang G. and Fix G. *An Analysis of the Finite Element Method.* Prentice-Hall, 1973.

[52] Temam R. *Navier-Stokes equations, Theory and Numerical Analysis.* North-Holland, 1984.

[53] Tomboulides A. G. *in preparation.* PhD thesis, Princeton University.

[54] Tomboulides A. G., Israeli M., and Karniadakis G. E. Efficient removal of boundary-divergence errors in time-splitting methods. *J. Sc. Comp.*, 4:291, 1989.

[55] Tomboulides A. G. and Karniadakis G. E. Outflow boundary conditions for viscous incompressible flows. In *Mini-Symposium on Outflow Boundary Conditions, Stanford, CA*, 1991.

[56] Wannier G. H. A contribution to the hydrodynamics of lubrication. *Quart. Appl. Math.*, 8:1, 1950.

[57] White F. M. *Viscous Fluid Flow.* McGraw-Hill, 1974.

[58] Yanenko N. *The Method of Fractional Steps.* Springer, 1971.

[59] Zanetti G. Hydrodynamics of lattice-gas automata. *Phys. Rev. A*, 40:1539, 1989.

9 Design of Incompressible Flow Solvers: Practical Aspects

Rainald Löhner

Summary

We describe a series of algorithms for the numerical simulation of incompressible flows. These algorithms are obtained by following a rational path from a list of design goals for practical incompressible flow solvers to their ultimate realization. Along the way, the important identity of artificial viscosity and pairs of different trial spaces for velocities and pressures is shown rigourously for the mini-element. Several numerical examples demonstrate the accuracy and versatility of the algorithms developed.

9.1 Introduction

The applications that require numerical simulations of incompressible flows may be grouped into two families:

- *Engineering design and optimization*: here the basic physics governing the flows to be simulated are relatively well understood, and the main requirement on the numerical methods employed is versatility, ease of use, and speed. Many configurations have to be simulated quickly, in order to develop or improve a new product. This implies that the whole process of simulating incompressible flow past an arbitrary, new configuration must take at most several days. Usually, the engineer desires a global figure, like lift and drag, as the end-product of a simulation.

- *Study of basic physics*: in this case, numerical simulations are used to obtain new insight into basic physical phenomena, like vortex merging and breakdown, or the transition to turbulence. The main requirement placed on the numerical methods employed is accuracy. The geometries for which these calculations are carried out are typically very simple (boxes, channels), and the time required to perform such a simulation plays a secondary role. Some of the runs performed to date have required hundreds of CRAY-hours. Usually, the physicist desires statistical data as the end-product of such a simulation.

Numerical simulations in both areas have reached a fairly mature state, as evidenced by an abundance of literature [Confs 1990]. Therefore, before going further, we must define the design goals for the incompressible flow solvers to be discussed here. We will concentrate on the first family of applications, i.e., design and optimization. Thus, we require:

- *Arbitrary geometries*: an engineer designing a new product does neither have the desire nor the time to go through the intellectual excercise of answering the question: will I be able to grid my new design idea? Therefore, he must be provided with tools that can quickly mesh any arbitrary domain. This implies naturally the use of *unstructured grids*.

- *Fast gridding*: an engineer needs the result immediately. How much time "immediately" is, is a matter of personality. What is certain however, is that if it takes more than a week to obtain a result, the engineer won't use numerical simulations as a design tool. Therefore, every aspect of a simulation must be expedited. Currently, the biggest bottleneck facing 3-D simulations is not CPU-time, but gridding time. There is no point in having a super-optimized 3-D incompressible Navier-Stokes solver if it takes 6 months to generate a mesh. The only automatic, fast grid generators currently available generate tetrahedral meshes (see Löhner and Parikh [1988], Baker [1989], Peraire et al. [1989]). Therefore, we will concentrate on *triangles and tetrahedra* for the choice of the spatial discretization.

- *Simple elements*: the engineer or technician performing the numerical simulations is typically more interested in lift and drag than mathematical elegance. Therefore, the mathematical complexity of the algorithm must be kept to a minimum. As an example, a piecewise linear, discontinuous pressure field is somewhat unnatural for someone unfamiliar with the LBB condition (see Gunzburger [1987]). The imposition of boundary conditions should be simple and straightforward. Thus, we are led to the use of *simple, low-order elements that have all the variables (velocities, pressure) at the same location*.

- *Unknown solutions*: an engineer will typically try to push his design to the limit. In the case of airfoils, separation will occur. This implies that although a steady flow may be desired, the physics dictate an unsteady flow. This is a very important consideration when designing a flow solver. Vast classes of problems have steady solutions, particularly for the low Reynolds number regime. Optimized flow solvers can be constructed for steady flow. However, these codes may predict a steady solution when in reality unsteadiness occurs. Moreover, the flow to be simulated may have regions where the flow is steady or quasi-steady, and other regions where the flow is unsteady. Thus, we are led to *flow solvers that can simulate steady and unsteady flows*.

- *Inviscid, incompressible flows*: in order to study trends, an engineer may wish to approximate a high Reynolds number flow by an inviscid flow. An inviscid flow will require a much coarser grid, which translates into fast turnaround. Therefore, the flow solver must also have the *ability to solve the incompressible Euler equations*.

Several other design goals are obvious, and are listed here without further explanation:

- In order to reduce software complexity and maintenance costs, the method should be the *same in 2-D as in 3-D*.

- The switch from laminar to turbulent viscosity should not involve a major change in the method.

- In order to be applicable to large-scale problems in 3-D, the method should lend itself for *easy implementation on parallel machines*.

9.2 Discretization in Time

In order to define the notation used we start by recalling the incompressible Navier-Stokes equations

$$\mathbf{v}_{,t} + \mathbf{v}\nabla\mathbf{v} + \nabla p = \nabla \cdot \sigma, \tag{9.2.1a}$$

$$\nabla \cdot \mathbf{v} = 0. \tag{9.2.1b}$$

Here p denotes the pressure, $\mathbf{v}$ the velocity vector and σ the stress-tensor, and both the pressure p and the stresses σ have been normalized by the (constant) density ρ. These equations are obtained from the more general, compressible flow case, as the sound-speed approaches an infinite value. This sound-speed is associated with the pressure, and forces the use of implicit time-marching schemes for the pressure. For practical reasons, it may also be advantageous to use implicit schemes for the advective and diffusive terms of equation 9.2.1a. As an example, consider a typical boundary layer grid with very elongated elements aligned with the attached flow. As soon as separation occurs, the explicit Courant-Friedrichs-Levy (CFL) criterion would impose time-step sizes that are orders of magnitude smaller than in the attached case (see Figure 9.1). In order to derive the discretization in time for the velocities, we start with the following Taylor-expansion

$$\Delta\mathbf{v} = \Delta t \mathbf{v}_{,t}\big|^{n} + \frac{\Delta t^2}{2}\mathbf{v}_{,tt}\big|^{n+\theta}. \tag{9.2.2}$$

Inserting equation 9.2.1a repeatedly into (9.2.2), using (9.2.1b), and ignoring any spatial

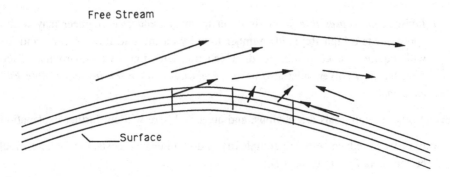

Figure 9.1: Separation point with typical boundary-layer grid: explicit schemes would imply very small timesteps in this situation.

derivatives of order higher than two yields

$$\Delta \mathbf{v} = \Delta t \left[-\mathbf{v} \nabla \mathbf{v} - \nabla p + \nabla \cdot \sigma \right] + \theta \Delta t \left[-\nabla \Delta p + \nabla \cdot \Delta \sigma \right]$$
$$+ \frac{\Delta t^2}{2} \left[(\mathbf{v} \nabla \mathbf{v}) \cdot \nabla \mathbf{v} + \nabla \mathbf{v} \otimes \mathbf{v} \nabla \mathbf{v} + \nabla p \nabla \mathbf{v} + \mathbf{v} \nabla \nabla p \right]^{n+\theta} . \tag{9.2.3}$$

Several observations should be made at this point:

- An additional factor appears in front of the streamline upwind diffusion (Brooks and Hughes [1982]) or balancing tensor diffusivity (Kelly et al.[1980]). Usually, only the advection-diffusion equation is studied. This assumes that the transport velocity field is steady. In the present case, the velocity field itself is being convected. This introduces the additional factor.

- Additional "mixed" velocity-pressure terms appear as a result of the consistent treatment of time-advancement for all terms of the Navier-Stokes equations. They may be interpreted as upwind-factors for the pressure.

In order to derive the discretization in time for the pressure, we re-state the divergence constraint as

$$c^{-2} p_{,t} + \nabla \cdot \mathbf{v} = 0, \tag{9.2.4}$$

where c is the speed of sound, and then proceed with the following Taylor-expansion

$$c^{-2} \Delta p = c^{-2} \Delta t \, p_{,t} |^n + c^{-2} \frac{\Delta t^2}{2} p_{,tt} |^{n+\theta} . \tag{9.2.5}$$

Inserting equation 9.2.4 repeatedly into (9.2.5), using (9.2.1b), ignoring any spatial derivatives of order higher than two, and taking the limit $c \to \infty$ yields

$$\Delta t \, \nabla \cdot \mathbf{v} = \frac{\Delta t^2}{2} \left[\nabla^2 p + \nabla \cdot \mathbf{v} \cdot \nabla \mathbf{v} \right]^{n+\theta} . \tag{9.2.6}$$

Observe that:

- A Laplacian "pressure-diffusion" appears naturally on the right-hand side. We will return to this Laplacian subsequently when we look for appropriate spatial discretizations.

- An additional "mixed" velocity-pressure term appears as a result of the consistent treatment of time-advancement for the divergence-equation.

The next task is to linearize the nonlinear terms in order to derive a timestepping procedure to advance the solution in time. Many different ways can be pursued. We just state the linearization employed here:

$$\left[\frac{1}{\Delta t} - \theta \Delta t \nabla \mathbf{v} \otimes \mathbf{v} \nabla - \theta \nabla \mathbf{D} \nabla\right] \Delta \mathbf{v} + \theta \nabla \Delta p =$$

$$- \mathbf{v} \cdot \nabla \mathbf{v} + \frac{\Delta t}{2} \nabla \mathbf{v} \otimes \mathbf{v} \nabla \mathbf{v} + \frac{\Delta t}{2} (\mathbf{v} \nabla \mathbf{v}) \cdot \nabla \mathbf{v} \qquad (9.2.7a)$$

$$+ \nabla \cdot \sigma - \nabla p + \frac{\Delta t}{2} (\nabla p \nabla \mathbf{v} + \mathbf{v} \nabla \nabla p)$$

$$\Delta t \nabla \cdot \mathbf{v} = \frac{\Delta t^2}{2} \left[\nabla^2 (p + \theta \Delta p) + \nabla \cdot \mathbf{v} \cdot \nabla \mathbf{v}\right] \qquad (9.2.7b)$$

with

$$\sigma = \mathbf{D} \nabla \mathbf{v}. \qquad (9.2.8)$$

The spatial discretization of this time-stepping scheme will result in the following coupled sytem of equations:

$$\begin{bmatrix} \mathbf{K}_{vv} & \mathbf{K}_{vp} \\ \mathbf{K}_{pv} & \mathbf{K}_{pp} \end{bmatrix} \begin{pmatrix} \Delta \mathbf{v} \\ \Delta \mathbf{p} \end{pmatrix} = \begin{pmatrix} \mathbf{r}_v \\ \mathbf{r}_p \end{pmatrix}. \qquad (9.2.9)$$

For practical 3-D problems, the direct solution of this coupled system of equations is beyond the reach of current supercomputer technology. The most common way to reduce the required storage and CPU-requirements is to decouple the system of equations into separate elliptic problems in the following iterative loop, sometimes referred to as *preconditioned Uzawa algorithm* (Gregoire et al. [1985]):

- U1. Given $\Delta \mathbf{p}$, compute $\Delta \mathbf{v}$:

$$\mathbf{K}_{vv} \Delta \mathbf{v} = \mathbf{r}_v - \mathbf{K}_{vp} \Delta \mathbf{p}. \qquad (9.2.10a)$$

- U2 Given $\Delta \mathbf{v}$, compute $\Delta \mathbf{p}$:

$$\left(-\mathbf{K}_{pv} \tilde{\mathbf{K}}_{vv}^{-1} \mathbf{K}_{vp} + \mathbf{K}_{pp}\right) \Delta \mathbf{p} = \mathbf{r}_p - \mathbf{K}_{pv} \tilde{\mathbf{K}}_{vv}^{-1} \mathbf{r}_v. \qquad (9.2.10b)$$

- U3 If not yet converged: goto U1.

Of course, little is gained unless an unexpensive approximation $\tilde{\mathbf{K}}_{vv}^{-1}$ to $\mathbf{K}_{vv}^{-1}$ can be found. Fortunately, such an approximation exists: simply take

$$\tilde{\mathbf{K}}_{vv}^{-1} = \mathbf{M}_l^{-1}, \qquad (9.2.11)$$

where $\mathbf{M}_l$ denotes the lumped mass-matrix. The reason why this approximation works is that in the limit, as the mesh size h tends to zero, Equations (9.2.10b, 9.2.11) will approximate the analytic result

$$\nabla^2 p = -\nabla \cdot \mathbf{v} \cdot \nabla \mathbf{v}, \qquad (9.2.12)$$

which is obtained by taking the divergence of the momentum equations (9.2.1a). Observe that:

- Many well-known schemes can be obtained as special cases of Equation (9.2.10). For example, projection schemes (Chorin [1968], Donea et al.[1982], Huffenus and Khaletzky [1984], Patera [1984], Kim and Moin [1985], Gresho and Chan [1990]), used commonly for transient problems, arise by taking only one iteration pass per timestep, most commonly with a starting guess for the pressure $\Delta p^0 = 0$.

- Unlike some projection schemes (e.g. Chorin [1968], Huffenus and Khaletzky [1984], Kim and Moin [1985]), this iterative scheme will always yield steady-state results that are independent of the time-step employed.

- Because of the special structure of $\mathbf{K}_{vv}$ that results from (9.2.7a), for laminar flows and some turbulence models the velocity increments $\Delta\mathbf{v}$ can be obtained from decoupled systems of equations, i.e., separate elliptic problems for each velocity component.

- The choice of $\mathbf{M}_l^{-1}$ to approximate $\mathbf{K}_{vv}^{-1}$ may not be appropriate if the advection operator is also integrated implicitly. In this case, the analytic decoupling alluded to in Equation (9.2.12) is no longer valid.

- We have observed in numerical experiments that the velocity increments $\Delta\mathbf{v}$ are obtained almost immediately in 1–2 iterative passes, whereas the pressure sometimes requires up to 10 passes to converge.

- In order to have high temporal accuracy for the advection terms, we always employ a consistent mass-matrix discretization in $\mathbf{K}_{vv}$. In doing so, we have not observed the loss of stability reported in Gresho and Chan [1990].

- For an in-depth discussion of the errors associated with projection schemes, see Gresho and Chan [1990].

Velocity	Pressure	Operator
p1	q0	(0,-1, 2,-1, 0)
p1	p1	(-1, 0, 2, 0,-1)
iso-p1	p1	(-1,-1, 4,-1,-1)

Table 9.1: Resulting operators for different velocity/pressure combinations.

9.3 Spatial Discretization

As stated before, we strive to perform the spatial approximation by using simple elements. If at all possible, all variables should be placed at the same location. As documented in the literature, the choice of location and polynomial order can be troublesome. The so-called Ladyzenskaya-Babuska-Brezzi (LBB) stability condition has to be met (Gunzburger [1987]). Instead of reviewing the procedures required to derive this condition, we look at the discrete Poisson-problem for the pressure (Equation (9.2.10b)). In there, the leading (first-order) matrix is of the form $\mathbf{K} = -\mathbf{K_{pv}M_l^{-1}K_{vp}}$. This is just the numerical equivalent of $\nabla^2 = \nabla \cdot \nabla$. Table 9.1 lists the resulting operators in 1-D for different choices of approximations of velocity and pressure. One can clearly see the unstable decoupling of nearest neighbors that occurs for the p1/p1 element. The p1/p0 element reproduces exactly the usual $(-1, 2, -1)$-Laplacian discretization, whereas the p1/p1+bubble (mini) element (see Thomasset [1981]), which in 1-D is the same as the p1/iso-p1 element, adds additional stabilizing terms to the unstable discretization resulting from the p1/p1 element. These stabilizing terms are equivalent to an artificial viscosity for the divergence equation, and are of the same form as in Equation (9.2.6). This equivalence can be shown to exist in 2-D and 3-D as well.

9.3.1 Equivalence of P1/Iso-P1 and Artificial Viscosity

In what follows, we will derive the equivalent artificial viscosity that the mini-element employs. By obtaining the exact value for the numerical viscosity that the mini-element adds to the continuity equation of the equivalent p1/p1-element, it is possible to construct better artificial viscosities. Particularly for stretched elements one can see how artificial viscosities should be constructed. We will perform the derivation for the usual Galerkin approximation of the steady-state form of the Navier-Stokes equations with upwinding of the advection terms:

$$\mathbf{v} \cdot \nabla \mathbf{v} + \nabla p = \frac{1}{Re} \nabla^2 \mathbf{v} + \frac{\Delta t}{2} \nabla \mathbf{v} \otimes \mathbf{v} \nabla \mathbf{v}. \tag{9.3.1}$$

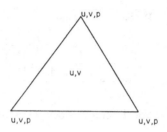

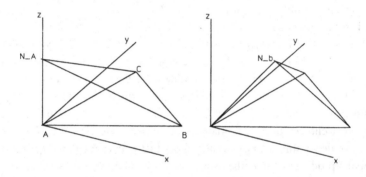

Figure 9.2: Notation used for the mini-element.

In order to simplify the analysis, we make the assumption that the velocity inside each p1-element is constant. This is equivalent to the use of 1-point quadrature for the integrals. For clarity, we define the following notation for the mini-element:

- N^i: the shape-functions associated with the corner points of the element, i.e., the equivalent p1-element shape-functions.

- N^b: the bubble-function associated with the center node; as is well known, at the center-node we only consider velocity degrees of freedom. As the bubble-function, we take the linear tent-function shown in Figure 9.2. This is not necessary, but simplifies the analysis somewhat.

- Nn: the number of nodes of the p1-element, i.e., $Nn = 3$ in 2-D and $Nn = 4$ in 3-D.

Let us now focus on the integrals associated with the bubble-function N^b. As these shape-functions disappear on the boundary, all boundary integrals that may arise due to integration by parts vanish identically. For the momentum equations, we obtain

$$\int_\Omega N^b v^l N^j_{,l} d\Omega \; v^m_j + \int_\Omega N^b N^i_{,m} d\Omega \; p_i =$$
$$- \frac{1}{Re} \int_\Omega N^b_{,l} N^j_{,l} d\Omega \; v^m_j - \frac{\Delta t}{2} \int_\Omega N^b_{,k} v^k v^l N^j_{,l} d\Omega \; v^m_j \,. \tag{9.3.2}$$

Evaluation of these integrals yields

$$K^{bi}_{lm} = \int_{\Omega_{el}} N^b_{,l} N^i_{,m} d\Omega = N^i_{,m} \int_{\Omega_{el}} N^b_{,l} d\Omega = 0 \,, \tag{9.3.3}$$

$$K^{bb}_{lm} = \int_{\Omega_{el}} N^b_{,l} N^b_{,m} d\Omega = NnA \sum_i N^i_{,l} N^i_{,m} \,, \tag{9.3.4}$$

$$G^{bi}_{l} = \int_{\Omega_{el}} N^b N^i_{,l} d\Omega = \frac{A}{Nn^2} N^i_{,l} \tag{9.3.5}$$

$$\int_{\Omega_{el}} N^b N^b_{,l} d\Omega = 0 \,. \tag{9.3.6}$$

Thus, from Equation (9.3.2) we obtain for the velocities at the center-node of the mini-element:

$$\left[\frac{1}{Re} K^{bb}_{ll} + \frac{\Delta t}{2} v^k v^l K^{bb}_{kl} \right] v^m_b = -v^l G^{bi}_{l} v^m_i - G^{bi}_{m} p_i \,. \tag{9.3.7}$$

Weighting the continuity equation with the p1 shape-functions N^i only yields

$$\int N^i N^j_{,m} d\Omega \; v^m_j = 0 \,. \tag{9.3.8}$$

These integrals are subdivided into the contributions from the p1-shape-functions N^i and the bubble-function N^b:

$$\int N^i N^j_{,m} d\Omega \; v^m_j = I_{p_1} + I_b \,, \tag{9.3.9}$$

with

$$I_b = \int_{\Omega_{el}} N^i N^b_{,m} d\Omega \; v^m_b = -\frac{A}{Nn^2} N^i_{,m} \; v^m_b \,. \tag{9.3.10}$$

Inserting (9.3.7) into (9.3.10) yields

$$I_b = \frac{A}{Nn^2} N^i_{,m} \frac{1}{\frac{1}{Re} K^{bb}_{ll} + \frac{\Delta t}{2} v^k v^l K^{bb}_{kl}} \left[v^l G^{bi}_{l} v^m_i + G^{bi}_{m} p_i \right] \,, \tag{9.3.11}$$

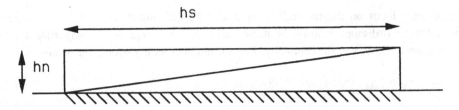

Figure 9.3: Typical boundary layer element.

or

$$I_b = A\,\beta\left[N^i_{,l}v^m N^j_{,m}v^l_j + N^i_{,l}N^j_{,l}p_j\right] \qquad (9.3.12)$$

with

$$\beta = \frac{1}{Nn^5\left[\frac{1}{Re}\sum_j |\nabla N^j|^2 + \frac{\Delta t}{2}\sum_j \left(\mathbf{v}\cdot\nabla N^j\right)^2\right]}. \qquad (9.3.13)$$

One can see that this same expression would have been obtained if instead of using the mini-element with equations (9.3.1,9.2.1b), we had used the p1-element with a modified continuity equation of the form:

$$\nabla\cdot\mathbf{v} = \nabla\cdot\beta\mathbf{v}\cdot\nabla\mathbf{v} + \nabla\beta\nabla p. \qquad (9.3.14)$$

But this is just Equation (9.2.7b) with $\beta = \frac{\Delta t}{2}$! Thus, the mini-element with the conventional Galerkin approximation is the same as the p1/p1-element with the Taylor-Galerkin approximation as far as the continuity equation is concerned. Let us now consider the case of highly stretched boundary layer elements. Denoting the streamwise element length by h_s and the normal element length by h_n, we assume that for accuracy, the cell Reynolds number $Re_c = h_n\,Re\ <\ 1$. This implies that the upwind-terms in Equation (9.3.13) can be ignored. Consider a triangle which has been obtained by subdivision of a rectangle into two, as shown in Figure 9.3. For such an element, we obtain

$$\beta = \frac{Re}{2Nn^5\left(\frac{1}{h_s^2} + \frac{1}{h_n^2}\right)}. \qquad (9.3.15)$$

Thus, for $h_s \gg h_n$

$$\beta = \frac{h_n^2\,Re}{2Nn^5} = 0.00206\,Re\,h_n^2. \qquad (9.3.16)$$

The exact value of the diffusion coefficient β depends on the "tent" function used for the center-node of the mini-element. Our numerical experience indicates that the values given by Equation (9.3.13) tend to give noisy pressure solutions. Therefore, in practice, a value ten times bigger than that given by Equation (9.3.13) is employed.

9.4 Elliptic Solvers

As shown before, the Uzawa algorithm requires the solution of several elliptic Poisson-type problems. The fastest solvers for this class of problems are unstructured multigrid solvers (Löhner and Morgan [1987]). They require good smoothers, as well as efficient intergrid transfer operators. In what follows, we describe a class of iterative solvers that lie between the complexity of multiple grids (multigrid), and the excessive memory requirements of direct solvers. Codes based on structured gridding have explored for a long time the useful properties of line-relaxation (Briley and McDonald [1977], Beam and Warming [1978], Rogers et al.[1985], Mansour and Hamed [1990]). Line-relaxation offers an economical way to circumvent directional stiffness by joining together neighbors of neighbors along the line. Thus, it offers a practical way to derive good preconditioners and smoothers. The concept of lines translates to snakes (Hassan et al. [1990]) in the context of an unstructured grid. The corresponding relaxation scheme to solve

$$\mathbf{K}\mathbf{u} = \mathbf{r}, \qquad (9.4.1)$$

where $\mathbf{K}$ is the matrix describing the elliptic problem, $\mathbf{u}$ the desired vector of unknowns, and $\mathbf{r}$ the right-hand side vector, becomes

$$\mathbf{K}_1\mathbf{K}_2\Delta\mathbf{u} = \mathbf{r} - \mathbf{K}\mathbf{u}. \qquad (9.4.2)$$

Because an unstructured grid usually does not possess an equal number of gridpoints along a certain direction, the resulting snakes may often exhibit folding (see Figure 9.4b). This implies that the information flow from the domain to the boundary may be slowed down considerably. This is not important for smoothers, but crucial for preconditioners. In the present context, we augment Equation (9.4.2) by a Preconditioned Conjugate Gradient (PCG) (Hestenes and Stiefel [1952]. Therefore, we seek a good preconditioner in the snake-relaxation. Wherever the snake folds, we reconnect it in the direction the snake is intended to continue. This gives rise to a more complex structure, which we call linelet (see Figure 9.4c). Whereas the storage requirements of snakes are fixed ($3N$, where N is the number of unknowns), the storage requirements of linelets depend on the structure of the mesh and the renumbering chosen. Consider the academic example shown in Figure 9.5. The linelet structure exhibited there may stem from a refined mesh region close to a curved wall. Tables 9.2–9.4 list the resulting extra storage and the ratio of snake-storage for symmetric matrices ($2N$) to linelet-storage for different orders of bifurcation. One can see that the extra amount of storage and CPU incurred by switching from snakes to linelets is indeed limited. For the general case, we have implemented a reverse Cuthill-McKee ordering (Cuthill and McKee [1969]) for the linelets. Practical calculations indicate $O(5N - 10N)$ storage requirements. This is deemed acceptable, as

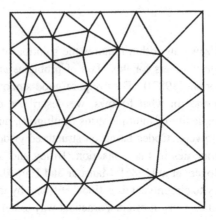

a) Original discretization

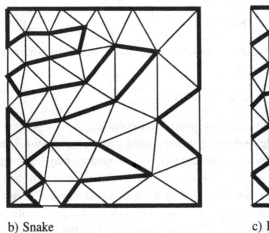

b) Snake   c) Linelets

Figure 9.4: Snake and Linelets for a 2-D discretization.

it takes much more than $6N$ operations to build a new right-hand side. As expected, the convergence rate of the PCG algorithm increases considerably when going from snakes to linelets.

9.5 Computation of Timesteps and Upwinding Factors

The allowable timestep for the elements is given by

$$\Delta t = r(Re_c) \cdot \Delta t_{CFL}, \qquad (9.5.1)$$

Level	Nr. Eqs./N	Add. Stor./N	Ratio	Ratio Symm.
1	3	1	0.333	1.167
2	7	5	0.714	1.357
3	15	17	1.133	1.567
4	31	49	1.581	1.790
5	63	129	2.048	2.024
6	127	321	2.528	2.264
7	255	769	3.016	2.508
8	511	1793	3.509	2.754
9	1023	4097	4.005	3.002
10	2047	9217	4.503	3.251

Table 9.2: Storage requirements for linelets; order of bifurcation: 2.

Level	Nr. Eqs./N	Add. Stor./N	Ratio	Ratio Symm.
1	4	2	0.500	1.250
2	13	14	1.077	1.538
3	40	68	1.700	1.850
4	121	284	2.347	2.174
5	364	1094	3.005	2.503
6	1093	4010	3.669	2.834
7	3280	14216	4.334	3.167
8	9841	49208	5.000	3.500
9	29524	167306	5.667	3.833
10	88573	560966	6.333	4.167

Table 9.3: Storage requirements for linelets; order of bifurcation: 3.

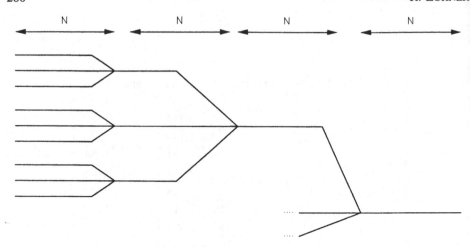

Figure 9.5: Linelet with order of bifurcation = 3.

Level	Nr. Eqs./N	Add. Stor./N	Ratio	Ratio Symm.
1	11	9	0.818	1.409
2	111	189	1.703	1.851
3	1111	2889	2.600	2.300
4	11111	38889	3.500	2.750
5	111111	488889	4.400	3.200

Table 9.4: Storage requirements for linelets; order of bifurcation: 10.

where $r(Re_c)$ denotes a reduction factor due to the element or cell Reynolds number Re_c, and Δt_{CFL} the allowable element timestep for the limit $Re \to \infty$. In what follows, we try to obtain a criterion akin to

$$\Delta t_{CFL} = \frac{\text{length covered}}{\text{velocity}} . \qquad (9.5.2)$$

For regular grids, this question seldomly arises. However, for large Reynolds number problems, one may have highly stretched grids close to the wall. Then, if the minimum element length is employed in Equation (9.5.2), a very conservative timestep Δt_{CFL} is computed. To obtain a more realistic representation of Equation (9.5.2), we make the observation that the element normals are related to the shape-function derivatives within an element as follows:

- direction of normal: $\mathbf{n}_i = \frac{\nabla N_i}{|\nabla N_i|}$;
- length of normal: $h_i = \frac{1}{|\nabla N_i|}$.

This implies that for each of the element normals, Equation (9.5.2) reduces to

$$\Delta t_i = \frac{h_i}{|\mathbf{v} \cdot \mathbf{n}_i|} = \frac{1}{|\mathbf{v} \cdot \nabla N_i|} . \tag{9.5.3}$$

The cell Reynolds number corresponding to this normal is given by:

$$Re_i = h_i Re = \frac{|\mathbf{v} \cdot \nabla N_i|}{|\nabla N_i|^2} Re . \tag{9.5.4}$$

Given this cell Reynolds number, we can compute the optimal upwinding-factor α for steady-state simulations:

$$\alpha_i = \coth\left(\frac{Re_i}{2}\right) - \frac{2}{Re_i} . \tag{9.5.5}$$

As the final timestep, we take the minimum timestep obtained from the normals of an element. On the other hand, for α we take the maximum value encountered over the normals of an element. As we employ implicit timestepping schemes, and the right-hand sides of Equations (9.2.7a,9.2.7b) depend on the timestep Δt, care has to be taken not to over-diffuse the solution. Therefore, we always limit the timestep taken for the construction of the right-hand sides by the optimal upwinding timestep:

$$\Delta t_{rhs} = \min(\Delta t, \Delta t_{opt}) . \tag{9.5.6}$$

This choice of timestep for the right-hand side will lead to steady-state solutions that are independent of the timestep-size chosen for the implicit solver. A von Neumann stability analysis indicates that for the matrices on the left-hand side the full value of the timestep should be employed.

9.6 Numerical Examples

9.6.1 Driven Cavity (2-D, viscous, steady-state, $Re = 10^2$)

This is a well-known test example. In order to compare different spatial discretizations, both the p1/iso-p1 and the p1/p1-element were employed. In both cases we employed 1,433 points and 2,712 elements. The results obtained are shown in Figures 9.6a–f. As one can see, the difference between the two elements is minimal. The locations of vortices are compared with the solutions obtained by Ghia et al. [1981] for a streamfunction-vorticity scheme on a 129×129 cartesian grid in Table 9.5. The location of the vortices for the present results were obtained graphically, with a maximum accuracy of ± 0.01. The comparison with the fine-grid results is extremely good. As this is a steady-state problem, local timesteps were employed to speed up convergence.

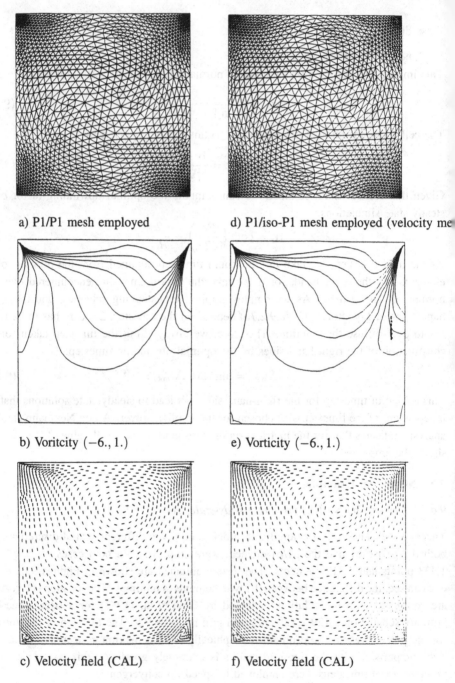

a) P1/P1 mesh employed

d) P1/iso-P1 mesh employed (velocity me

b) Voritcity $(-6., 1.)$

e) Vorticity $(-6., 1.)$

c) Velocity field (CAL)

f) Velocity field (CAL)

Figure 9.6: Driven cavity: $Re = 10^2$.

Vortex	p1/p1	p1/iso-p1	Ghia et al.
cc	0.620, 0.743	0.620, 0.743	0.6170, 0.7344
bl	0.035, 0.037	0.035, 0.037	0.0313, 0.0391
br	0.940, 0.064	0.940, 0.064	0.9453, 0.0625

Table 9.5: $Re = 10^2$: location of vortices.

Vortex	p1/p1	Ghia et al.
cc	0.520, 0.530	0.5117, 0.5333
bl1	0.067, 0.164	0.0586, 0.1641
br1	0.830, 0.060	0.7656, 0.0586
ul1	0.067, 0.933	0.0703, 0.9141
bl2	0.015, 0.016	0.0156, 0.0195
br2	0.970, 0.025	0.9336, 0.0625

Table 9.6: $Re = 10^4$: location of vortices.

9.6.2 Driven Cavity (2-D, viscous, steady-state, $Re = 10^4$)

The problem statement is the same as before, but the Reynolds number has now been increased to $Re = 10^4$. The grid employed, shown in Figures 9.7a,b, was constructed by inspection of the results obtained by Ghia et al. on a 257×257 cartesian grid. The minimum element length prescribed was $\delta = 0.03$, while the maximum stretching ratio for the elements was set to 1:10. The total number of gridpoints is 9,054. The results obtained using p1/p1-elements are shown in Figures 9.7c–f. The locations of vortices are compared with the solutions obtained by Ghia et al. in Table 9.6. Again, the locations of the vortices for the present results were obtained graphically, with an accuracy of ±0.01. Except for the corner sub-vortices, the comparison with the fine grid solutions is again very good.

9.6.3 Circular Arc Cascade (2-D, viscous, steady-state, $Re = 10^3$)

The problem set-up and the boundary conditions were taken from Mansour and Hamed [1990]. The thickness to chord ratio was set to 0.2 and the pitch to chord ratio was 2.0. The grid employed, consisting of 5,948 p1/p1 elements and shown in Figure 9.8a, was taylored to be similar to that used in Mansour and Hamed [1990]. The results

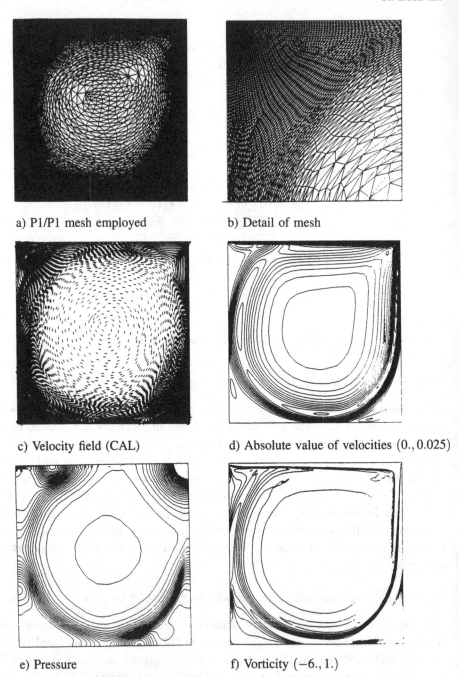

a) P1/P1 mesh employed   b) Detail of mesh

c) Velocity field (CAL)   d) Absolute value of velocities $(0., 0.025)$

e) Pressure   f) Vorticity $(-6., 1.)$

Figure 9.7: Driven cavity: $Re = 10^4$.

Re-nr.	Timestep	Steps/Cycle	Num. Str.	Exp. Str.
100	0.05	120	0.167	0.167
140	0.05	113	0.177	0.177
1000	0.05	96	0.208	0.208

Table 9.7: Comparison of Strouhal-numbers.

obtained after 200 steps, corresponding to a residual decrease of 4 orders of magnitude, are shown in Figures 9.8b–d. Flow separation occurs at about 85% of the chord length. These results are almost graphically undistinguishable from those reported in Mansour and Hamed [1990].

9.6.4 Circular Cylinder (2-D, viscous, transient)

This is another well-known test example. We performed a parametric study for the von Karman vortex street observed experimentally (Schlichting [1979]). The numerically observed Strouhal-numbers are compared with the experimental values in Table 9.7. As one can see, the agreement is exceptional. Figure 9.9 shows one example of this study, for a Reynolds number of $Re = 10^3$. The mesh close to the cylinder is shown in Figure 9.9a, the local and global pressure fields in Figure 9.9b,c and the velocity field close to the cylinder in Figure 9.9d.

9.6.5 Airfoil (2-D, inviscid, steady-state)

This case was used to test the capability of the methodology outlined above to predict inviscid incompressible flows. The angle of attack was set to $\alpha = 5^o$. The mesh, displayed in Figure 9.10a, consisted of 3,199 points and 6,196 p1/iso-p1 elements. The solution obtained is shown in Figure 9.10b.

9.6.6 Airfoil (2-D, viscous, transient, $Re = 10^4$)

After obtaining the solution for the inviscid case, the resulting wake-line was used to construct a local, structured c-mesh for the expected boundary layer. The boundary layer thickness was estimated from boundary-layer theory. The resulting mesh, consisting of 10,909 points and 21,516 p1/iso-p1 elements, is shown in Figure 9.11a. While the angle of attack was kept at $\alpha = 5^o$, the Reynolds number was set to $Re = 10^5$. The resulting unsteady flowfield is shown in Figures 9.11b–e. The separation point occurs at 35% of the chord-length, and is in good agreement with experimental data.

a) MESH , NELEM= 5948 , NPOIN= 3072

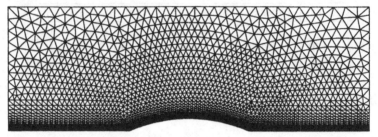

b) ABS(VEL), MIN= 0.00E+00 , MAX= 1.30E+00 , DUC= 1.00E-01

c) PRESSURE, MIN= 7.75E-01 , MAX= 1.45E+00 , DUC= 2.50E-02

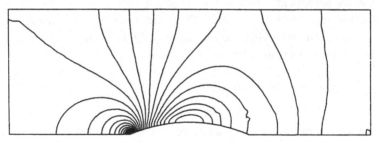

d) VORTICI., MIN=-7.60E+01 , MAX= 5.00E+00 , DUC= 3.00E+00

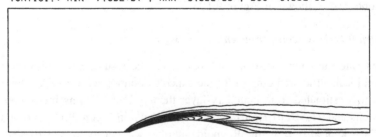

Figure 9.8: Circular arc cascade.

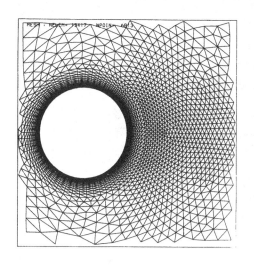

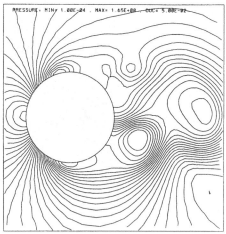

a) Mesh close to the cylinder   b) Pressure close to the cylinder

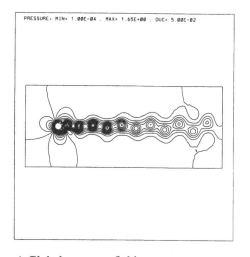

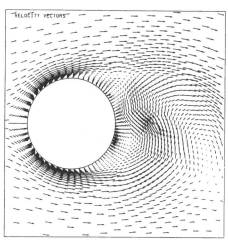

c) Global pressure field   d) Velocity field close to the cylinder

Figure 9.9: von Karman vortex shedding from a circular cylinder.

MESH   NELEM= 6196 . NPOIN=   3199

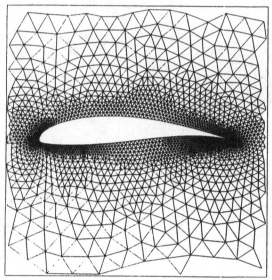

a) Velocity mesh employed

PRESSURE   MIN= 0.15E+00 . MAX= 0.15E+01 . DUC= 0.70E-01

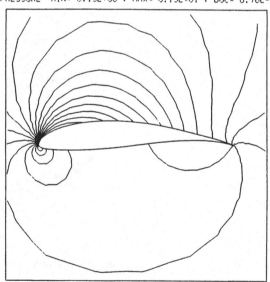

b) Pressure contours

Figure 9.10: Inviscid flow past an airfoil.

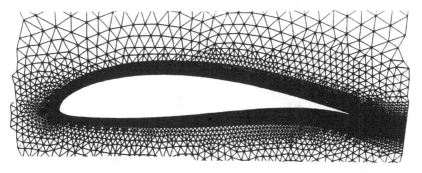

a) Velocity mesh employed

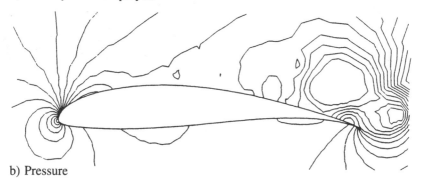

b) Pressure

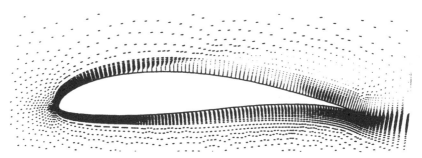

c) Velocity field

Figure 9.11: Flow past airfoil: $Re = 10^5$.

9.6.7 Inlet Flow in a Channel (3-D, viscous, steady-state, $Re = 100$)

The surface triangulation of the volumetric mesh is shown in Figure 9.12a. The mesh consisted of 30,070 tetrahedra and 6,481 points. Figures 9.12b–e show the surface mesh,

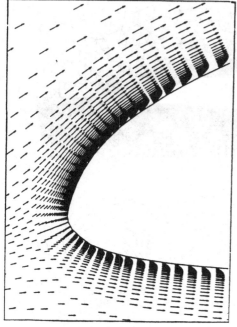

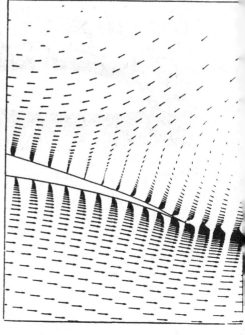

d) Leading edge detail   e) Trailing edge detail

Figure 9.11: *cont.*

pressure, absolute value of the velocities and velocities on one of the outside surfaces. As expected, after an initial run-up distance, the flow settles into the expected channel flow pattern with linear pressure decrease.

9.7   Conclusions

We have described several numerical techniques that have proven successful when trying to solve incompressible flow problems in practice. Because the numerical simulation of incompressible flows has reached a mature state, it seems reasonable to incorporate in more depth the requirements and aspects that arise in engineering and design practice. Therefore, we established in the introductory section a list of requirements deemed important when trying to solve quickly practical problems.

Although perhaps mathematically not very elegant, the resulting algorithms meet these requirements, as evidenced by the diverse numerical examples presented. The addition of an artificial viscosity term for the continuity equation was shown to be identical to

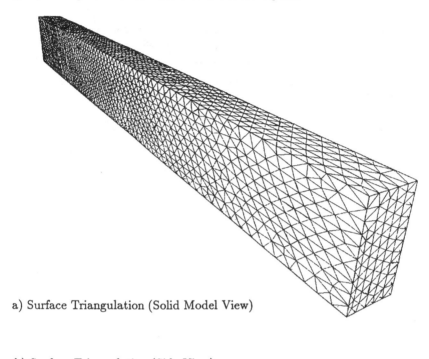

a) Surface Triangulation (Solid Model View)

b) Surface Triangulation (Side View)

c) Pressure (Side View)

d) Absolute Value of the Velocities (Side View)

e) Velocities (Side View)

Figure 9.12: Inlet flow in a 3-D channel.

the mini-element. The suspicion arises that many mixed elements with different shape-function spaces introduce some form of numerical viscosity, be it of second or fourth order. A more thorough investigation of this question remains a task for the future.

Other developments will center on the incorporation of turbulence models, and further optimization of the algorithms in 3-D.

Acknowledgements

This work was supported by ONR under Grant N00014-90-J-1416, with Dr. Spiro Lekoudis as the technical monitor, and the Laboratory for Computational Physics and Fluid Dynamics of the Naval Research Laboratory.

References

[1] Baker, T. J. (1989). "Developments and Trends in Three-Dimensional Mesh Generation," *Appl. Num. Math.* **5**, 275–304.

[2] Beam, R. M. and Warming, R. F. (1978). "An Implicit Finite Difference Algorithm for Hyperbolic Systems in Conservation-Law Form," *J. Comp. Phys.* **22**, 87–110.

[3] Briley, W. R., and McDonald, H. (1977). "Solution of the Multi-Dimensional Compressible Navier-Stokes Equations by a Generalized Implicit Method," *J. Comp. Phys.* **21**, 372–397.

[4] Brooks, A. N. and Hughes, T. J. R. (1982). "Streamline Upwind/Petrov Galerkin Formulations for Convection Dominated Flows with Particular Emphasis on the Incompressible Navier-Stokes Equations," *Comp. Meth. Appl. Mech. Eng.* **32**, 199–259.

[5] Chorin, A. J. (1968). "Numerical Solution of the Navier-Stokes Equations," *Math. Comp.* **22**, 745–762.

[6] Confs (1990). See the following Conference Series: *Finite Elements in Fluids* **I-VII**, J. Wiley & Sons, *Int. Conf. Num. Meth. Fluid Dyn.* **I-XI**, Spinger Lecture Notes in Physics, *AIAA CFD Conf.* **I-IX**, AIAA CP, *Num. Meth. Laminar and Turbulent Flow*, Pinerige Press, and others.

[7] Cuthill, E., and McKee, J. (1969). "Reducing the Bandwidth of Sparse Symmetric Matrices," *Proc. ACM Nat. Conf.*, New York 1969, 157–172.

[8] Donea, J., Giuliani, S., Laval, H., and Quartapelle, L. (1982). "Solution of the Unsteady Navier-Stokes Equations by a Fractional Step Method," *Comp. Meth. Appl. Mech. Eng.* **30**, 53–73.

[9] Ghia, U., Ghia, K. G., and Shin, C. T. (1981, October). "Solution of the Incompressible Navier-Stokes Equations by Coupled Strongly-Implicit Multigrid Method," *Proc. Symp. Multigrid Methods*, NASA Ames.

[10] Gregoire, J. P., Benque, J. P., Lasbleiz, P. and Goussebaile, J. (1985). "3-D Industrial Flow Calculations by Finite Element Method," *Springer Lecture Notes in Physics* **218**, 245–249.

[11] Gresho, P. M., and Chan, S. T. (1990). "On the Theory of Semi-Implicit Projection Methods for Vicous Incompressible Flows and its Implementation via a Finite Element Method That Introduces a Nearly-Consistent Mass Matrix," *Int. J. Num. Meth. Fluids* to appear.

[12] Gunzburger, M. D. (1987). "Mathematical Aspects of Finite Element Methods for Incompressible Viscous Flows," *Finite Elements: Theory and Application* (Dwoyer, Hussaini and Voigt eds.), Springer Verlag, 124–150.

[13] Hassan, O., Morgan, K., and Peraire, J. (1990). "An Implicit Finite Element Method for High Speed Flows," AIAA-90-0402.

[14] Hestenes, M. and Stiefel, E. (1952). "Methods of Conjugate Gradients for Solving Linear Systems," *J. Res. Nat Bur. Standards* **49**, 409–436.

[15] Huffenus, J. D. and Khaletzky, D. (1984). "A Finite Element Method to Solve the Navier-Stokes Equations Using the Method of Characteristics," *Int. J. Num. Meth. Fluids* **4**, 247–269.

[16] Kelly, D. W., Nakazawa, S., Zienkiewicz, O. C., and Heinrich, J. C. (1980). "A Note on Anisotropic Balancing Dissipation in Finite Element Approximation to Convection Diffusion Problems," *Int. J. Num. Meth. Eng.* **15**, 1705–1711.

[17] Kim, J. and Moin, P. (1985). "Application of a Fractional-Step Method to Incompressible Navier-Stokes Equations," *J. Comp. Phys.* **59**, 308–323.

[18] Löhner, R. and Morgan, K. (1987). "An Unstructured Multigrid Method for Elliptic Problems," *Int. J. Num. Meth. Eng.* **24**, 101–115.

[19] Löhner, R. and Parikh, P. (1988). "Three-Dimensional Grid Generation by the Advancing Front Method," *Int. J. Num. Meth. Fluids* **8**, 1135–1149.

[20] Mansour, M. L. and Hamed, A. (1990). "Implicit Solution of the Incompressible Navier-Stokes Equations on a Non-Staggered Grid," *J. Comp. Phys.* **86**, 147–167.

[21] Patera, A. T. (1984). "A Spectral Element Method for Fluid Dynamics: Laminar Flow in a Channel Expansion," *J. Comp. Phys.* **54**, 468–488.

[22] Peraire, J., Morgan, K., and Peiro, J. (1990). "Unstructured Finite Element Mesh Generation and Adaptive Procedures for CFD," *AGARD-CP-464*, 18.

[23] Rogers, S. D., Kwak, D., and Kaul, U. (1985). "On the Accuracy of the Pseudocompressibility Method in Solving the Incompressible Navier-Stokes Equations," AIAA-85-1689.

[24] Sani, R., Gresho, P. M., Lee, R. L., and Griffiths, D. F. (1981a). "The Cause and Cure (?) of the Spurious Pressures Generated by Certain FEM Soltions of the Incompressible Navier Stokes Equations, Part 1," *Int. J. Num. Meth. Fluids* **1**, 17–43.

[25] Sani, R., Gresho, P. M., Lee, R. L., Griffiths, D. F., and Engleman, M. (1981b). "The Cause and Cure (?) of the Spurious Pressures Generated by Certain FEM Soltions of the Incompressible Navier Stokes Equations, Part 2," *Int. J. Num. Meth. Fluids* **1**, 171–204.

[26] Schlichting, H. (1979). *Boundary Layer Theory*, McGraw-Hill.

[27] Taylor, C. and P. Hood, P. (1973). "A Numerical Solution of the Navier-Stokes Equations Using the Finite Element Method," *Comp. Fluids* **1**, 73–100.

[28] Thomasset, F. (1981). *Implementation of Finite Element Methods for Navier-Stokes Equations*, Springer-Verlag.

10 The Covolume Approach to Computing Incompressible Flows

R. A. Nicolaides

10.1 Introduction

This article contains a summary account of covolume methods for incompressible flows. Covolume methods are a recently developed way to solve both compressible and incompresssible flow problems on unstructured meshes. The general idea is to use complementary pairs of control volumes to discretize flux, circulation and other expressions which occur in the governing equations. These complementary volumes (*covolumes* for short) are related by an orthogonality property which is a basic feature of the covolume approach. One of the simplest mesh configurations which is suitable is the Delaunay-Voronoi mesh pair. This is introduced in the next section. After that we proceed through div-curl systems to the stationary Stokes equations and the Navier-Stokes equations. We will show that for uniform meshes the covolume equations for the stationary Stokes equations specialize to the MAC (staggered mesh) scheme, and that the MAC scheme itself is actually equivalent to a velocity-vorticity scheme. Some numerical results are presented in the last section.

Since this article is intended only as an overview, we will present most of the results in a two dimensional setting. Almost all of the ideas and techniques do generalize nicely to three dimensions but are harder to visualize than in two dimensions. Given our limited aims it would be inappropriate to present proofs of most of the mathematical results. We will refer to the original sources for these and other details.

One of the reasons for introducing covolume methods is to find lower order methods for viscous flows which are free of "spurious mode" problems. Spurious modes are a common feature of low order discretizations of flow problems, both compressible and incompressible. It can be proved for some low order finite element schemes that convergence of the pressure (at least) to the exact solution occurs arbitrarily slowly as the mesh size approaches zero (Boland and Nicolaides [1984 and 1985]). The presence of "weakly" spurious modes is responsible for this behavior. A brief introduction to this topic is contained in Appendix 10.A to this article.

295

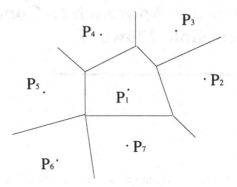

Figure 10.1: The Voronoi polygons of $P_1 - P_7$.

10.2 Covolume Meshes

In this section we will summarize the properties of a particular type of dual mesh system. These meshes have an orthogonality property which is exploited by covolume schemes. The best known example of the type of mesh used in covolume schemes is the Voronoi-Delaunay pair. We will introduce this first. Section 10.2.5 mentions orthogonal mesh pairs which are not Voronoi-Delaunay systems. Riedinger et al [1988] contains references to many of the basic papers in the Voronoi-Delaunay field.

10.2.1 Voronoi Diagrams

Suppose we have N nodes x_i $i = 1, 2, \ldots, N$ in the plane. It makes sense to ask for the set of all points in the plane which are closer to a particular node than to the others. This set of points is the *Voronoi* (or *Dirichlet*) *region* associated with the node. For a node x_j the Voronoi region P_j is the set of points

$$P_j := \{x \in \Re^2;\ |x_j - x| < |x_k - x|,\ k \neq j\}$$

Figure 10.1 illustrates the idea. This figure illustrates the general fact that the Voronoi regions are convex polygons whose sides bisect the lines joining certain pairs of vertices. Convexity follows because each Voronoi region is formed as the intersection of half spaces.

10.2.2 Delaunay Triangulations

Two nodes are said to be *adjacent* if their Voronoi regions share a common boundary segment. We can make a dual graph by joining all pairs of adjacent nodes by straight

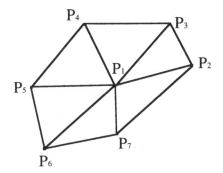

Figure 10.2: Delaunay triangulation of $P_1 - P_7$.

lines. It turns out that the resulting graph is a triangulation (usually – see below) of the convex hull of the nodes. It is called the (or a) Delaunay triangulation of the nodes. Figure 10.2 shows the Delaunay triangulation of the set of nodes of Figure 10.1. It follows from elementary geometry that the triangles and polygons are complementary to each other in the sense that the edges of either set of such domains are perpendicular to the faces of the other. Stated like this, the same property is true in three dimensions. The tetrahedra and polyhedra in three dimensions are regarded as *complementary volumes* or *covolumes* for short.

In some cases (for example four nodes at the vertices of a square) the Delaunay triangulation can contain figures other than triangles. If a strict triangulation is needed these figures can themselves be triangulated. Such triangulations can be done in more than one way so that uniqueness of the triangulation can be lost. A "generic" triangulations is one which has no degeneracy of this type. In general, the Delaunay triangulation can be an equivalence class of triangulations.

A Delaunay triangulation may be characterized in several ways. One way is by the easily shown property that the circumcircle of any triangle (circumsphere of any tetrahedron in three dimensions) contains no nodes other than the ones on its boundary. This gives another interpretation of the degenerate case just mentioned. In a degenerate case, the "triangulation" contains cyclic figures other than triangles the simplest being cyclic quadrilaterals.

10.2.3 Locally Equiangular Meshes

There is another characterization which is very useful in two dimensions. This approach is not suitable for three dimensions however.

Suppose given a bounded polygonal planar domain Ω and consider triangulations of Ω based on a fixed set of N nodes including the boundary nodes. In any triangulation, we can locate quadrilaterals made up from two adjacent triangles. In general they may be convex or nonconvex. Restricting attention to convex quadrilaterals the common triangle side will form a diagonal of the quadrilateral. This diagonal may join either the pair of corners of the quadrilateral whose angle sum is greater than or less than 180 degrees. For a cyclic quadrilateral both angle sums are the same. A triangulation is called *locally equiangular* if, for every such convex quadrilateral the diagonal joins the corners with angle sum at least equal to 180 degrees. There is no need for Ω to be convex in this definition.

A Delaunay triangulation is locally equiangular. This may be easily proved using the characterization given at the end of the last section. Then the uniqueness of a generic Delaunay triagulation implies that local equiangularity characterizes the Delaunay triangulation when Ω is convex.

10.2.4 Nonconvex Domains

The duality definition of the Delaunay mesh cannnot be used when the domain Ω is not convex because the mesh necessarily triangulates the convex hull of the nodes. In the nonconvex case the (triangulation) mesh segments which define the boundary are assumed to be given. Then we may simply define a Delaunay mesh to be a triangulation which has the circumcircle property mentioned at the end of Section 10.2.2, or, in two dimensions, which is locally equiangular. If Ω is convex this will produce the standard Delaunay triangulation. The dual of this triangulation which is obtained by joining the circumcenters of adjacent triangles is related to the Voronoi diagram. In fact, the dual figures associated with the *interior* nodes are just the Voronoi figures of those nodes.

In viscous flow computations the dual cells at the boundaries are used for discretization. Figure 10.3 shows the situation. The cells are formed from the dual edges drawn to the midpoints of the boundary segments and the intercepted part of the boundary itself. These are *hybrid* cells in the sense that their boundaries are part primal and part dual mesh edges. The boundary of the dual cell is ABCDEFOA in the figure. It is not convex although it is a simple (not self intersecting) polygon.

10.2.5 Arbitrary Triangulations

Given an arbitrary triangulation of Ω, a dual triangulation can be constructed by joining the circumcenters of adjacent triangles. In particular, the triangles surrounding an interior node will have their circumcenters joined and a closed path will be associated with the node. If the region enclosed in the path is convex, it must be the Voronoi region of

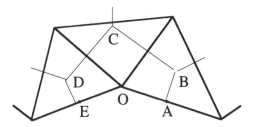

Figure 10.3: A boundary covolume.

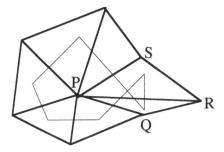

Figure 10.4: A "fishy" covolume.

the node. However, in general this path can be non simple. This is illustrated in Figure 10.4.

In line with Section 10.2.3 note that by exchanging diagonals in the convex quadrilateral PQRS we do obtain a simple and convex cell around the node as Figure 10.5 shows.

It is worth knowing that in two dimensions only two kinds of dual cells are possible. They are convex cells or non simple cells. Simple nonconvex dual cells do not occur in two dimensions. This is clearly true if the triangulation contains only acute angled triangles. In fact in that case the triangulation is locally equiangular so that the interior covolumes are Voronoi regions. The general result can be proved by elementary geometry.

If the triangulation contains obtuse angled triangles it may or may not be a Delaunay mesh, and if it is not it must contain at least one nonsimple dual cell. These more general dual meshes can still be used for covolume discretizations since the orthogonality property still holds. However they require more detailed programming than the simple case.

In three dimensions it is difficult to visualize the structure of the covolumes which are derived from an arbitrary triangulation. It is not known whether the faces of the

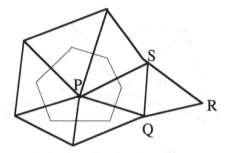

Figure 10.5: The effect of diagonal exchange.

covolumes can be simple but not convex. The orthogonality property does hold however. In the Voronoi situation the covolumes are convex and therefore so are their faces.

10.3   Planar div curl Systems

In this section we will consider the covolume discretization of the system

$$\begin{aligned} \operatorname{div} \mathbf{u} &= 0 \\ \operatorname{curl} \mathbf{u} &= \omega \\ \mathbf{u}.\mathbf{n}|_\Gamma &= f. \end{aligned} \qquad (10.3.1)$$

The main points can be illustrated using a simply connected domain Ω. The extra features of the multiply connected case are treated in Nicolaides [1991a]. We will first show how the discrete flux and circulation are defined. Then we will outline a related discrete vector field theory leading to a discrete Helmholtz decomposition. This theory can be used to prove existence and solvability for the discrete equations. The main error estimate also follows easily from the theory. The proofs of the results stated in this section are given in Nicolaides [1991a] which contains additional results. For analogous three dimensional results see Nicolaides and Wu [1991b].

10.3.1   Discretization

We will work with a Voronoi-Delaunay mesh system although this is not strictly necessary. A positive direction on each edge σ of the triangulation is defined to be from low to high node number. The positive normal direction to the edge is such that it and the positive direction are oriented like the (right handed) coordinate system. The covolume scheme approximates the quantities $\mathbf{u}.\mathbf{n}$. The discrete equations are obtained as follows. First, we integrate the equation (10.3.2a) over each triangle and use the divergence theorem. The resulting equation for a general triangle τ with boundary $\partial\tau$ is

$$\int_{\partial\tau} \mathbf{u}.\mathbf{n}\, ds = 0$$

which is approximated by

$$\sum_{i\in\partial\tau} u_i h_i = 0$$

where u_i denotes an approximation to $\mathbf{u}.\mathbf{n}$ on the edge labelled i and where h_i denotes the corresponding edge length signed positively if the normal points out of the triangle and signed negatively otherwise.

In a similar way the equation (10.3.2b) may be integrated over each interior covolume and Stokes theorem applied. Denoting by τ' a general interior covolume with boundary $\partial\tau'$ the result is

$$\int_{\partial\tau'} \mathbf{u}.\mathbf{t}\, ds = \int_{\tau'} \omega\, dx\, dy$$

where $\partial\tau$ is described counterclockwise. This equation is approximated by

$$\sum_{j\in\partial\tau'} u_j h'_j = \int_{\tau'} \omega\, dx\, dy$$

where h'_j denotes the corresponding dual edge length signed positively if the dual edge is directed in the positive direction of the integration path and negatively otherwise.

The boundary equation (10.3.2c) is approximated by defining boundary values u_k using

$$u_k := \frac{1}{|\sigma_k|}\int_{\sigma_k} f\, ds \qquad (10.3.2)$$

if the normal to the boundary edge σ_k points outside Ω and the negative of this quantity otherwise.

The same approximate velocity components are used in these discretizations. This is unlike a conventional approach where we could expect to find both velocity components appearing in connection with a single control volume. In contrast to this the covolume method uses just a single velocity component and two control volumes. It is apparent that the orthogonality of the mesh edges is the property which makes this possible. Use of only one component of the velocity field is the primary feature of the covolume approach for isotropic problems. Anisotropic problems can require the use of both tangential and normal components, see Hu and Nicolaides [1991].

10.3.2   Solvability

The div curl system (10.3.2) can be consistent only if

$$\int_\Gamma f \, ds = 0 \qquad\qquad (10.3.3)$$

We will assume this property since otherwise there can be no solution to the equations. It is reasonable to expect a similar property to show up in the discrete problem. It arises in the following way. First we will count equations and unknowns. Suppose that the triangulation has N nodes, T triangles and E edges. The number of edges and nodes on the boundary Γ is the same since the boundary is closed. Altogether we have E unknowns and $T + N$ equations including the boundary equations. According to Euler's formula applied to the triangulation we have $E = T + N + 1$. Thus there is one more equation than the number of unknowns and we expect a consistency condition to be necessary for solvability of the linear equations. This condition turns out to be the expected one, that

$$\sum_{\sigma_k \in \Gamma} u_k h_k = 0$$

with the standard sign convention for h_k. Substituting from (10.3.2) we obtain consistency in the discrete equations if and only if (10.3.3) holds; this means that discrete consistency follows if the continuous problem is consistent. In fact provided the div curl equations are consistent, the discrete equations do have exactly one solution.

10.3.3   Discrete Field Theory (1)

In addition to the flux and circulation other vector field operations can be defined and used to obtain discrete analogs of continuous results. First we will introduce two potentials defined respectively on the nodes of the primal and dual meshes. These will turn out to be analogous to the stream function and velocity potential, and we will use the notation Ψ and Φ for them. Ψ is defined on the primal nodes, including the boundary nodes. Φ is defined on dual nodes which coincide with circumcenters of the triangles. Sometimes it is necessary to also define Φ on the midpoints of boundary edges. This will be mentioned explicitly where it occurs.

Difference operators R and G are defined relative to the positive directions on the primal and dual mesh edges by

$$R\Psi \quad := \quad \frac{\Psi_i - \Psi_j}{h_{i,j}}$$

$$G\Phi \quad := \quad \frac{\Phi_k - \Phi_l}{h'_{k,l}}$$

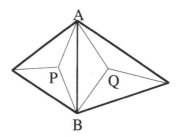

Figure 10.6: A kite domain $APBQ$.

In this, i and j are the numbers of two adjacent primal nodes and the double subscript on h denotes the distance between them. Similar conventions apply in the second definition.

We have the following potential type results:

i. $Du = 0$ *if and only if* $\exists \Psi$ *such that* $u = R\Psi$.

ii. $Cu = 0$ *if and only if* $\exists \Phi$ *such that* $u = G\Phi$.

The sufficiency of the conditions is easily seen to be true by substitution.

We will introduce an inner product for E dimensional vectors. In the primal triangulation join each circumcenter to the vertices of its triangle. Any edge of the triangulation now has two associated triangles whose vertices are the nodes which define the edge and the circumcenters of the triangles which share the edge. Figure 10.6 illustrates this. The two new triangles are APB and ABQ. $PBQA$ is not a rhombus in general, although PQ does orthogonally bisect AB. $PBAQ$ does have the shape of a kite. Its area is $hh'/2$ and this is the weight which is placed on the edges in the inner product. Denoting these areas by w_i the inner product is defined by

$$[u, v] = \sum_{\sigma_j} u_j v_j w_j$$

and the corresponding norm by

$$\|u\|_W^2 := \sum_{\sigma_j} u_j^2 w_j \,.$$

In terms of the inner product there are two useful analogs of integration by parts formulas. These are

$$[u, G\Phi] = (Du, \Phi) \qquad u|_\Gamma = 0$$

and

$$[u, R\Psi] = (Cu, \Psi) \qquad \Psi|_\Gamma = 0 \,.$$

The first formula is analogous to

$$\int_\Omega \mathbf{u}.\nabla\Phi\,dx\,dy = -\int_\Omega \Phi\operatorname{div}\mathbf{u}\,dx\,dy \qquad \mathbf{u}.\mathbf{n}|_\Gamma = 0$$

and there is a corresponding interpretation for the second.

10.3.4 Discrete Vector Fields (2)

The most important result of the discrete field theory is an analog of the Helmholtz decomposition of a vector field. Helmholtz's decomposition is essentially the statement that an arbitrary smooth vector field can be decomposed into the sum of a divergence free part and a part with zero curl. The decomposition is unique once suitable boundary conditions are prescribed on the components. The analog of this within the covolume framework is the following.

Any set of components u defined on the edges of the triangulation and with zero values on Γ can be represented in exactly one way in the form $u = z + w$ where $Dz = 0$, $Cw = 0$ and w and z are both zero on Γ.

This theorem has many uses. One use is for estimating the error in the covolume approximation to the div curl problem.

10.3.5 Discretization Error

Let $u^{(n)}$ denote the E dimensional vector with components

$$u_i^{(n)} = \frac{1}{h_i}\int_{\sigma_i}\mathbf{u}.\mathbf{n}\,ds\,.$$

This is the mean of $\mathbf{u}.\mathbf{n}$ on the ith edge of the triangulation. The error estimate for the covolume approximation to the div curl problem is

$$\|u - u^{(n)}\|_W \le C\max(h, h')|\mathbf{u}|_{1,\Omega}$$

where h and h' denote the largest edge lengths in the primal and dual meshes and on the right we have the usual Sobolev 1-seminorm.

If we introduce the stream function Ψ for the continuous problem (recall that div $u = 0$) and the piecewise linear interpolant $\hat{\Psi}$ for the discrete stream function ($Du = 0$) then from the last estimate we can derive the following estimate:

$$\|\hat{\Psi} - \Psi\|_{1,\Omega} \le C\max(h, h')\|\Psi\|_{2,\Omega}\,. \tag{10.3.4}$$

There are some points about this estimate which are worth noting. If we substitute $\mathbf{u} = \operatorname{curl}\Psi$ into the equation (10.3.2b) and define Ψ to be zero at one point on Γ then it follows that Ψ solves a Dirichlet problem. If we apply similar transformation at the discrete level then we obtain a discretization of this Dirichlet problem. The resulting discretization of Poisson's equation is at least as old as McNeal [1953]. It is usually obtained directly by approximating the integrated form of Poisson's equation

$$\int_{\partial V} \frac{\partial \Psi}{\partial n} \, ds = \int_V g \, dx \, dy$$

in each interior covolume. The estimate (10.3.4) shows that this covolume discretization for the Dirichlet problem has the same accuracy as piecewise linear finite elements for the usual norm. The discrete Laplacian generated by this procedure occurs again in Section 10.4.6.

If we reverse the roles of div and curl in the equations (10.3.2) so that the curl equation becomes homogeneous, then a parallel calculation to the one above gives an approximation to the Neumann problem which does not seem to be well known. Also, although the coefficient matrix of the Dirichlet problem coincides with the finite element matrix on the same mesh the Neumann matrix does not seem to be derivable from any standard finite element approach.

10.4 Stationary Stokes Equations

In this section we will begin with a method for discretization of the viscous terms in the Stokes equations. This turns out to involve the vorticity through the definition of the vector Laplacian. Discretization of the remaining terms in the equations is based on the ideas already introduced in the previous sections. We will specialize the discrete equations to a uniform Cartesian mesh and show how the standard MAC equations result. Following that, an equivalence between the MAC discretization and a velocity vorticity formulation is established. The results will be presented for two dimensions but extend nicely to three dimensions. The results for this section are based on Nicolaides [1989], Nicolaides [1991b] and Choudhury and Nicolaides [1990].

10.4.1   Stokes Equations

We will consider the stationary Stokes equations

$$
\begin{aligned}
\Delta \mathbf{u} - \nabla p &= \mathbf{f} \\
\operatorname{div} \mathbf{u} &= 0 \\
\mathbf{u}|_\Gamma &= 0 \,.
\end{aligned}
$$

Since we are working with only the normal components of the velocity field it makes sense to satisfy the corresponding component of the momentum equation. This means

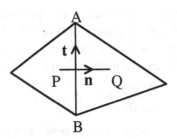

Figure 10.7: Normal and tangential directions.

that we should approximate the equation

$$\mathbf{n}.(\Delta \mathbf{u} - \nabla p) = \mathbf{n}.\mathbf{f} \qquad (10.4.1)$$

where $\mathbf{n}$ denotes the normal to an edge of the triangulation. We will apply this equation to every edge of the triangulation which has at most one endpoint on Γ.

10.4.2   Viscous Term

The vector Laplacian term appearing in equation (10.4.1) is equivalent to

$$\mathbf{n}.(\text{grad div } \mathbf{u} - \text{curl curl } \mathbf{u}) \qquad (10.4.2)$$

using the definition of the vector Laplacian. Figure 10.7 shows an edge AB of the triangulation and its shared triangles. P and Q are the circumcenters of their triangles and the directions n and t are indicated. Then (10.4.2) is equivalent to

$$\frac{\partial}{\partial n}(\text{div } \mathbf{u}) - \frac{\partial}{\partial t}(\text{curl } \mathbf{u})$$

where the differentiations are along the directions shown in the figure. It follows that to approximate this expression we require only approximations to div $\mathbf{u}$ at P and Q and to curl $\mathbf{u}$ at A and B. Then we can use simple differencing to approximate the derivatives. The divergences associated with the triangles can be approximated by $\hat{D}u|_P$ and $\hat{D}u|_Q$ where the $\hat{}$ denotes normalization by triangle area, and the curls may be approximated by $\hat{C}$ where this time $\hat{}$ denotes normalization by covolume area. Since Cu is currently defined only for interior nodes, we have to extend it to allow for the case when a node of the edge in question is on Γ.

The method for this is illustrated in Figure 10.8. The node B is a boundary node. To compute a circulation for B we use the circulation path $ABQPCA$. The new point

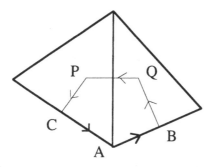

Figure 10.8: A boundary circulation path.

arising here is the requirement of velocity components on the boundary segments CA and AB. These are supplied by the tangential boundary conditions. The components along BQ and PC are provided by the normal boundary conditions. Normalizing the circulation by the hybrid covolume area gives the approximation to the curl at A.

The matrix of circulation operators extended to the boundary nodes by this method is denoted by C_b and by $\hat{C}_b$ when normalized by covolume areas.

It should be mentioned that in one sense we do not need to difference the first term in (10.4.2) since it is zero by incompressibility. However, there are reasons for including it in the discretization. For example its presence can enhance the performance of iterative solvers, particularly when the current iteration's velocity approximation is not solenoidal (even though it will become solenoidal in the iteration limit).

10.4.3 Discrete System

The discrete approximation to the Stokes equations may now be written as the linear system of equations

$$(G\hat{D}u - R\hat{C}_b u) - Gp' = F$$
$$\hat{D}u = 0$$
$$u|_\Gamma = 0.$$

The operators appearing have been defined above. Only the boundary condition on the normal component of **u** appears explicitly here. The tangential component appears implicitly in the computation of $\hat{C}_b$ as we showed in the previous section.

Apart from an additive constant in the pressure, these equations have a unique solution. This can be proved using the discrete Helmholtz decomposition of Section 10.3.3. The details may be found in Nicolaides [1991b].

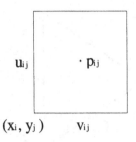

u_{ij} $\cdot\, p_{ij}$

(x_i, y_j) v_{ij}

Figure 10.9: Notation for rectangular meshes.

10.4.4 The MAC Scheme

We will summarize the basic equations of the well known MAC discretization in this section confining attention to the stationary Stokes equations. A recommended reference for MAC techniques is Hirt [1979].

The domain here is the square $0 < x < 1, 0 < y < 1$ which is overlaid with a uniform square mesh of side h. The variables used are the horizontal velocity components defined at the midpoint of the vertical mesh segments, the vertical velocity components defined at the midpoints of the horizontal mesh segments, and the pressures defined at the centers of the mesh cells. These variables are indexed by the method shown in Figure 10.9.

The approximation for the divergence equation is

$$\frac{u_{i+1,j} - u_{i,j}}{h} + \frac{v_{i,j+1} - v_{i,j}}{h} = 0. \tag{10.4.3}$$

The x momentum equation is approximated by

$$\frac{u_{i+1,j} - 2u_{i,j} + u_{i-1,j}}{h^2} + \frac{u_{i,j+1} - 2u_{i,j} + u_{i,j-1}}{h^2} - \frac{p_{i,j} - p_{i-1,j}}{h} = f_{x;i,j} \tag{10.4.4}$$

and the approximation for the y momentum equation is similar. At boundaries these equations call for velocity components at positions outside Ω. For example, near to the boundary $y = 1$ the x momentum equation is

$$\frac{u_{i+1,n-1} - 2u_{i,n-1} + u_{i-1,n-1}}{h^2} + \frac{u_{i,n} - 2u_{i,n-1} + u_{i,n-2}}{h^2} - \frac{p_{i,n-1} - p_{i-1,n-1}}{h} = f_{x;i,n-1}$$

and $u_{i,n}$ is outside $\overline{\Omega}$. The standard remedy is to define $u_{i,n}$ through the equation

$$\frac{u_{i,n} + u_{i,n-1}}{2} := u(x_i, 1)$$

where the right side is given by the boundary conditions.

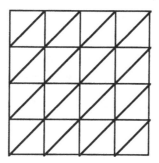

Figure 10.10: A standard triangulation with $n = 4$.

10.4.5 Equivalent Covolume Scheme

A covolume scheme producing equations which are equivalent to these MAC equations, including the boundary equations, can be obtained as follows. We use a standard triangulation of Ω illustrated for the case $n = 4$ in Figure 10.10.

Since the triangles are right angled their circumcenters are opposite the right angles and the circumcenters of paired triangles coincide. In the covolume scheme, pressures are defined at the circumcenters of the triangles which therefore coincide with the MAC pressure locations.

The velocity components of the covolume scheme are defined to be normal to the triangle edges and directed such that relative to the horizontal and vertical edges they are the x and y velocity components.

To obtain the MAC equation (10.4.3) consider the pair of adjacent triangles from the covolume mesh shown in Figure 10.11 with a convention for naming the variables. The two flux equations of the covolume approximation are

$$
\begin{aligned}
hu_{i+1,j} - hv_{i,j} - h\sqrt{2}w_{i,j} &= 0 \\
-hu_{i,j} + hv_{i,j+1} + h\sqrt{2}w_{i,j} &= 0 .
\end{aligned}
$$

Adding these gives

$$
h(u_{i+1,j} - u_{i,j}) + h(v_{i,j+1} - v_{i,j}) = 0 \tag{10.4.5}
$$

which is equivalent to (10.4.3).

To recover (10.4.4) from the covolume equations we use the covolumes enclosed by broken lines in Figure 10.12.

The covolume momentum equation for $u_{i,j}$ uses the equations given in Sections 10.4.2

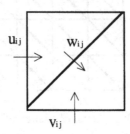

Figure 10.11: The "diagonal" velocity component.

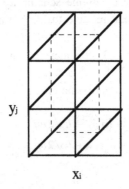

Figure 10.12: Covolumes for MAC scheme.

and 10.4.3. The equation is

$$\frac{2(-hu_{i,j} + hv_{i,j+1} + h\sqrt{2}w_{i,j})}{h^3} - \frac{2(hu_{i,j} - hv_{i-1,j} - h\sqrt{2}w_{i-1,j})}{h^3}$$

$$-\frac{(hv_{i,j+1} - hu_{i,j+1} - hv_{i-1,j+1} + hu_{i,j})}{h^3} + \frac{(hv_{i,j} - hu_{i,j} - hv_{i-1,j} + u_{i,j-1})}{h^3}$$

$$-\frac{p_{i,j} - p_{i-1,j}}{h} = f_{x;i,j}$$

and we may eliminate the barred terms using

$$h\sqrt{2}w_{i,j} = hu_{i+1,j} - hv_{i,j}$$

$$h\sqrt{2}w_{i-1,j} = hu_{i-1,j} - hv_{i-1,j+1}.$$

This results in the equation

$$\frac{u_{i+1,j} - 2u_{i,j} + u_{i-1,j}}{h^2} + \frac{u_{i,j+1} - 2u_{i,j} + u_{i,j-1}}{h^2}$$

$$-Q + \frac{p_{i,j} - p_{i-1,j}}{h} = f_{x;i,j}$$

where

$$Q = h^{-2}(u_{i+1,j} - u_{i,j} + u_{i-1,j} - u_{i,j} - v_{i,j+1} - v_{i,j} - v_{i-1,j+1} + v_{i-1,j}).$$

Using (10.4.5) it follows that Q is zero so that the equation is equivalent to (10.4.4) as we wanted to show. A similar calculation permits the boundary equation to be derived from the associated covolume equation. It follows that the covolume method is a genuine extension of the MAC scheme as defined above.

10.4.6 Velocity Vorticity Interpretation

In this section we will consider the time dependent Stokes equations,

$$\frac{\partial \mathbf{u}}{\partial t} = \Delta \mathbf{u} - \nabla p + f$$

$$\operatorname{div} \mathbf{u} = 0$$

$$\mathbf{u}|_{\Gamma} = 0.$$

together with suitable initial conditions. The discrete system corresponding to these equations is

$$u_t = (G\hat{D}u - R\hat{C}_b u) - Gp' + F$$

$$\hat{D}u = 0$$

$$u|_{\Gamma} = 0.$$

The subscript denotes forward time differencing. Backward time differencing could also be used. Defining

$$\omega' := \hat{C}_b u$$

and multiplying the discrete momentum equation by C it follows (Section 10.3.3) that

$$\begin{aligned} \omega'_t &= -CR\omega' + CF \\ &= L_h\omega' + CF \end{aligned}$$

where L_h denotes the discrete Laplacian which appeared previously in Section 10.3.5.

This shows that we can replace the primitive variable system with the vorticity velocity system

$$\begin{aligned} \omega'_t = L_h\omega' + CF \qquad\qquad \omega|_\Gamma &= \hat{C}_b u|_\Gamma \\ Du &= 0 \\ \hat{C}u &= \omega' \\ u|_\Gamma &= 0. \end{aligned}$$

The first pair of equations are the vorticity transport equation and the vorticity boundary condition. The next three are the velocity equations. This equation is exactly what we would obtain if we directly discretized the continuous velocity vorticity equations by the covolume method without using the primitive variable equations. Figure 10.13 diagrams these relationships. The horizontal lines denote transformations between primitive variable and velocity vorticity formulations. The vertical lines denote covolume discretizations. The main conclusion is that in the covolume framework, just as in the continuous framework, the two formulations are equivalent in the sense that the solution of either formulation provides the solution to the other. This property is not usual in discretizations of incompressible flows.

As an application we may consider the MAC scheme. It follows that the MAC scheme can equally well be regarded as a discretization of the velocity vorticity form of the flow problem. We could pursue this further and introduce a stream function (Section 10.3.3, (i)). Then we could show the equivalence of the classical stream function vorticity method with the primitive variable MAC scheme. It follows that the only advantages of any of these formulations over the others lie in the convenience of solving the discrete equations systems. The numerical approximations are mathematically identical for the same quantities.

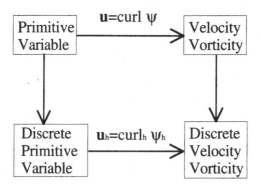

Figure 10.13: Vorticity tranformations.

10.4.7 Error Estimate

A natural norm for analysis of the covolume approximation is

$$\|u\|_{1,h}^2 := \sum_i (\hat{D}u)_i^2 A_i + \sum_j (\hat{C}_b u)_j^2 A_j'$$

where A_i and A_i' denote respectively triangle and covolume areas. The covolume sum is extended over all covolumes including those on Γ. This norm is a discrete form of the Sobolev $\mathbf{H}_0^1(\Omega)$ norm. If $Du = 0$ it reduces to a discrete $L^2(\Omega)$ norm of the (discrete) vorticity.

For quantities defined on the triangle circumcenters a natural discrete form of the $L^2(\Omega)$ norm is

$$\|p\|_A^2 := \sum_i p_i^2 A_i$$

where the sum is over all triangles of the mesh.

For any covolume A_j we define the mean vorticity of the solution u by

$$\overline{\omega}_j = \frac{1}{A_j'} \int_{A_j} \omega \, dx \, dy \,.$$

The approximate vorticity is denoted by

$$\omega_j' := \hat{C}_b u|_j$$

where u denotes the solution of the discrete Stokes equations. Similarly, for each triangle we define the mean of the pressure p by

$$\bar{p}|_i := \frac{1}{A_i} \int_{A_i} p \, dx \, dy \, .$$

Using these notations we have the following estimate for the errors of the uniform mesh MAC scheme (Nicolaides [1991b]).

$$\|u^{(n)} - u\|_{1,h} \equiv \|\bar{\omega} - \omega'\|_{A'} \leq Ch(|\mathbf{u}|_2 + |p|_2)$$
$$\|\bar{p} - p'\|_A \leq Ch(|\mathbf{u}|_2 + |p|_2)$$

where the norms on the right are the ordinary Sobolev $\mathbf{H}^2$ and H^2 seminorms. These estimates are obtained under the regularity requirement $\omega \in H^2(\Omega)$. It is probable that the $\mathbf{L}^2$ error in the velocity is second order in h but this has not been proved.

10.5 Navier-Stokes Equations

The new issue which occurs with the full equations is the convection term. The three major ways to express this term (conservative form, total pressure form and the usual nonconservative form) lead to different covolume discretizations. However all of them involve the tangential velocity component relative to the triangle (or tetrahedral) edge (or face). So far we have not needed to use these components but it seems to be necessary in the context of the convective terms. We will begin this section by considering how the tangential components can be obtained. The results are of independent interest, since there are several other situations where tangential components are needed. An example is in computing the dissipation function in connection with the energy equation. That is another case where tangential components are apparently essential. It is to be expected that tangential components should be needed for some purposes. Even in the Cartesian setting they are required for the convection terms. However in contrast to that case, in the covolume setting it is not obvious which of several options should be used.

10.5.1   Elementwise Extension of u

To begin we will construct a constant vector field in a triangle which has prescribed (constant) normal components on its edges. Since we have only two parameters at our disposal and three conditions to satisfy there must be some restriction on the normal components for a solution to exist. A necessary and sufficient condition is that the discrete flux of the normal components is zero. Specifically, we have the following result.

There is a constant vector $\mathbf{v}$ *such that* $\mathbf{v}.\mathbf{n}_i = u_i$ $i = 1, 2, 3$ *if and only if*

$$u_1 h_1 + u_2 h_2 + u_3 h_3 = 0$$

where $\mathbf{n}_i$ $i = 1, 2, 3$ *denote the unit normals to the triangle sides.*

The simple proof of this is given in Nicolaides [1989].

In practice any pair of the above equations can be solved for the components of $\mathbf{v}$. In this way the discrete incompressibility condition ensures that we can always extend the field components as a constant vector into the interior of the triangle. There is a similar result for three dimensions.

Although all of the results in the following sections assume that $\mathbf{v}$ is computed in this way, there is another approach due to Hall et al [1991] which is of interest. In this technique, a vector $\mathbf{v}^*$ is defined as

$$\mathbf{v}^* := \frac{1}{3} \sum_{i=1}^{3} (u_i \mathbf{n}_i + v_i \mathbf{t}_i)$$

and the coefficients $\mathbf{v}_j$ are then determined from the linear equations

$$\mathbf{v}^*.\mathbf{t}_j = v_j \qquad j = 1, 2, 3 \,.$$

Hall et al [1991] report good results using this in conjunction with their weighting for the convection terms.

10.5.2 Tangential Components

Figure 10.14 shows two adjacent triangles labelled L and R. In each triangle a constant vector field is defined and denoted respectively by $\mathbf{v}_L$ and $\mathbf{v}_R$. Normally these vectors will be computed by the method Section 10.5.1. The tangential field component along the common edge AB must be computed from the left and right fields $\mathbf{v}_L$ and $\mathbf{v}_R$. Let $\mathbf{n}$ denote the unit normal to AB directed from L to R and let $\mathbf{t}$ denote the unit tangent directed from A to B. There are several acceptable methods to define the tangential component but all of them use a linear combination of the tangential components of $\mathbf{v}_L$ and $\mathbf{v}_R$. We will consider these methods individually.

Define $v_L := \mathbf{v}_L.\mathbf{t}$ and $v_R := \mathbf{v}_R.\mathbf{t}$ and let v denote the tangential component along AB. The simplest choice for v is

$$v := \frac{1}{2}(v_L + v_R)\,. \tag{10.5.1}$$

A partial justification for this can be given as follows. The discontinuity of the velocity field across AB implies the existence of a vortex sheet there. Choosing the tangential

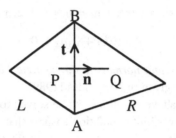

Figure 10.14: Left and right triangles.

velocity of the sheet to be the average of the velocities on either side is consistent with the
tangential velocity of the sheet which is induced by the vorticity. This is not a complete
explanation since, for example, it ignores the induced velocities of the other edges. On
the other hand, since u is continuous these effects presumably become negligible when
the mesh size approaches zero. The error estimate stated in the next section uses this
method to define tangential velocities.

The next method defines v by

$$v := \frac{1}{h'}(h'_L v_L + h'_R v_R)$$

where $h' := h_L + h_R$. This definition has the property that the discrete flux out of any
covolume is zero. In other words, this extension preserves the solenoidal property of the
original u components. We will prove this statement.

Referring to Figure 10.15 we will compute the flux out of the covolume surrounding
the node at B.

The flux across PQ is $-h'_L v_L - h'_R v_R$, continuing with our earlier notation. The total
flux out of the covolume consists of a sum of similar terms. The contribution to this
sum from the triangle ABC is given by $-\mathbf{v}_R.(h'_R \mathbf{t}_{AB} + h'_S \mathbf{t}_{CB})$ where h'_S denotes QM
in Figure 10.16 which shows another view of the triangle ABC. Note that $\mathbf{v}_R$ is the
vector interpolant for this triangle.

Next, we will show that $h'_R \mathbf{t}_{AB} + h'_S \mathbf{t}_{CB} = (h_{AC}/2)\mathbf{n}_{AC}$ where h_{AC} denotes the length
of the edge AC and $\mathbf{n}_{AC}$ denotes its outward pointing normal. In fact, referring to
Figure 10.16 this results from applying the equations

$$0 = \int_\tau \nabla(1)\, dx\, dy = \int_{\partial\tau} \mathbf{n}\, ds$$

to the triangle QMN and noting that the sides of LMN and ABC are parallel while
QM is perpendicular to BC and QN is perpendicular to AB.

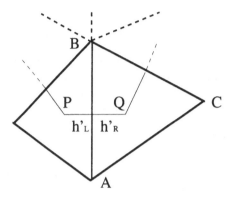

Figure 10.15: Notation for tangential computation.

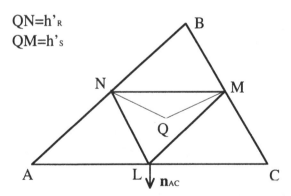

Figure 10.16: Proving the zero flux property.

Summing the resulting terms over the triangles simply gives the flux out of the patch of triangles which meet at B and this sum is zero by discrete incompressibility of u. The result follows.

10.5.3   Convective Discretization

There are several forms of the convective terms that are used in practice and each of them can be discretized. We will consider the (momentum) conservative form

$$\text{div}\,\mathbf{u} \otimes \mathbf{u}$$

and the total pressure form

$$\omega(-v, u)$$

where we recall that in the total pressure form a modified pressure given by

$$p + \frac{1}{2}(u^2 + v^2)$$

is used, and where $\mathbf{u} := (u, v)$.

Recalling that the normal component of these vectors is required let n denote the constant direction normal to a side of a given triangle T with area A. Then we approximate

$$
\begin{aligned}
\text{div}\,\mathbf{u} \otimes \mathbf{u} &\approx \tfrac{1}{A} \int_{\partial T} (\mathbf{u}.\mathbf{n})(\mathbf{u}.\mathbf{m})\, ds \\
&\approx \tfrac{1}{A} \sum_i (\mathbf{u}.\mathbf{n})_i (\mathbf{u}.\mathbf{m}_i) h_i
\end{aligned}
\qquad (10.5.2)
$$

where the divergence theorem was used and where $\mathbf{m}$ denotes the outer normal to ∂T and the sum is over the triangle edges. It is for evaluating the quantities $(\mathbf{u}.\mathbf{n})_i$ that the tangential field components are used since only at the edge of T with normal $\mathbf{n}$ is it explicitly known. The procedure is to substitute the values of $(\mathbf{u}.\mathbf{n})_i$ into the last equation above, having first computed an approximation to $\mathbf{u}$ using the already known normal component and the tangential component computed by the method of Section 10.5.2. Considering these approximations to hold at triangle circumcenters, we have to weight them suitably to obtain value at the midpoints of edges. Linear interpolation or some other weighting can be used. Hall et al [1991] use a relative weighting by triangle areas.

For the total pressure form the calculations are simpler since $\mathbf{n}.(-v, u)$ is just the negative of the tangential velocity component. Using v' to denote the tangential velocity component computed in Section 10.5.2 the approximation is

$$\mathbf{n}.[\omega(-v, u)] \approx -v'\overline{\omega} \qquad (10.5.3)$$

where $\overline{\omega}$ denotes the average of the two values of ω at the endpoints of the edge whose normal is $\mathbf{n}$.

For the uniform mesh case corresponding to the estimates in Section 10.4.7 we have the following which is proved in Nicolaides and Wu [1991c]:

$$\|\bar{\omega} - \omega'\|_{A'} \leq K(\mathbf{u}, p, \mathbf{f})h$$
$$\|\bar{p} - p'\|_A \leq K(\mathbf{u}, p, \mathbf{f})h$$

where K is independent of h. The regularity conditions for this are $\mathbf{u} \in \mathbf{H}^2(\Omega)$, $\omega \in H^2(\Omega)$, and $p \in H^2(\Omega)$. As in the linear case, this estimate is effectively for a discrete $\mathbf{H}^1$ norm of the error. It is likely that the velocity error is second order (see Section 10.6.2) but this has not been proved.

The total pressure discretization lets us extend the commutative diagram of Section 10.4 to the nonlinear case. This follows from taking the circulation of the discrete momentum equation around covolumes just as in Section 10.4. For a given covolume the result is a sum of the form

$$\sum_j \bar{\omega}_j v'_j h'_j$$

which is what would result for the convective term if we integrated the vorticity transport equation

$$\omega_t + \text{div}(\omega\mathbf{u}) = \nu\Delta\omega$$

over the covolume. This implies that the commutative diagram of Section 10.4 is valid in this sense even for the nonlinear equations.

Another consequence is that the covolume discretization conserves vorticity to the extent that the *equivalent* discretization of the vorticity transport equation is conservative.

10.6 Hamel Flow

This section describes the results of applying the covolume method to a viscous flow for which there is a similarity solution. The flow is bounded by two parallel vertical segments and two segments at an angle to one another. The flow is assumed to be radial into the (virtual) sink at the junction of the nonparallel segments. This flow was first analysed by Hamel although numerous later contributions exist. The assumption of radial flow enables the problem to be reduced to the solution of a second order nonlinear boundary value problem.

It is possible to make some progress with this equation using elliptic functions, but the result is not as explicit as one would like for comparison purposes with numerical approximations. For that reason, it was solved numerically using central differences and Newton's method on a mesh with 1000 nodes. This gives us explicit numerical values for assessing the accuracy of the covolume approximations. The Hamel problem exhibits a boundary layer at sufficiently high Reynolds numbers, but the mesh always contained a substantial number of nodes inside the boundary layer. We were satisfied that the solution was satisfactorily resolved for the Reynolds numbers reported below. Thus, we will refer to these numerical solutions as "exact" even though in reality they were computed on a (very fine) mesh and are merely highly accurate.

10.6.1   The Similarity Solution

The original problem considered by Hamel was a steady flow between two plane walls meeting at an angle 2α. Assuming the walls to meet at the origin O (where there is a singularity in both velocity and pressure) we assume the velocity components in polar coordinates to be

$$u_r = \frac{1}{r}F(\Theta)$$
$$u_\Theta = 0$$

and substitute them into the momentum equations from which we then eliminate the pressure. The result of this is the equation

$$2FF' + \nu F''' + 4\nu F' = 0.$$

Integrating this once and defining $\nu G = F$ we obtain

$$G'' + G^2 + 4G = \frac{c}{\nu^2}$$
$$G(-\alpha) = G(\alpha) = 0$$

where c is a constant of integration and the boundary conditions represent the no slip conditions.

A suitable Reynolds number for the converging flow problem is $R := -\alpha F(0)/\nu$. Adjusting c changes $F(0)$ and hence the Reynolds number.

The momentum equations can also be integrated explicitly to obtain the formula for pressure

$$p = \frac{2\nu}{r}u_r - \frac{c}{2r^2} + p_0$$

where p_0 is chosen to make p have zero mean over the domain.

10.6.2   Numerical Results

In this section we will show the results of some calculations using the choices (10.5.1) and (10.5.2) for the convection terms. We have repeated the calculations using the choices

mesh	A	B	C	D	E
h	0.36	0.18	0.09	0.045	0.0225
$\|p - p'\|_A$	0.35d+01	0.76d+00	0.21d+00	0.69d-01	0.38d-01
$\|u - u^{(n)}\|_W$	0.27d+01	0.38d+00	0.68d-01	0.15d-01	0.29d-02
$|\omega' - \overline{\omega}|_{A'}$	0.29d+02	0.10d+02	0.33d+01	0.12d+01	0.39d+00

Table 10.1: Summary of test results to determine rate of covergence.

(10.5.1) and (10.5.3), but only minor differences in the results were found. These and other calculations are given with more details in Nicolaides and Wu [1991a].

Figures 10.17 and 10.18 show two meshes which were used to obtain the results given below. Symmetry of the flow about the centerline of the channel was used in the calculations. The second mesh differs from the first only in having four additional mesh lines in the radial direction. Figures 10.19 and 10.20 show the x and y velocity components at a fixed station along the channel for $Re = 272.9$ which corresponds to $\nu = .01$ and $c = 26.95$. In the figure captions $\rho = 1/\nu$. The high gradients at the upper wall are indicative of the more pronounced boundary layer which forms at higher Reynolds numbers. The coarser mesh has no difficulty in resolving the velocity in this case. This is not quite true for the pressure in Figure 10.21. However, the slight oscillation in the pressure disappears under the refinement to the finer mesh in Figure 10.22. This shows that the pressure oscillation was a resolution problem and not a stability problem.

A similar set of figures is shown for the case of $Re = 408.8$. This time, Figure 10.21 shows an oscillation in the x velocity component, which once again vanishes under refinement as the same figure shows. Similar remarks apply to the y component in Figure 10.24 although this time the coarser mesh oscillations are of considerably smaller magnitude. The coarse mesh pressure oscillations are extensive in Figure 10.25, but once again disappear completely under the refinement.

Table 10.1 summarizes the results of some tests to determine the rate of convergence of the covolume approximations. Figure 10.27 shows a starting mesh which was uniformly refined four times to produce the mesh sequence. The Reynolds number here was 61.

The pressure errors are divided by a factor of about 3 with each mesh subdivision except for the last. The corresponding factor for the velocities is at least 4 and for the vorticity it is close to 3. The pressure and vorticity factors are considerably better than the factor of 2 predicted by the theory of Section 10.5.3. The computed orders of convergence are about $h^{1.5}$ for both pressure and vorticity.

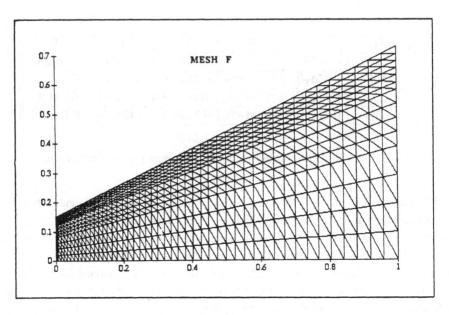

Figure 10.17: The coarse mesh.

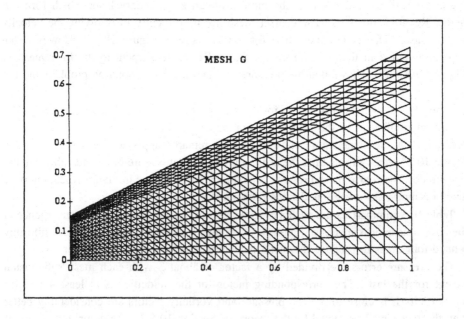

Figure 10.18: The fine mesh.

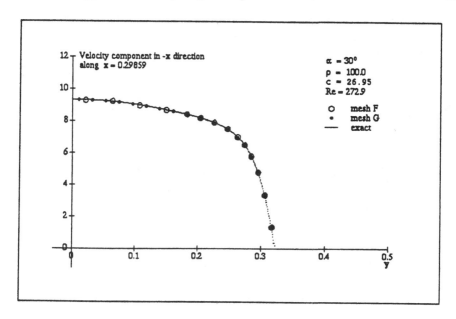

Figure 10.19: u along $x = 0.29859$, $Re = 272.9$.

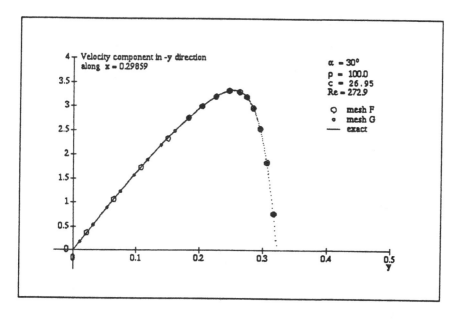

Figure 10.20: v along $x = 0.29859$, $Re = 272.9$.

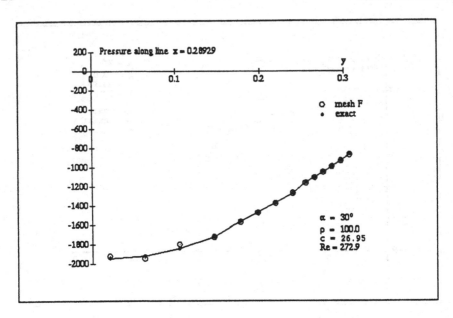

Figure 10.21: p along $x = 0.29859$ (coarse mesh), $Re = 272.9$.

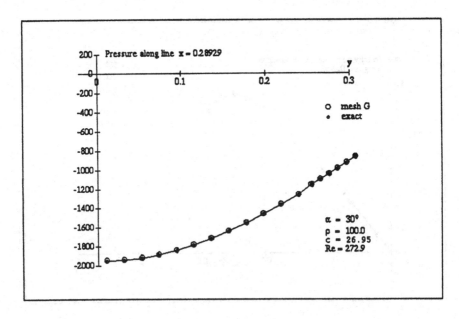

Figure 10.22: p along $x = 0.29859$ (fine mesh), $Re = 272.9$.

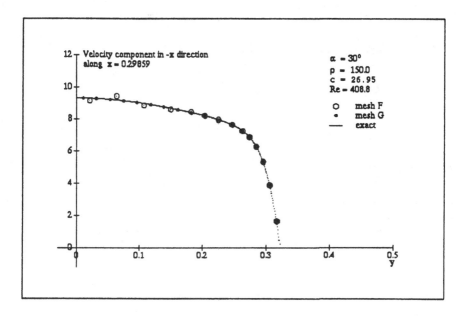

Figure 10.23: u along $x = 0.29859$, $Re = 408.8$.

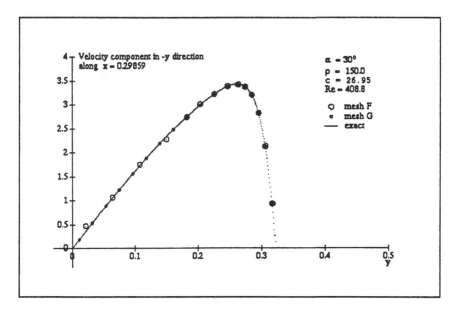

Figure 10.24: v along $x = 0.29859$, $Re = 408.8$.

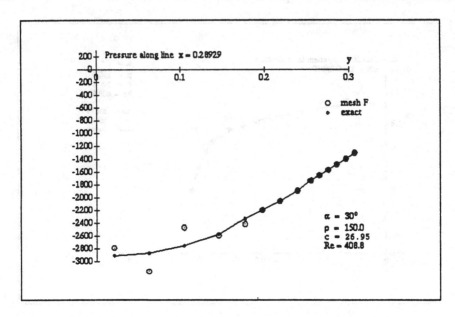

Figure 10.25: p along $x = 0.29859$ (coarse mesh), $Re = 408.8$.

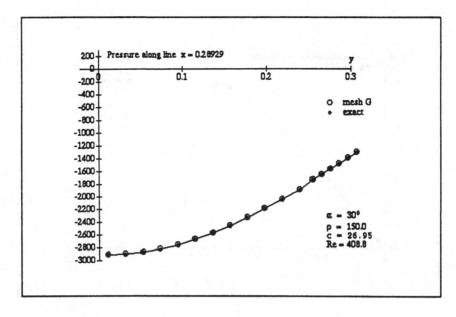

Figure 10.26: p along $x = 0.29859$ (fine mesh), $Re = 408.8$.

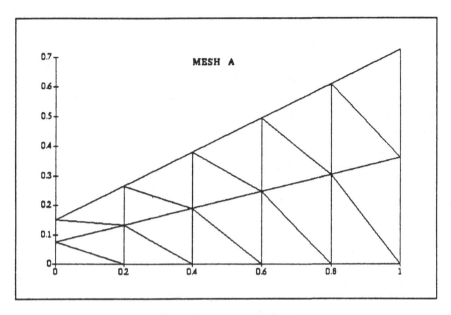

Figure 10.27: Initial mesh.

The vorticity errors in the last row of Table 10.1 appear to be rather large, but the norm of the exact vorticity is itself rather large, being around 600 in both its maximum value and its L^2 norm. Looked at in the relative sense the vorticity errors are therefore quite reasonable.

Appendix 10.A: On Spurious Modes

Spurious modes are a common phenomenon in CFD, occuring in both compressible and incompressible flow algorithms. The most prevalent example of a spurious mode is the infamous "checkerboard mode." If it occurs in an algorithm, the algorithm designer usually explicitly rules it inadmissible (by orthogonality in finite elements) or adds new terms *ad hoc* such as second and fourth order dissipation terms so that it no longer is a solution of the discrete equations. In this appendix we will concentrate on a finite element situation where the checkerboard mode is eliminated by the first approach. The interesting fact is that this is not enough to ensure a good approximation from the finite element scheme. The basic difficulty is that there are other "weakly" spurious modes which, while being oscillatory like the checkerboard mode, do not make their presence felt in such an obvious way. Nevertheless, they are present and can be activated in the right combination of circumstances. The main aim of this appendix is to show an example of a weakly spurious mode and to state a rigorous result about how such modes may produce erroneous solutions on finite meshes.

Rigorous results are available for analysis of finite elements in the incompressible context. The finite element case is special because of the wealth of relevant mathematical tools which have been accumulated. It has not been possible to analyse spurious modes for general finite volume and finite difference methods in the same depth. However, it is quite possible that in these environments too, weakly spurious modes remain after eliminating explicit modes like the checkerboard.

The technical mathematical difficulties which surround the topic of spurious modes tend to make the subject somewhat obscure. That is a pity since the subject is of direct practical interest even if a little subtle. We have chosen a slightly different way to explain the results here in the hope of making the results more accessible and useful.

10.A.1   Spurious Modes

We will begin by putting the checkerboard mode into the setting which we will use below for the weakly spurious mode. Fundamentally, the issue is one of bounding the solutions to equations by their data. To illustrate this, we will consider the equation

$$\text{div}\,\mathbf{u} = q \qquad \text{in } \Omega$$
$$\mathbf{u}|_\Gamma = 0 \tag{10.A.1}$$

where Ω is a bounded domain and Γ denotes its boundary. Rather than a single equation it pays to consider the entire class of equations with data q from some function space. It is standard to take q in the class $L^2(\Omega)$ of square integrable functions with zero average over Ω and to look for solutions $\mathbf{u}$ in the space $\mathbf{H}_0^1(\Omega)$ of functions with square integrable first derivatives. The norms in these spaces are defined respectively by

$$\|q\|_0^2 := \int_\Omega q^2 d\Omega$$

$$|\mathbf{u}|_1^2 := \int_\Omega |\nabla \mathbf{u}|^2 d\Omega$$

In terms of these spaces, the equation (10.A.1) is said to be well posed if there is a constant K such that for each q we can find a solution u satisfying

$$|\mathbf{u}|_1 \leq K\|q\|_0. \tag{10.A.2}$$

It is important that the same constant K can be used for every q. Other descriptions used for this property are (i) that the solutions are bounded by their data, and (ii) that the solutions depend continuously on their data. The last idea is a very useful one. It expresses the idea that the perturbation in a solution caused by a perturbation of the data is bounded by K times the magnitude of the perturbation. It could be difficult in practice to compute a solution to an equation which lacks this property. In the particular case of equation (10.A.1) we will refer to the well posedness of the problem as *div stability*.

Proofs of (10.A.2) may be found in the standard books Girault and Raviart [1986], Ladyzhenskaya [1969] and Temam [1984].

We encounter a discrete form of the div stability condition in setting up the finite element error estimates for the incompressible flow problem. To be specific, we will take the standard continuous piecewise bilinear velocity space V_h coupled with the piecewise constant pressure space S_h. The domain Ω is the square $-1 < x < 1$, $-1 < y < 1$ which is decomposed into $2n \times 2n$ subsquares of side $h := 1/n$. The weak form for the incompressibility constraint is

$$\int_\Omega s_h \operatorname{div} \mathbf{u}_h \, dx \, dy = 0 \quad \forall s_h \in S_h \,.$$

By choosing s_h in turn to be the standard basis functions for S_h, we obtain after integration a matrix representation of this equation which we may denote by

$$\mathrm{h}^2 \operatorname{div}_{\mathrm{h}} u_h = 0 \,.$$

The left hand side is evaluated elementwise by integration and is expressed as a linear combination of values of u_h at the element corners. For the element with lower left corner $(x_i, y_j) := (ih, jh)$ the value is

$$h(\overline{u}_{i+1,j+1/2} - \overline{u}_{i,j+1/2}) + h(\overline{v}_{i+1/2,j+1} - \overline{v}_{i+1/2,j})$$

where u and v denote approximate velocity values and barred quantities denote their averages.

It turns out that we must have a div stability result for div_h in order to obtain error estimate for the approximation. This condition is analogous to (10.A.2) and says that there must be a constant K such that given any q_h there is a solution to

$$\operatorname{div}_h \mathbf{u}_h = q_h \qquad\qquad (10.\mathrm{A}.3)$$

satisfying

$$|\mathbf{u}_h|_1 \le K \|q_h\|_0 \qquad\qquad (10.\mathrm{A}.4)$$

where K is independent not only of q_h but also of h. The bad case is when K tends to infinity as h tends to zero since then there is no bound on the effect of a perturbation in the data on the solution.

Next we will check the div stability when q_h is the checkerboard mode. For this we assume there is at least one solution to (10.A.3). Multiplying the equation by q_h and integrating over Ω we obtain

$$\|q_h\|_0^2 = \int_\Omega q_h \, \mathrm{div}_h \, \mathbf{u}_h \, dx \, dy = \int_\Omega q_h \, \mathrm{div} \, \mathbf{u}_h \, dx \, dy = 0 = 0|u_h|_1^2 \,.$$

The last equality shows that K for the checkerboard is effectively infinite. This means that so long as the checkerboard mode is in the pressure space S_h we cannot have a well posed problem (and we cannot obtain satisfactory error estimates). In this particular case the problem is easy to fix by redefining the pressure space to be all piecewise constants which are orthogonal to the checkerboard mode.

10.A.2 Weakly Spurious Modes

We will introduce square macroelements made up of four subsquares of the mesh. The entire mesh is the union of square macroelements of this type with centroids at (Ih, Jh), $I, J = \pm 1, \pm 3, \ldots, \pm(2n-1)$. With each macroelement we associate a local checkerboard function with the top left value equal to $+1$ and denoted by $\mathrm{alt}_{h,IJ}$ where $\mathrm{alt}_{h,IJ}$ is zero outside macroelement (I, J). Now we define

$$q_h := \sum_{IJ} I \mathrm{alt}_{j,IJ} \,. \tag{10.A.5}$$

Note that q_h has zero average over Ω.

Assuming that there is a solution to equation (10.A.3) with this particular q_h we will check div stability. As before, multiplying (10.A.3) by q_h and integrating we obtain

$$\int_\Omega q_h \, \mathrm{div} \, \mathbf{u}_h \, dx \, dy = \|q_h\|_0^2 \,.$$

The integral on the left can be estimated after a calculation (Boland and Nicolaides [1984] and [1985]) and the result is

$$\left| \int_\Omega q_h \, \mathrm{div} \, \mathbf{u}_h \, dx \, dy \right| \leq h\sqrt{3} \|q_h\|_0 |\mathbf{u}_h|_1 \,.$$

Now it follows that

$$|\mathbf{u}_h|_1 \geq \frac{1}{h\sqrt{3}} \|q_h\|_0$$

which implies that the best K we can hope to get is $O(h^{-1})$. This means that even though we have eliminated the checkerboard mode, we still do not have a well posed equation since in the limit the "stability constant" $K(h)$ tends to infinity. In this sense, (10.A.5) is a weakly spurious mode.

It is actually possible to implement a "post-processing" operation on a numerical solution which eliminates the contamination resulting from this and related spurious modes (see e.g. Boland and Nicolaides [1985]). However the post-processing operation is problem specific, and we do not know how to define it on, say, an arbitrary isoparametric mesh or for most other schemes with stability problems.

It is quite easy to establish a connection with the usual definition of div-stability. The usual definition requires that there should exist a constant $\gamma \geq 0$ and independent of h and q_h such that

$$\sup_{|\mathbf{u}_h|_1 \leq 1} \int_\Omega q_h \operatorname{div} \mathbf{u}_h \, dx \, dy \geq \gamma \|q_h\|_0 \, .$$

This implies that (10.A.4) holds with $K = 1/\gamma$. To show this, multiply (10.A.3) by $\operatorname{div} \mathbf{u}_h$ and integrate to get

$$\sup_{|\mathbf{u}_h|_1 \leq 1} \int_\Omega q_h \operatorname{div} \mathbf{u}_h \, dx \, dy \quad \leq \quad \frac{1}{|\mathbf{u}_h|_1} \int_\Omega (\operatorname{div} \mathbf{u}_h)^2 \, dx \, dy$$

$$= \quad \frac{1}{|\mathbf{u}_h|_1} \|q_h\|_0^2 \, .$$

from which we obtain

$$\gamma \|q_h\|_0 \leq \frac{1}{|\mathbf{u}_h|_1} \|q_h\|_0$$

which is equivalent to (10.A.4) as we wanted to show. It follows that the usual definition of div stability is violated for the bilinear-constant pair since we have shown that (10.A.5) is false for these elements.

10.A.3   Further Comments

We have seen that certain technical requirements are not satisfied by the bilinear constant element pair, but that does not necessarily mean that we will see any practical manifestation of inaccuracies. On the other hand it would seem to be essential to independently check (by a *stable* method) any unexpected result from a computation performed using bilinear-constant elements. Of course the possibility of computed results containing unrevealed errors remains.

More can be said about the occurence of stability errors in the bilinear-constant context. In fact, it is possible to construct an $\mathbf{L}^2(\Omega)$ function $\mathbf{f}$ such that the standard bilinear-constant approximations to the stationary Stokes equations behave in a basically unacceptable way as h tends to zero. We will quote a specific result from Boland and

Nicolaides [1985] to illustrate this. This result shows that the pressure error can converge to zero slower than any positive power of h. The earlier definitions of Ω, h and the triangulations of are still in force. Also we are dealing with homogeneous boundary conditions, incompressible flow and body forces denoted by $\mathbf{f}$.

Given any $0 < \beta < 1$ there is a constant K_β and a body force $\mathbf{f}_\beta \in L^2(\Omega)$ such that the pressure error of the bilinear-constant element pair satisfies

$$\|p - p^h\|_0 \geq K_\beta h^\beta$$

for sufficiently small h.

This is a more direct result than the well posedness results above, since it directly asserts that a case exists where convergence of p^h is close to nonexistent. This is significant since the well posedness results by themselves do not necessarily prove that the computations can be unreliable. An example of an $\mathbf{f}_\beta$ with the property of the above theorem is constructed in Boland and Nicolaides [1985] (Theorem 6.1). The construction is somewhat involved and will not be given here. It is possible to argue that such complicated body forces do not occur in practice. Two counterarguments can be given to this. First, the $\mathbf{f}_\beta$ which is constructed is a valid candidate in the context of the generally accepted mathematical theory of the Stokes equations. If one accepts this theory, then presumably one accepts its consequences although the existence of unsatisfactory consequences might suggest that the theory needs more refinement. Second, and more tentatively, we could presumably interpret the convection terms as some sort of a body force for the Stokes equation. If that were the case, then it would probably not be a good idea to assume too much about its form.

Although we have concentrated on the bilinear-constant elements, it is easy to devise many other element pairs with weakly spurious modes. Unfortunately, it seems that weakly spurious modes are almost a generic property of low order finite element schemes for viscous flows. Although there are some convergent low order finite elements, they tend to be somewhat contrived or violate some reasonable requirement such as mass conservation or balancing of the velocity and pressure errors in the error estimates. For a recent survey of div stability results see Gunzburger [1989] as well as the standard reference Girault and Raviart [1986] and also Pironneau [1990].

References

[1] Boland J. M. and Nicolaides R. A. [1984]. "On the stability of bilinear-constant velocity pressure finite elements," Numer. Math. **44**, 219–222.

[2] Boland J. M. and Nicolaides R. A. [1985]. "Stable and semistable low order finite elements for viscous flows," SIAM J. Numer. An. **22**, 474–492.

[3] Choudhury S. and Nicolaides R. A. [1990]. "Discretization of incompressible vorticity-velocity equations on triangular meshes," Int. Jnl. Num. Meth. in Fluids. **11**, 6.

[4] Girault V. and Raviart P-A. [1986]. *Finite element methods for Navier-Stokes equations*, Springer-Verlag.

[5] Gunzburger M. D. [1989]. *Finite Element Methods for Viscous Incompressible Viscous Flows*, Academic Press.

[6] Hall C. A., Cavendish J. C., and Frey W. H. [1991]. "The dual variable method for solving fluid flow difference equations on Delaunay triangulations," Comp. Fluids, **20**, 2, 145–164.

[7] Hirt C. W. [1979]. "Simplified solution algorithms for fluid flow problems," Numerical Methods for Partial Differential Equations, ed. S. V. Parter, Academic Press.

[8] Hu X. and Nicolaides R. A. [1991]. "Covolume techniques for anisotropic media," Numer. Math., **61**, 215–234.

[9] Ladyzhenskaya O. A. [1969]. *The Mathematical Theory of Viscous Incompressible Flow*, Gordon and Breach.

[10] MacNeal R. H. [1953]. "An asymmetrical finite difference network," Quart. Appl. Math. **11**, 295–310.

[11] Nicolaides R. A. [1989]. "Flow discretization by complementary volume schemes," AIAA paper 89-1978. Proceedings of 9th AIAA CFD Mtg. Buffalo NY.

[12] Nicolaides R. A. [1991a]. "Direct discretization of planar div-curl problems," SIAM Jnl. Num. An., **29**, 1, 32–56.

[13] Nicolaides R. A. [1991b]. "Analysis and convergence of the MAC scheme. I The linear problem," SIAM Jnl. Num. An. (to appear, 1992).

[14] Nicolaides R. A. and Wu X. [1991a]. "Numerical solution of the Hamel problem by a covolume method," *Advances in CFD* ed. W. G .Habashi and M. M. Hafez, (to appear).

[15] Nicolaides R. A. and Wu X. [1991b]. "Solution of three dimensional div-curl systems by a covolume method," Math. Comp. (to be submitted).

[16] Nicolaides R. A. and Wu X. [1991c]. "Analysis and convergence of the MAC scheme. II The Navier-Stokes equations," Math. Comp. (submitted).

[17] Pironneau O. [1990]. *Finite Elements for Fluids*, John Wiley.

[18] Riedinger R., Habar M., Oelhafen P., and Guntherodt H. J. [1988]. "About the Delaunay-Voronoi tesselation," Jnl. Comp. Phys. **74**, 61–72.

[19] Temam, R. [1984]. *Navier-Stokes Equations*, North-Holland.

11 Vortex Methods: An Introduction and Survey of Selected Research Topics

Elbridge Gerry Puckett

11.1 Introduction

Vortex methods are a type of numerical method for approximating the solution of the incompressible Euler or Navier-Stokes equations. In general, vortex methods are characterized by the following three features.

1. The underlying discretization is of the vorticity field, rather than the velocity field. Usually this discretization is Lagrangian in nature and frequently it consists of a collection of particles which carry concentrations of vorticity.

2. An approximate velocity field is recovered from the discretized vorticity field via a formula analogous to the Biot-Savart law in electromagnetism.

3. The vorticity field is then evolved in time according to this velocity field.

In the past two decades a number of different numerical methods for computing the motion of an incompressible fluid have been proposed that have the above features. In this article we consider a class of such methods which are based on the work of Chorin [1973, 1978, 1980, and 1982]. Members of this class are related by the manner in which a vorticity field in an inviscid, incompressible flow is discretized and subsequently evolved. It is common practice to use the term *vortex method* or *the vortex method* to refer to a member of this class when it is used to model the incompressible Euler equations. One can modify the vortex method and use it to model the incompressible Navier-Stokes equations by adding a random walk. This is known as *the random vortex method*. One can also replace the random walk by some non-random technique for solving the diffusion equation. Such methods are generally referred to as *deterministic vortex methods*.

The vortex method can be regarded as a discretization of the equations of fluid motion in vorticity form rather than in the usual velocity-pressure form. This has the advantage of eliminating the pressure from the number of dependent variables to be computed. In addition, as a consequence of the "Biot-Savart Law," the velocity can be written as a linear superposition of basic velocity functions, plus possible corrections for boundary conditions. These velocity functions arise as the convolution of an integral kernel with a concentrated mass of vorticity, usually referred to as a *vortex* or *vortex blob*. This

335

representation of the velocity field as a linear combination of identical velocity functions is both mathematically appealing and relatively straightforward to program. Furthermore, the nature of this discretization is such that diffusive errors due to the numerical method are small. This facilitates relatively inexpensive computations of inviscid or slightly viscous flows – especially when the vorticity is concentrated in small regions of the computational domain.

One of the goals of this article is to provide a reader who is familiar with computational fluid dynamics but not necessarily vortex methods a clear and fairly comprehensive introduction to the subject. We have included a short review of the mathematical theory of vortex methods, especially as it relates to cutoff functions and the influence that these functions have on the convergence rate of the method. We have also included (without proof) the statements of several theorems that have been proved regarding the accuracy and rate of convergence of various vortex method approximations to Euler and Navier-Stokes flow.

In addition, we have endeavored to describe several currently active areas of research in the field. This includes recent work on vorticity boundary conditions and numerical methods for implementing these boundary conditions. We also discuss deterministic vortex methods and describe several of these methods in detail. We briefly mention the recent work on point vortex methods, especially concerning the convergence of the point vortex method to solutions of the incompressible Euler equations in two and three dimensions. Finally, we describe several techniques for reducing the cost of a vortex method computation from $O(N^2)$ to $O(N \log N)$ or even $O(N)$ where N is the number of basic velocity functions in the computation. In particular, we describe the two dimensional versions of the *method of local corrections* and the *fast multipole method* in some detail. It is our hope that we have presented this material in sufficient detail that an interested reader would be able to program many of these methods from our description of them.

This article is *not* a survey of all vortex methods. It is intended to be a short intro-duction to the field with a brief discussion of selected research topics of interest to the author. Consequently, there has been no attempt to refer to all researchers in the field or all "important work." In fact, there are many aspects of the field which we have not discussed at all. For example, there is no discussion of the vast body of literature concerning the vortex method approximation of a vortex sheet (e.g., see Bernard and Chorin [1973] or Krasny [1986a, 1986b and 1987]) or of the technique of contour dy-namics (e.g., Zabusky, Hughes, and Roberts [1979] or Zabusky and Overman [1983]) and related ideas (e.g., Buttke [1990]). On the other hand, we have made a concerted effort to provide as complete a bibliography as possible in those areas that are discussed.

We apologize in advance for any omission or oversight that we have made in this regard. For a more thorough review of vortex methods and more extensive bibliographies the interested reader is referred to the excellent review articles of Beale and Majda [1984], Hald [1991], Leonard [1981 and 1985], and Sethian [1991]. The author has benefited tremendously from reading these articles.

In the following section we review the equations of incompressible fluid flow and introduce various formulations of these equations that will be useful in the subsequent discussion. In Section 11.3 we describe two and three dimensional versions of the vortex method for approximating solutions of the incompressible Euler equations in unbounded regions Ω of $I\!R^n$. This allows us to ignore the issue of satisfying the no-flow boundary condition on $\partial\Omega$. We discuss the modifications necessary in order to do so in Section 11.4. In Section 11.5 we introduce the random vortex method for approximating solutions of the Navier-Stokes equations in unbounded regions Ω of $I\!R^n$. We also discuss current research into alternatives to the random walk; i.e., deterministic vortex methods. If Ω has a solid boundary, then one must have a mechanism for approximately satisfying the no-slip boundary condition on $\partial\Omega$. In Section 11.6 we describe a technique for doing this called the *vortex sheet method*. We also mention some recent work on alternatives to this method. Section 11.6 also contains a short discussion on parameter selection for a hybrid random vortex / vortex sheet method computation. At the present this is probably the most widely used type of vortex method for computing viscous flows with boundaries in "engineering" applications. Finally, in Section 11.7 we end with a discussion of "fast vortex methods." This includes a detailed description of two techniques that have been introduced in the past decade: the method of local corrections and the fast multipole method.

The author is indebted to a number of individuals for freely discussing their work and making suggestions during the writing of this article. He would especially like to thank Bill Ashurst, Chris Anderson, Tom Beale, Alexandre Chorin, Claude Greengard, Leslie Greengard, Ole Hald, and Steve Roberts for reading portions of the manuscript.

11.2 The Equations of Incompressible Fluid Flow in Vorticity Form

In this section we derive the Euler and Navier-Stokes equations in vorticity form from the usual velocity-pressure formulation of the equations. We also discuss the issue of boundary conditions, especially recent work on vorticity boundary conditions for the Navier-Stokes equations. We end with a brief introduction to the notion of fractional step methods. A more thorough account of the role that vorticity plays in the mathematical theory of fluid flow may be found in Majda [1986].

11.2.1   The Euler Equations

First we consider the incompressible Euler equations. Let $\Omega \subset I\!R^n$ with $n = 2$ or 3. The incompressible Euler equations are

$$\mathbf{u}_t + (\mathbf{u} \cdot \nabla)\mathbf{u} = -\nabla p, \tag{11.2.1a}$$

$$\nabla \cdot \mathbf{u} = 0, \tag{11.2.1b}$$

$$\mathbf{u}(\mathbf{x}, 0) = \mathbf{u}^0(\mathbf{x}), \tag{11.2.1c}$$

$$\mathbf{u} \cdot \boldsymbol{\eta} = 0 \quad \text{on} \quad \partial\Omega, \tag{11.2.1d}$$

where $\mathbf{u}$ is the velocity, p is the pressure, and $\boldsymbol{\eta}$ is a unit vector normal to $\partial\Omega$.[1] We let $\mathbf{x} = (x, y)^T$, $\mathbf{u} = (u, v)^T$ when $n = 2$ and $\mathbf{x} = (x, y, z)^T$, $\mathbf{u} = (u, v, w)^T$ when $n = 3$. Note that equation (11.2.1a) is a vector equation with n components. Equations (11.2.1a–11.2.1d) describe the flow of an incompressible, inviscid fluid in Ω with initial data $\mathbf{u}^0$. We assume that the flow has constant density $\rho = 1$. The boundary condition (11.2.1d) is commonly called the *no-flow boundary condition*.

The Free-Space Problem   During the subsequent discussion we will have occasion to consider the case when $\Omega = I\!R^n$. In this event it is understood that the boundary condition (11.2.1d) has been replaced by

$$\mathbf{u}(\mathbf{x}) \to 0 \quad \text{as} \quad \mathbf{x} \to \infty. \tag{11.2.2}$$

This is sometimes called the *free-space problem* and (11.2.2) is often referred to as the *free-space boundary condition*.

The vorticity ω is defined by

$$\omega = \nabla \times \mathbf{u}. \tag{11.2.3}$$

In two dimensions the vorticity vector $\omega = (0, 0, \partial_x v - \partial_y u)^T$ points in the direction perpendicular to the (x, y) plane and is usually considered a scalar, $\omega = \partial_x v - \partial_y u$.

Suppose $\Omega = I\!R^n$ and consider the following problem. Given the vorticity $\omega(\mathbf{x}, t)$ at time t, reconstruct the velocity $\mathbf{u}(\mathbf{x}, t)$ such that $\mathbf{u}$ satisfies (11.2.1b), (11.2.2) and (11.2.3). We can do so by finding a function ψ such that

$$\mathbf{u} = \nabla \times \psi. \tag{11.2.4}$$

This function can be found by solving

$$\Delta\psi = -\omega, \tag{11.2.5a}$$

$$\nabla\psi(\mathbf{x}) \to 0 \quad \text{as} \quad \mathbf{x} \to \infty. \tag{11.2.5b}$$

---

[1] We denote vector quantities in bold face type.

Note that $\mathbf{u}$ automatically satisfies (11.2.1b) since $\nabla \cdot (\nabla \times \mathbf{f}) = 0$ for any function $\mathbf{f}$. The function ψ is called the *stream function*. As with the vorticity, in two dimensions ψ points in the direction perpendicular to the (x, y) plane and is generally thought of as a scalar, ψ.

Let G denote the Green's function for the Poisson equation (11.2.5a) in $\mathbb{R}^n$. Then we have

$$\psi(\mathbf{x}, t) = \int_{\mathbb{R}^n} G(\mathbf{x} - \mathbf{x}') \, \omega(\mathbf{x}', t) \, d\mathbf{x}' = (G * \omega)(\mathbf{x}, t), \qquad (11.2.6)$$

where $*$ denotes convolution. Since $\partial_x(G * \omega) = (\partial_x G) * \omega$ and similarly for $\partial_y(G * \omega)$ and $\partial_z(G * \omega)$ we find from (11.2.4) and (11.2.6) that

$$\mathbf{u}(\mathbf{x}, t) = \int_{\mathbb{R}^n} (\nabla \times G)(\mathbf{x} - \mathbf{x}') \, \omega(\mathbf{x}', t) \, d\mathbf{x}' = \int_{\mathbb{R}^n} \mathbf{K}(\mathbf{x} - \mathbf{x}') \, \omega(\mathbf{x}', t) \, d\mathbf{x}' = (\mathbf{K} * \omega)(\mathbf{x}, t)$$
$$(11.2.7)$$

where $\mathbf{K} = \nabla \times G$. Equation (11.2.7) is frequently referred to as the "Biot-Savart law" in analogy with the well known formula of that name from electrostatics. In two dimensions $\mathbf{K}$ is a vector,

$$\mathbf{K}(\mathbf{x}) = \frac{1}{2\pi |\mathbf{x}|^2} (-y, x)^T, \qquad (11.2.8)$$

which has a singularity of order $O(|\mathbf{x}|^{-1})$ at $\mathbf{x} = \mathbf{0}$. In three dimensions $\mathbf{K}$ is a matrix,

$$\mathbf{K}(\mathbf{x}) = \frac{1}{4\pi |\mathbf{x}|^3} \begin{pmatrix} 0 & z & -y \\ -z & 0 & x \\ y & -x & 0 \end{pmatrix}, \qquad (11.2.9)$$

with a singularity of order $O(|\mathbf{x}|^{-2})$ at $\mathbf{x} = \mathbf{0}$. The singularity in $\mathbf{K}$ has played a central role in the mathematical theory of vortex methods.

Note that if ω has compact support or decays sufficiently fast as $\mathbf{x} \to \infty$, then $\mathbf{u}$ given by (11.2.7) automatically satisfies the free-space boundary condition (11.2.2). Thus, when $\Omega = \mathbb{R}^n$, we have found a way to write the velocity $\mathbf{u}$ in terms of the vorticity ω.

The No-Flow Boundary Condition When $\Omega \subset \mathbb{R}^n$ but $\Omega \neq \mathbb{R}^n$ the function $\mathbf{u}$ given by (11.2.7) will not necessarily satisfy the no-flow boundary condition (11.2.1d) on $\partial \Omega$. We can rectify this situation by writing the velocity $\mathbf{u}$ as the sum of a "rotational" flow $\mathbf{u}_\omega$ and an irrotational (vorticity free) flow $\mathbf{u}_p$ as follows. Given the vorticity field $\omega(\mathbf{x})$ in Ω let $\mathbf{u}_\omega = \mathbf{K} * \omega$ (where we define $\omega(\mathbf{x}) = 0$ for $\mathbf{x} \notin \Omega$ for the purposes of the integration over $\mathbb{R}^n$ in (11.2.7)). In general, $\mathbf{u}_\omega$ will fail to satisfy (11.2.1d) on $\partial \Omega$. We seek a scalar function ϕ in Ω such that

$$\Delta \phi = 0 \qquad (11.2.10a)$$

$$\frac{\partial \phi}{\partial \eta} = -\mathbf{u}_\omega \cdot \eta \qquad \text{on} \quad \partial\Omega. \tag{11.2.10b}$$

In other words, we seek a solution to the Laplace equation (11.2.10a) in Ω with Neumann boundary conditions (11.2.10b). The theory of elliptic equations (e.g., see Garabedian [1986]) guarantees the existence of a solution ϕ under reasonable restrictions on $\partial\Omega$ and the boundary data.

The *potential flow* $\mathbf{u}_p$ defined by

$$\mathbf{u}_p(\mathbf{x}) = \nabla \phi(\mathbf{x}), \tag{11.2.11}$$

satisfies (11.2.1b) since $\nabla \cdot \mathbf{u}_p = \Delta \phi = 0$. Also, $\nabla \times \mathbf{u}_p = 0$ since $\nabla \times \nabla f = 0$ for any scalar function f. Now consider the velocity field given by,

$$\mathbf{u}(\mathbf{x}) = \mathbf{u}_\omega(\mathbf{x}) + \mathbf{u}_p(\mathbf{x}). \tag{11.2.12}$$

It is apparent from the foregoing that $\mathbf{u}$ satisfies $\nabla \cdot \mathbf{u} = 0$ and $\nabla \times \mathbf{u} = \omega$. Furthermore, it follows from (11.2.10b), (11.2.11) and (11.2.12) that

$$\mathbf{u} \cdot \eta = \mathbf{u}_\omega \cdot \eta - \mathbf{u}_\omega \cdot \eta = 0 \qquad \text{on} \quad \partial\Omega.$$

Hence $\mathbf{u} = \mathbf{u}_\omega + \mathbf{u}_p$ satisfies (11.2.1b,d) and (11.2.3) as desired.

In a vortex method computation we construct an approximation $\tilde{\mathbf{u}}$ to the exact solution $\mathbf{u}$ of (11.2.1a–d) in precisely this manner: Given an approximation $\tilde{\omega}$ to the vorticity field ω we take $\tilde{\mathbf{u}}_\omega = \mathbf{K} * \tilde{\omega}$ to be our approximation to $\mathbf{u}_\omega$. Then, in order to satisfy the no-flow boundary condition (11.2.1d), we add to $\tilde{\mathbf{u}}_\omega$ an approximation $\tilde{\mathbf{u}}_p$ to the potential flow $\mathbf{u}_p$ by solving the Neumann problem (11.2.10a–b) with boundary data $-\tilde{\mathbf{u}}_\omega \cdot \eta$. In Section 11.4 we will discuss various techniques for computing $\tilde{\mathbf{u}}_p$.

11.2.2   The Euler Equations in Vorticity Form

By taking the curl of equations (11.2.1a,c) we obtain the vorticity form of the Euler equations,

$$\frac{D\omega}{Dt} = (\omega \cdot \nabla)\mathbf{u}, \tag{11.2.13a}$$

$$\nabla \cdot \mathbf{u} = 0, \tag{11.2.13b}$$

$$\omega = \nabla \times \mathbf{u}, \tag{11.2.13c}$$

$$\omega(\mathbf{x}, 0) = \omega^0(\mathbf{x}), \tag{11.2.13d}$$

$$\mathbf{u}(\mathbf{x}, t) \cdot \eta(\mathbf{x}) = \mathbf{0} \qquad \text{for} \quad \mathbf{x} \in \partial\Omega. \tag{11.2.13e}$$

Here $D/Dt \equiv \partial_t + (u \cdot \nabla)$ is the so-called *material derivative*. Note that by taking the curl of (11.2.1a) we have eliminated the pressure from the set of equations to be solved. This greatly simplifies the task of finding approximations to **u**. Furthermore, from (11.2.7), (11.2.10a–b) and (11.2.12) we see that the velocity depends linearly on the vorticity. The vortex method takes advantage of this fact by discretizing the vorticity field as a linear superposition of patches of vorticity. Each vortex patch induces a velocity field via (11.2.7) that is easy to compute, and the total velocity field is simply the sum of these velocities plus a correction for boundary conditions.

11.2.3 The Navier-Stokes Equations

We now turn to the Navier-Stokes equations. Let ν denote the kinematic viscosity. In vorticity form the Navier-Stokes equations are given by,

$$\frac{D\omega}{Dt} = (\omega \cdot \nabla)\mathbf{u} + \nu \Delta \omega \,, \tag{11.2.14a}$$

$$\nabla \cdot \mathbf{u} = 0 \,, \tag{11.2.14b}$$

$$\omega = \nabla \times \mathbf{u} \,, \tag{11.2.14c}$$

$$\omega(\mathbf{x}, 0) = \omega^0(\mathbf{x}) \,, \tag{11.2.14d}$$

$$\mathbf{u} = \mathbf{0} \quad \text{on} \quad \partial \Omega \,. \tag{11.2.14e}$$

These equations describe the flow of a viscous, incompressible fluid in Ω with initial data ω^0. They can be derived from the usual velocity-pressure form of the Navier-Stokes equations in the same way we derived equations (11.2.13a–e) from (11.2.1a–d).

From a computational point of view these equations present two additional difficulties over the Euler equations. The first difficulty is that of accurately modeling the diffusive effects of the term $\nu \Delta \omega$ on the right hand side of (11.2.14a). In particular, for very small ν solutions of (11.2.14a–e) can have important features which vary on spatial scales that are $O(\sqrt{\nu})$. This leads to the requirement that a numerical method must resolve spacial features of the flow that are $O(\sqrt{\nu})$. For example, in a finite difference computation this requirement would lead to the need for $O(\sqrt{\nu^{-1}})$ grid points per spacial dimension (e.g., see Henshaw, Kreiss, and Reyna [1991]). This can become prohibitively expensive as $\nu \to 0$. We will return to this issue in our discussion of the random vortex method in Section 11.5.

The other difficulty that has been introduced is the addition of a second boundary condition

$$\mathbf{u}(\mathbf{x}, t) \cdot \tau(\mathbf{x}) = 0 \qquad \text{for} \qquad \mathbf{x} \in \partial \Omega \,, \tag{11.2.15}$$

where τ is a unit vector tangent to $\partial\Omega$. This is usually referred to as the *no-slip boundary condition* (or sometimes, the *stick boundary condition*). This boundary condition has a very important physical interpretation: it accounts for the creation of vorticity at $\partial\Omega$. In fact, it can be shown that in an incompressible flow this is the only means by which vorticity is created (e.g. see Chorin and Marsden [1979], Schlichting [1979], or White [1974]).

It is not immediately apparent how best to approximate the boundary condition (11.2.15) within the context of a vortex method computation. Chorin [1973 and 1978] has devised several methods in which concentrations of vorticity – i.e. vortices – are added at points on $\partial\Omega$ with strengths chosen so that the resulting velocity field approximately cancels $\tilde{\mathbf{u}} \cdot \tau$ on $\partial\Omega$ where $\tilde{\mathbf{u}}$ is a numerical approximation to (11.2.12). These vortices are subsequently allowed to diffuse into Ω and participate in the flow. The process of diffusing the vortices into the flow is accomplished by letting each vortex undergo a random walk. This method has the attraction of mimicking the physical process of vorticity creation at $\partial\Omega$. In one form or another this idea is at the basis of almost all methods currently being used to approximate (11.2.15) in vortex method computations.

There has been much research into clarifying the mathematical principles underlying Chorin's idea – as well as into finding alternatives to it. We briefly discuss some of these efforts here. Let Ω be a compact subset of $I\!\!R^2$ which lies in the interior of a closed curve $\partial\Omega$. Let $\mathbf{u}$ be an incompressible velocity field in Ω that satisfies the no-flow boundary condition (11.2.1d) on $\partial\Omega$ and let $\omega = \nabla \times \mathbf{u}$ be the corresponding vorticity field. The stream function ψ associated with $\mathbf{u}$ may be found by solving

$$\Delta\psi = -\omega, \tag{11.2.16a}$$

$$\psi = 0 \quad \text{on} \quad \partial\Omega. \tag{11.2.16b}$$

Therefore ψ satisfies (11.2.4),

$$\mathbf{u}(\mathbf{x}) = (\psi_y(\mathbf{x}), -\psi_x(\mathbf{x})), \tag{11.2.17}$$

and (11.2.1d),

$$\mathbf{u} \cdot \eta = 0 \quad \text{on} \quad \partial\Omega.$$

Now suppose that $\mathbf{u}$ also satisfies the no-slip boundary condition (11.2.15) on $\partial\Omega$. Then we must have

$$\frac{\partial\psi}{\partial\eta} = 0 \quad \text{on} \quad \partial\Omega. \tag{11.2.18}$$

This constitutes a second boundary condition on ψ – one in addition to (11.2.16b). In other words, the problem of finding ψ from ω is now over determined.

Quartapelle and Valz-Gris [1981] considered the question "What are the conditions on the vorticity so that the stream function ψ constructed by solving (11.2.16a,b) also satisfies the boundary condition (11.2.18)?" Their answer was that ω must be orthogonal to all harmonic functions η in Ω. In other words, that

$$\int_\Omega \omega(\mathbf{x})\, \eta(\mathbf{x})\, d\mathbf{x} = 0\,, \tag{11.2.19}$$

for all η harmonic in Ω. Thus, Quartapelle and Valz-Gris have replaced the boundary condition (11.2.15) by a constraint on the evolution on the vorticity (11.2.19). They implemented this constraint numerically by advancing the solution of (11.2.14a) one time step without regard for this constraint and then projecting the result down onto the component which satisfies (11.2.19). A detailed implementation of this algorithm is presented in Quartapelle [1981].

Anderson [1989] studied the connection between Chorin's [1978] algorithm and the Quartapelle and Valz-Gris algorithm. He was able to show that Quartapelle and Valz-Gris' constraint on the vorticity was equivalent to altering the vorticity at $\partial\Omega$ by exactly the same amount Chorin added at $\partial\Omega$.

It is natural to inquire about the possibility of finding a boundary condition on the vorticity ω consistent with the boundary condition (11.2.14e) on $\mathbf{u}$. In this regard Cottet [1988a] has shown that for domains $\Omega \subseteq I\!\!R^2$ the following system of equations are equivalent to the Navier-Stokes equations (11.2.14a–e) (see also Mas-Gallic [1990]),

$$\frac{D\omega}{Dt} = \nu\Delta\omega\,, \tag{11.2.20a}$$

$$\omega(\mathbf{x}, 0) = \omega^0(\mathbf{x})\,, \tag{11.2.20b}$$

$$\omega = \nabla \times \mathbf{u} \qquad \text{on} \quad \partial\Omega\,, \tag{11.2.20c}$$

$$\Delta\mathbf{u} = -\nabla \times \omega\,, \tag{11.2.20d}$$

$$\mathbf{u} = \mathbf{0} \qquad \text{on} \quad \partial\Omega\,. \tag{11.2.20e}$$

One can regard these equations as two coupled subsystems (11.2.20a–c) and (11.2.20d–e). The first system is an evolution equation for the vorticity (11.2.20a) with initial condition (11.2.20b) and boundary condition (11.2.20c). The second system is Poisson's equation (11.2.20d) with Dirichlet boundary data (11.2.20e). Both Cottet [1988a] and Mas-Gallic [1990] have proposed numerical methods based on these equations for solving the Navier-Stokes equations in $I\!\!R^2$. One of the essential differences between these algorithms and Chorin's algorithm is that the algorithms of Cottet and Mas-Gallic do not use a random walk. Such methods are commonly referred to as *deterministic vortex methods*. We will return to this topic in Section 11.5.

We also mention that Hou and Wetton [1990] have constructed a finite difference method for approximating solutions of the Navier-Stokes equations in vorticity-stream formulation which incorporate a vorticity boundary condition. They have proved the convergence of this method to solutions of the Navier-Stokes equations. This is currently a very active area of research. We refer the interested reader to Anderson [1989], Cottet [1988a], Hou and Wetton [1990], Mas-Gallic [1990], Quartapelle [1981], and Quartapelle and Valz-Gris [1981] for further details and references.

11.2.4   The Diffusion Equation

Solutions of the Navier-Stokes equations (11.2.14a–e) may be approximated by alternately solving two distinct systems of partial differential equations for small time steps Δt; with the solution of the first system being the initial data for the second and so on. The first of these systems is the Euler equations (11.2.13a–e). The second is the *diffusion* or *heat equation*,

$$\omega_t = \nu \Delta \omega \,, \tag{11.2.21a}$$

$$\mathbf{u} = \mathbf{0} \quad \text{on} \quad \partial \Omega \,. \tag{11.2.21b}$$

Note that this is not the usual boundary value problem for the heat equation – the boundary data is prescribed for the velocity $\mathbf{u}$ rather than the vorticity ω. This is simply a reflection of the fact that we are using the boundary condition (11.2.14e) on the velocity with the evolution equations (11.2.14a–d) for the vorticity. It is natural to inquire if one can replace (11.2.21b) with an equivalent boundary condition on the vorticity. This is essentially the same question that is being addressed in the work on vorticity boundary conditions for equations (11.2.14a–d).

Equation (11.2.21a) describes the rate at which the vorticity is diffusing in the flow. In many instances it is possible to arrange the computation so that if the no-flow boundary condition (11.2.1d) is satisfied by the initial data to (11.2.21a,b), then the solution of (11.2.21a) will satisfy (11.2.1d) at later times as well. Thus, the boundary condition usually associated with the solution of (11.2.21a) is the no-slip boundary condition (11.2.15). This boundary condition may be regarded as a source term that appears during the solution of (11.2.21a) (e.g., see Puckett [1989a], p. 310).

11.2.5   Fractional Step Methods

The idea of solving the Euler equations for a small time step Δt and then the heat equation for the same time step and using the result as an approximation to the solution of the Navier-Stokes equations at time Δt is sometimes called *viscous splitting*. This is a special case of a more general technique known as *operator splitting*. Viscous splitting for the Navier-Stokes equations in $\Omega = I\!R^n$ for $n = 2, 3$ with the free space boundary condition (11.2.2) has been given a rigorous justification by Beale and Majda [1981], Ebin and Marsden [1970], and Ying [1990]. The problem is much more difficult when boundaries are included. Work on this problem appears in Benfatto and Pulverenti [1983] and Alessandrini, Douglis, and Fabes [1983].

Numerical methods based on operator splitting are called *fractional step* methods. Such methods are applicable to a wide variety of partial differential equations. In particular, the random vortex method is a fractional step method for approximating solutions of the Navier-Stokes equations. The first step consists of a vortex method approximation to the solution of the Euler equations (11.2.13a–e) in which the no-flow boundary condition is satisfied by adding a potential flow. The second step is a random walk approximation to the solution of the heat equation (11.2.21a) in which the no-slip boundary condition (11.2.15) is satisfied by creating vortex elements on $\partial\Omega$. The initial data for one step is the result obtained from the previous step.

11.3   The Vortex Method

The vortex method is a numerical method for approximating solutions of the incompressible Euler equations (11.2.1a–d). It is a *particle method* in which fluid particles carrying concentrations of vorticity are followed as their positions and concentrations evolve with the motion of the fluid. Fundamental to all particle methods is the notion of the fluid flow map $\mathbf{x} : \Omega \times [0, T] \to \Omega$ defined so that $\mathbf{x}(\alpha, t)$ is the trajectory of the fluid particle which at time $t = 0$ is at the point α. For fixed α this trajectory may be found by solving the ordinary differential equation,

$$\frac{d\mathbf{x}}{dt}(\alpha, t) = \mathbf{u}(\mathbf{x}(\alpha, t), t), \tag{11.3.1a}$$

$$\mathbf{x}(\alpha, 0) = \alpha. \tag{11.3.1b}$$

In this section we will assume that $\Omega = I\!R^n$ and substitute the free-space boundary condition (11.2.2) for the no-flow boundary condition (11.2.1d). In this case $\mathbf{u}$ is a solution of the incompressible Euler equations (11.2.1a–c) with boundary condition (11.2.2). Therefore, by (11.2.7) we have

$$\mathbf{u}(\mathbf{x}, t) = \int_{I\!R^n} \mathbf{K}(\mathbf{x} - \mathbf{x}')\, \omega(\mathbf{x}', t)\, d\mathbf{x}' = \int_{I\!R^n} \mathbf{K}(\mathbf{x} - \mathbf{x}(\alpha', t))\, \omega(\mathbf{x}(\alpha', t), t)\, d\alpha'.$$

The last equality follows from the fact that $\nabla \cdot \mathbf{u} = 0$ and hence, that the Jacobian of the transformation $\alpha' \to \mathbf{x}(\alpha', t)$ equals 1 (e.g., see Chorin and Marsden [1990], p. 10).

Thus, (11.3.1a,b) can be rewritten as

$$\frac{d\mathbf{x}}{dt}(\alpha, t) = \int_{\mathbb{R}^n} \mathbf{K}(\mathbf{x}(\alpha, t) - \mathbf{x}(\alpha', t))\, \omega(\mathbf{x}(\alpha', t), t)\, d\alpha', \qquad (11.3.2a)$$

$$\mathbf{x}(\alpha, 0) = \alpha. \qquad (11.3.2b)$$

Thus, if the fluid flow map $\alpha \to \mathbf{x}(\alpha, t)$ and the vorticity field $\omega(\mathbf{x}, t)$ are known, then the velocity at later times may be written as an integral over the domain at the initial time $t = 0$. In two dimensions the vorticity is constant on particle paths $\alpha \to \mathbf{x}(\alpha, t)$ and this leads to an especially simple representation of the velocity field at later times.

11.3.1   Two Dimensions

As a consequence of the scalar character of the vorticity in two dimensions the two dimensional version of the vortex method is simpler than its three dimensional counterpart. Therefore we begin by describing the vortex method in $\mathbb{R}^2$. In two dimensions the fluid velocity $\mathbf{u}$ is perpendicular to the vorticity

$$(\omega \cdot \nabla)\mathbf{u} = 0.$$

Hence, the right hand side of (11.2.13a) is zero,

$$\frac{D\omega}{Dt} = 0.$$

This equation states that the vorticity is transported passively by the flow. In other words, that the vorticity is constant along particle trajectories,

$$\frac{d}{dt}\omega(\mathbf{x}(\alpha, t), t) = 0,$$

or equivalently,

$$\omega(\mathbf{x}(\alpha, t), t) = \omega^0(\alpha). \qquad (11.3.3)$$

This is the most significant difference between the fluid flow equations in two and three dimensions. In general, when $n = 3$ we have $(\omega \cdot \nabla)\mathbf{u} \neq 0$. This provides a mechanism for the vorticity to change as it moves with the flow. In particular, the vorticity may now stretch and fold. It is generally believed that this is the mechanism by which turbulence is created and driven (e.g., see Tennekes and Lumley [1972]). This explains why the phrase "two dimensional turbulence" is sometimes considered an oxymoron.

Using (11.3.3) we can rewrite (11.3.2a,b) as

$$\frac{d\mathbf{x}}{dt}(\alpha, t) = \int_{\mathbb{R}^n} \mathbf{K}(\mathbf{x}(\alpha, t) - \mathbf{x}(\alpha', 0))\, \omega^0(\alpha')\, d\alpha', \qquad (11.3.4a)$$

$$\mathbf{x}(\alpha, 0) = \alpha. \qquad (11.3.4b)$$

Thus, given the fluid flow map $\alpha \to \mathbf{x}(\alpha, t)$, we see that the velocity can be expressed as an integral over the vorticity field at time $t = 0$. The equivalence of the Lagrangian equations (11.3.4a,b) and the Euler equations (11.2.1a–c) with boundary condition (11.2.2) is discussed in McGrath [1968].

In the vortex method the solution of (11.3.4a,b) is approximated by discretizing the integral on the right hand side of (11.3.4a) and solving the resulting equations for a finite number of particles. Consider a rectangular grid with mesh width h that covers the support of $\omega^0 = \nabla \times \mathbf{u}^0$.[2] Let α_i, $i = 1, \ldots, N$, denote the centers of the grid cells and let $\omega_i = \omega(\alpha_i)$. If we denote by $\tilde{\mathbf{x}}_i(t)$ the trajectory of the particle that starts at α_i and evolves according to this discretized system, then an obvious approximation to (11.3.4a,b) consists of solving the following system of N ordinary differential equations,

$$\frac{d\tilde{\mathbf{x}}_i}{dt}(t) = \sum_{j \neq i} \mathbf{K}(\tilde{\mathbf{x}}_i(t) - \tilde{\mathbf{x}}_j(t))\, \omega_j\, h^2 , \tag{11.3.5a}$$

$$\tilde{\mathbf{x}}_i(0) = \alpha_i . \tag{11.3.5b}$$

The numerical solution of this system is known as the *point vortex method*. This method was originally used by Rosenhead [1931] for computing the roll-up of a vortex sheet. It is apparent from (11.2.8) that whenever two particle trajectories approach each other, the velocity that each induces on the other goes to infinity. For this reason it was generally believed that the point vortex method was unstable and would not converge to solutions of the incompressible Euler equations. However, Goodman, Hou, and Lowengrub [1990a] have recently proved that the 2-d point vortex method does in fact converge to solutions of (11.3.4a,b). We will briefly discuss this and related work in Section 11.3.5.

Chorin [1973] introduced the idea of replacing (11.3.5a,b) by

$$\frac{d\tilde{\mathbf{x}}_i}{dt}(t) = \sum_{j \neq i} \mathbf{K}_\delta(\tilde{\mathbf{x}}_i(t) - \tilde{\mathbf{x}}_j(t))\, \omega_j\, h^2 , \tag{11.3.6a}$$

$$\tilde{\mathbf{x}}_i(0) = \alpha_i . \tag{11.3.6b}$$

where the new kernel $\mathbf{K}_\delta$ is close to $\mathbf{K}$ except at the origin where $\mathbf{K}_\delta$ is bounded. From a theoretical point of view Hald has shown that it is convenient to obtain $\mathbf{K}_\delta$ by convoluting $\mathbf{K}$ with a *smoothing function* f_δ. So $\mathbf{K}_\delta = \mathbf{K} * f_\delta$ where $f_\delta : I\!\!R^2 \to I\!\!R$ is a scalar function on $I\!\!R^2$ which is obtained from a fixed scalar function $f : I\!\!R^2 \to I\!\!R$ of integral one via the relation

$$f_\delta(\mathbf{x}) = \frac{1}{\delta^2} f(\mathbf{x}/\delta) . \tag{11.3.7}$$

---

[2]For simplicity we always assume that ω^0 has compact support. With suitable modifications one can also handle initial vorticity distributions ω^0 that are periodic or that have unbounded support but decay sufficiently fast at infinity.

If we compare (11.3.6a) with (11.3.1a), we find that solving the system (11.3.6a,b) amounts to replacing the exact velocity $\mathbf{u}(\mathbf{x}, t)$ in (11.3.1a) by the approximate velocity

$$\tilde{\mathbf{u}}(\mathbf{x}, t) = \sum_i \mathbf{K}_\delta(\mathbf{x} - \tilde{\mathbf{x}}_i(t)) \, \omega_i \, h^2 \,, \tag{11.3.8}$$

and computing the N approximate particle trajectories $\tilde{\mathbf{x}}_i$ that start at the points α_i. The discretization presented here is essentially a trapezoid rule approximation to the right hand side of (11.3.4a). Nicolaides [1986] and Chiu and Nicolaides [1988] have studied other discretizations of this integral.

Since by definition $\mathbf{K}_\delta = \mathbf{K} * f_\delta$ and since $\nabla \times (\mathbf{K} * g) = g$ for any scalar function g which decays sufficiently fast as $|\mathbf{x}| \to \infty$, it follows from taking the curl of equation (11.3.8) that our approximation to the vorticity field is

$$\tilde{\omega}(\mathbf{x}, t) = \sum_i f_\delta(\mathbf{x} - \tilde{\mathbf{x}}_i(t)) \, \omega_i \, h^2 \,. \tag{11.3.9}$$

The ith term on the right hand side of (11.3.9) is referred to as the ith *vortex* or ith *vortex blob*, $\tilde{\mathbf{x}}_i(t)$ is its position at time t and ω_i is its *strength* or *weight*.

The smoothness of $\mathbf{K}_\delta$ depends on f_δ. We can choose f_δ so that the kernel $\mathbf{K}_\delta$ will be bounded or even continuous and differentiable at $\mathbf{x} = 0$. Chorin [1973] originally used

$$f_\delta(\mathbf{x}) = \begin{cases} (2\pi|\mathbf{x}|\delta)^{-1} & |\mathbf{x}| < \delta, \\ 0 & |\mathbf{x}| \geq \delta. \end{cases} \tag{11.3.10}$$

This yields a velocity of the form (11.3.8) with

$$\mathbf{K}_\delta(\mathbf{x}) = (-y, \, x)^T \begin{cases} (2\pi|\mathbf{x}|\delta)^{-1} & |\mathbf{x}| < \delta, \\ (2\pi|\mathbf{x}|^2)^{-1} & |\mathbf{x}| \geq \delta. \end{cases} \tag{11.3.11}$$

This has probably been the most widely used velocity kernel in two dimensional "engineering" implementations of the vortex method. By comparing (11.3.11) with (11.2.8) we see that $\mathbf{K}_\delta(\mathbf{x}) = \mathbf{K}(\mathbf{x})$ for $\mathbf{x} \geq \delta$ but that $\mathbf{K}_\delta$ remains bounded as $\mathbf{x} \to 0$. Thus, the effect of f_δ is to "cutoff" the infinite velocity at the origin. For this reason the function f_δ is also referred to as the *cutoff function* and δ is called the *cutoff radius*. For f_δ given by (11.3.10) it is apparent from (11.3.9) that the vorticity due to a vortex blob at $\mathbf{x}_i$ is now concentrated in a circular region or "core" about $\mathbf{x}_i$. Hence f_δ is sometimes also called a *core function*.

Usually – but not always – f is chosen to be radially symmetric: $f(\mathbf{x}) = f(r)$ where $r = |\mathbf{x}|$. It is sometimes the case, especially for higher order cutoffs, that f does not have compact support. In these cases the name cutoff radius for δ no longer quite makes sense. For this reason some authors prefer to refer to δ as the *cutoff parameter*. Assuming that the exact solution $\mathbf{u}(\mathbf{x}, t)$ of (11.2.1a–c) with boundary condition (11.2.2) is sufficiently smooth, then the properties of f will determine the accuracy of the vortex method approximation to $\mathbf{u}$. We discuss cutoff functions further in Section 11.3.3 below.

Suppose that for some $R_0 > 0$ we have $\mathbf{K}_\delta(\mathbf{x}) = \mathbf{K}(\mathbf{x})$ whenever $|\mathbf{x}| \geq R_0$. (In Section 11.3.3 we will show that this is true for radially symmetric f with compact support.) Let C be any circle of radius $R > R_0$ centered at $\tilde{\mathbf{x}}_i(t)$. Then the circulation Γ_i about the ith vortex is

$$
\begin{aligned}
\Gamma_i &= \oint_C \mathbf{K}_\delta(\mathbf{x} - \tilde{\mathbf{x}}_i(t))\, \omega_i\, h^2 \cdot d\mathbf{s} \\
&= \frac{\omega_i h^2}{2\pi} \oint_C \frac{(-y, x)}{|\mathbf{x}|^2} \cdot d\mathbf{s} \\
&= \frac{\omega_i h^2}{2\pi} \int_0^{2\pi} (-\sin\theta, \cos\theta) \cdot (-\sin\theta, \cos\theta)\, d\theta \\
&= \omega_i h^2 \,.
\end{aligned}
\tag{11.3.12}
$$

This formula is useful for determining the vortex strengths when vortices can enter the computation by some mechanism other than as part of the initial discretization of the vorticity field. This occurs most often within the context of satisfying the no-slip boundary condition (11.2.15) on $\partial\Omega$. We will discuss such techniques in Section 11.6. It is also possible to design a vortex method in which vortices are created at a flame front. Such algorithms have been studied by Ashurst and McMurtry [1989], Ghoniem, Chorin, and Oppenheim [1982] and Sethian [1984].

Equations (11.3.6a,b) are a system of N ordinary differential equations that describe the approximate particle trajectories $\tilde{\mathbf{x}}_i(t)$, each of which is continuous in the variable t. One hopes that for appropriately chosen f_δ these approximate particle trajectories remain close to the exact particle trajectories $\mathbf{x}(\alpha_i, t)$. This turns out to be true and can be rigorously proven as we shall see in Section 11.3.4.

In an actual computation the trajectories $\tilde{\mathbf{x}}_i(t)$ are further approximated by discretizing 11.3.6a,b) in time. One can do this by employing almost any numerical ODE solver. For example, let $\tilde{\mathbf{x}}_i^k$ denote the time discretized approximation to $\tilde{\mathbf{x}}_i(t)$ at time $t_k = k\Delta t$ and let

$$
\tilde{\mathbf{u}}^k(\mathbf{x}) = \sum_i \mathbf{K}_\delta(\mathbf{x} - \tilde{\mathbf{x}}_i^k)\, \omega_i\, h^2 \,,
\tag{11.3.13}
$$

denote the approximate velocity field. If we use the first order Euler's method to approximate (11.3.6a,b) in time, then the new particle positions at time $t_{k+1} = (k+1)\Delta t$ are given by,

$$
\tilde{\mathbf{x}}_i^{k+1} = \tilde{\mathbf{x}}_i^k + \Delta t\, \tilde{\mathbf{u}}^k(\tilde{\mathbf{x}}_i^k) \,.
\tag{11.3.14}
$$

Thus, since the vorticity is constant on particle paths, the approximate velocity field at time t_{k+1} is now,

$$\tilde{\mathbf{u}}^{k+1}(\mathbf{x}) = \sum_i \mathbf{K}_\delta(\mathbf{x} - \tilde{\mathbf{x}}_i^{k+1})\, \omega_i\, h^2\,.$$

Higher order time discretizations have been studied theoretically by Anderson and Greengard [1985] and Hald [1987]. Examples of computations which involve a high order time discretization may be found in Baden and Puckett [1990], Cheer [1989], Nordmark [1991], Sethian [1984], and Sethian and Ghoniem [1988].

Note that since the vortex strengths ω_i have not changed on the approximate particle paths $\tilde{\mathbf{x}}_i^k \to \tilde{\mathbf{x}}_i^{k+1}$ we are approximately satisfying (11.2.13a) (or equivalently (11.3.3)). Furthermore, since $\tilde{\mathbf{u}}^{k+1} = \mathbf{K} * \tilde{\omega}^{k+1}$ explicitly satisfies (11.2.13b,c), the new velocity field $\tilde{\mathbf{u}}^{k+1}$ is a consistent approximation to the exact solution of (11.2.13a–c) at time t_{k+1} with initial data $\tilde{\omega}^k$ at time $t_k = k\Delta t$. Finally, for smoothing functions f_δ which have compact support or which decay sufficiently fast at infinity, the approximate velocity field $\tilde{\mathbf{u}}^k$ given by (11.3.13) will automatically satisfy the free space boundary condition (11.2.2).

A Remark Concerning Notation   We denote the exact trajectory of the particle which at time $t = 0$ is at the point $\boldsymbol{\alpha}$ by $\mathbf{x}(\boldsymbol{\alpha}, t)$. In other words, $\mathbf{x}(\boldsymbol{\alpha}, t)$ is a solution of the exact Lagrangian equations (11.3.4a,b). If we are considering a finite number of exact particle trajectories, say starting at the points $\boldsymbol{\alpha}_i$, $i = 1, \ldots, N$, then we simply write $\mathbf{x}_i(t)$. So the $\mathbf{x}_i(t)$ denote solutions of the finite set of ODEs

$$\frac{d\mathbf{x}_i}{dt}(t) = \int_{\mathbb{R}^n} \mathbf{K}(\mathbf{x}_i(t) - \mathbf{x}(\boldsymbol{\alpha}, t))\, \omega^0(\boldsymbol{\alpha})\, d\boldsymbol{\alpha}\,,$$

$$\mathbf{x}_i(0) = \boldsymbol{\alpha}_i\,,$$

for $i = 1, \ldots, N$, where the flow map $\mathbf{x}(\boldsymbol{\alpha}, t)$ is assumed to be known.

We denote a (time continuous) approximation of the ith particle trajectory $\mathbf{x}_i(t)$ by $\tilde{\mathbf{x}}_i(t)$. The paths $\tilde{\mathbf{x}}_i(t)$ are determined by solving the ODEs (11.3.6a,b). We denote a time discretized approximation to the $\tilde{\mathbf{x}}_i(t)$ at time t_k by $\tilde{\mathbf{x}}_i^k$. For example, if we use the first order Euler's method, then the $\tilde{\mathbf{x}}_i^k$ are advanced in time according to (11.3.14) with $\tilde{\mathbf{u}}^k$ defined by (11.3.13). We will employ a similar notation for the three dimensional versions of these quantities.

The purpose of distinguishing between the time continuous approximation $\tilde{\mathbf{x}}_i(t)$ to $\mathbf{x}_i(t)$ and the time discretized version $\tilde{\mathbf{x}}_i^k$ is as follows. Historically the first convergence proofs for the vortex method were for the time continuous case. In other words, it was shown that the $\tilde{\mathbf{x}}_i(t)$ converge to the $\mathbf{x}_i(t)$ and that the corresponding velocity field $\tilde{\mathbf{u}}(\mathbf{x}, t)$ converges to the exact velocity field $\mathbf{u}(\mathbf{x}, t)$ as the discretization of the right hand side of (11.3.2a) is refined. Subsequently, it was also proved that the $\tilde{\mathbf{x}}_i^k$ converge to $\mathbf{x}_i(t_k)$ and that $\tilde{\mathbf{u}}^k(\mathbf{x})$ converges to $\mathbf{u}(\mathbf{x}, t_k)$ in the appropriate limit. However many workers continue to state and prove theorems for only time continuous versions of the vortex method. In particular, there has not yet been published a convergence proof for the time discretized version of the random vortex method in three dimensions. By emphasizing the distinction between the time discretized and time continuous versions of the vortex method we will be able to give precise statements of some of these theorems. It is also our hope that by making the distinction clear here this will assist readers that go on to read other papers in the field.

We have neglected one essential issue in our discussion of the vortex method in two dimensions. This is the problem of satisfying the no-flow boundary condition (11.2.1d) in a computation. We will discuss this in Section 11.4. We now describe the vortex method for approximating solutions of (11.2.1a–c) with boundary condition (11.2.2) in three dimensions.

11.3.2 Three Dimensions

Recall that in $I\!R^3$ the vorticity is a vector which may point in any direction. This implies that the right hand side of (11.2.13a) is in general nonzero. This term is usually associated with the "stretching" of vorticity. One can think of (11.2.13a) as an evolution equation for the vorticity in which the vorticity is stretched by the flow as it is transported along particle trajectories. To see this note that by (11.2.13a) the evolution of vorticity that is carried with the point $\mathbf{x}(\alpha, t)$ is given by

$$\frac{d\omega}{dt}(\mathbf{x}(\alpha,t),t) = (\omega(\mathbf{x}(\alpha,t),t) \cdot \nabla)\,\mathbf{u}(\mathbf{x}(\alpha,t),t)\,. \tag{11.3.15}$$

Hence the stretching term $(\omega \cdot \nabla)\mathbf{u}$ governs the rate of change of the vorticity as it is convected with the flow.

A three dimensional vortex method also involves the discretization of (11.3.2a,b). However, now the vorticity $\omega(\mathbf{x}, t)$ is a vector quantity and it is no longer constant along particle paths. As in two dimensions we discretize the right hand side of (11.3.2a) by approximating the velocity field by a sum of the form

$$\tilde{\mathbf{u}}(\mathbf{x}, t) = \sum_i \mathbf{K}_\delta(\mathbf{x} - \tilde{\mathbf{x}}_i(t))\,\tilde{\omega}_i(t)\,h^3\,. \tag{11.3.16}$$

Here $\tilde{\omega}_i(t)$ is an approximation of $\omega(\tilde{\mathbf{x}}_i(t), t)$, $\mathbf{K}_\delta = \mathbf{K} * f_\delta$ with $\mathbf{K}$ given by (11.2.9), and f_δ is a smoothing function of the form

$$f_\delta(\mathbf{x}) = \frac{1}{\delta^3}f(\mathbf{x}/\delta)\,, \tag{11.3.17}$$

for some $f : I\!\!R^3 \to R$ of integral one. However, unlike the two dimensional case, we must now find a way to update the vortex strengths $\omega_i(t)$. A variety of techniques have been proposed for doing this. Essentially three dimensional vortex methods are distinguished from one another by the manner in which they discretize equation (11.3.15).

In this article we will be primarily concerned with a method due to Anderson which was first described in Anderson and Greengard [1985]. In this method one uses (11.3.16) to find an expression for $\nabla \mathbf{u} \equiv (\nabla u, \nabla v, \nabla w)$ of the form

$$\nabla \tilde{\mathbf{u}}(\mathbf{x}, t) = \nabla \sum_j \mathbf{K}_\delta(\mathbf{x} - \tilde{\mathbf{x}}_j(t)) \, \tilde{\omega}_j(t) \, h^3 \,. \tag{11.3.18}$$

Since $\mathbf{K}_\delta$ is a known function of $\mathbf{x}$, the right hand side of (11.3.18) can be explicitly computed. This suggests the following system of $2N$ ordinary differential equations for the vortex positions and strengths

$$\frac{d\tilde{\mathbf{x}}_i}{dt}(t) = \sum_j \mathbf{K}_\delta(\tilde{\mathbf{x}}_i(t) - \tilde{\mathbf{x}}_j(t)) \, \tilde{\omega}_j(t) \, h^3 \,, \tag{11.3.19a}$$

$$\tilde{\mathbf{x}}_i(0) = \alpha_i \,, \tag{11.3.19b}$$

$$\frac{d\tilde{\omega}_i}{dt}(t) = (\tilde{\omega}_i(t) \cdot \nabla) \sum_j \mathbf{K}_\delta(\tilde{\mathbf{x}}_i(t) - \tilde{\mathbf{x}}_j(t)) \, \tilde{\omega}_j(t) \, h^3 \,, \tag{11.3.20a}$$

$$\tilde{\omega}_i(0) = \omega^0(\alpha_i) \,. \tag{11.3.20b}$$

Note that equations (11.3.20a,b) describe the evolution of the vortex strengths $\tilde{\omega}_i(t)$ in time rather than the evolution of the vorticity field itself. Nevertheless, it is natural to expect that the solution of these equations converges to the exact solution of the 3D incompressible Euler equations in the limit as h, $\delta \to 0$. This has been proved by Beale [1986] and Cottet [1988b] under certain reasonable assumptions on the smoothness of the underlying flow field and the choice of the smoothing function f_δ. In addition, Long [1990] has proved that a version of the random vortex method based on this method converges to solutions of the 3-D Navier-Stokes equations. We will discuss these results further in Sections 11.3.5 and 11.5.3.

A Method of Beale and Majda An alternative discretization of (11.3.15) was proposed by Beale and Majda [1982a]. It is based on the fact that equation (11.2.13a) implies

$$\omega(\mathbf{x}(\alpha, t), t) = (\nabla_\alpha \mathbf{x}(\alpha, t)) \, \omega^0(\alpha) \,,$$

(e.g., see Chorin and Marsden [1990]). This suggests the following method for updating the vortex strengths

$$\tilde{\omega}_i(t) = (\nabla_\alpha^h \tilde{\mathbf{x}}(\alpha_i, t)) \cdot \omega^0(\alpha_i), \qquad i = 1, \dots, N, \qquad (11.3.21)$$

where ∇_α^h denotes some (high order accurate) finite difference approximation to the gradient of the fluid flow map which is defined on the grid determined by the vortex positions at time $t = 0$. It has been proved by Beale and Majda [1982a and 1982b] and Anderson and Greengard [1985] that the solution of the coupled system of equations (11.3.19a,b) and (11.3.21) also converges to the solution of the 3-d incompressible Euler equations in the appropriate limit.

The Vortex Filament Method We mention one other approach, originally proposed by Chorin [1980 and 1982], which is sometimes called the *vortex filament method*. In this method the vorticity field is represented as a collection of vectors. Each vector is represented by two points, considered to be at the head and tail of the vector, and the magnitude of the vorticity is proportional to the distance between these two points. The evolution of this vorticity field is approximated by allowing these points to flow with the velocity field induced by this approximate vorticity field via (11.2.7). The stretching term is automatically accounted for by the movement of these two points in relation to one another. (In an actual computation one usually divides a vector into two smaller vectors when the distance between the two points becomes large.) Greengard [1986] observed that the vortex filament method is essentially equivalent to the method of Beale and Majda and used this observation to prove that the method converges to solutions of the incompressible Euler equations under the appropriate conditions. The interested reader is also referred to the work of Anderson and Greengard [1985 and 1989] for further information concerning the vortex filament method.

These are just a few examples of a wide variety methods that have been suggested for updating the vortex strengths in three dimensions. We refer the reader to the literature for a more detailed discussion of this topic: Ashurst and Meiburg [1988], Anderson and Greengard [1985], Beale and Majda [1982a and 1982b], Fishelov [1990a], Greengard [1986], Knio and Ghoniem [1990] and Leonard [1980 and 1985]. We also remark that Beale, Eydeland, and Turkington [1991] have conducted numerical studies to compare several of the algorithms described above.

In an actual computation one must discretize the solution of (11.3.19a,b) and (11.3.20a,b) (or (11.3.21)) in time. As in two dimensions let $\tilde{\mathbf{x}}_i^k$ denote the time discretized approximation to $\tilde{\mathbf{x}}_i(t_k)$ and $\tilde{\mathbf{u}}^k$ denote the approximation to $\tilde{\mathbf{u}}(\mathbf{x}, t_k)$. The time discretized velocity field at time t_k may therefore be written as

$$\tilde{\mathbf{u}}^k(\mathbf{x}) = \sum_i \mathbf{K}_\delta(\mathbf{x} - \tilde{\mathbf{x}}_i^k)\, \tilde{\omega}_i^k\, h^3\,.$$

If we wish to approximate (11.3.19a,b) to first order in time, then the particle positions are advanced according to (11.3.14), just as in the two dimensional case. Similarly, a first order time discretization of (11.3.20a,b) is given by

$$\omega_i^{k+1} = \omega_i^k + \Delta t\,(\omega_i^k \cdot \nabla)\,\tilde{\mathbf{u}}^k(\mathbf{x}_i^k)\,. \tag{11.3.22}$$

A Remark Concerning the Approximate Vorticity Field in Three Dimensions   In analogy with the two dimensional case it is natural to let the approximate vorticity field be defined by

$$\mathbf{vort}(\mathbf{x}, t) = \sum_i f_\delta(\mathbf{x} - \tilde{\mathbf{x}}_i(t))\,\tilde{\omega}_i(t)\, h^3\,. \tag{11.3.23}$$

However, in general this function fails to be divergence free. (In two dimensions the fact that the vorticity is a vector pointing in a direction perpendicular to the x, y-plane guarantees that (11.3.9) is divergence free.) It is apparent then that $\mathbf{vort} \neq \nabla \times \tilde{\mathbf{u}}$. Instead one can define

$$\tilde{\omega}(\mathbf{x}, t) = \nabla \times \tilde{\mathbf{u}}(\mathbf{x}, t) = \sum_i \zeta_\delta(\mathbf{x} - \tilde{\mathbf{x}}_i^k)\,\tilde{\omega}_i^k\, h^3\,, \tag{11.3.24}$$

where $\zeta_\delta \equiv \nabla \times \mathbf{K}_\delta$. (Hald [1991] has taken the trouble to actually write out ζ_δ.)

As Beale [1986] has pointed out, $\tilde{\omega}$ is simply the L^2 projection of **vort** onto the space of divergence free vector fields. In particular, the velocity field induced by (11.3.23) is identical to that induced by (11.3.24). Since in an actual computation one does not need to explicitly evaluate the vorticity field in order to evolve the flow field, the question of which of the above expressions represents the computed vorticity field is largely academic. However, if one is interested in computing values of the vorticity field in addition to the velocity (say, for example, in order to plot the vorticity field) or if one is interested in proving that the computed vorticity field converges to the exact one, then it may very well be appropriate to use (11.3.24) rather than (11.3.23). We refer the interested reader to the comments of Anderson and Greengard [1989, p. 1129] and Beale [1986, p. 405] for a more detailed discussion of this issue.

11.3.3   Initial Conditions

We now briefly discuss how to choose the initial vortex positions and strengths. Suppose we are given an initial vorticity field $\omega^0(\mathbf{x})$ with support contained in a bounded set $\Omega_0 \subseteq I\!R^n$. We wish to approximate ω^0 by a sum of the form (11.3.9), (11.3.23), or (11.3.24). We begin by creating a square grid of side h that covers Ω_0. We let the initial vortex positions $\tilde{\mathbf{x}}_i^0 = \tilde{\mathbf{x}}_i(0)$ be the centers of the grid cells and the initial particle strengths be

$$\omega_i^0 = \omega(\tilde{\mathbf{x}}_i^0) . \qquad (11.3.25)$$

This is the type of initial condition that is assumed in most proofs that the vortex method (*respectively* random vortex method) converges to solutions of Euler's (*respectively* Navier-Stokes) equations. (However in many computations - especially viscous ones – vorticity may introduced into the flow via other mechanisms. We will discuss this further in Section 11.5.2 below.) In this context the grid spacing h plays a central role in the accuracy of the method. It, together with the cutoff radius δ, determines the accuracy of the initial discretization and as well as the accuracy of the subsequent approximation $\tilde{\mathbf{u}}^k$ to the flow field at later times.

It is also possible to choose the initial particle strengths so that

$$\omega_i^0 h^n = \int_{B_i} \omega(\mathbf{x}) \, d\mathbf{x} , \qquad (11.3.26)$$

where B_i denotes the ith grid cell. For example, Hald [1979] used this initial condition in his convergence proof. In fact, (11.3.26) may be preferable to (11.3.25) when the initial data is not very smooth. We also note that Anderson and Greengard [1985], Nicolaides [1986], and Chiu and Nicolaides [1988] have studied how to choose the vortex positions and strengths on irregular grids in a manner that will preserve the overall accuracy of the method.

11.3.4 Cutoff Functions

We now turn to a discussion the cutoff function f_δ. As we have already noted the properties of f, together with the smoothness of the underlying flow field, will determine the accuracy of the vortex method. In this section we discuss some of the properties necessary for an accurate smoothing function and describe some that have been proposed. In the following section we state some of the convergence results that have been obtained by researchers in the field. These results will illustrate how the choice of cutoff function influences the rate of convergence. Sections 11.3.4 and 11.3.5 are loosely based on Hald's [1991] review article.

In what follows C denotes a generic constant, D denotes differentiation, $\beta = (\beta_1, \ldots, \beta_n)$ is a multi-index the components of which are non-negative integers, and $|\beta| = \beta_1 + \ldots + \beta_n$. So for example, when $n = 2$ we have,

$$\mathbf{x}^\beta = x^{\beta_1} y^{\beta_2} , \qquad \text{and} \qquad D^\beta f = \frac{\partial^{|\beta|} f}{\partial_x^{\beta_1} \partial_y^{\beta_2}} ,$$

for any function $f : \mathbb{R}^2 \to \mathbb{R}$.

We remind the reader that given $f : \mathbb{R}^n \to \mathbb{R}$, the function f_δ is defined by (11.3.7) in two dimensions or (11.3.17) in three dimensions. There have been a variety of assumptions placed on the smoothing function f_δ in the literature. For example, in the first convergence proof of any kind for the vortex method Hald and Del Prete [1978] assumed that f_δ was chosen so that $\mathbf{K}_\delta(\mathbf{x}) = \mathbf{K} * f_\delta(\mathbf{x})$ is continuous for $\mathbf{x} \neq 0$ and that for all $|\beta| \leq 2$ the derivatives of $\mathbf{K}_\delta$ satisfy

$$\left| D^\beta \mathbf{K}_\delta(\mathbf{x}) \right| \leq \begin{cases} C \delta^{-1} |\mathbf{x}|^{-|\beta|} & 0 < |\mathbf{x}| \leq \delta, \\ C |\mathbf{x}|^{-|\beta|-1} & \delta < |\mathbf{x}|. \end{cases}$$

(For example, Chorin's cutoff (11.3.11) satisfies these assumptions.) Given the above assumptions on $\mathbf{K}_\delta$ Hald and Del Prete were able to prove first order convergence of the two dimensional vortex method for short times.

By strengthening these assumptions Hald [1979] was subsequently able to establish second order convergence of the two dimensional vortex method for arbitrarily long times. In this work Hald made the following assumptions on the smoothing function: f_δ is radially symmetric $f_\delta(\mathbf{x}) = f_\delta(r)$ where $r = |\mathbf{x}|$, $f_\delta(r)$ vanishes for $r \geq \delta$, $f_\delta(r)$ is twice continuously differentiable with respect to r, f_δ'' is continuously differentiable except possibly at $r = \delta$, $f_\delta'(0) = 0$,

$$\int_0^\delta r f_\delta(r) \, dr = \frac{1}{2\pi},$$

$$\int_0^\delta r^3 f_\delta(r) \, dr = 0,$$

and for $p \leq 3$ the derivatives of f_δ can be estimated by

$$\left| \frac{d^p}{dr^p} f_\delta(r) \right| \leq \begin{cases} C \delta^{-2-p} & r \leq \delta, \\ 0 & r > \delta. \end{cases}$$

In order to obtain higher order convergence still more more stringent requirements on f_δ are necessary. The first general theory of high order convergence for the vortex method is due to Beale and Majda [1982b]. In this paper the authors defined a class of smoothing functions called $FeS^{-L,p}$ and proved that for $f \in FeS^{-L,p}$ both two and three dimensional versions of the vortex method converge to solutions of the Euler equations at an arbitrarily high rate. The order of convergence depends only on the smoothness of the flow and the appropriate choice of L and p. (The parameters L and p here correspond to the parameters L and m defined below.) Cottet [1982] and Raviart [1985] have considered similar classes in their work. However, rather than compare various classes of smoothing functions we pick one and briefly discuss the nature of the various requirements. The following class, called $M^{L,m}$, is due to Anderson and Greengard [1985].[3]

Definition: Let $f : I\!\!R^n \to I\!\!R$ be a scalar function defined on $I\!\!R^n$ for $n = 2$ or 3. For $L \geq 3$ we say that the function $f \in M^{L,m}$ if f is L times continuously differentiable and satisfies the following properties,

$$\int_{I\!\!R^n} f(\mathbf{x})\,d\mathbf{x} = 1\,, \tag{11.3.27}$$

$$\int_{I\!\!R^n} \mathbf{x}^\beta f(\mathbf{x})\,d\mathbf{x} = 0\,, \qquad \text{for all } 1 \leq |\beta| \leq m - 1\,, \tag{11.3.28}$$

$$\int_{I\!\!R^n} |\mathbf{x}|^m\,|f(\mathbf{x})|\,d\mathbf{x} < \infty\,, \tag{11.3.29}$$

$$|\mathbf{x}|^{n+|\beta|}\,|D^\beta f(\mathbf{x})| \leq C\,, \qquad \text{for all } |\beta| \leq L\,, \tag{11.3.30}$$

$$|\mathbf{x}|^{m+n+2}\,|f(\mathbf{x})| \leq C\,. \tag{11.3.31}$$

Given $f \in M^{L,m}$ we define

$$f_\delta(\mathbf{x}) = \frac{1}{\delta^n} f(\mathbf{x}/\delta)\,. \tag{11.3.32}$$

Functions f_δ with $f \in M^{L,m}$ are called mth order cutoff functions. The justification for this terminology will become apparent in the next section.

From (11.3.27) and (11.3.32) it follows that the family of functions $\{f_\delta : \delta > 0\}$ is an approximation to the identity (e.g., see Folland [1976, p. 17]). In other words, for all $g \in \mathcal{L}^p$, with $1 \leq p < \infty$,

$$f_\delta * g \to g \qquad \text{as} \qquad \delta \to 0\,,$$

in the $\mathcal{L}^p$ norm. Thus we may regard f_δ as an approximation to the Dirac delta function.

Condition (11.3.28) is what has become known as the *moment condition*. It states that all moments of the function f up to order $m - 1$ vanish. It is natural to choose f to be radially symmetric. In this case all of the odd moments of f vanish by symmetry and it follows that m is even. In particular, if f is radially symmetric and no moment conditions are imposed, then $m = 2$. Conditions (11.3.27–11.3.29) ensure that replacing the singular kernel $\mathbf{K}$ by the smoothed kernel $\mathbf{K}_\delta = \mathbf{K} * f_\delta$ in (11.2.7) causes a change in the velocity which is $O(\delta^m)$ (e.g., see Anderson and Greengard [1985, p. 422]).

---

[3]The M stands for "mollifier."

Conditions (11.3.29–11.3.31) govern the decay of f at infinity. These are essentially technical requirements that are used in the convergence theory to ensure that the remainder in certain integrals is easy to estimate. A variety of similar conditions have been considered in the literature. For instance, Beale and Majda [1982b] originally considered bounds on the Fourier transform of f rather than (11.3.29–11.3.31). In general, one needs f to be "smooth and rapidly decreasing." For example, in Beale and Majda [1985] the authors replace (11.3.29–11.3.31) with

$$|D^\beta f(\mathbf{x})| \leq C\left(1 + |\mathbf{x}|^2\right)^{-j}, \qquad (11.3.33)$$

for all multi-indices β and every integer j. Conditions (11.3.27–11.3.28) together with (11.3.33) are perhaps the simplest requirements one can place on f. However (11.3.33) is a bit more restrictive than is needed in the theory. Finally we note that if one only considers functions f with compact support, then (11.3.29–11.3.31) can be omitted altogether. Choosing f to have compact support may also be desirable for other reasons which we discuss at the end of this section.

For radially symmetric functions $f(\mathbf{x}) = f(r)$, one can show that

$$\mathbf{K}_\delta(\mathbf{x}) = \mathbf{K}(\mathbf{x}) \int_{|\mathbf{y}| \leq r} f_\delta(\mathbf{y})\, d\mathbf{y} = \mathbf{K}(\mathbf{x})\, F(r), \qquad (11.3.34)$$

(e.g., see Hald [1979], p. 735). The function $F(r) = 2\pi \int_0^r s f_\delta(s)\, ds$ is called the the *shape factor* or sometimes (somewhat confusingly) the *cutoff function*. It is apparent from (11.3.27) and (11.3.34) that if f has compact support, say $f(r) = 0$ for $r > a$, then $\mathbf{K}_\delta$ agrees with $\mathbf{K}$ for $r > a\delta$. We also note that if f_δ is radially symmetric, then $\mathbf{K}_\delta$ can be constructed from f_δ and $\mathbf{K}$ by using (11.3.34).

A wide variety of explicit cutoff functions of various orders of accuracy have appeared in the literature. We list a few of these here. Beale and Majda [1985] studied cutoffs based on the Gaussian. For example, in two dimensions

$$f_2(r) = \frac{1}{\pi}\, e^{-r^2}. \qquad (11.3.35)$$

is a second order cutoff. (Here we have used the subscript 2 to indicate that this cutoff is second order.) The authors then demonstrate how to obtain a fourth order cutoff by taking a linear combination of two Gaussians with different scalings

$$f_4(r) = c_1\, f_2(r) + c_2\, f_2\left(\frac{r}{a}\right), \qquad (11.3.36)$$

where a is arbitrary subject only to $a \neq 1$. The constants c_1 and c_2 can readily be determined by noting that f_4 must satisfy (11.3.27–11.3.28) with $m = 4$. For example, if we choose a such that $a^2 = 2$, then

$$f_4(r) = e^{-r^2} - \frac{1}{2} e^{-r^2/2}.$$

Similarly, one can derive sixth order cutoffs by taking a sum of the form $c_1 f_4(r) + c_2 f_4(r/a)$. A typical function of this form is

$$f_6(r) = c_1 e^{-r^2} + c_2 e^{-r^2/2} - c_3 e^{-r^2/4}.$$

Beale and Majda [1985] also constructed two dimensional cutoff functions of the form

$$f_m(r) = \frac{1}{\pi} P_m(r^2) e^{-r^2},$$

where P_m is some polynomial that must be determined. Applying (11.3.27) and the moment conditions (11.3.28) to functions of this form one finds that $P_m = Q_m - Q'_m$ where the Q_m are the Laguerre polynomials normalized so that the constant term is 1. For example,

$$P_2(r) = 1,$$

$$P_4(r) = 2 - r,$$

$$P_6(r) = 3 - 3r + \frac{r^2}{2},$$

$$P_8(r) = 4 - 6r + 2r^2 - \frac{r^3}{6}.$$

Both Beale and Majda [1985] and Beale [1986] have also considered high order cutoffs in three dimensions. The analogue of (11.3.35) in three dimensions is

$$f_2(r) = \frac{3}{4\pi} e^{-r^3}.$$

Beale and Majda [1985] demonstrate how to derive higher order 3D cutoffs from lower order ones by considering sums analogous to (11.3.36). They also consider sums of the form

$$f_{m+2}(r) = c_1 f_m(r) + c_2 r f'_m(r).$$

Starting with a second order 3D cutoff of the form

$$f_2(r) = \frac{3}{4\pi} \left\{ (1 + r^6)^{-\frac{1}{2}} - r^6 (1 + r^6)^{-\frac{3}{2}} \right\},$$

Beale [1986] has used this relation to derive the following fourth and sixth order 3D cutoffs,

$$f_4(r) = \frac{3}{4\pi} \left\{ (1 + r^6)^{-\frac{1}{2}} + \left(\frac{3}{2} - r^6\right) (1 + r^6)^{-\frac{3}{2}} + \frac{9r^6}{2} (1 + r^6)^{-\frac{5}{2}} \right\},$$

$$f_6(r) = \frac{3}{4\pi} \left\{ (1 + r^6)^{-\frac{1}{2}} - r^6 (1 + r^6)^{-\frac{3}{2}} + \frac{27}{8} (1 + r^6)^{-\frac{5}{2}} - \frac{135r^6}{8} (1 + r^6)^{-\frac{7}{2}} \right\}.$$

Hald [1987] (see also Hald [1991]) has devised the following two dimensional infinite order cutoff

$$f_\infty(r) = \frac{4}{45\pi r^3} \left\{ 16 J_3(4r) - 10 J_3(2r) + J_3(r) \right\}, \tag{11.3.37}$$

where J_3 is the Bessel function of order 3. This is one of the simplest members of a class of infinite order cutoffs based on Bessel functions which are discussed by Nordmark [1991]. This cutoff falls outside the classes of high order cutoffs studied by Anderson and Greengard [1985], Beale and Majda [1982b and 1985], and Cottet [1982]. In particular, f given by (11.3.37) fails to satisfy the moment condition (11.3.28) for $m = 4$ and conditions (11.3.29–11.3.31) for $L = m = 2$. The function simply decays too slowly. Nevertheless, by using new arguments Hald [1987] has succeeded in establishing the high order convergence of vortex methods with these cutoffs. We also remark that Hald [1987] has shown that there are no infinite order cutoffs with compact support.

Nordmark [1991] has derived the following eighth order cutoff which has compact support

$$f(r) = \begin{cases} \frac{52}{\pi} \left[1 - 21r^2 + 105r^4 - 140r^6 \right] \left[1 - r^2 \right]^9 & r \le 1, \\ 0 & r > 1. \end{cases} \tag{11.3.38}$$

In his PhD thesis Nordmark [1988] conducted extensive numerical tests to compare (11.3.37) and (11.3.38). These two functions are amazingly similar. Their graphs – both in physical and Fourier space – are nearly indistinguishable from one another.

In closing we note that since f is radially symmetric in all of the above examples, then $\mathbf{K}_\delta$ can be constructed from f_δ and $\mathbf{K}$ with the aid of (11.3.34). Explicit formulas for $\mathbf{K}_\delta$ corresponding to these examples also appear in Beale and Majda [1985], Hald [1987 and 1991], and Nordmark [1991].

From a computational point of view there are several factors which must be taken into account that do not enter into the convergence theory. These are:

(i) How efficiently can $\mathbf{K}_\delta$ be computed?

(ii) Does f_δ have compact support?

The first consideration obviously determines the speed of the computation. In particular, since the cost of evaluating the velocity $\tilde{\mathbf{u}}$ given by (11.3.13) or (11.3.16) at each of the N vortex positions is $O(N^2)$, the cost of computing $\mathbf{K}_\delta$ can have a very real effect on the speed of the overall computation. A variety of techniques have been devised to reduce the $O(N^2)$ cost of the velocity computation to $O(N \log N)$ or even $O(N)$. (Some of these ideas are discussed in Section 11.7.) However these techniques usually assume that $\mathbf{K}_\delta(\mathbf{x}) = \mathbf{K}(\mathbf{x})$ for $\mathbf{x}$ sufficiently far from the center of the vortex. Hence, cutoff functions with large or unbounded support become problematic. This may also be the case when one wishes to satisfy a boundary condition of the form (11.2.1d) on $\partial\Omega$. In this regard we remark that Merriman [1991] has done some interesting work on the relation between cutoff functions and boundary conditions.

We refer the reader to the articles of Beale and Majda [1985], Hald [1991], and Nordmark [1991] for a more detailed discussion concerning the construction of high order smoothing functions. Numerical experiments to compare various cutoff functions may be found in Beale and Majda [1985], Hald and del Prete [1978], Nordmark [1991], and Perlman [1985].

11.3.5 The Convergence of the Vortex Method to Solutions of the Euler Equations

We will now state (without proof) several generic theorems which establish the validity of the vortex method approximation to solutions of the incompressible Euler equations. In the following discussion we will measure the error in both a discrete and continuous p-norm over bounded subsets of $\mathbb{R}^n$. For any function g defined on the points $\alpha_i \in \mathbb{R}^n$, $i = 1, \ldots, N$ we let

$$\|g\|_p = \left(\sum_i |g(\alpha_i)|^p h^n \right)^{\frac{1}{p}}.$$

Let B denote any arbitrary bounded set in $\mathbb{R}^n$. For any integrable function g defined on B we let

$$\|g\|_{L^p(B)} = \left(\int_B |g(\mathbf{x})|^p \, d\mathbf{x} \right)^{\frac{1}{p}}.$$

In the following discussion all statements concern the free space problem (11.2.1a–c) with boundary condition (11.2.2).

Before proceeding we remind the reader of the following. In two dimensions it has been shown that for smooth initial data solutions of the incompressible Euler equations exist and remain smooth for all time (e.g., see Kato [1967] or McGrath [1968]). In three dimensions however the existence of smooth solutions to these equations has only been established for short time (e.g., see Kato [1972]). In fact, many researchers believe that in three dimensions singularities can form in finite time, irrespective of the smoothness of the initial data. It has been hypothesized that this phenomenon is linked to the onset of turbulence. In this regard we note that Beale, Kato, and Majda [1984] have shown that if singularities do form in a solution of the 3D Euler equations, then the maximum norm of the vorticity necessarily grows without bound.

In light of the above, the most one can hope to prove in three dimensions is that the vortex method converges to the solution of the Euler equations as long as the solution remains smooth. In other words, at the present one cannot hope to prove that the numerical method converges to a solution of the 3D Euler equation for arbitrarily large times t.

The next two theorems concern the continuous time version of the vortex method. In both of these theorems one can take p to be any $1 \leq p < \infty$. The following result can, with some modifications, be found in Beale and Majda [1982b], Cottet [1982], Anderson and Greengard [1985], and Raviart [1985]. The statement as it appears here is a synopsis of these results due to Hald [1991].

Theorem 11.3.1 (Convergence in Two Dimensions) *Assume that the solution* **u** *of the incompressible Euler equations is smooth enough and that the initial vorticity field has compact support. Let* $f \in M^{L,m}$ *with* L *sufficiently large and* $m \geq 2$. *Set* $\delta = ch^q$ *with* $0 < q < 1$. *Then there exists a constant* C *such that for all* h *sufficiently small we have*

(i) convergence of particle paths

$$\max_{0 \leq t \leq T} \|\tilde{\mathbf{x}}_i(t) - \mathbf{x}_i(t)\|_p \leq C\left[\delta^m + \left(\frac{h}{\delta}\right)^L \delta\right],$$

(ii) convergence of particle velocities

$$\max_{0 \leq t \leq T} \|\tilde{\mathbf{u}}(\tilde{\mathbf{x}}_i(t), t) - \mathbf{u}(\mathbf{x}_i(t), t)\|_p \leq C\left[\delta^m + \left(\frac{h}{\delta}\right)^L \delta\right],$$

(iii) convergence of the velocity field

$$\max_{0 \leq t \leq T} \|\tilde{\mathbf{u}}(\mathbf{x}, t) - \mathbf{u}(\mathbf{x}, t)\|_{L^p(B)} \leq C\left[\delta^m + \left(\frac{h}{\delta}\right)^L \delta\right].$$

Note that in the above theorem we have not stated a result analogous (ii) and (iii) for the vorticity field. Proofs that the approximate vorticity field converges to the exact vorticity field in two dimensions do appear in the literature however. For example, Hald and Del Prete [1978] prove a statement analogous to (iii) for the vorticity in the maximum norm for short times and for a specific choice of cutoff function. The general understanding of the theory as well as the analytical tools available have improved considerably since their proof. In fact, the current state of the theory is such that, if everything is smooth enough, then obtaining bounds for the vorticity from statements of the form (i–iii) above should be relatively straightforward. However, most workers do not bother to explicitly state and prove such results. In particular, we know of no explicit statement in the literature analogous to (ii–iii) for the vorticity field in two dimensions.

In three dimensions there is a notable exception to this trend. The following theorem is due to Beale [1986]. (See also Cottet [1988b].) What is meant by $\tilde{\omega}$ here is the function defined by (11.3.24)

$$\tilde{\omega}(\mathbf{x}, t) = \sum_i \zeta_\delta(\mathbf{x} - \mathbf{x}_i(t)) \, \omega_i(t) \, h^3 .$$

Theorem 11.3.2 (Convergence in Three Dimensions) *Assume that the solution* $\mathbf{u}$ *of the incompressible Euler equations exists and and remains smooth on the interval* $0 \leq t \leq T$ *and that the initial vorticity field has compact support. Let* $f \in M^{L,m}$ *with* L *sufficiently large and* $m \geq 4$.[4] *Set* $\delta = ch^q$ *with* $\frac{1}{3} < q < 1$. *Then there exists a constant* C *such that for all* h *sufficiently small we have*

(i) convergence of particle paths

$$\max_{0 \leq t \leq T} \|\tilde{\mathbf{x}}_i(t) - \mathbf{x}_i(t)\|_p \leq Ch^{mq},$$

(ii) convergence of particle velocities

$$\max_{0 \leq t \leq T} \|\tilde{\mathbf{u}}(\tilde{\mathbf{x}}_i(t), t) - \mathbf{u}(\mathbf{x}_i(t), t)\|_p \leq Ch^{mq},$$

(iii) convergence of the velocity field

$$\max_{0 \leq t \leq T} \|\tilde{\mathbf{u}}(\mathbf{x}, t) - \mathbf{u}(\mathbf{x}, t)\|_{L^p(B)} \leq Ch^{mq},$$

(iv) convergence of particle vorticities

$$\max_{0 \leq t \leq T} \|\tilde{\boldsymbol{\omega}}(\tilde{\mathbf{x}}_i(t), t) - \boldsymbol{\omega}(\mathbf{x}_i(t), t)\|_p \leq Ch^{mq-1},$$

(v) convergence of the vorticity field

$$\max_{0 \leq t \leq T} \|\tilde{\boldsymbol{\omega}}(\mathbf{x}, t) - \boldsymbol{\omega}(\mathbf{x}, t)\|_{L^p(B)} \leq Ch^{mq-1}.$$

It is apparent from these two theorems that an essential requirement of the theory is that

$$\frac{h}{\delta} \to 0, \qquad \text{as} \quad h, \delta \to 0.$$

In other words, the mesh length h must go to zero faster than the cutoff parameter δ. This implies that the vortex cores tend to overlap at an increasing rate as h, $\delta \to 0$.

Since $\delta = ch^q$, in Theorem 11.3.2 we see that the error in (i–iii) is of order δ^m. This is the motivation for calling $f \in M^{L,m}$ an mth order cutoff. If in Theorem 11.3.1 the constant L is such that

$$L > \frac{(m-1)q}{(1-q)}, \tag{11.3.39}$$

then it is easy to see that the convergence rate is also $O(h^{mq}) = O(\delta^m)$. If we choose q close to 1, then such errors are essentially of order h^m. However, by (11.3.39) this

---

[4]Cottet [1988b] has shown that it is sufficient to take $m \geq 2$.

means that $L \sim (1 - q)^{-1}$ must be very large. Since one of the assumptions (which we have not stated) in both of these theorems is that the flow field is at least C^{L+n+1}, this will only work for very smooth flow fields.

From a computational point of view it is evident that choosing q close to 1 in Theorem 2 yields the fastest rate of convergence. However this is based on the assumption that the underlying flow field is very smooth. Roughly speaking, for $f \in M^{L,m}$ if the flow is at least C^{L+n+1}, then the optimal choice for δ will be $\delta = h^{1-\epsilon}$ where $\epsilon \ll 1$ (e.g., see Beale and Majda [1982b], Cottet [1982 and 1991], Anderson and Greengard [1985] and Raviart [1985]). On the other hand, if the flow has only a small number of derivatives and the cutoff is of high order, then choosing $\delta = h^{1/2}$ seems to be best (e.g., see Hald [1979]). In particular, this holds for infinite order cutoffs (e.g., see Hald [1987]).

In practice one usually takes h to be as small as possible and then chooses δ based on heuristic considerations. For an example: How does the solution look? How does the method perform on a simple test problem with a given choice of parameters? For example, see Sethian and Ghoniem's [1988] comprehensive study of the flow past a backward facing step. We also remark that numerical experiments appear to indicate that a larger δ yields a more accurate solution at late times. We refer the interested reader to Beale, Eydeland, and Turkington [1991], Beale and Majda [1985], Hald and Del Prete [1978], Nordmark [1991], and Perlman [1985] for a more detailed discussion of these issues.

Convergence of the Time Discretized Method   In practice a vortex method computation also involves the discretization of (11.3.6a,b) or (11.3.19a,b)–(11.3.20a,b) in time. Anderson and Greengard [1985] were the first to prove that the time discretized vortex method converges to solutions of the incompressible Euler equations. They studied the modified Euler method and an explicit class of multi-step schemes that includes the Adams-Bashforth methods. Their proof for the modified Euler method can be extended to include other second order Runge-Kutta methods. The treatment of higher-order Runge-Kutta methods is more difficult. Hald [1987] has proved the convergence of the classical fourth order Runge-Kutta method for vortex methods with infinite order cutoffs. The following theorem is a synopsis of Anderson and Greengard's result.

Theorem 11.3.3 (Convergence of the Time Discretized Vortex Method) *Suppose that the hypotheses of Theorem 11.3.1 (respectively Theorem 11.3.2) hold for $n = 2$ (respectively $n = 3$) and that the fluid flow map $\mathbf{x} : \mathbb{R}^{n+1} \to \mathbb{R}^n$ is $s + 1$ times continuously differentiable. Assume further that the ODE solver is a member of the class of multi-step methods considered by Anderson and Greengard [1985] and that it has a local truncation error which is order Δt^{r+1}. Let $T = k_0 \Delta t$ and for $0 \leq k \leq k_0$ let $\mathbf{x}_i^k = \mathbf{x}_i(k\Delta t)$. Then* *there exists a constant C such that for all h, Δt sufficiently small we have*

(i) convergence of particle paths

$$\max_{0 \le k \le k_0} \|\tilde{\mathbf{x}}_i^k - \mathbf{x}_i^k\|_2 \le C \left[\delta^m + \left(\frac{h}{\delta}\right)^L \delta + h^s + \Delta t^r \right]. \tag{11.3.40}$$

11.3.6   The Point Vortex Method

If we let f_δ be the Dirac delta function, $f_\delta(\mathbf{x}) = \delta(\mathbf{x})$, then in $\mathbb{R}^2$ the approximate vorticity (11.3.9) becomes a collection of point masses

$$\tilde{\omega}(\mathbf{x}, t) = \sum_i \delta(\mathbf{x} - \mathbf{x}_i(t))\, \omega_i(t)\, h^2\,,$$

and the approximate velocity becomes

$$\tilde{\mathbf{u}}(\mathbf{x}, t) = \sum_i \mathbf{K}(\mathbf{x} - \mathbf{x}_i(t))\omega_i(t)h^2\,.$$

This is the "classical" *point vortex method*. As we have already noted, the singularity in $\mathbf{K}$ gives rise to an arbitrarily large velocity as $\mathbf{x} \to \mathbf{x}_i$. For this reason it was generally believed that the point vortex method was unstable and hence would not converge to solutions of the incompressible Euler equations. However, Goodman, Hou, and Lowengrub [1990a] have recently proved that the point vortex method does in fact converge to solutions of two dimensional Euler equations. The key idea in their proof is that for smooth flow the smoothness of the flow map and its inverse guarantee that for finite times two particles will remain sufficiently far apart. In fact, for any finite time two neighboring particles which are initially separated by a distance h will remain separated by a distance which is $O(h)$. Thus, the point vortex method in essence has a cutoff which is asymptotically $O(h)$ for any finite time. The authors also prove that 2D point vortex method is second order. This follows from the interesting fact that certain errors in the stability estimate cancel because the Biot-Savart kernel $\mathbf{K}$ is odd. (Beale [1986] used this fact to improve his accuracy estimates for a 3D vortex blob method.)

Hou and his colleagues have also been able to prove that three dimensional point vortex methods converge to solutions of the Euler equations for sufficiently smooth flows. It turns out that the analytical tools one needs to prove this depend on how one discretizes the stretching term (11.3.15). Hou and Lowengrub [1990] have used an argument similar to the one used in two dimensions to prove the convergence of a method in which the stretching term is approximated by a Lagrangian finite differencing such as in equation (11.3.21). When the stretching term is "grid-free" as in (11.3.20a,b) the analysis is somewhat harder. Cottet, Goodman and Hou [1991] have used several additional ideas to establish the convergence of a three dimensional point vortex method of this type.

As Hou [1991] has pointed out, the convergence analysis does not necessarily imply that the point vortex method is practical without smoothing or some form of desingularization. The theory states that for any given time T there is a constant $h_0(T)$ such that for all $h \leq h_0$ the method is stable and converges. However, h_0 may be impractically small. Furthermore, given a fixed discretization of the initial vorticity field, there will be a critical time t_c beyond which two neighboring particles may become so close that the stability analysis no longer applies.

These observations have lead Hou and his co-workers to consider several possible remedies. Hou [1992] has proposed a method for "desingularizing" the integral kernel **K** that seems to work well. Hou, Lowengrub, and Shelley [1991] have studied a regridding technique for the desingularized point vortex method. The basic idea behind a regridding algorithm is to stop the computation when the error begins to rise and reinitialize the vortex positions and strengths in a manner that will maintain a given level of accuracy. Regridding algorithms and other techniques for improving the long time accuracy of vortex methods have also been studied by Beale [1988] and Nordmark [1991]. Hald [1991] has reviewed some of these efforts.

For further discussion concerning the point vortex method and related topics see Hou [1991].

11.4   The No-Flow Boundary Condition

We now discuss the modifications to the computed flow field that are required in order to satisfy the no-flow boundary condition (11.2.1d) on $\partial\Omega$. Assume that we have a solution $\mathbf{u}_\omega$ of (11.2.1a–c) that fails to satisfy (11.2.1d) on $\partial\Omega$. We seek a scalar function ϕ such that

$$\Delta\phi = 0 \qquad \text{in } \Omega, \qquad (11.4.1a)$$

$$\nabla\phi \cdot \boldsymbol{\eta} = -\mathbf{u}_\omega \cdot \boldsymbol{\eta}, \qquad \text{on } \partial\Omega. \qquad (11.4.1b)$$

If we define $\mathbf{u}_p \equiv \nabla\phi$ then, as we have shown in Section 11.2.1, the velocity field $\mathbf{u} \equiv \mathbf{u}_\omega + \mathbf{u}_p$ satisfies (11.2.1a–d). The velocity $\mathbf{u}_p$ is called a *potential flow* (e.g., see Chorin and Marsden [1990]).

Numerically we attempt to perform a discrete analog of the continuous solution just described. In general, at the end of each time step we have some function $\mathbf{u}_\omega$ which fails to satisfy (11.2.1d). The problem is to find ϕ such that ϕ satisfies (11.4.1a,b). In other words we wish to solve Laplace's equation for ϕ subject to Neumann boundary conditions. In some instances this can be done exactly by using the method of images or some similar idea. In other instances it is necessary to determine a numerical approximation $\tilde{\phi}$ to ϕ. In either case, effective numerical techniques for determining $\mathbf{u}_p$ or an approximation $\tilde{\mathbf{u}}_p$ may often be found in the scientific literature. We describe several simple examples here.

To begin, note that the solution of the potential flow problem is dictated in large part by the shape of the domain. For example, let $n = 2$ and suppose that Ω is the upper half plane $\{(x, y) : y > 0\}$. If $\mathbf{u}_\omega$ is the velocity due to a collection of N vortices at (x_i, y_i) with strengths ω_i, then it is a simple matter to check that the potential flow $\mathbf{u}_p$ we seek is precisely the flow due to a collection of N vortices at $(x_i, -y_i)$ with strengths $-\omega_i$. Here we have employed the well known *method of images* (e.g., see Garabedian [1986, pp. 246–247]) to find a solution of (11.4.1a,b) in the half plane $y > 0$.

As another example consider flow past a circle of radius R centered at the origin in two dimensions. So $\Omega = \{\mathbf{x} : |\mathbf{x}| > R\}$ and $\partial\Omega = \{\mathbf{x} : |\mathbf{x}| = R\}$. Suppose $\mathbf{u}_\omega$ is the flow induced by a collection of vortices at $\mathbf{x}_i \in \Omega$ with strengths ω_i, $i = 1, \ldots, N$. Let

$$\mathbf{x}_i' = \frac{R^2}{|\mathbf{x}_i|^2} \, \mathbf{x}_i$$

be the radial image of the point $\mathbf{x}_i$ with respect to the circle $\partial\Omega$. Then a potential flow $\mathbf{u}_p$ which cancels the normal component of $\mathbf{u}_\omega$ on $\partial\Omega$ is the flow due to a collection of vortices with positions $\mathbf{x}_i'$ and strengths $-\omega_i$, $i = 1, \ldots, N$, (e.g., see Garabedian [1986, pp. 247–249]). However, the total circulation about $\partial\Omega$,

$$\Gamma_\Omega = \oint_C \mathbf{u}_\omega \cdot \mathbf{ds} \,, \tag{11.4.2}$$

has now changed. Here C is any circle of radius $r > \max |\mathbf{x}_i|$ centered at the origin. Another potential flow that satisfies

$$\mathbf{u}_p \cdot \boldsymbol{\eta} = -\mathbf{u}_\omega \cdot \boldsymbol{\eta}$$

and yet leaves (11.4.2) unchanged is the flow due to the N radial images at positions $\mathbf{x}_i'$ and strengths $-\omega_i$ together with N additional vortices, all located at the origin, and having strengths ω_i.

This illustrates an important fact about the potential flow. Sometimes it is necessary to consider other properties of the flow field, such as total circulation, when choosing $\mathbf{u}_p$. Also note that, in general, ϕ is not unique. Given any solution ϕ of (11.4.1a,b), the function $\phi + c$ for any constant c, is also a solution of (11.4.1a,b).

For more general domains the method of images is not always practical, or necessarily even possible. In these cases one can resort to one of several different strategies. Sometimes it is possible to conformally map Ω onto another domain Ω' for which the method of images will work. However there are potential pitfalls with this approach.

368 E. G. PUCKETT

Large numerical errors can occur as a result of the conformal mapping. For example, this might occur in regions where $\partial\Omega$ is not differentiable, such as at a corner. Accurate numerical algorithms for conformal mapping are an active area of research and results in this area will ultimately impact vortex methods (e.g., see Howell [1990] and Howell and Trefethen [1990].)

Another approach to solving the potential flow problem is to place a grid over the computational domain and use a fast Poisson solver on the grid. In this case one must be careful to ensure that the introduction of the grid does not also introduce the very types of diffusive errors that the vortex method was designed to eliminate. For example, see the method of Mayo [1985]. We will discuss this issue further in Section 11.7. One can also use a multigrid method or a finite element method to solve the potential flow problem (11.4.1a,b).

Finally we mention that there have been a number of recent advances in the fast numerical solution of integral equations. Thus an effective way to solve (11.4.1a,b) might be to reformulate (11.4.1a,b) as an integral equation and use a fast solver to obtain an approximation to the potential flow $\mathbf{u}_p$. These techniques have the advantage of being applicable to a wide variety of domains. We refer interested reader to Greenbaum, Greengard and McFadden [1991], Greengard [1988], and Rokhlin [1985] for further details.

11.5 Vortex Method Approximations to Solutions of the Navier-Stokes Equations

We now consider vortex methods for computing viscous flow. We begin by describing the granddaddy of all such methods, the *random vortex method*. As its name connotes, the random vortex method is essentially a vortex method with the addition of a random approximation to the viscous terms in the Navier-Stokes equations. After describing this method we state several theorems that have been proved concerning its convergence to solutions of the Navier-Stokes equations. We then describe several vortex methods in which the random walk has been replaced by a non-random approximation to the viscous terms in the Navier-Stokes equations. We also state a convergence theorem that has been proved for one of these methods. We close with a few remarks regarding random versus non-random vortex methods and a brief discussion of some of the important open questions in the field.

11.5.1 The Random Vortex Method

The random vortex method was first introduced by Chorin [1973]. We will now describe this method as it is used to approximate solutions of the Navier-Stokes equations (11.2.14a–d) in $\Omega = I\!\!R^n$ with the free space boundary condition (11.2.2). In Section

6 we will discuss the modifications to this algorithm that are required when Ω is a bounded domain with boundary $\partial\Omega$ upon which the boundary condition (11.2.14e) must be satisfied.

In practice one is usually interested in determining the velocity field $\tilde{\mathbf{u}}^k = \mathbf{K} * \tilde{\omega}^k$ at some time $t_k = k\Delta t$ rather than in just determining the corresponding vorticity field $\tilde{\omega}^k$. Most theoretical work on the random vortex method has been concerned with estimating the error between the approximate and exact velocity field, rather than the error between the corresponding vorticity fields. For example, it has been proved that this velocity field converges to the exact velocity field in the appropriate limit. (In fact, at the time of this writing, no analogous statement concerning the vorticity field has yet been proved.) However, since the pressure term is eliminated from the momentum equations when they are written in vorticity form, we prefer to introduce the random vortex method as a fractional step method for solving (11.2.14a–d) for the approximate vorticity field $\tilde{\omega}^k$.

Let

$$\tilde{\omega}^k(\mathbf{x}) = \sum_{j=1}^{N} f_\delta(\mathbf{x} - \tilde{\mathbf{x}}_j^k)\,\Gamma_j^k\,, \tag{11.5.1}$$

where it is understood that $\Gamma_j^k = \omega_j h^2$ when $n = 2$ and $\Gamma_j^k = \omega_j^k h^3$ when $n = 3$. We wish to emphasize that when $n = 3$ one can just as well replace (11.5.1) with (11.3.24); the details of the computation remain the same. The random vortex method is a fractional step method. The first step consists of approximating the solution of the Euler equations (11.2.13a–c) at time $t_{k+1} = t_k + \Delta t$ with initial data (11.5.1) given at time t_k. This is accomplished by updating the vortex positions $\tilde{\mathbf{x}}_j^k$ and strengths Γ_j^k as described in Sections 11.3.1 and 11.3.2. This yields an intermediate vorticity field of the form,

$$\tilde{\omega}^{k+\frac{1}{2}}(\mathbf{x}) = \sum_{j} f_\delta(\mathbf{x} - \tilde{\mathbf{x}}_j^{k+\frac{1}{2}})\,\Gamma_j^{k+1}\,. \tag{11.5.2}$$

Here $\tilde{\mathbf{x}}_j^{k+\frac{1}{2}}$ simply denotes the particle positions at the end of the first of two steps; i.e., as a result of (11.3.14) or perhaps some higher order ODE solver. If $n = 2$, then $\Gamma_j^{k+1} = \Gamma_j^k$. Otherwise, we assume that $\Gamma_j^{k+1} = \omega_j^{k+1} h^3$ where ω_j^{k+1} is given by (11.3.22) or some higher order time discretization of (11.3.20a).

The second fractional step is an approximate solution of the heat equation,

$$\omega_t = \nu \Delta \omega\,, \tag{11.5.3}$$

at time t_{k+1} with initial data (11.5.2) at time t_k. This is accomplished by letting each vortex undergo a random walk,

$$\tilde{\mathbf{x}}_j^{k+1} = \tilde{\mathbf{x}}_j^{k+\frac{1}{2}} + \boldsymbol{\eta}_j\,,$$

where $\eta_j = (\eta_j^1, \ldots, \eta_j^n)$ and the $\{\eta_j^i : i = 1, \ldots, n, \ j = 1, \ldots, N\}$ are independent, Gaussian distributed random numbers with mean 0 and variance $2\nu\Delta t$. The approximate vorticity field at time t_{k+1} is therefore,

$$\tilde{\omega}^{k+1}(\mathbf{x}) = \sum_j f_\delta(\mathbf{x} - \tilde{\mathbf{x}}_j^{k+1})\,\Gamma_j^{k+1}\,.$$

Why random walk the vortices? Suppose one wishes to solve the diffusion equation (11.5.3) in $\Omega = I\!R^n$ with initial data (11.5.2). The exact solution at time t is given by,

$$\begin{aligned}
\omega(\mathbf{x}, t) &= (4\pi\nu t)^{-n/2} \int_{I\!R^n} e^{-(\mathbf{x}-\boldsymbol{\eta})^2/4\nu t}\, \tilde{\omega}^{k+\frac{1}{2}}(\boldsymbol{\eta})\, d\boldsymbol{\eta} \\
&= (4\pi\nu t)^{-n/2} \int_{I\!R^n} \left(\sum_j f_\delta(\mathbf{x} - (\tilde{\mathbf{x}}_j^{k+\frac{1}{2}} + \boldsymbol{\eta}))\, \Gamma_j^{k+1} \right) e^{-\boldsymbol{\eta}^2/4\nu t}\, d\boldsymbol{\eta} \\
&= E\left[\sum_j f_\delta(\mathbf{x} - (\tilde{\mathbf{x}}_j^{k+\frac{1}{2}} + \boldsymbol{\eta}))\, \Gamma_j^{k+1} \right],
\end{aligned}$$

where E denotes expectation over Gaussian random variables $\boldsymbol{\eta}$ on $I\!R^n$ with mean 0 and variance $2\nu t$. Thus, the exact solution at time t_{k+1} of the diffusion equation (11.5.3) with initial data (11.5.2) at time t_k is *precisely* the expected value of the function obtained from (11.5.2) by letting each of the vortex positions $\tilde{\mathbf{x}}_j^{k+\frac{1}{2}}$ undergo an independent random walk with mean 0 and variance $2\nu\Delta t$.

Millanazo and Saffman [1977] and Roberts [1985] have studied the accuracy of the random vortex method in two dimensions by using it to approximate a known radially symmetric solution of the Navier-Stokes equations. These studies indicate that the rate of convergence is $O(\sqrt{\nu/N})$ where N is the number of vortices and ν is the kinematic viscosity. In $I\!R^2$ this corresponds to a rate which is first order in h. To see this suppose that at time $t = 0$ the support of the vorticity $\Omega_0 \subset I\!R^2$ is covered by N equal cells of side $h = \sqrt{N}^{-1}$. Then the observed convergence rate for the random vortex method is $O(\sqrt{N}^{-1}) = O(h)$. Similarly, one can show that in $I\!R^3$ a rate that is $O(\sqrt{N}^{-1})$ corresponds to a rate that is $O(h^{3/2})$.

11.5.2   More on Initial Conditions

In practice the random vortex method tends to be used in one of two ways. One way is to prescribe an initial vorticity field ω^0 which is compactly supported in some unbounded domain such as $\Omega = I\!R^n$. In this case the initial vortex positions and weights may be determined as described in Section 11.3.3. However, in many engineering applications no initial vorticity field is specified. The initial condition often consists of a bounded domain Ω in which a potential flow that satisfies the no-flow boundary condition on $\partial\Omega$ is given. As the computation progresses vorticity is allowed to enter the flow via some vortex creation mechanism designed to mimic the physical creation of vorticity at $\partial\Omega$. We discuss techniques for doing this in Section 11.6. At the current time there is no complete theory for the convergence of such a computation to the exact solutions of the Navier-Stokes equations.

11.5.3 Convergence of the Random Vortex Method

Since its introduction there has been considerable interest in proving that the random vortex method converges to solutions of the Navier-Stokes equations. However, it has not been until the last decade that this fact has been rigorously established. We will now briefly review what has been proved to date about the random vortex method.

Marchiorio and Pulverenti [1982] were the first establish a convergence theorem for the random vortex method. They showed that solutions of the stochastic differential equation which is the continuous time version of the random vortex method converge to solutions of the Navier-Stokes equations in $I\!\!R^2$. However, their result does not exhibit a rate of convergence in terms of the parameters h, δ, Δt, and ν which is desirable from a computational point of view.

Subsequently, Goodman [1985] proved that a time discretized version of the random vortex method converges to solutions of the Navier-Stokes equations in $I\!\!R^2$. In particular, Goodman was able to show that for any fixed $t_k = k\Delta t$,

$$\sup_{\mathbf{x}\in B} |\tilde{\mathbf{u}}^k(\mathbf{x}) - \mathbf{u}(\mathbf{x},t_k)| \leq C\,N^{-\frac{1}{4}}\,\log N$$

with high probability. Here B is an arbitrary bounded set in $I\!\!R^2$. Thus, Goodman established a convergence rate which is $O(N^{-\frac{1}{4}}\log N)$. This is essentially one half order slower than the rate of $N^{-\frac{1}{2}}$ which one might expect based on the work of Millanazo and Saffman [1977] and Roberts [1985] that was mentioned above.

Long [1988 and 1990] has been able to establish a faster rate of convergence for the *time continuous* version of the random vortex method in both two and three dimensions. We state a synopsis of Long's results here (see also Hald [1991]). In the following it is assumed that the initial vorticity field has support contained in a bounded set $\Omega_0 \subset I\!\!R^n$ and that the initial discretization of the vorticity field is as described in Section 11.3.4. In particular, we assume that Ω_0 has been covered with a square grid of side $h = O(N^{-1/n})$, for $n = 2$ or 3, that the initial vortex positions $\tilde{\mathbf{x}}_j(0)$ are taken to be the centers of the grid points, and that the initial vortex strengths are of the form $\Gamma_j(0) = \omega^0(\tilde{\mathbf{x}}_j(0))\,h^n$. In both of the following theorems p is any real number that satisfies $1 \leq p < \infty$.

Theorem 11.5.1 (Convergence in Two Dimensions) *Suppose that ω^0 is supported inside a bounded set $\Omega_0 \subset \mathbb{R}^2$ and that ω^0 smooth enough that the solution ω of the Navier-Stokes equations (11.2.14a–d) with initial data ω^0 and boundary condition (11.2.2) exists and is sufficiently smooth on the interval $0 \leq t \leq T$. Let $\mathbf{u} = \mathbf{K} * \omega$ be the associated velocity field. Let $f \in M^{L,m}$ with L sufficiently large and $m \geq 2$. Set $\delta = ch^q$ with $0 < q < 1$ and let B be a bounded set in $\mathbb{R}^2$. Then there exists a constant C such that for all h sufficiently small we have*

(i) convergence of the particle paths

$$\max_{0 \leq t \leq T} \|\tilde{\mathbf{x}}_j(t) - \mathbf{x}_j(t)\|_p \leq C[\delta^m + (\frac{h}{\delta})^L \delta + h|\log h|]$$

(ii) convergence of the particle velocities

$$\max_{0 \leq t \leq T} \|\tilde{\mathbf{u}}(\tilde{\mathbf{x}}_j(t), t) - \mathbf{u}(\mathbf{x}_j(t), t)\|_p \leq C\, [\delta^m + (\frac{h}{\delta})^L \delta + h|\log h|]$$

(iii) convergence of the velocity field

$$\max_{0 \leq t \leq T} \|\tilde{\mathbf{u}}(\mathbf{x}(t), t) - \mathbf{u}(\mathbf{x}, t)\|_{L^p(B)} \leq C[\delta^m + (\frac{h}{\delta})^L \delta + h|\log h|]$$

with probability $1 - o(h)$.

Note that the computed velocity field $\tilde{\mathbf{u}}$ is a random quantity, whereas the velocity $\mathbf{u}$ is a deterministic quantity. This theorem says that the error between them is small with high probability. In other words, if we run the problem just once with h sufficiently small, then we should expect an accurate solution.

The proof of this theorem may be found in Long [1988]. The statement as it appears here is a synopsis due to Hald [1991]. Long's convergence rate is nearly optimal in the sense that for q close to 1 (i.e. δ close to h) the dominant term in the error is,

$$O(h\,|\log h|) = O(\sqrt{N}^{-1} \log N) \approx O(\sqrt{N}^{-1}).$$

We note that in the absence of some variance reduction technique one cannot hope to obtain a rate that is better than $O(\sqrt{N}^{-1})$. (See Handscomb and Hammersley [1964] or Maltz and Hitzl [1979] for a justification of this statement and a discussion of variance reduction techniques.)

Long has also established the convergence of the random vortex method in $\mathbb{R}^3$.

Theorem 11.5.2 (Convergence in Three Dimensions) *Suppose that ω^0 is supported inside a bounded set $\Omega_0 \subset \mathbb{R}^3$ and that ω^0 smooth enough that the solution ω of the Navier-Stokes equations (11.2.14a–d) with initial data ω^0 and boundary condition (11.2.2) exists and is sufficiently smooth on the interval $0 \le t \le T$. Let $\mathbf{u} = \mathbf{K} * \omega$ be the associated velocity field. Let $f \in M^{L,m}$ with L sufficiently large and $m \ge 4$. Set $\delta = ch^q$ with $0 < q < \frac{3}{5}$. Let B be a bounded set in $\mathbb{R}^3$. Then there exists a constant C such that for all h sufficiently small we have*

(i) convergence of the particle paths

$$\max_{0 \le t \le T} \|\tilde{\mathbf{x}}_j(t) - \mathbf{x}_j(t)\|_p \le C \left[\delta^m + h \left(\frac{h}{\delta}\right)^{1/2} |\log h| \right] ,$$

(ii) convergence of the particle velocities

$$\max_{0 \le t \le T} \|\tilde{\mathbf{u}}(\tilde{\mathbf{x}}_j(t), t) - \mathbf{u}(\mathbf{x}_j(t), t)\|_p \le C \left[\delta^m + h \left(\frac{h}{\delta}\right)^{1/2} |\log h| \right] ,$$

(iii) convergence of the velocity field

$$\max_{0 \le t \le T} \|\tilde{\mathbf{u}}(\mathbf{x}, t) - \mathbf{u}(\mathbf{x}, t)\|_{L^p(B)} \le C \left[\delta^m + h \left(\frac{h}{\delta}\right)^{1/2} |\log h| \right] ,$$

with probability $1 - o(h)$.

The proof of this theorem may be found in Long [1990]. To the author's knowledge at the present time it is the only convergence proof for the random vortex method in $\mathbb{R}^3$.

In both of the above theorems the constant C depends only on T, k, m, p, q, the diameter of B, the diameter of Ω_0 and the bounds for a finite number of derivatives of the velocity field $\mathbf{u}(\mathbf{x}, t)$. For simplicity we have actually stated a somewhat weaker form of Long's results. Long [1988 and 1990] can replace the $1 - o(h)$ probability of having a high error by a term of the form $1 - o(h^\gamma)$ where the constant γ can be chosen to be as large as one likes. However, the constant C on the right hand sides of the error bounds depends on γ and, in general, will increase as γ increases.

At the present time all convergence theorems for the random vortex method are for $\Omega = \mathbb{R}^n$ with the free space boundary condition (11.2.2). The convergence of the method to solutions of the Navier-Stokes equations in bounded domains $\Omega = \mathbb{R}^n$ with the boundary condition (11.2.14e) on $\partial\Omega$ is still an open and difficult problem. We will discuss these issues further in Section 11.6.

11.5.4 Deterministic Vortex Methods

During the past decade much research has been devoted to improving the convergence rate of the random vortex method. For example, Chang [1988] has applied high order numerical techniques that he developed for the integration of stochastic differential equations (Chang [1987]) to a random vortex method computation. However the vast majority of these efforts have been directed at finding a non-random technique for replacing the random walk algorithm with a higher order approximation to solutions of the diffusion equation (11.5.3). These methods are generically called *deterministic vortex methods*. This is a very active area of research and there are still many open questions in the field. We will now briefly illustrate some ideas that are currently being considered.

A very natural and perhaps one of the first examples of a deterministic vortex method is what has become known as the *core spreading algorithm*. (See Kuwahara and Takami [1973] or Leonard [1980] for details.) In the two dimensional (continuous time) version of this algorithm the approximate vorticity has the form

$$\tilde{\omega}(\mathbf{x}, t) = \sum_j f_t(\mathbf{x} - \tilde{\mathbf{x}}_j(t))\, \Gamma_j, \tag{11.5.4}$$

where the core function f_t now depends on the time t rather than on a constant cutoff parameter δ. The function $f_t(\mathbf{x})$ is chosen to be the solution at time t of the heat equation with initial data f_0 at time $t = 0$,

$$f_t(\mathbf{x}) = (G_t * f_0)(\mathbf{x}) = \frac{1}{4\pi\nu t} \int e^{-(\mathbf{x}-\mathbf{y})^2/4\nu t}\, f_0(\mathbf{y})\, d\mathbf{y}.$$

Here G_t denotes the heat kernel in $\mathbb{R}^2$,

$$G_t(\mathbf{x}) = \frac{1}{4\pi\nu t}\, e^{-\mathbf{x}^2/4\nu t}. \tag{11.5.5}$$

If one chooses f_0 to be the Dirac delta function $f_0(\mathbf{x}) = \delta(\mathbf{x})$, then

$$f_t(\mathbf{x}) = \frac{1}{4\pi\nu t} e^{-\mathbf{x}^2/4\nu t}.$$

In other words, f_t is a Gaussian smoothing function but now the "mass" of the core spreads out in time at a rate proportional to $\sqrt{\nu t}$. In general, f_0 can be taken to be any one of the two dimensional smoothing functions discussed in Section 11.3.4.

If we let $\mathbf{K}_t = \mathbf{K} * f_t$, then the approximate vorticity (11.5.4) yields an approximate velocity field of the form,

$$\tilde{\mathbf{u}}(\mathbf{x}, t) = \sum_j \mathbf{K}_t(\mathbf{x} - \tilde{\mathbf{x}}_j(t))\, \Gamma_j. \tag{11.5.6}$$

One might expect the velocity field $\tilde{\mathbf{u}}$ to converge to a solution $\mathbf{u}$ of the Navier-Stokes equations. However Greengard [1985] has shown that it converges to a solution of the wrong equation! We briefly outline his argument here.

Let $\mathbf{X}(\alpha, t)$ denote the solution of the Lagrangian equations

$$\frac{d\mathbf{X}}{dt}(\alpha, t) = \int \mathbf{K}_t(\mathbf{X}(\alpha, t) - \mathbf{X}(\alpha', t))\, \omega^0(\alpha')\, d\alpha', \qquad (11.5.7a)$$

$$\mathbf{X}(\alpha, 0) = \alpha. \qquad (11.5.7b)$$

where ω^0 is the initial vorticity field. (Compare this with equations (11.3.4a,b).) Let $\mathbf{U}$ denote the velocity field corresponding to (11.5.7a,b),

$$\mathbf{U}(\mathbf{x}, t) = \int \mathbf{K}_t(\mathbf{x} - \mathbf{X}(\alpha, t))\, \omega^0(\alpha)\, d\alpha. \qquad (11.5.8)$$

So (11.5.7a) can be rewritten as $(d\mathbf{X}/dt)(\alpha, t) = \mathbf{U}(\mathbf{X}(\alpha, t), t)$. Greengard first showed that for appropriately chosen f_0 solutions of

$$\frac{d\tilde{\mathbf{x}}_i}{dt}(t) = \sum_j \mathbf{K}_t(\tilde{\mathbf{x}}_i(t) - \tilde{\mathbf{x}}_j(t))\, \Gamma_j, \qquad (11.5.9a)$$

$$\tilde{\mathbf{x}}_i(0) = \alpha_i. \qquad (11.5.9b)$$

converge to solutions of (11.5.7a,b). In other words he showed that $\tilde{\mathbf{u}} \to \mathbf{U}$ where $\tilde{\mathbf{u}}$ is given by (11.5.6) and $\mathbf{U}$ is given by (11.5.8).

Thus, it suffices to compare $\mathbf{U}$ to solutions $\mathbf{u}$ of the Navier-Stokes equations. Let $\xi = \nabla \times \mathbf{U}$ denote the vorticity field corresponding to $\mathbf{U}$. We compare ξ to solutions $\omega = \nabla \times \mathbf{u}$ of (11.2.14a–d) with initial data ω^0 and boundary condition (11.2.2). Let ζ denote the passive transport of ω^0 in the velocity field $\mathbf{U}$. In other words, $\zeta(\mathbf{X}(\alpha, t), t) = \omega^0(\alpha)$ and hence,

$$\partial_t \zeta + \mathbf{U} \cdot \nabla \zeta = 0.$$

Changing variables $\alpha \to \mathbf{X}(\alpha, t)$ in (11.5.8) yields $\mathbf{U} = \mathbf{K}_t * \zeta = \mathbf{K} * (G_t * \zeta)$. Hence $\xi = \nabla \times \mathbf{U} = G_t * \zeta$.

Now the exact vorticity ω satisfies

$$\frac{\partial \omega}{\partial t} + \mathbf{u} \cdot \nabla \omega = \nu \Delta \omega. \qquad (11.5.10)$$

But if we differentiate $\xi = G_t * \zeta$ with respect to t and use the fact that $\partial_t(G_t * \zeta) = \nu \Delta(G_t * \zeta) + G_t * (\partial_t \zeta)$ we find that ξ satisfies

$$\frac{\partial \xi}{\partial t} + G_t * (\mathbf{U} \cdot \nabla \zeta) = \nu \Delta \xi. \qquad (11.5.11)$$

By differentiating (11.5.10) and (11.5.11) once again with respect to t one can show that the difference $\omega_t - \xi_t$ satisfies an ODE in t which is not identically zero unless the initial data ω^0 is radially symmetric. Hence, the exact solution ω of (11.5.10) with initial data ω^0 is not identical to the solution ξ of (11.5.11) with initial data ω^0 *unless the initial vorticity field is radially symmetric*. From (11.5.11) we see that the core spreading algorithm accurately approximates the diffusion of vorticity but that the vorticity is advected by an averaged velocity field rather than by the local velocity. We should emphasize that it follows from Greengard's proof that the core spreading algorithm *does* approximate the correct equations when the vorticity field is radially symmetric.

We now describe a deterministic vortex method in $I\!R^2$ due to Cottet and Mas-Gallic [1990] that *has* been shown to converge to solutions of the Navier-Stokes equations. Let ω^0 denote an initial vorticity distribution and assume that the support of ω^0 is contained in some bounded set $\Omega_0 \subset I\!R^2$. Assume that a square grid of side h has been placed over Ω_0 and let

$$\tilde{\omega}^0(\mathbf{x}) = \sum_j \delta(\mathbf{x} - \tilde{\mathbf{x}}_j^0)\, \tilde{\omega}_j^0\, h^2 \,,$$

be an approximation to ω^0 where, as usual, $\mathbf{x}_j^0$ is the center of the jth grid cell and $\tilde{\omega}_j^0 = \omega^0(\tilde{\mathbf{x}}_j^0)$.

In general, given the approximate particle positions $\tilde{\mathbf{x}}_j^k$ and their weights $\tilde{\omega}_j^k$ at time $t_k = k\Delta t$, the approximate vorticity field is given by,

$$\tilde{\omega}^k(\mathbf{x}) = \sum_j \delta(\mathbf{x} - \tilde{\mathbf{x}}_j^k)\, \tilde{\omega}_j^k\, h^2 \,.$$

The approximation to the velocity field in the interval $t \in [t_k, t_{k+1}]$ is given by,

$$\tilde{\mathbf{u}}(\mathbf{x}, t) = \sum_j (\mathbf{K} * G_{\Delta t})(\mathbf{x} - \tilde{\mathbf{x}}_j^k(t))\, \tilde{\omega}_j^k\, h^2 \,,$$

where $G_{\Delta t}$ is the heat kernel (11.5.5) with $t = \Delta t$. To obtain the approximate particle paths $\tilde{\mathbf{x}}_i(t)$ for $t \in [t_k, t_k + \Delta t]$ we solve the following N ordinary differential equations,

$$\frac{d\tilde{\mathbf{x}}_i}{dt}(t) = \tilde{\mathbf{u}}(\tilde{\mathbf{x}}_i(t), t)\,, \qquad t \in [t_k, t_{k+1}]\,, \qquad (11.5.12a)$$

$$\tilde{\mathbf{x}}_i(t_k) = \tilde{\mathbf{x}}_i^k\,. \qquad (11.5.12b)$$

This is simply a standard vortex method in the interval $[t_k, t_{k+1}]$ with $\mathbf{K}_\delta = \mathbf{K} * f_\delta$ replaced by $\mathbf{K} * G_{\Delta t}$. Now set $\tilde{\mathbf{x}}_j^{k+1} = \tilde{\mathbf{x}}_j(t_{k+1})$ to obtain an intermediate vorticity field of the form,

$$\tilde{\omega}^{k+\frac{1}{2}}(\mathbf{x}) = \sum_j \delta(\mathbf{x} - \tilde{\mathbf{x}}_j^{k+1})\, \tilde{\omega}_j^k\, h^2 \,. \qquad (11.5.13)$$

Finally, to update the vortex strengths $\tilde{\omega}_i^k \to \tilde{\omega}_i^{k+1}$ one explicitly solves the diffusion equation (11.5.3) with initial data (11.5.13) to obtain,

$$\tilde{\omega}_i^{k+1} = (G_{\Delta t} * \tilde{\omega}^{k+\frac{1}{2}})(\mathbf{x}_i^{k+1}) = \sum_j G_{\Delta t}(\mathbf{x}_i^{k+1} - \mathbf{x}_j^{k+1})\, \tilde{\omega}_j^k\, h^2 .$$

Note that in the description of the algorithm given here the particle paths $\tilde{\mathbf{x}}_i(t)$, being solutions of the ODEs (11.5.12a,b), are smooth functions of t whereas the vortex strengths $\tilde{\omega}_i^k$ are piecewise constant functions of t. Cottet and Mas-Gallic have proven that this version of the algorithm converges to solutions of the Navier-Stokes equations. Of course in actual practice one discretizes the system (11.5.12a,b) in time to obtain piecewise constant approximations to the $\tilde{\mathbf{x}}_i(t)$.

In the two dimensional version of the random vortex method the vortex strengths remain constant and the diffusion of vorticity is approximated by having the particle positions take a random walk with appropriate mean and variance. Here, on the other hand, the diffusion of vorticity is approximated by changing the particle strengths in a manner consistent with the exact solution of the heat equation. Also note that in Cottet and Mas-Gallic's method the approximate vorticity field is a collection of point vortices, but that the approximate velocity field is smoothed by convoluting it with the heat kernel $G_{\Delta t}$. This is not the same as the core spreading algorithm however. In the core spreading algorithm the smoothing function $f_t = G_t * f_0$ spreads out in time. In Cottet and Mas-Gallic's algorithm Δt plays the role of the cutoff parameter δ and is fixed for all time.

Cottet and Mas-Gallic [1990] have proven that this algorithm converges to solutions of the Navier-Stokes equations. We state their main result here. Let $\omega(\mathbf{x}, t)$ denote the exact solution of the Navier-Stokes equations (11.2.14a–d) in $\mathbb{R}^2$ with initial data ω^0 and boundary condition (11.2.2). Let $\mathbf{u}(\mathbf{x}, t) = (\mathbf{K} * \omega)(\mathbf{x}, t)$ and let $W^{m,p}(\mathbb{R}^2)$ denote the space of all functions $f : \mathbb{R}^2 \to \mathbb{R}$ such that $D^\alpha f \in L^p$ for all multi-indices α with $|\alpha| \le m$. Then we have the following,

Theorem 11.5.3 (Cottet and Mas-Gallic) *Assume that* $\omega^0 \in W^{m,\infty}(\mathbb{R}^2) \cap W^{m,1}(\mathbb{R}^2)$ *for any* $m \ge 0$ *and that* h *and* Δt *satisfy*

$$h \le C_0\, (\nu \Delta t)^{1/2+s} \tag{11.5.14}$$

where C_0 *and* s *are arbitrary positive constants. Then given* $T > 0$ *and* $\nu_0 > 0$ *there exists a constant* C_1 *depending only on* ω^0 *and* T *such that for* Δt *small enough,*

$$\max_{0 \le t \le T} \|\mathbf{u}(\cdot, t) - \tilde{\mathbf{u}}(\cdot, t)\|_p \le C_1\, \nu\, \Delta t$$

for all $2 < p < \infty$ *and* $\nu \le \nu_0$.

Note that since the solution of the heat equation $G_{\Delta t} * \tilde{\omega}^{k+\frac{1}{2}}$ has infinite support - even though $\tilde{\omega}^{k+\frac{1}{2}}$ does not – the total amount of vorticity in the flow is *not* conserved by this algorithm,

$$\int_{\mathbb{R}^2} \omega^0(\mathbf{x})\, d\mathbf{x} \approx \sum \tilde{\omega}_j^0 h^2 > \sum \tilde{\omega}_j^1 h^2 > \ldots > \sum \tilde{\omega}_j^k h^2 > \ldots .$$

(Nevertheless the method converges to solutions of the Navier-Stokes equations.) Choquin and Huberson [1989] have suggested the following modification which does conserve the total amount of vorticity in the flow,

$$\tilde{\omega}_i^{k+1} = \tilde{\omega}_i^k + \sum_j G_{\Delta t}(\tilde{\mathbf{x}}_i^{k+1} - \tilde{\mathbf{x}}_j^{k+1})\,(\tilde{\omega}_j^k - \tilde{\omega}_i^k) h^2$$

This is just one example of a class of deterministic vortex methods that have been considered by Cottet, Mas-Gallic, and their colleagues. We refer the interested reader to Choquin and Lucquin-Desreux [1988], Choquin and Huberson [1989], Cottet [1990 and 1991], Cottet and Mas-Gallic [1990], Lucquin-Desreux [1987], and Mas-Gallic [1990] for further details and references.

Fishelov [1990b] has studied a deterministic vortex method in which the diffusion of vorticity is also approximated by adjusting the particle weights rather than their positions. We briefly describe the three dimensional, time continuous version of this algorithm here. Suppose we are given an approximate vorticity distribution of the form

$$\tilde{\omega}(\mathbf{x}, t) = \sum_j f_\delta(\mathbf{x} - \tilde{\mathbf{x}}_j(t))\, \tilde{\omega}_j(t)\, h^3\,,$$

where f_δ is some smoothing function. Then the associated velocity field is given by

$$\tilde{\mathbf{u}}(\mathbf{x}, t) = \sum_j \mathbf{K}_\delta(\mathbf{x} - \tilde{\mathbf{x}}_j^k)\, \tilde{\omega}_j(t)\, h^3\,,$$

with $\mathbf{K}_\delta = \mathbf{K} * f_\delta$. Consider the diffusive term $\nu \Delta \omega$ of equation (11.2.14a). Fishelov has made the observation that one can evaluate this term explicitly,

$$\nu \Delta \tilde{\omega}(\mathbf{x}, t) = \nu \sum_j \Delta f_\delta(\mathbf{x} - \tilde{\mathbf{x}}_j(t))\, \tilde{\omega}_j(t)\, h^3\,,$$

provided that f_δ is chosen so that an explicit representation of Δf_δ can be found. This gives rise to the following coupled system of ordinary differential equations,

$$\frac{d\tilde{\mathbf{x}}_i}{dt}(t) = \sum_j \mathbf{K}_\delta(\tilde{\mathbf{x}}_i(t) - \tilde{\mathbf{x}}_j(t))\, \tilde{\omega}_j(t)\, h^3\,, \qquad (11.5.15a)$$

$$\tilde{\mathbf{x}}_i(0) = \mathbf{x}_i^0\,, \qquad (11.5.15b)$$

and

$$\frac{d\tilde{\omega}_i}{dt}(t) = (\tilde{\omega}_i(t) \cdot \nabla_x) \sum_j \mathbf{K}_\delta(\tilde{\mathbf{x}}_i(t) - \tilde{\mathbf{x}}_j(t)) \, \tilde{\omega}_j(t) \, h^3$$

$$+ \, \nu \sum_j \Delta f_\delta(\tilde{\mathbf{x}}_i(t) - \tilde{\mathbf{x}}_j(t)) \, \tilde{\omega}_j(t) \, h^3 \,, \qquad (11.5.16a)$$

$$\tilde{\omega}_i(0) = \omega^0(\mathbf{x}_i^0) \,. \qquad (11.5.16b)$$

Equation (11.5.15a,b) is our usual vortex method approximation to

$$\omega_t + (\mathbf{u} \cdot \nabla) \, \omega = 0 \,,$$

while equation (11.5.16a,b) is an approximation to

$$\omega_t = (\omega \cdot \nabla) \mathbf{u} + \nu \Delta \omega \,.$$

One expects that, given the appropriate relationship between h and δ, the solution of equations (11.5.15a,b) and (11.5.16a,b) will converge to a function that satisfies the Navier-Stokes equations

$$\omega_t + (\mathbf{u} \cdot \nabla) \, \omega = (\omega \cdot \nabla) \mathbf{u} + \nu \Delta \omega \,, \qquad (11.5.17)$$

in the limit as $h, \delta \to 0$. Fishelov [1990b] has done some analysis of the accuracy and stability of this algorithm, but at the present time its convergence to solutions of (11.2.14a–d) or to the associated velocity field $\mathbf{u} = \mathbf{K} * \omega$ remains open.

This is by no means an exhaustive list of the methods that have been developed as an alternative to the random walk. It is simply a representative sample chosen on the basis of the author's knowledge and experience. We refer the reader to the references listed above for further information on this topic.

11.5.5 A Few Remarks

We close this section with a few comments concerning what we believe are the important issues in the area of vortex methods for viscous flows. It is widely believed that in order to resolve the smallest relevant scales in a viscous flow a grid-based numerical method must satisfy $\Delta x = O(\sqrt{\nu})$ where Δx is the grid spacing in each space direction (e.g., see Henshaw, Kreiss, and Reyna [1991]). Thus, in order to maintain a given level of accuracy in n space dimensions, the computational cost will increase like $O(\nu^{-n/2})$ as $\nu \to 0$.

The random vortex method was specifically designed to circumvent this problem (e.g., see Chorin [1973]). The claim that is generally made with regards to this method – and vortex methods in general – is that because it is a particle method the "diffusive" errors found in grid-based methods are largely eliminated. Furthermore, since the particles naturally concentrate in regions of greatest vorticity, it is also claimed that vortex methods have a sort of a built in adaptivity – concentrating computational cost in regions of greatest interest. If these claims are true, then the computational cost required to maintain a fixed level of accuracy should not increase as $\nu \to 0$. However, these claims have never been rigorously substantiated. Now that the convergence of the random vortex method to solutions of the Navier-Stokes equations has been established it is time to carefully investigate the actual dependence of the error on the viscosity ν.

Similarly, it is often claimed that various deterministic vortex methods are superior to the random vortex method because they have a higher-order convergence rate. This is certainly the case for fixed ν. However, a very important issue that has yet to be addressed is "How does the error in a deterministic vortex method depend on ν?" Or equivalently, "How does the computational cost of a deterministic vortex method depend on ν?" For example, if the condition relating Δt, h, and ν in (11.5.14) cannot be eliminated, then – for ν small enough – it will be less expensive to use the random vortex method to compute a given flow to some fixed level of accuracy.

Finally, we remark that it may very well turn out to be the case that the answer to the above questions depends on the number of space dimensions n. For example, it is entirely possible that some restriction similar to (11.5.14) is necessary in three dimensions but not in two. It will be of great interest to have the answer to these questions in both two and three dimensions.

11.6 The No-Slip Boundary Condition

We now turn to a discussion of numerical techniques for satisfying the no-slip boundary condition (11.2.15) in a random vortex method computation. Chorin [1973] originally satisfied (11.2.15) by creating vortices at grid points on $\partial\Omega$ such that the velocity induced by these vortices cancelled the tangential component of the velocity along $\partial\Omega$. After further study Chorin [1978] proposed another particle creation algorithm known as the *vortex sheet method*. This method is based on approximating solutions of the Prandtl boundary layer equations in a thin region adjacent to $\partial\Omega$. The vortex sheet method has probably been the most widely used technique for satisfying the no-slip boundary condition in random vortex method computations (e.g., see Baden and Puckett [1990], Cheer [1983 and 1989], Choi, Humphrey, and Sherman [1988], Chorin [1980], Fishelov [1990a], Ghoniem, Chorin, and Oppenheim [1982], Sethian [1984], Sethian and Ghoniem

[1988], Summers, Hanson, and Wilson [1985], Tiemroth [1986], and Zhu [1989]). For this reason we will describe it in some detail below. Chorin's original vortex creation algorithm has not been as widely used. However, Anderson *et al.* [1990] have recently used a variant of Chorin's technique to study certain problems associated with vortex shedding past a cylinder. See also Leonard [1980] for a discussion of Chorin's original idea and references to other workers who have used it.

At the time of this writing no one has proved that the random vortex method together with some algorithm for satisfying the no-slip boundary condition converges to solutions of the Navier-Stokes equations. A complete analysis of the vortex sheet method is also an open problem. Some progress has been made in this area however. Marchiorio and Pulverenti [1983] constructed a sequence of stochastic processes which are similar to the vortex sheet method and showed that they converge to solutions of the Prandtl boundary layer equations. However their proof contains no convergence rates or error estimates that are of computational value. Hald [1986] has proved the convergence of a random walk algorithm with particle creation to solutions of a convection-diffusion equation. Puckett [1989] has proved a consistency result for the random walk and particle creation portion of the vortex sheet method. However, the stability and an analysis of the "inviscid" part of the algorithm remain open problems.

There has also been progress in related areas. For example, Mas-Gallic [1990] has recently proved the convergence of a two dimensional deterministic vortex method with boundary conditions. Also, as we mentioned earlier, Hou and Wetton [1990] have proved the convergence of a finite difference method for approximating solutions of the Navier-Stokes equations in vorticity-stream formulation with a vorticity boundary condition. We also refer the interested reader to Ghoniem and Sherman [1985] for an interesting discussion of random walk methods and particle creation algorithms. However, the issue of how best to satisfy the no-slip boundary condition in a vortex method computation is an important and largely unanswered question.

11.6.1   *The Vortex Sheet Method*

Let Ω denote a domain containing a viscous, incompressible fluid which has a solid boundary $\partial\Omega$ upon which the no-slip boundary condition (11.2.15) must be satisfied. In a hybrid random vortex / vortex sheet method computation the computational domain is divided into two regions: an interior (or exterior) region Ω_{NS} located away from $\partial\Omega$ and a *sheet layer* Ω_{Pr} located adjacent to $\partial\Omega$. In Ω_{NS} one uses the random vortex method to approximate solutions of the incompressible Navier-Stokes equations, while in Ω_{Pr} one uses the vortex sheet method to approximate solutions of the Prandtl boundary layer equations. We use the term sheet layer to distinguish the *computational* boundary layer from the *physical* boundary layer. While the justification for dividing the computational domain into the two regions Ω_{NS} and Ω_{Pr} is soundly based on the theory of boundary layers (e.g., see Schlichting [1979]), it is sometimes the case that the sheet method is used in regions where the underlying assumptions implicit in the use of the Prandtl equations are in doubt. However, even though many aspects of the combined random vortex / vortex sheet algorithm remain to be rigorously justified – such as the use of the vortex sheet method near points of separation – we emphasize that it has been successfully used by a number of workers to model a wide variety of flows.

Both the random vortex method and the vortex sheet method are particle methods. The particles carry concentrations of vorticity; the velocity field within each of the regions is uniquely determined by the particle positions and their concentrations together with the appropriate boundary conditions. Both methods are fractional step methods. One of the fractional steps consists of evolving the particles and their concentrations in this velocity field. The other step consists of letting the particle positions undergo a random walk to account for the diffusive effects of viscosity.

In Ω_{NS} the particles are called *vortices* or *vortex blobs*. In Ω_{Pr} they are called *vortex sheets*. The no-flow boundary condition is satisfied on $\partial\Omega$ by imposing a potential flow on Ω_{NS} which cancels the normal component of the velocity due to the vortices. The no-slip boundary condition is satisfied by creating vortex sheets on $\partial\Omega$ which subsequently participate in the flow. The two solutions are matched by converting sheets that leave the sheet layer into vortices with the same circulation, converting vortices that enter the sheet layer into sheets with the same circulation, and letting the velocity at infinity in the Prandtl equations be the tangential component of the velocity on $\partial\Omega$ due to the vortices in Ω_{NS}. The sheet creation process and subsequent movement of the sheets into the interior of the flow mimics the physical process of vorticity creation at a boundary and is one of the attractive features of this numerical method.

For simplicity we describe the vortex sheet method in $I\!\!R^2$. The generalization to $n = 3$ is straightforward (e.g., see Chorin [1980] or Fishelov [1990a]). Let (x, y) denote coordinates that are parallel and perpendicular to the boundary respectively. Let (u, v) denote the corresponding velocity components, ω the vorticity, and ν the kinematic viscosity. Assume that the boundary is located at $y = 0$ and let $U_\infty(x, t)$ denote the "velocity at infinity" which is imposed on the flow from outside the boundary layer. In vorticity formulation the Prandtl equations are

$$\omega_t + u\omega_x + v\omega_y = \nu\,\omega_{yy}, \tag{11.6.1a}$$

$$\omega = -u_y, \tag{11.6.1b}$$

$$u_x + v_y = 0, \tag{11.6.1c}$$

$$u(x, 0, t) = 0, \tag{11.6.1d}$$

$$v(x, 0, t) = 0, \qquad (11.6.1e)$$

$$\lim_{y \to \infty} u(x, y, t) = U_\infty(x, t). \qquad (11.6.1f)$$

Note that in the limiting process ($\nu \to 0$) by which one derives the Prandtl equations from the Navier-Stokes equations the vorticity $\omega = v_x - u_y$ has become $\omega = -u_y$ since $v_x = O(\sqrt{\nu} u_y)$.

In the vortex sheet method the vorticity at time $t = k\Delta t$ is approximated by a sum of linear concentrations of vorticity,

$$\tilde{\omega}^k(x, y) = \sum_j \omega_j b_l(x - x_j^k) \delta(y_j^k - y). \qquad (11.6.2)$$

Each term of the sum in (11.6.2) is referred to as a *vortex sheet*. The jth sheet has center (x_j^k, y_j^k) and *strength* or *weight* ω_j. Here δ is the Dirac delta function, and $b_l = b(x/l)$ is what we refer to as the *smoothing* or *cutoff function* in analogy with the vortex method. The most commonly used cutoff is the *hat* or *tent* function originally proposed by Chorin [1978],

$$b(x) = \begin{cases} 1 - |x| & |x| \leq 1, \\ 0 & \text{otherwise.} \end{cases} \qquad (11.6.3)$$

The parameter l is often referred to as the *sheet length* – even though the support of b_l is typically of length nl for some integer $n \geq 2$. Since b_l has finite support and since $\delta(y_j - y)$ is 0 for $y \neq y_j$, we see that the jth sheet is simply a line segment parallel to the boundary which carries a delta function concentration of vorticity. For b_l with b defined by (11.6.3) each sheet has length $2l$ and the vorticity concentration varies linearly along the length of the sheet – having a value of ω_j at the center and 0 at the ends. We briefly discuss other possible choices for b_l at the end of this section.

We can use (11.6.1b) and (11.6.1f) to write the tangential velocity in terms of the vorticity,

$$u(x, y, t) = U_\infty(x, t) + \int_y^\infty \omega(x, s, t) ds. \qquad (11.6.4)$$

Our approximation to u at time $k\Delta t$ is determined by (11.6.2) and (11.6.4),

$$\tilde{u}^k(x, y) = U_\infty(x, k\Delta t) + \sum_j \omega_j b_l(x - x_j^k) H(y_j^k - y), \qquad (11.6.5)$$

where $H(y)$ is the Heaviside function

$$H(y) = \begin{cases} 1 & y \geq 0, \\ 0 & \text{otherwise.} \end{cases}$$

From (11.6.5) we see that as one crosses the jth sheet in the vertical direction there is a jump in the tangential velocity $\tilde{u}$ of size $\omega_j b_l(x - x_j)$. This is the motivation for referring to the computational elements as vortex sheets. To find the velocity component normal to the boundary we first use (11.6.1c) and (11.6.1e) to write

$$v(x, y, t) = - \int_0^y u_x(x, s, t) ds \,. \tag{11.6.6}$$

Then, by approximating u_x with a centered divided difference, we obtain our approximation to v,

$$\tilde{v}^k(x, y) = -\partial_x U_\infty(x, t) y - \frac{1}{l} \sum_j \omega_j (b_l(x^+ - x_j^k) - b_l(x^- - x_j^k)) \min(y, y_j^k) \,. \tag{11.6.7}$$

where $x^+ = x + l/2$ and $x^- = x - l/2$.

Since $\tilde{u}^k$ and $\tilde{v}^k$ were constructed using (11.6.4) and (11.6.6) respectively the velocity field $(\tilde{u}^k, \tilde{v}^k)$ automatically satisfies equations (11.6.1b,c) and the boundary conditions (11.6.1e,f). Furthermore, given U_∞, this velocity field is completely determined by the sheet positions (x_j^k, y_j^k) and their strengths ω_j.

The vortex sheet method is a fractional step method. The first step is the numerical solution of the convective part of equation (11.6.1a)

$$\omega_t + u\omega_x + v\omega_y = 0 \,. \tag{11.6.8}$$

The second step is the numerical solution of the diffusive part of (11.6.1a)

$$\omega_t = \nu\omega_{yy} \tag{11.6.9}$$

subject to the no-slip boundary condition (11.6.1d).

Given an approximation $(\tilde{u}^k, \tilde{v}^k)$ to the velocity field at the kth time step the velocity at the next time step is determined as follows. We first evaluate $(\tilde{u}^k, \tilde{v}^k)$ at the center of each sheet. Denote the velocity at the center of the jth sheet by $(\tilde{u}_j^k, \tilde{v}_j^k)$. Our numerical approximation to (11.6.8) is found by moving the center of each sheet one time step of length Δt in this direction to obtain

$$(x_j^{k+\frac{1}{2}}, y_j^{k+\frac{1}{2}}) = (x_j^k, y_j^k) + \Delta t(\tilde{u}_j^k, \tilde{v}_j^k) \,. \tag{11.6.10}$$

In general, the sheet positions given by (11.6.10) induce a non-zero tangential velocity on the boundary which we denote by $\tilde{u}^{k+\frac{1}{2}}(x, 0)$. In order to approximately satisfy the no-slip boundary condition (11.6.1d) we create sheets on the boundary. Let $a_r, r = 1, \ldots, M$

denote equally spaced grid points at $y = 0$ with grid spacing l: $a_{r+1} - a_r = l$, let $u_r = \tilde{u}^{k+\frac{1}{2}}(a_r, 0)$, and let ω_{max} denote a computational parameter called the *maximum sheet strength*. Then for each r we create $q_r = [|u_r|/\omega_{max}]$ sheets with centers $(a_r, 0)$ and strengths $-\text{sign}(u_r)\,\omega_{max}$. Here $[x]$ denotes the greatest integer less than or equal to x.

The numerical solution of the diffusion equation (11.6.9) is found by letting all sheets – new and old – undergo a random walk in the y direction, reflecting those that go below the boundary. Therefore, the new sheet positions at time $(k + 1)\Delta t$ are given by

$$(x_j^{k+1}, y_j^{k+1}) = (x_j^{k+\frac{1}{2}}, |y_j^{k+\frac{1}{2}} + \eta_j|)$$

where the η_j are independent, Gaussian distributed random numbers with mean 0 and variance $2\nu\Delta t$.

We wish to make several comments regarding the sheet creation algorithm here. First, note that in our presentation of the algorithm all sheets have magnitude ω_{max} and that we create no sheets at a_r when $|u_r| < \omega_{max}$. Hence, the no-slip boundary condition is satisfied at a_r only up to order ω_{max}. Originally Chorin [1978 and 1980] created sheets at the rth grid point whenever $|u_r| \geq \epsilon$ for some $\epsilon \ll \omega_{max}$ such that $\omega_j \leq \omega_{max}$ for all j and the sum of the strengths of these sheets exactly cancel u_r. For example, ϵ might be chosen to be on the order of the computer's round off error. However, this algorithm creates more sheets than the one described above, and since the work required to compute $(\tilde{u}_j^k, \tilde{v}_j^k)$ at the center of each sheet is $O(lN^2)$ where N is the number of sheets in the flow, this greatly increases the computational cost of the algorithm. Furthermore, numerical experiments to compare the two sheet creation algorithms conducted by Puckett [1989a] and Zhu [1989] have shown that there is no tangible increase in the accuracy of the numerical approximation when this latter, more expensive, sheet creation algorithm is used.

The second point we would like to make concerns the manner in which the no-slip boundary condition is satisfied and its relation to the cutoff function b_l. As noted above, $\tilde{u}^{k+\frac{1}{2}}(x, 0)$ is, in general, non-zero. Ideally one would like to add some function to $\tilde{u}^{k+\frac{1}{2}}$ which can be represented by the sum of sheets, and which cancels $\tilde{u}^{k+\frac{1}{2}}(x, 0)$ at all points x on the boundary but leaves $\tilde{u}^{k+\frac{1}{2}}(x, y)$ unchanged for $y > 0$. In other words, we wish to find some function of the form $\sum \omega_j b_l(x - x_j)\text{H}(y_j - y)$ such that

$$\tilde{u}^{k+\frac{1}{2}}(x, y) + \sum_j \omega_j\, b_l(x - x_j)\, \text{H}(y_j - y) = \begin{cases} \tilde{u}^{k+\frac{1}{2}}(x, y) & y > 0, \\ 0 & y = 0. \end{cases}$$

In general this is not possible. However, one can find ω_j and (x_j, y_j) so that this holds exactly for $y > 0$ and within $O(l)$ for $y = 0$. For example, when b is defined by (11.6.3) choosing (x_j, y_j) to be the grid points $(a_r, 0)$ reduces the problem to that of finding the coefficients of a piecewise linear interpolant to $-\tilde{u}^{k+\frac{1}{2}}(x, 0)$ with node points at the a_r (e.g., see Schultz [1973]). In other words, we wish to find coefficients c_r such that

$$\sum_r c_r b_l(x - a_r) \approx -\tilde{u}^{k+\frac{1}{2}}(x, 0). \qquad (11.6.11)$$

For the piecewise linear basis functions b_l (i.e., with b given by (11.6.3)) it turns out that $c_r = -u_r$ is the correct choice, since then the left hand side of (11.6.11) is the usual piecewise linear interpolant of $-\tilde{u}^{k+\frac{1}{2}}$ at $y \equiv 0$. In actual practice we approximate the left hand side of this expression by creating q_r sheets at each point $x = a_r$ with strengths $\omega_r = \pm\omega_{max}$ such that $q_r\omega_r \approx -u_r$.

This idea can be generalized to make use of higher order interpolation procedures. For example, one can replace b_l with a basis function for cubic splines (e.g., see Schultz [1973] or DeBoor [1978]). In Puckett [1987] the author studied the effect that this type of smoothing function has on the accuracy and rate of convergence of the vortex sheet method. The interested reader is referred there for further details.

Several studies have been made of the accuracy with which the vortex sheet method approximates solutions of the Prandtl equations. The method was used to approximate Blasius flow in both Chorin [1978] and Puckett [1989a] and to approximate Falkner-Skan flow in Summers [1989]. In particular, Puckett [1989a] contains an extensive tabulation of the error in approximating Blasius flow as a function of the computational parameters Δt, l, and ω_{max} while Summers [1989] examines the computed solutions for a family of flows, some of which contain stagnation points or separation points.

11.6.2   Choosing the Computational Parameters

There are three computational parameters in the vortex sheet method: the time step Δt, the sheet length l, and the maximum sheet strength ω_{max}. The only generally agreed upon constraint that these parameters must satisfy is the so called "CFL" condition,

$$\Delta t\, U_{\max} \leq l \qquad (11.6.12)$$

where $U_{\max} = \max U_\infty(x)$. The justification usually given for (11.6.12) is that one wants to ensure that sheets move downstream at a rate of no more that one grid point per time step. This is an *accuracy condition* (as opposed to a stability condition) which ensures that information propagating in the streamwise direction will influence all features in the flow which are at least $O(l)$.

We also propose another accuracy condition,

$$\Delta t\, \omega_{max} \leq Cl^2 \qquad (11.6.13)$$

where C is a constant with dimensions $1/L$ and L is a typical length scale. (For example, L might be the length of the boundary.) This condition is a consequence of requiring that the degree with which we refine u as a function of y be of the same order as the degree with which we refine features in the streamwise direction, $O(U_{max}/\omega_{max}) = O(L/l)$, and then using (11.6.12). Note that since

$$\frac{d}{dx}\omega_j b_l(x - x_j) = O(\omega_{max}/l),$$

sheets induce local (non-physical) streamwise gradients in $\tilde{u}$ which are $O(\omega_{max}/l)$. Condition (11.6.13) relates the size of these gradients to the ratio $l/\Delta t$.

We wish the circulation Γ about a vortex element to remain the same when a sheet leaves the sheet layer and becomes a vortex or vice-versa. If b_l is the piecewise linear smoothing function with b given by (11.6.3), then this implies that

$$|\Gamma_j| = l\,\omega_{max}, \tag{11.6.14}$$

where Γ_j is given by (11.3.12). Equation (11.6.14) serves to relate the computational parameters h and ω_j discussed in Sections 11.3 and 11.5 to the sheet length l and maximum sheet strength ω_{max}.

In a hybrid random vortex method / vortex sheet method computation there are two computational parameters which still must be chosen: the cutoff radius δ and the sheet layer thickness ϵ. We would like to relate the cutoff radius δ to the vortex sheet parameters ω_{max} and l. Let us assume that we are using Chorin's cutoff function (11.3.10). Then the velocity kernel is given by (11.3.11). We seek δ so that a vortex at the edge of the boundary layer and its image (with respect to the boundary) with opposite sign will induce the same tangential velocity on the boundary as a sheet with the same position and strength. (Note that the sheet does not require an image since, by (11.6.6) and (11.6.7), the sheets satisfy the no-flow boundary condition (11.6.1e) exactly at $y = 0$.) If we set

$$\delta = \frac{l}{\pi},$$

then we find that for $|x| < \delta$ a vortex at (x, y), together with its image at $(x, -y)$ will induce the same tangential velocity on the boundary at $(x, 0)$ as a sheet with center (x, y).

Recall that the random walks have standard deviation $\sqrt{2\nu\Delta t}$. One wishes to avoid having random walks which travel the length of the sheet layer in one time step and this principle is generally taken into account when choosing the sheet layer thickness ϵ. Usually ϵ is taken to be $\epsilon = C\sqrt{\nu\Delta t}$ for some constant C. Typically $C = 2$ or 3. This yields a boundary layer which has the appropriate scale, $O(\sqrt{\nu})$ (see Schlichting

[1979]). We reiterate however that ϵ represents the thickness of a numerical boundary layer which should be distinguished from the physical boundary layer.

11.7   Fast Vortex Methods

The velocity at the point $\mathbf{x}$ due to N vortices with positions $\mathbf{x}_j$ and weights Γ_j is given by

$$\tilde{\mathbf{u}}(\mathbf{x}, t) = \sum_{j=1}^{N} \mathbf{K}_\delta(\mathbf{x} - \mathbf{x}_j)\Gamma_j. \qquad (11.7.1)$$

The cost of evaluating this function at a single point $\mathbf{x}$ is $O(N)$ operations. In a vortex method computation one needs to evaluate (11.7.1) at each of the N positions $\mathbf{x}_j$ at least once per time step. (Several evaluations per time step are required for higher order time discretizations.) Therefore, the cost of computing one time step of the vortex method is $O(N^2)$ and this cost can become prohibitive as $N \to \infty$.

Similar considerations pertain to astronomical calculations of the force of gravity due to a large number of stars acting on one another or in computational models of plasmas in which electrons and ions interact. The underlying theme in each of these fields is that there are N objects, a potential function associated with each object, and one wishes to calculate the force on each object induced by the other $N - 1$ objects. The problem of evaluating the force due to N objects at each of the N objects is generically referred to as the "N-body problem". The direct evaluation of a sum of the type shown in (11.7.1) at each of the positions $\mathbf{x}_j$ is often referred to as the *direct method*. Techniques for reducing the cost of an N-body computation that arise in one field will often have applications to other fields. Thus there has been much cross-fertilization in this area of vortex methods. We refer the interested reader to the review article of L. Greengard [1990] for a discussion of the underlying similarities between these problems and for further references.

There are a variety of approximation techniques available for reducing the cost of computing (11.7.1) to $O(N \log N)$ or even $O(N)$. Because of their importance to the practical implementation of vortex methods – especially in three dimensions – the study of such fast vortex methods is currently an active area of research. We will briefly outline several ideas that have been introduced in the last decade for speeding up vortex method computations in two dimensions and indicate the direction that some of this research is taking in three dimensions.

For the purposes of illustrating some of the ideas that can be used to reduce the cost of the direct method we will assume that $\Omega = I\!R^2$. The stream function ψ associated with a *point* vortex at $\mathbf{x}_j$ with strength $\Gamma_j = \omega_j h^2$ is the solution of Poisson's equation,

$$\Delta \psi = -\omega \qquad (11.7.2)$$

where

$$\omega(\mathbf{x}) = \delta(\mathbf{x} - \mathbf{x}_j)\Gamma_j, \qquad (11.7.3)$$

and δ is the Dirac delta function. Hence, ψ is the fundamental solution of Poisson's equation in $I\!R^2$, centered at the point $\mathbf{x}_j$, and multiplied by the vortex strength Γ_j,

$$\psi(\mathbf{x}) = -\frac{1}{2\pi} \log(|\mathbf{x} - \mathbf{x}_j|)\,\Gamma_j\,. \qquad (11.7.4)$$

The velocity $\tilde{\mathbf{u}}_j$ due to the vortex at $\mathbf{x}_j$ with strength Γ_j may be found by differentiating ψ,

$$\tilde{\mathbf{u}}_j(\mathbf{x}) = \nabla^\perp \psi(\mathbf{x}) \equiv (\psi_y(\mathbf{x}), -\psi_x(\mathbf{x}))\,. \qquad (11.7.5)$$

Thus, if the velocity is thought of as a force field, then the stream function ψ plays the role of its potential. Analogously, in electrostatics the potential associated with a point charge at $\mathbf{x}_j$ with charge Γ_j is also given by (11.7.4) but now the electrostatic field due to this charge is given by

$$\mathbf{E}(\mathbf{x}) = -\nabla \psi(\mathbf{x}) = -(\psi_x(\mathbf{x}), \psi_y(\mathbf{x}))$$

One can readily see that computing the velocity due to a point vortex or the electrostatic field due to a point charge are equivalent problems. Similar analogies hold for the gravitational field due to a body at $\mathbf{x}_j$.

11.7.1 The Vortex-in-Cell Method

In essence the velocity field due to N vortices can be found by solving a linear elliptic problem. This observation can be exploited to reduce the cost of the computation by employing fast algorithms for the solution of elliptic problems. For example, one can place a grid over the support of the vorticity Ω_0, interpolate the values of the vorticity ω onto the grid,[5] solve (11.7.2) on the grid for ψ, and use divided differences to obtain an approximation to $\tilde{\mathbf{u}} = (\psi_y, -\psi_x)$.

This procedure yields values of the velocity field on the grid. To find the value of the velocity at a vortex center $\mathbf{x}_j$ we can interpolate the values of the velocity from the grid onto $\mathbf{x}_j$. Thus, if we have $N_g = N_x \times N_y$ grid points and an $O(N_g \log N_g)$ method for solving (11.7.2) on the grid, then the cost of evaluating the velocity at each

---

[5]When dealing with point vortices or "point charges" one can use a weighted area rule for the interpolation. (See e.g., Hockney and Eastwood [1981] or Leonard [1980].)

of the N vortices is now $O(N + N_g \log N_g)$. This idea was probably first introduced by Birdsall and Fuss [1969] within the context of computing plasma simulations. It is the simplest example of a "fast vortex method" and is commonly known as the *vortex-in-cell* method, or sometimes the *cloud-in-cell* method. See Baker [1979] and Christiansen [1973] for numerical applications of this technique to vortex method computations or see Leonard [1980] for a more detailed review of these ideas. Cottet [1987] has proven the convergence of a vortex-in-cell method applied to a vortex blob method.

The problem with the vortex-in-cell method is a loss of accuracy due to the approximate solution of (11.7.2) on the grid, the two interpolation steps, and the divided difference approximation to $\tilde{\mathbf{u}}$. For example, suppose that we use the vortex-in-cell method to approximate the velocity field due to a single point vortex at $\mathbf{x}_j$; i.e., to solve (11.7.2) with the right hand side given by (11.7.3). A close examination of the error in this procedure will show that the error in approximating $\tilde{\mathbf{u}}_j$ at some point $\mathbf{x}$ increases as the distance between $\mathbf{x}$ and $\mathbf{x}_j$ decreases. This is because the constants in the error bounds depend on ψ and its derivatives and these become unbounded as $\mathbf{x} \to \mathbf{x}_j$. Thus, the order of the error will remain the same for all $\mathbf{x}$, but the actual computed error will increase substantially as $\mathbf{x}$ approaches $\mathbf{x}_j$.

This loss of accuracy as $\mathbf{x} \to \mathbf{x}_j$ can be ameliorated by only considering the influence of "far away" vortices when approximating the velocity field at a point $\mathbf{x}$ from values on the grid. For vortices near $\mathbf{x}$ the velocity at $\mathbf{x}$ due to those vortices is computed exactly

$$\tilde{\mathbf{u}}^{nearby}(\mathbf{x}, t) = \sum_{|\mathbf{x}-\mathbf{x}_j|<C} \mathbf{K}_\delta(\mathbf{x} - \mathbf{x}_j)\Gamma_j . \qquad (11.7.6)$$

Note that this sum is only over those vortices "near" $\mathbf{x}$. Methods of this type have been used for plasma simulations by Hockney, Goel and Eastwood [1974] as well as others. They are commonly called particle-particle/particle-mesh (P^3M) methods. See Hockney and Eastwood [1981] for a review of this and similar methods.

In some sense all "accurate" fast vortex methods have a radius C of the form that appears in equation (11.7.6). Outside this radius the velocity field due to a vortex may be accurately calculated using some fast approximation technique; inside this radius the velocity field must be computed directly.

11.7.2   The Method of Local Corrections

An interesting variant of the particle-particle/particle-mesh idea, known as the method of local corrections, was proposed by Anderson [1986]. The method of local corrections was designed so that it could be used with high order accurate velocity kernels such as those described in Section 11.3.4. It is based on the observation that the velocity field due to a point vortex is harmonic away from the center of the vortex,

$$\Delta \tilde{\mathbf{u}}_j = 0 \qquad \text{for} \qquad \mathbf{x} \neq \mathbf{x}_j,$$

where $\tilde{\mathbf{u}}_j = \mathbf{K}(\mathbf{x} - \mathbf{x}_j)\Gamma_j$ is the velocity field due to a point vortex centered at $\mathbf{x}_j$ (c.f. equation (11.7.5)).

Suppose that we are given a grid with mesh spacing h_1 and an rth order accurate approximation to the Laplacian on this grid denoted by Δ^{h_1}. (It is not necessary that this grid be identical to the grid – with spacing denoted by h - that is used to determine the initial data. In fact, optimum accuracy and efficiency may very well be achieved when $h_1 \neq h$.) Rather than solving (11.7.2) for the stream function ψ, Anderson solves a Poisson equation for the velocity field $\tilde{\mathbf{u}}$,

$$\Delta \tilde{\mathbf{u}}^{h_1} = \sum_j \mathbf{g}_{D_j} \qquad (11.7.7)$$

where the right hand side is the sum of approximations to $\Delta \tilde{\mathbf{u}}_j$ on the grid

$$\mathbf{g}_{D_j}(\mathbf{x}) = \begin{cases} \Delta^{h_1} \tilde{\mathbf{u}}_j & |\mathbf{x} - \mathbf{x}_j| < D, \\ 0 & |\mathbf{x} - \mathbf{x}_j| \geq D. \end{cases} \qquad (11.7.8)$$

Since $\Delta \tilde{\mathbf{u}} = 0$ for $\mathbf{x} \neq \mathbf{x}_j$, we can choose D so that $\mathbf{g}_{D_j}$ is an rth order accurate approximation to $\Delta \tilde{\mathbf{u}}_j$ on the grid. Anderson calls the constant D the *spreading radius*. The point of using the approximation $\mathbf{g}_{D_j}$ to $\Delta \tilde{\mathbf{u}}_j$ is that for small D the right hand side of (11.7.7) can be computed in only $O(N)$ operations rather than in the $O(NN_g)$ operations required to approximate $\Delta \tilde{\mathbf{u}}_j$ at each of the N_g grid points.

Once the right hand side of (11.7.7) is available we use our favorite fast Poisson solver to approximate the solution of (11.7.7) on the grid. This results in an approximation $\tilde{\mathbf{u}}^{h_1}$ on the grid to the velocity field $\tilde{\mathbf{u}}$ given by (11.7.1). In particular, note that by solving for $\tilde{\mathbf{u}}^{h_1}$ directly we have eliminated the error due to approximating $\tilde{\mathbf{u}} = (\psi_y, -\psi_x)$ with divided differences of ψ on the grid. Instead we now must compute the $\mathbf{g}_{D_j}$. Numerical experiments by Anderson [1985], Baden [1987], and Buttke [1991] have demonstrated that for $D = h_1$ or $D = 2h_1$ the approximation $\tilde{\mathbf{u}}^{h_1}$ to $\tilde{\mathbf{u}}$ on the grid is highly accurate.

The approximation $\tilde{\mathbf{u}}^{h_1}$ is next interpolated back onto the vortex positions. In the interpolation step Anderson cleverly exploited the fact that in two dimensions the two components of the velocity field $\tilde{\mathbf{u}} = (\tilde{u}, \tilde{v})$ due to the sum of point vortices centered at points $\mathbf{x}_j$ are the real and complex parts of a complex analytic function,

$$F(\mathbf{x}) = \tilde{u}(z) - i\tilde{v}(z). \qquad (11.7.9)$$

Here we have identified $I\!\!R^2$ with the complex plane C: $\mathbf{x} = (x, y) \rightarrow z = x + iy$. Thus, one can use Lagrange interpolation in the complex plane to obtain a fourth order interpolation at only four points. In contrast, a bilinear interpolation at four points for each of the velocity components separately would only be second order. Lagrange interpolation at more points will result in higher order interpolation formulas for the velocity. However it should be noted that because this idea is based on a representation of the velocity as a function in the complex plane it does not generalize easily to three dimensions. In fact, the task of finding high order interpolation formulas is an important issue when generalizing the method of local corrections to three dimensions (e.g., see Buttke [1991]).

The last, and very essential part of this procedure is to *correct* the velocity field $\tilde{\mathbf{u}}^{h_1}$ before interpolating it back onto the individual vortices. This is accomplished as follows. Suppose we wish to calculate the velocity at $\mathbf{x}$ by interpolating $\tilde{\mathbf{u}}^{h_1}$ from the points α_i, $i = 1, \ldots, m$ on the grid. Let the vortices which satisfy $|\mathbf{x} - \mathbf{x}_j| < C$ be the "nearby" vortices. (Anderson calls the distance C the *correction radius*.) For each i we evaluate the velocity at α_i induced by all nearby vortices and subtract this velocity from $\tilde{\mathbf{u}}^{h_1}(\alpha_i)$. The resulting values of $\tilde{\mathbf{u}}^{h_1}$ at the α_i are the ones used to interpolate the velocity onto the point $\mathbf{x}$. Finally, the exact velocities due to the vortices near $\mathbf{x}$ – i.e., the expression in (11.7.6) – are added to this value of the velocity at $\mathbf{x}$ to obtain our complete approximation to $\tilde{\mathbf{u}}$ at $\mathbf{x}$.

It is important to note that all of the steps in this procedure but the very last are performed as if the vortices at the points $\mathbf{x}_j$ are *point vortices*. Only when the velocity at $\mathbf{x}$ is corrected by (11.7.6) is the velocity field due to a vortex blob at the $\mathbf{x}_j$ used. If the smoothing function f_δ has compact support, say within a disk of radius R, then the velocity field due to a vortex blob at $\mathbf{x}_j$ and a point vortex at $\mathbf{x}_j$ are identical at distances greater than R from $\mathbf{x}_j$. Thus the method of local corrections will accurately represent the influence of the core function f_δ provided that $R \leq C$.

As with the vortex-in-cell method the cost of the method of local corrections is also $O(N + N_g \log N_g)$. It is somewhat more complex to program than a vanilla vortex-in-cell method, but considerably more accurate. Applications of this technique to both vector and parallel computers have been studied by Baden [1987] and a version for viscous flow with solid boundaries has been implemented by Baden and Puckett [1990]. We also note that Mayo [1985] employed similar ideas in the solution of Laplace's equation in irregular regions.

11.7.3   Tree Codes, Multipole Expansions, and the Fast Multipole Method

We now turn to a somewhat different set of techniques for reducing the cost of directly calculating (11.7.1) in two dimensions. These techniques rely on representing the velocity field due to a cluster of vortices as a truncated power series in z^{-1} which is valid sufficiently far from the cluster. (Once again we are identifying $I\!R^2$ with the complex plane C.) By combining these expansions with a hierarchical data structure which consists of nested boxes covering the computational domain, one can devise an algorithm of arbitrary accuracy which is $O(N \log N)$. Several clever modifications of this idea due to Greengard and Rokhlin [1987] can be made to reduce the cost of such a method to $O(N)$. We outline the methods as they would be applied to a point vortex computation and indicate at the end the modifications necessary for working with vortex blobs. The following discussion is loosely based on the review article of Greengard [1990]. For further details and a description of these methods in a more general setting see Greengard and Rokhlin [1987] or Greengard [1990].

The Tree Data Structure For the sake of simplicity let us assume that there are $N = 4^\gamma$ vortices uniformly distributed in a square box of side 1 which we will call Ω. We begin by describing the hierarchical data structure. Let us call the initial box level 0. Now subdivide this box into 4 equal boxes of side $1/2$ and call this level 1. In general, level l is obtained from level $l-1$ by subdividing each box at level l into 4 smaller boxes of equal size. Continue this procedure until at the lowest level of refinement we have N boxes. Thus, at level 0 there is 1 box, at level 1 there are 4 boxes and, in general, at the lth level there are 4^l boxes. Given a box A at level $l-1$, the boxes at level l which are obtained by subdividing box A are called the *children* of box A. A box is the called the *parent* of its children.

Definition 1 Two boxes are said to be *near neighbors* if they are at the same level of refinement and share a boundary point. (A box is a near neighbor of itself.)

Definition 2 Two boxes are said to be *well separated* if they are at the same level of refinement and not near neighbors.

Definition 3 With each box A is associated an *interaction list* which consists of all of the children of the near neighbors of A's parent that are well separated from A.

The interaction list is the key concept here. As we shall see, it enables one to organize the computation in an efficient manner. Note that at levels 0 and 1 all boxes are near neighbors of each other and hence each box's interaction list is empty. Note also that at any level of refinement the maximum size of the interaction list is 27. (It is smaller for boxes that lie on the boundary of the computational domain.)

The purpose of this tree of boxes is twofold. First, it provides a mechanism for determining which vortices are well separated from a given box A and thus amenable to a far field approximation of the velocity due to the vortices in box A. Second, it provides a mechanism, namely the interaction list, for keeping track of those vortices that have not yet had their interactions with the vortices in box A computed.

Multipole Expansions   We now turn to the issue of fast far field approximations for computing the velocity between vortices in boxes that are well separated from one another. In order to describe this approximation it is again helpful to identify $I\!R^2$ with the complex plane C: $\mathbf{x} = (x, y) \rightarrow z = x + iy$ and view the velocity as the real and imaginary parts of the function F in the complex plane given by (11.7.9). If we substitute the expression for $\mathbf{K}$ given by (11.2.8) into (11.7.1) (with $\mathbf{K}_\delta$ replaced by $\mathbf{K}$ since we are considering point vortices) we find that the velocity field due to a cluster of m^6 point vortices is

$$\tilde{\mathbf{u}}(\mathbf{x}, t) = -\frac{1}{2\pi} \sum_{j=1}^{m} \Gamma_j \frac{(y - y_j, x_j - x)}{(x - x_j)^2 + (y - y_j)^2} \quad .$$

Hence, F has the form

$$F(z) = \frac{1}{2\pi i} \sum_{j=1}^{m} \Gamma_j \frac{1}{(z - z_j)}$$

where we have identified $\mathbf{x}_j \in I\!R^2$ with $z_j \in C$. Note that F is an analytic function for $z \neq z_j$.

Now consider a cluster of vortices located at points z_j all of which are contained in a box A with center z_A such that $|z_j - z_A| < R$ for all j. Let z be such that $|z - z_A| > 2R$. Then $F(z)$ can be written as a *multipole expansion* about the point $z = z_A$,

$$F(z) = \sum_{k=0}^{\infty} \frac{a_k}{(z - z_A)^{k+1}} \tag{11.7.10}$$

where the a_k are given by

$$a_k = \frac{1}{2\pi i} \sum_{j=1}^{m} \Gamma_j (z_j - z_A)^k \qquad k = 0, 1, 2, \dots \quad . \tag{11.7.11}$$

Note that the cost of forming each coefficient a_k is $3 \cdot m$ operations. (The $2\pi i$ can be divided into the strengths Γ_j initially to avoid needless divisions later.) Hence, the cost of forming the coefficients for the first p terms of (11.7.10) is $O(mp)$.

We can derive bounds for the error in using the first p terms of (11.7.10) as follows. Let

$$c = \frac{R}{|z - z_A|} \quad \text{and} \quad W = \frac{1}{2\pi} \sum_{j=1}^{m} |\Gamma_j|. \tag{11.7.12}$$

---

[6]Note the use of the integer m rather than N. This is a notational convenience so that we may later consider the cost of taking many such expansions when the total number of vortices is $N > m$.

Then

$$
\left| F(z) - \sum_{k=0}^{p} \frac{a_k}{(z - z_A)^{k+1}} \right| = \left| \sum_{k=p+1}^{\infty} \frac{a_k}{(z - z_A)^{k+1}} \right|
$$

$$
\leq W \sum_{k=p+1}^{\infty} \frac{R^k}{|z - z_A|^{k+1}}
$$

$$
\leq \frac{W}{R} \sum_{k=p+1}^{\infty} \frac{R^{k+1}}{|z - z_A|^{k+1}}
$$

$$
\leq \frac{W}{R} \left| \frac{R}{z - z_A} \right|^{p+2} \sum_{k=0}^{\infty} \frac{R^k}{|z - z_A|^k} \tag{11.7.13}
$$

$$
\leq \frac{W}{R} \left| \frac{R}{z - z_A} \right|^{p+2} \frac{1}{1 - c}
$$

$$
\leq \frac{W}{R} \frac{1}{c^{-1} - 1} \, c^{p+1}
$$

$$
\leq \frac{W}{R} \left(\frac{1}{2} \right)^{p+1}
$$

since, by assumption, $c < 1/2$.

In practice R is half the diagonal length of a box and hence $R \to 0$ as the level of refinement increases. However one can control the size of W/R as follows. In a typical (two dimensional) vortex method computation Ω is divided into $N = 4^\gamma$ squares each of side $h = 1/\sqrt{N} = 2^{-\gamma}$. Let z_j, $j = 1, \ldots, N$ denote the centers of the squares. Let $\omega(z)$ denote the vorticity at time $t = 0$ and let $\|\omega\|_\infty$ denote the sup norm of $\omega(z)$ in Ω. The initial data consists of N vortices, with positions z_j and strengths $\Gamma_j = \omega(z_j) h^2$.

At any given level l there are 4^l boxes and hence, there are $m_l = N/4^l = 4^{\gamma-l} = 2^{2(\gamma-l)}$ vortices per box. Furthermore, the side of a box at level l has length 2^{-l} and hence the distance from the center of a box to any corner is

$$
\left| F(z) - \sum_{k=0}^{p} \frac{a_k}{(z - z_A)^{k+1}} \right| \leq \frac{W}{R_l} \left(\frac{1}{2} \right)^{p+1}
$$

$$
= \frac{1}{2\pi} \frac{2}{\sqrt{2}} 2^l \sum_{j=1}^{m_l} |\Gamma_j| \left(\frac{1}{2} \right)^{p+1}
$$

$$
\leq \frac{1}{\sqrt{2\pi}} 2^l \, m_l \, \|\omega\|_\infty \, h^2 \left(\frac{1}{2} \right)^{p+1} \tag{11.7.14}
$$

$$
= \frac{1}{\sqrt{2\pi}} 2^{2\gamma-l} \, \|\omega\|_\infty \, \frac{1}{N} \left(\frac{1}{2} \right)^{p+1}
$$

$$\leq \frac{1}{\sqrt{2\pi}} \, \|\omega\|_\infty \left(\frac{1}{2}\right)^{p+1}.$$

since $2^{2\gamma-l} \leq N$ for all l. This estimate holds at all levels of refinement l.

Tree Codes One can combine the tree data structure and the multipole expansions (11.7.10) to obtain an $O(N \log N)$ algorithm for approximating (11.7.1) at each of the N vortices as follows. For the time being let us assume that $p = 2$.

Starting at level 2 form the multipole expansion coefficients given by (11.7.11) for each box A and use them to evaluate the velocity at all vortices which are contained in A's interaction list.[7] (Since there are no well separated boxes at levels 0 or 1 there is no need to perform this step at these levels.) So far we have calculated the influence of the vortices in box A on all the other vortices except those contained in A's near neighbors. To compute these interactions we recursively repeat the above procedure at successive levels of refinement. At each level calculate the coefficients (11.7.11) for every box A and evaluate the resulting multipole expansion at each vortex in A's interaction list. At the last level we also evaluate the interactions between each box and its nearest neighbors *directly*.

The total amount of work in this procedure is $O(N \log N)$. To see this first note that since there are N vortices and each vortex contributes to $p = 2$ expansion coefficients, it takes $3 \cdot p \cdot N = 6 \cdot N$ operations to form the coefficients (11.7.11) at each level. From the point of view of a vortex in box A at level l the cost of evaluating its velocity due to the expansions in the boxes in its interaction list is at most $(3 \cdot p + 1) \cdot 27 = 7 \cdot 27$ operations since there are at most 27 members of the interaction list and it takes 7 operations to evaluate the expansion (11.7.10) at a point z when $p = 2$. (The term $(z - z_A)^{-1}$ need only be evaluated once for each z.) Since there are N vortices the cost of evaluating all of the expansions at a given level is therefore $7 \cdot 27 \cdot N$. Thus, the cost of this procedure at any given level is $7 \cdot 27 \cdot N + 6 \cdot N = 195 \cdot N$ operations. At the last level there are at most 8 nearest neighbors and 1 vortex per box (by assumption the vortices are uniformly distributed in the computational domain). Thus, the cost of evaluating the direct interactions at the last level is $8N$ operations. (Self interactions are not computed.) Finally, there are $\log_4 N$ levels so the total amount of work is $O(N + N \log N) = O(N \log N)$ as claimed.

Of course, in general the vortices are not uniformly distributed in the computational domain. In practice one can implement an adaptive version of this algorithm that at each level checks every box to see if it contains any vortices. If not, this box is pruned

---

[7]For convenience we will often refer to the vortices contained in the boxes of A's interaction list as being in A's interaction list.

from the tree structure and ignored at subsequent levels. For most practical problems the running time of this algorithm will remain $O(N \log N)$. However the constant of proportionality will depend on the distribution of vortices in Ω. See Carrier, Greengard and Rokhlin [1988] for an example of such an algorithm.

Algorithms of this type have been used extensively in astrophysical calculations and are commonly called *tree codes*. See Apple [1985] or Barnes and Hut [1986] for a more complete discussion of tree codes. By taking $p > 2$ we will have a more accurate algorithm. This is essentially the approach taken by van Dommelen and Rundensteiner [1989]. Their algorithm is still $O(N \log N)$ but now the constant of proportionality depends on the precision desired; i.e., on p.

The Fast Multipole Method Greengard and Rokhlin [1987] have made several observations that allows one to construct an algorithm with $O(N)$ running time. These ideas are similar to those used by Rokhlin [1985] for the rapid solution of integral equations. We will briefly describe the Greengard-Rokhlin algorithm here. The interested reader is referred to Greengard and Rokhlin [1987] for further details and for proofs of the statements made below.

To begin, suppose we are given the multipole expansion (11.7.10) for $F(z)$ about the center z_A of some box A. Then we can expand F in a multipole expansion about the center z_P of A's parent P,

$$F(z) = \sum_{l=0}^{\infty} \frac{b_l}{(z - z_P)^{k+1}} .$$

This expansion will be valid to the accuracy stated in (11.7.13) provided that it is only evaluated at points z in boxes that are well separated from P. The coefficients b_l may be formed from the (already known) a_k,

$$b_l = \sum_{k=0}^{l} \binom{l}{k} a_k (z_A - z_P)^{l-k} \qquad (11.7.15)$$

where $\binom{l}{k}$ are the binomial coefficients. Thus, if the coefficients a_k, $k = 1, \ldots, p$ of the first p terms of each child's multipole expansions are known, then one can obtain the coefficients b_l, $l = 1, \ldots, p$ of the parent's multipole expansion in $O(p^2)$ operations.

Next note that the multipole expansion (11.7.10) of $F(z)$ is analytic outside the circle of radius R centered at z_A. Therefore, $F(z)$ can be expanded in a Taylor series about any point z_B that satisfies $|z_B - z_A| > R$. In particular, suppose z_A and z_B are the centers of boxes A and B and that A and B are well separated from one another; i.e.,

$|z_B - z_A| > 2R$, where R is the half diagonal of a box. Then $F(z)$ can be expanded in a Taylor series about z_B that is valid in a circle of radius R centered at z_B,

$$F(z) = \sum_{l=0}^{\infty} c_l (z - z_B)^l .$$

Greengard and Rokhlin call this the *local expansion* in B of the velocity field due to the vortices at the $z_j \in A$. The important thing to notice here is that the coefficients c_l may also be written in terms of the coefficients a_k,

$$c_l = \frac{1}{(z_A - z_B)^{l+1}} \sum_{k=0}^{\infty} \binom{l+k}{k} \frac{a_k}{(z_A - z_B)^k} (-1)^{k+1}$$

In practice, to obtain pth order accuracy one truncates the series on the right hand side of this expression after p terms. Thus, given the multipole expansion coefficients a_k for $k = 1, \ldots, p$, we can compute pth order approximations $\tilde{c}_l$ to the local expansion coefficients c_l, $l = 1, \ldots, p$ of the form

$$\tilde{c}_l = \frac{1}{(z_A - z_B)^{l+1}} \sum_{k=0}^{p} \binom{l+k}{k} \frac{a_k}{(z_A - z_B)^k} (-1)^{k+1} \qquad (11.7.16)$$

in $O(p^2)$ operations.

Finally, we have a formula similar to (11.7.15) for shifting the local expansion about the center of a given box A onto the centers of A's children. Let z_A denote the center of box A at level $l - 1$ and let z_C denote the center of one of A's children. Then

$$\sum_{l=0}^{p} c_l(z - z_A)^l = \sum_{l=0}^{p} \left(\sum_{k=l}^{p} c_l \binom{k}{l} (z_C - z_A)^{k-l} \right) (z - z_C)^l . \qquad (11.7.17)$$

Note that this formula is exact. The left hand side is simply a polynomial in powers of $(z - z_A)$ and the right hand side is simply the same polynomial in powers of $(z - z_C)$.

We are now in position to describe the fast multipole method. The method consists of two passes, beginning with an upward pass from the finest level of boxes up to level 2. This is the coarsest level that still has boxes with nonempty interaction lists. We then retrace our steps going back down through the tree hierarchy, until at the finest level of refinement each box has a local expansion that represents the velocity field due to all of the vortices except near neighbors.

Upward Pass: Starting at the finest level we form the coefficients (11.7.11) for every box. For each parent P at level l we take the coefficients formed at level $l+1$ for each of

P's children and using (11.7.15) shift them on to P's center. By adding the contributions from each of P's children we form the coefficients of P's multipole expansion. At each level we save the coefficients associated with each box for the later use.

Downward Pass: Given a box A at level 2 we use (11.7.16) to convert the coefficients of the multipole expansions associated with each box in A's interaction list to a local expansion about A's center and add these coefficients together. Before moving to the next level down we use (11.7.17) to shift these coefficients onto the center of each of A's children. We then recursively repeat the above procedure: In each box A at level l we have the coefficients of a local expansion that represents the velocity field due to all of the vortices in boxes that are well separated from A's parent. To these coefficients we add the coefficients for the local expansions in A's interaction list. At the end of this procedure each box at the finest level will contain the coefficients of a local expansion that represents the velocity field due to all of the vortices except its nearest neighbors – their contribution to the velocity field will be evaluated directly. Thus, to obtain the velocity at each vortex we evaluate the velocity field given by the appropriate local expansion and add to it the exact velocity due to its nearest neighbors.

The total cost of the above procedure is $O(Np^2)$. To see this first note that the cost of forming all of the local expansion coefficients at the finest level is $O(Np)$. The cost of shifting the multipole coefficients from a child onto its parent is $O(p^2)$ and since this must be done at most N times (there are at most N children) the cost of the remainder of the upward pass is $O(Np^2)$. Similarly, one can show that the cost of the downward pass is also $O(Np^2)$. At the finest level the cost of evaluating the appropriate local expansion at each vortex is $O(Np)$ and the cost of computing the nearest neighbor interactions directly is $O(N)$.

Some care should be taken in the programming of this algorithm to achieve the maximum possible savings. We refer the interested reader to Greengard and Rokhlin [1988] for details. Greengard and Rokhlin [1987] have also studied the implementation of various boundary conditions in conjunction with the fast multipole method. We also remark that Greengard [1988] has devised a fast technique along similar lines for computing the flow due to vortices in a channel.

Just as with the method of local corrections we can use this algorithm to evaluate the velocity field due to a collection of vortex blobs. For example, if the core functions have support contained in a region of radius R, then we must choose the maximum level of boxes so that at all levels $|\mathbf{x}_i - \mathbf{x}_j| < R$ implies that the vortices at $\mathbf{x}_i$ and $\mathbf{x}_j$ are nearest neighbors. It is conceivable that one might be able to find core functions with large or even infinite support, but which induce a velocity that is sufficiently close to that induced by a point vortex at distances greater than R from the vortex center. For such core functions both the method of local corrections and fast multipole expansions might then be applicable. However the author is not aware of any published work concerning this issue.

11.7.4   Fast Vortex Methods in Three Dimensions

The number of vortices that are required to retain a given level of accuracy, of say $O(h^r)$, increases from $O(h^{-2})$ in two dimensions to $O(h^{-3})$ in three. Therefore, fast algorithms are especially crucial in three dimensional vortex method computations. The generalization to three dimensions of the ideas illustrated above is currently an active area of research. In what follows we will simply mention recent work and give the appropriate references.

Buneman, Couet and Leonard [1981] have studied a three dimensional version of the vortex-in-cell method. A discussion of some of the salient ideas may be found in the review articles of Leonard [1980 and 1985]. Greengard and Rokhlin [1988a] have implemented a three dimensional version of their fast multipole method which is based on spherical harmonics. Buttke [1991] has also studied the fast multipole method in three dimensions. He has concluded that in three dimensions it is best to eliminate the hierarchy of expansions. This results in a method which is $O(N^{3/2})$ rather than $O(N \log N)$ or $O(N)$. Buttke presents evidence that for many three dimensional applications this technique will be less expensive than the full fast multipole method. However it should be noted that Greengard and Rokhlin [1988b] have devised strategies which will significantly reduce the cost of the 3D multipole algorithm.

Buttke [1991] also describes a three dimensional implementation of the method of local corrections and compares its performance to his version of the fast multipole method. Almgren, Buttke, and Colella [1991] have also implemented a three dimensional version of the method of local corrections. Their implementation includes an adaptive mesh refinement algorithm based on a multigrid method.

Finally we note that Anderson [1990] has designed a method which shares many of the features of the fast multipole method. However Anderson's method is based on Poisson's formula for representing solutions of Laplace's equation as a boundary integral over a disk or sphere rather than on multipole or spherical harmonic expansions. According to Anderson this technique has the advantage that the computer program for the method in two dimensions can easily be modified for three dimensions. He also discusses how to exploit the essential structure of a multigrid code to quickly develop a version of this method in either two or three dimensions.

References

[1] Alessandrini, G., Douglis, A., and Fabes, E. [1983]. An Approximate Layering Method for the Navier-Stokes Equations in Bounded Cylinders. *Ann. Mat. Pura Appl.* **135**, 329–347.

[2] Almgren, A., Buttke, T. and Colella, P. [1991]. A Fast Vortex Method in Three Dimensions. *Proceedings of the 10th AIAA Computational Fluid Dynamics Conference.* Honolulu, HI, 10 pages.

[3] Anderson, C. R. [1985]. A Method of Local Corrections for Computing the Velocity Field Due to a Distribution of Vortex Blobs. *J. Comp. Phys.* **61**, 111–123.

[4] Anderson, C. R. [1986]. A Vortex Method for Flows with Slight Density Variations. *J. Comp. Phys.* **62**, 417–432.

[5] Anderson, C. R. [1988]. Observations on Vorticity Creation Boundary Conditions. *Mathematical Aspects of Vortex Dynamics* (Ed. by R. E. Caflisch), S.I.A.M., Philadelphia, 144–159.

[6] Anderson, C. R. [1989]. Vorticity Boundary Conditions and Boundary Vorticity Generation for Two-Dimensional Viscous Incompressible Flows. *J. Comp. Phys.* **80**, 72–97.

[7] Anderson, C. R. [1990]. The Fast Multipole Method without Multipoles. *UCLA Computational and Applied Mathematics*, Report **90-14**, 1–28.

[8] Anderson, C. R. and Greengard, C. [1985]. On Vortex Methods. *SIAM J. Numer. Anal.* **22**, 413–440.

[9] Anderson, C. R. and Greengard, C. [1989]. The Vortex Ring Merger Problem at Infinite Reynolds Number. *Commun. Pure Appl. Math.* **42**, 1123–1139.

[10] Anderson, C. R., Greengard, C., Greengard, L. and Rokhlin, V. [1990]. On the Accurate Calculation of Vortex Shedding. *Phys. Fluids A* **2**, 883–885.

[11] Appel, A. W. [1985]. An Efficient Program for Many Body Simulations. *SIAM J. Sci. Stat. Comput.* **6**, 85–103.

[12] Ashurst, W. T. and Meiburg, E. [1988]. Three Dimensional Shear Layers Via Vortex Dynamics. *J. Fluid Mech.* **180**, 87–116.

[13] Ashurst, W. T. and McMurtry, P. A. [1989]. Flame Generation of Vorticity: Dipoles from Monopoles. *Combust. Sci. Tech.* **66**, 17–37.

[14] Baden, S. B. [1987]. Run Time Partitioning of Scientific Continuum Calculations Running on Multiprocessors. *Ph.D Thesis*, U. C. Berkeley Computer Science Dept.

[15] Baden, S. B. [1988]. Very Large Vortex Calculations in Two Dimensions. *Vortex Methods* (Ed. by C. Anderson and C. Greengard), Lecture Notes in Mathematics **1360**, Springer Verlag, New York, 96–120.

[16] Baden, S. B. and Puckett, E. G. [1990]. A Fast Vortex Method for Computing 2D Viscous Flow. *J. Comp. Phys.* **91**, 278–297.

[17] Baker, G. R. [1979]. The "Cloud-in-Cell" Technique Applied to the Roll Up of Vortex Sheets. *J. Comp. Phys.* **31**, 76–95.

[18] Barnes, J. and Hut, P. [1986]. A Hierarchical $O(N \log N)$ Force Calculation Algorithm. *Nature* **324**, 446–449.

[19] Beale, J. T. [1986]. A Convergent 3-D Vortex Method with Grid-Free Stretching. *Math. Comp.* **46**, 401–424.

[20] Beale, J. T. [1988]. On the Accuracy of Vortex Methods at Large Times. *Computational Fluid Dynamics and Reacting Gas Flows* (Ed. by B. Engquist et al.), Springer Verlag, New York, 19–32.

[21] Beale, J. T., Eydeland, A., and Turkington, B. [1991]. Numerical Tests of 3-D Vortex Methods using a Vortex Ring with Swirl. *Vortex Dynamics and Vortex Methods* (Ed. by C. Anderson and C. Greengard), Lectures in Applied Mathematics **28**, American Mathematical Society, Providence, RI, 1–9.

[22] Beale, J. T., Kato, T., and Majda, A. [1984]. Remarks on the Breakdown of Smooth Solutions for the 3-D Euler Equations. *Commun. Math. Phys.* **94**, 61–66.

[23] Beale, J. T. and Majda, A. [1981]. Rates of Convergence for Viscous Splitting of the Navier-Stokes Equations. *Math. Comp.* **37**, 243–259.

[24] Beale, J. T. and Majda, A. [1982a]. Vortex Methods I: Convergence in Three Dimensions. *Math. Comp.* **39**, 1–27.

[25] Beale, J. T. and Majda, A. [1982b]. Vortex Methods II: High Order Accuracy in Two and Three Dimensions. *Math. Comp.* **39**, 29–52.

[26] Beale, J. T. and Majda, A. [1984]. Vortex Methods for Fluid Flow in Two or Three Dimensions. *Contemp. Math.* **28**, 221–229.

[27] Beale, J. T. and Majda, A. [1985]. High Order Accurate Vortex Methods with Explicit Velocity Kernels. *J. Comp. Phys.* **58**, 188–208.

[28] Benfatto, G. and Pulvirenti, M. [1983]. A Diffusion Process Associated to the Prandtl Equations. *J. Funct. Anal.* **52**, 330–343.

[29] Birdsall, C. K. and Fuss, D. [1969]. Clouds-in-Clouds, Clouds-in-Cells, Physics for Many-Body Plasma Simulations. *J. Comp. Phys.* **3**, 494–511.

[30] Buneman, O. [1973]. Subgrid Resolution of Flows and Force Fields. *J. Comp. Phys.* **11**, 250–268.

[31] Buneman, O., Couet, B. and Leonard, A [1981]. Simulation of Three Dimensional Incompressible Flows with a Vortex-in-Cell Method. *J. Comp. Phys.* **39**, 305–328.

[32] Buttke, T. F. [1990]. A Fast Adaptive Vortex Method for Patches of Constant Vorticity in Two Dimensions. *J. Comp. Phys.* **89**, 161–186.

[33] Buttke, T. F. [1991]. Fast Vortex Method Methods in Three Dimensions. *Vortex Dynamics and Vortex Methods* (Ed. by C. Anderson and C. Greengard), Lectures in Applied Mathematics **28**, American Mathematical Society, Providence, RI, 51–66.

[34] Caflisch, R. [1988]. Mathematical Analysis of Vortex Dynamics. *Mathematical Aspects of Vortex Dynamics* (Ed. by R. E. Caflisch), S.I.A.M., Philadelphia, 1–24.

[35] Carrier, J., Greengard, L. and Rokhlin, V. [1988] A Fast Adaptive Multipole Algorithm for Particle Simulations. *SIAM J. Sci. Stat. Comput.* **9**, 669–686.

[36] Chang, C. C. [1987]. Numerical Solution of Stochastic Differential Differential Equations with Constant Coefficients. *Math. Comp.* **49**, 523–542.

[37] Chang, C. C. [1988]. Random Vortex Methods for the Navier-Stokes Equations. *J. Comp. Phys.* **76**, 281–300.

[38] Cheer, A. Y. [1983]. A Study of Incompressible 2-D Vortex Flow Past a Circular Cylinder. *SIAM J. Sci. Stat. Comput.* **4**, 685–705.

[39] Cheer, A. Y. [1989]. Unsteady Separated Wake Behind an Impulsively Started Cylinder. *J. Fluid Mech.* **201**, 485–505.

[40] Chiu, C. and Nicolaides, R. A. [1988]. Convergence of a Higher-Order Vortex Method for Two-Dimensional Euler Equations. *Math. Comp.* **51**, 507–534.

[41] Choi, Y., Humphrey, J. A. C. and Sherman, F. S. [1988]. Random Vortex Simulation of Transient Wall-Driven Flow in a Rectangular Enclosure. *J. Comp. Phys.* **75**, 359–383.

[42] Choquin, J. P. and Lucquin-Desreux, B. [1988]. Accuracy of a Deterministic Particle Method for the Navier-Stokes Equations. *Inter. J. Num. Meth. Fluids* **8**, 1439–1458.

[43] Choquin, J. P. and Huberson, S. [1989]. Particle Simulations of Viscous Flows. *preprint* 27 pages.

[44] Chorin, A. J. [1973]. Numerical Study of Slightly Viscous Flow. *J. Fluid Mech.* **57**, 785–796.

[45] Chorin, A. J. [1978]. Vortex Sheet Approximation of Boundary Layers. *J. Comp. Phys.* **27**, 428–442.

[46] Chorin, A. J. [1980]. Vortex Models and Boundary Layer Instability. *SIAM J. Sci. Stat. Comput.* **1**, 1–21.

[47] Chorin, A. J. [1982]. The Evolution of a Turbulent Vortex. *Commun. Math. Phys.* **35**, 517–535.

[48] Chorin, A. J. and Bernard, P. S. [1973]. Discretization of a Vortex Sheet with an Example of Roll-up. *J. Comp. Phys.* **13**, 423–429.

[49] Chorin, A. J. and Marsden, J. E. [1990]. *A Mathematical Introduction to Fluid Mechanics*, Springer-Verlag, New York.

[50] Christiansen, J. P. [1973]. Numerical Simulation of Hydrodynamics by the Method of Point Vortices. *J. Comp. Phys.* **13**, 363–379.

[51] Cottet, G. H. [1982]. Methodes Particulaires pour l'equation d'Euler dans le plan. *These de 3eme cycle*, l'Universite Pierre et Marie Curie, Paris VI.

[52] Cottet, G. H. [1987]. Convergence of a Vortex in Cell Method for the Two Dimensional Euler Equations. *Math. Comput.* **49**, 407–425.

[53] Cottet, G. H. [1988a]. Boundary Conditions and Deterministic Vortex Methods for the Navier-Stokes Equations. *Mathematical Aspects of Vortex Dynamics* (Ed. by R. E. Caflisch), S.I.A.M., Philadelphia, 128–143.

[54] Cottet, G. H. [1988b]. A New Approach to the Analysis of Vortex Methods in Two and Three Dimensions. *Ann. Inst. H. Poincare Anal. non Lineaire* **5**, 227–285.

[55] Cottet, G. H. [1990]. A Particle-Grid Superposition Method for the Navier-Stokes Equations. *J. Comp. Phys.* **89**, 301–318.

[56] Cottet, G. H. [1991]. Large Time Behaviour for Deterministic Particle Approximations to the Navier-Stokes Equations. *Math. Comput.* **56**, 45–60.

[57] Cottet, G. H., Goodman, J. and Hou, T. Y. [1991]. Convergence of the Grid-Free Point Vortex Method for the Three Dimensional Euler Equations. *SIAM J. Numer. Anal.* **28**, 291–307.

[58] Cottet, G. H. and Mas-Gallic, S. [1990]. A Particle Method to Solve the Navier-Stokes System. *Numer. Math.* **57**, 805–827.

[59] De Boor, C. [1978]. *A Practical Guide to Splines*, Springer-Verlag, New York.

[60] Ebin, D. and Marsden, J. E. [1970]. Groups of Diffeomorphisms and the Motion of an Incompressible Fluid. *Ann. of Math.* **92**, 102–163.

[61] Fishelov, D. [1990a]. Vortex Methods for Slightly Viscous Three Dimensional Flow. *SIAM J. Sci. Stat. Comput.* **11**, 399–424.

[62] Fishelov, D. [1990b]. A New Vortex Scheme for Viscous Flows. *J. Comp. Phys.* **86**, 211–224.

[63] Folland, G. B. [1976]. *A Introduction to Partial Differential Equations*, Princeton University Press, Princeton, New Jersey.

[64] Garabedian, P. R. [1986]. *Partial Differential Equations* Chelsea, New York.

[65] Ghoniem, A. F., Chorin, A. J. and Oppenheim, A. K. [1982]. Numerical Modelling of Turbulent Flow in a Combustion Tunnel. *Phil. Trans. R. Soc. Lond. A* **304**, 303–325.

[66] Ghoniem, A. F. and Sherman, F. S. [1985]. Grid Free Simulation of Diffusion Using Random Walk Methods. *J. Comp. Phys.* **61**, 1–37.

[67] Goodman, J. [1987]. Convergence of the Random Vortex Method. *Commun. Pure Appl. Math.* **40**, 189–220.

[68] Goodman, J., Hou, T. Y. and Lowengrub, J. [1990]. The Convergence of the Point Vortex Method for the 2-D Euler Equations. *Commun. Pure Appl. Math.* **43**, 415–430.

[69] Goodman, J. and Hou, T. Y. [1991]. New Stability Estimates for the 2-D Vortex Method. *Commun. Pure Appl. Math.* **44**, 1015–1031.

[70] Greenbaum, A., Greengard, L. and McFadden, G. B. [1991] Laplace's Equation and the Dirichlet-Neumann Map in Multiply Connected Domains. NYU Research Report, DOE/ER/25053-5 27 pages.

[71] Greengard, C. [1985]. The Core Spreading Vortex Method Approximates the Wrong Equation. *J. Comput. Phys.* **61**, 345–347.

[72] Greengard, C. [1986]. Convergence of the Vortex Filament Method. *Math. Comp.* **47**, 387–398.

[73] Greengard, L. [1988]. Potential Flow in Channels. *SIAM J. Sci. Stat. Comput.* **11**, 603–620.

[74] Greengard, L. [1990]. The Numerical Solution of the N-Body Problem. *Computers in Physics*, Mar/Apr, 142–152.

[75] Greengard, L. and Gropp, W. D. [1990]. A Parallel Version of the Fast Multipole Method. *Computers Math. Applic.* **20**, 63–71.

[76] Greengard, L. and Rokhlin, V. [1987]. A Fast Algorithm for Particle Simulations. *J. Comput. Phys.* **73**, 325–348.

[77] Greengard, L. and Rokhlin, V. [1988a] The Rapid Evaluation of Potential Fields in Three Dimensions. *Vortex Methods* (Ed. by C. Anderson and C. Greengard), Lecture Notes in Mathematics **1360**, Springer Verlag, New York. 121–141.

[78] Greengard, L. and Rokhlin, V. [1988b] On the Efficient Implementation of the Fast Multipole Algorithm. Research Report, YALEDU/DCS/RR-602, 17 pages.

[79] Jolles, A. and Huberson, S. [1989]. Numerical Simulation of Navier-Stokes Equations by Means of Particle/ Mesh Methods. *LIMSI Research Report* **89-4**, 1–47.

[80] Hald, O. H. [1979]. Convergence of Vortex Methods for Euler's Equations II. *SIAM J. Numer. Anal.* **16**, 726–755.

[81] Hald, O. H. [1986]. Convergence of a Random Method with Creation of Vorticity. *SIAM J. Sci. Stat. Comput.* **7**, 1373–1386.

[82] Hald, O. H. [1987]. Convergence of Vortex Methods for Euler's Equations III. *SIAM J. Numer. Anal.* **24**, 538–582.

[83] Hald, O. H. [1991]. Convergence of Vortex Methods. *Vortex Methods and Vortex Motion* (Ed. by K. Gustafsson and J. Sethian), SIAM, Philadelphia, 33–58.

[84] Hald, O. H. and del Prete, V . M. [1978]. Convergence of Vortex Methods for Euler's Equations. *Math. Comp.* **32**, 791–809.

[85] Handscomb, D. C. and Hammersley, J. M. [1964]. Monte Carlo Methods. Methuen & Co. Ltd., London.

[86] Henshaw, W., Kreiss, H. O. and Reyna, L. G. [1991]. On Smallest Scale Estimates and a Comparison of the Vortex Method to the Pseudo-spectral Method. *Vortex Dynamics and Vortex Methods* (Ed. by C. Anderson and C. Greengard), Lectures in Applied Mathematics **28**, American Mathematical Society, Providence, RI, 303–325.

[87] Hockney, R. W., Goel, S. P. and Eastwood, J. W. [1974]. Quiet High-Resolution Computer Models of a Plasma. *J. Comp. Phys.* **14**, 148–158.

[88] Hockney, R. W., and Eastwood, J. W. [1981]. *Computer Simulations Using Particles* McGraw Hill, New York.

[89] Hou, T. Y. [1992]. A New Desingularization Method for Vortex Methods. *Math. Comp.* (to appear).

[90] Hou, T. Y. [1991b]. A Survey on Convergence Analysis for Point Vortex Methods. *Vortex Dynamics and Vortex Methods* (Ed. by C. Anderson and C. Greengard), Lectures in Applied Mathematics **28**, American Mathematical Society, Providence, RI, 327–339.

[91] Hou, T. Y. and Lowengrub, J. [1990]. Convergence of a Point Vortex Method for the 3-D Euler Equations. *Commun. Pure Appl. Math.* **43**, 965–981.

[92] Hou, T. Y., Lowengrub, J. and Shelly, M. [1991]. Exact Desingularization and Local Regridding for Vortex Methods. *Vortex Dynamics and Vortex Methods* (Ed. by C. Anderson and C. Greengard), Lectures in Applied Mathematics **28**, American Mathematical Society, Providence, RI, 341–362.

[93] Hou, T. Y., and Wetton, B. T. R. [1990]. Convergence of a Finite Difference Scheme for the Navier-Stokes Equations using Vorticity Boundary Conditions.*SIAM J. Numer. Anal.* (to appear).

[94] Howell, L. H. [1990]. Computation of Conformal Maps by Modified Schwarz-Christoffel Transformations. *Ph.D Thesis*, MIT Math. Dept.

[95] Howell, L. H. and Trefethen, L. N. [1990]. A Modified Schwarz-Christoffel Transformation for Elongated Regions. *SIAM J. Sci. Stat. Comput.* **11**, 928–949.

[96] Kato, T. [1967] On Classical Solutions of the Two-Dimensional Non-Stationary Euler Equations. *Arch. Rat. Mech. Anal.* **25**, 188–200.

[97] Kato, T. [1972] Nonstationary Flows of Viscous and Ideal Fluids in R^3. *J. Funct. Anal.* **9**, 296–305.

[98] Knio, O. and Ghoniem, A. F. [1990]. Numerical Study of a Three Dimensional Vortex Method. *J. Comp. Phys.* **86**, 75–106.

[99] Krasny, R. [1986a]. Desingularization of Periodic Vortex Sheet Roll-up. *J. Comput. Phys.* **65**, 292–289.

[100] Krasny, R. [1986b]. A Study of Singularity Formation in a Vortex Sheet by the Point Vortex Approximation. *J. Fluid Mech.* **167**, 65–93.

[101] Krasny, R. [1987]. Computation of Vortex Sheet Roll-Up in the Trefftz Plane. J. Fluid Mech. **184**, 123–155.

[102] Kuwahara, K. and Takami, H. [1973] Numerical Studies of Two-Dimensional Vortex Motion by a System of Points J. Phys. Soc. Japan, **34**, 247–253

[103] Leonard, A. [1980]. Vortex Methods for Flow Simulation. *J. Comp. Phys.* **37**, 289–335.

[104] Leonard, A. [1985]. Computing Three Dimensional Flows with Vortex Elements. *Ann. Rev. Fluid Mech.* **17**, 523–559.

[105] Long, D. G. [1988]. Convergence of the Random Vortex Method in Two Dimensions. *J. A.M.S.* **4**, 779–804.

[106] Long, D. G. [1990]. Convergence of the Random Vortex Method in Three Dimensions. *Lawrence Berkeley Laboratory Report*, LBL-29832, 1–63.

[107] Lucquin-Desreux, B. [1987]. Particle Approximation of the Two-Dimensional Navier-Stokes Equations. *Rech. Aerosp.* **4**, 1–12.

[108] Majda, A. [1986] Vorticity and the Mathematical Theory of Incompressible Flow. *Commun. Pure Appl. Math.* **19**, (special volume) S187-S220.

[109] Majda, A. [1988] Vortex Dynamics: Numerical Analysis, Scientific Computing, and Mathematical Theory. *ICIAM '87: Proceedings of the First International Congress of Industrial and Applied Mathematics.* (Paris 1987) SIAM, 153–182.

[110] Maltz, F. H. and Hitzl, D. L. [1979]. Variance Reduction in Monte-Carlo Computations Using Multi-Dimensional Hermite Polynomials. *J. Comp. Phys.* **32**, 345–376.

[111] Marchioro, C. and Pulvirenti M. [1982] Hydrodynamics in Two Dimensions and Vortex Theory. *Commun. Math. Phys.* **84**, 483–503.

406   E. G. PUCKETT

[112] Marchioro, C. and Pulvirenti M. [1983] A Diffusion Process Associated to the Prandtl Equation. *J. Func. Anal.* **52**, 330–343.

[113] Marsden, J. E. and Tromba, I. [1981]. *Vector Calculus*, Freeman and Co., San Francisco.

[114] Mas-Gallic, S. [1990]. Deterministic Particle Method: Diffusion and Boundary Conditions. *Universite Pierre et Marie Curie, Centre National de la Recherche Scientifique*, R90029, 1–27.

[115] Mayo, A. [1985]. Fast High Order Accurate Solution of Laplace's Equation on Irregular Domains. *SIAM J. Sci. Stat. Comput.* **6**, 144–157.

[116] McGrath, F. J. [1968] On Classical Solutions of the Two-Dimensional Non-Stationary Euler Equations. *Arch. Rational Mech. Anal.* **27**, 329–348.

[117] Merriman, B. [1991]. Particle Approximation. *Vortex Dynamics and Vortex Methods* (Ed. by C. Anderson and C. Greengard), Lectures in Applied Mathematics **28**, American Mathematical Society, Providence, RI, 481–546.

[118] Milinazzo, F. and Saffman, P. G. [1977]. The Calculation of Large Reynolds Number Two-Dimensional Flow Using Discrete Vortices with Random Walk. *J. Comp. Phys.* **23**, 380–392.

[119] Nicolaides, R. A. [1986]. Construction of a High Order Accurate Vortex and Particle Methods. *Appld. Numer. Math.* **2**, 313–320.

[120] Nordmark, H. O. [1988]. Higher Order Vortex Methods with Rezoning. *Ph.D Thesis*, U. C. Berkeley Math. Dept.

[121] Nordmark, H. O. [1991]. Rezoning for Higher Order Vortex Methods. *J. Comp. Phys.* **97**, 366–397.

[122] Perlman, M. [1985]. On the Accuracy of Vortex Methods. *J. Comp.Phys.* **59**, 200–223.

[123] Puckett, E. G. [1987]. A Study of Two Random Walk Methods and Their Rates of Convergence. *Ph.D Thesis*, U. C. Berkeley Math. Dept.

[124] Puckett, E. G. [1989a]. A Study of the Vortex Sheet Method and Its Rate of Convergence. *SIAM J. Sci. Stat. Comput.* **10**, 298–327.

[125] Puckett, E. G. [1989b]. Convergence of a Random Particle Method to Solutions of the Kolmogorov Equation $u_t = \nu u_{xx} + u(1 - u)$. *Math. Comp.* **52**, 615–645.

[126] Raviart, P. A. [1985]. An Analysis of Particle Methods. *Numerical Methods in Fluid Dynamics* (Ed. by F. Brezzi), Lecture Notes in Mathematics **1127**, Springer Verlag, New York, 245–324.

[127] Quartapelle, L. [1981]. Vorticity Conditioning in the Computation of Two Dimensional Viscous Flows. *J. Comp. Phys.* **40**, 453–477.

[128] Quartapelle, L. and Valz-Griz, F. [1981]. Projection Conditions on the Vorticity in Viscous Incompressible Flows. *Int. J. Numer. Methods Fluids.* **1**, 129–144.

[129] Roberts, S. G. [1985]. Accuracy of the Random Vortex Methods for a Problem with Non-smooth Initial Conditions. *J. Comp. Phys.* **58**, 29–43.

[130] Roberts, S. G. [1989]. Convergence of a Random Walk Method for the Burgers Equation. *Math. Comp.* **52**, 647–673.

[131] Rokhlin, V [1985] Rapid Solution of Integral Equations of Classical Potential Theory. *J. Comput. Phys.* **60**, 187–207.

[132] Rosenhead, L. [1931]. The Formation of Vortices from a Surface of Discontinuity. *Proc. R. Soc. Lon.,* A **134**, 170–192.

[133] Schlichting, H. [1979]. *Boundary-Layer Theory*, McGraw Hill, New York.

[134] Schultz, M. H. [1973]. *Spline Analysis*, Prentice-Hall Inc., Englewood Cliffs, NJ

[135] Sethian, J. A. [1984]. Turbulent Combustion in Open and Closed Vessels. *J. Comp. Phys.* **54**, 425–456.

[136] Sethian, J. A. [1991]. A Brief Overview of Vortex Methods. *Vortex Methods and Vortex Motion* (Ed. by K. Gustafasson and J. Sethian), SIAM, Philadelphia, 1–32.

[137] Sethian, J. A. and Ghoniem, A. F. [1988]. Validation Study of Vortex Methods. *J. Comp. Phys.* **74**, 283–317.

[138] Siggia, E. D. and Aref, H. [1981]. Point Vortex Simulation of the Inverse Energy Cascade in Two-Dimensional Turbulence. *Phys. Fluids* **24**, 171–173.

[139] Summers, D. M. [1989]. A Random Vortex Simulation of Falkner-Skan Boundary Layer Flow. *J. Comp. Phys.* **85**, 86–103.

[140] Summers, D. M., Hanson, T. and Wilson, C. B. [1985]. A Random Vortex Simulation of Wind Flow Over a Building. *Int. J. Num. Meth. Fluids* **5**, 849–871.

[141] van Dommelen, L. and Rundensteiner, E. A. [1989]. Fast Adaptive Summation of Point Forces in the Two Dimensional Poisson Equation. *J. Comp. Phys.* **83**, 126–147.

[142] Tennekes, H. and Lumley, J. L. [1972]. *A First Course in Turbulence*, MIT Press, Cambridge, MA.

[143] Tiemroth, E. [1986]. The Simulation of the Viscous Flow Around a Cylinder by the Random Vortex Method. *Ph.D Thesis*, U. C. Berkeley Naval Arch. Dept.

[144] White, F. M. [1974]. *Viscous Fluid Flow*, McGraw Hill, New York.

[145] Ying, L. A. [1990]. Viscous Splitting for the Unbounded Problem of the Navier-Stokes Equations. *Math. Comp.* **55**, 89–113.

[146] Zabusky, N. J., Hughes, M. H. and Roberts, K. V. [1979]. Contour Dynamics for the Euler Equations in Two Dimensions. *J. Comp. Phys.* **30**, 96–106.

[147] Zabusky, N. J., and Overman, E. A. [1983]. Regularization and Contour Dynamical Algorithms I. Tangential Regularization. *J. Comp. Phys.* **52**, 351–373.

[148] Zhu, J. [1989]. An Elliptical Vortex Method. *Ph.D Thesis*, NYU Courant Institute.

12 New Emerging Methods in Numerical Analysis: Applications to Fluid Mechanics

Roger Temam

Summary

Our object in this article is to present some aspects of new emerging methods in numerical analysis and their application to Computational Fluid Dynamics. These methods stem from Dynamical Systems Theory.

The attractor describing a turbulent flow is approximated by smooth manifolds and by projecting the Navier-Stokes equations onto these manifolds we obtain new algorithms, the Inertial Projections. These algorithms have proven to be stable, efficient and well suited for long time integration of the equations. They can be implemented with all forms of spatial discretizations, spectral methods, finite elements and finite differences (possibly also wavelets).

12.1 Introduction

Algorithms that have been introduced at a time of scarce computing resources may not be well adapted to supercomputing and to the more difficult problems that are tackled at the present time or that we foresee for the near future.

For incompressible fluid mechanics (or thermal convection), by using the full Navier-Stokes equations (NSE) we can, at present time, compute flows at the onset of turbulence: in particular the permanent regime is not anymore a stationary flow. The flow can be time periodic if a Hopf bifurcation has occurred or the permanent regime can be an even more complicated flow. Examples of time dependent flows have been numerically computed in the case of the driven cavity (see e.g. [15], [16], [20], [4] and [34]) and for other types of flows (see e.g. [14], [33] and [30]). Section 12.2 of this article is a brief survey, for the CFD practitioner, of some basic and relevant concepts in Dynamical Systems Theory (behavior for large time of the solutions of the NSE).

Although low dimensional turbulence (chaos) has been observed in some flows it is likely that a fine resolution and a large number of parameters are necessary for most turbulent flows, in particular those close to or related to industrial applications. There is some evidence (see Section 12.3) that when a large number of unknowns is used, it is not desirable nor economical to treat all the unknowns in the same way. *We rather*

409

advocate here the decomposition of the set of unknowns into two or several arrays of unknowns that have different orders of magnitude and that should be therefore treated differently. It is interesting that some theoretical questions in dynamical systems theory (approximations of the attractor) can help for such computational problems; namely they provide a methodology for the treatment of different arrays of unknowns with different orders of magnitude.

The decomposition of the unknowns into two (or more) arrays corresponds in fluid mechanics to the decomposition of the velocity vector field into large and small eddies, i.e.,

$$\mathbf{u} = \mathbf{y} + \mathbf{z}, \tag{12.1.1}$$

or

$$\mathbf{u} = \mathbf{y} + \mathbf{z}_1 + \cdots + \mathbf{z}_r. \tag{12.1.2}$$

In Section 12.4 we describe appropriate decompositions of the vector field $\mathbf{u}$ into small and large eddies, like (12.1.1), for spectral methods, finite elements and finite differences. These decompositions are essential for the construction of Approximate Inertial Manifolds: these are smooth manifolds approximating the attractor which have typically an equation of the form (in case (12.1.1)):

$$\mathbf{z} = \mathbf{\Phi}(\mathbf{y}). \tag{12.1.3}$$

When the law (12.1.3) is enforced, the small and large eddies $\{\mathbf{y}, \mathbf{z}\}$ are treated differently, the small ones being slaved by the large ones through (12.1.3). We call Inertial Projections any numerical method that amounts to projecting the equations on an approximate inertial manifold like (12.1.3).

In Section 12.5 we discuss the Nonlinear Galerkin Method: this is an inertial projection method in the context of (pseudo-) spectral methods. It has been already successfully applied to the computation of the Navier-Stokes equations in space dimension 2 and 3. The method has proved to be very efficient in this case. Its utilization has resulted in an increased stability and in a significant reduction in CPU computing time.

Finally in Section 12.6 we discuss the IU method (Incremental Unknowns Method). With finite differences, using the concept of incremental unknowns (see [37]), one can obtain a decomposition of $\mathbf{u}$ into small and large eddies as in (12.1.1). The corresponding inertial projection is described in Section 12.6.

12.2 Computing Turbulent Flows Using the Navier-Stokes Equations

A turbulent flow can be a forced flow or a non-forced flow. In the latter case and assuming that the fluid fills a bounded domain of the space, turbulence decays due to viscosity and the fluid tends to rest, during a time of order, at most, $0(Re)$, Re being the Reynolds number and the time being nondimensionalized. Here we will be interested in forced flows, where turbulence is maintained so that the fluid never comes to rest. Furthermore we will assume that the forcing is time independent which corresponds to most laboratory experiments in fluid mechanics; interesting examples of time dependent forcings may occur in problems related to the control of fluids (see e.g. [1], [19], [33]) but this question will not be addressed here.

The study of the long time behavior of a forced flow is closely related to turbulence and this raises difficult problems in fluid mechanics and mathematics. In order to describe the relation between long time behavior and turbulence (and computing), let us recall one of the typical experiments in fluid mechanics, the flow past a sphere.

A sphere is embedded in an incompressible fluid with velocity $\mathbf{U}_\infty$ at infinity. Most commonly the sphere lies in a wind tunnel, the fluid is air and $\mathbf{U}_\infty$ is the velocity at the entrance of the wind tunnel, $\mathbf{U}_\infty$ is the forcing. The permanent regime which is observed after the initial transient period when the wind tunnel is started depends on $\mathbf{U}_\infty$ or, alternatively on the nondimensional Reynolds number

$$Re = \frac{\nu \ U_\infty}{R},$$

ν the kinematic viscosity, R the radius of the sphere, $U_\infty =\mid \mathbf{U}_\infty \mid$ the modulus of $\mathbf{U}_\infty$. In a practical experiment the apparatus (ν and R) remains unchanged, and we increase Re by increasing the speed U_∞.

In the nonforced case, $U_\infty = 0$, then whatever the initial distribution of velocities around the sphere, the flow will come to rest after a certain time. Figure 12.1 shows the permanent regimes that are observed for increasing values of U_∞ (or Re). If Re is sufficiently small say $Re \leq Re_1$ the flow becomes steady; there is actually a unique stationary solution $\mathbf{u_s}$ satisfying the required boundary conditions and whatever the initial distribution of velocities, the flow converges to $\mathbf{u_s}$ after a certain time ($t \rightarrow \infty$). Furthermore this flow is laminar (Figure 12.1 (a)).

If Re is increased, say $Re_1 \leq Re \leq Re_2$, then again the flow becomes stationary after the initial transient period (Figure 12.1 (b)). The stationary solution is however more complicated than before; in particular von Kármán vortices have appeared behind the sphere. From the mathematical point of view the stationary solution is not unique anymore, there has been a bifurcation of stationary solutions.

In the next step, $Re_2 \leq Re \leq Re_3$, after a Hopf bifurcation has occurred, the flow never becomes stationary again; in fact the permanent regime is a time periodic one. Here the von Kármán vortices move periodically to the right and vanish, while reappearing on the left in a seemingly time-periodic manner. If U_∞ is larger, say $Re_3 \leq Re \leq Re_4$,

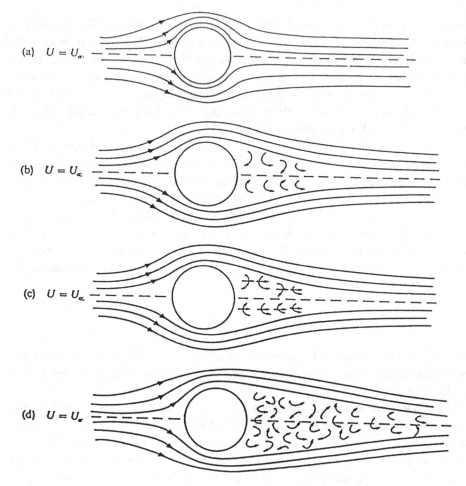

Figure 12.1: Flow past a sphere.

the displacement of the von Kármán vortices to the right is less regular and corresponds to a seemingly quasi-periodic flow. Finally if U_∞ is sufficiently large, $Re > Re_4$, (Figure 12.1 (d)) a fully turbulent completely unstructured flow appears in the wake of the sphere. The flow is nonstationary for all time and can be described only, from the mathematical point of view, by the attractor that it defines.

Although the above description is a simplified one, it is typical of many flows, laboratory flows (driven cavity, Poiseuille, Couette-Taylor, Bénard problem ...) as well as industrial problems. The present computing power allows us to compute nonstationary flows i.e., permanent regimes that are time dependent while the driving forces are time independent. Describing and obtaining a nonstationary flow (permanent regime) is more difficult than describing and obtaining a stationary solution. This poses new challenges in computing. Indeed a stationary flow is described by a single velocity vector field and the convergence to it is exponential in time. While a time dependent solution is described by a family of vector fields and the time convergence to such a solution may not be exponential and may even be slow (see [38]).

Another difficulty in computing turbulent flows is that the permanent regime could depend on the initial data [38]. For a given apparatus (ν and the geometry fixed) and for given driving forces, the permanent regime that we observe depends on the initial data. From the theoretical viewpoint there is a global attractor in the phase space that encompasses all the permanent regimes; the global attractor comprises "smaller" attractors that possess their individual basin of attraction: for a given initial distribution of velocities, the flow converges to the permanent regime (attractor) assigned by its basin of attraction.

Similarly if other parameters are perturbed (e.g. the shape of the domain filled by the fluid, the boundary velocity or the volume forces) the permanent regime that is observed does not depend only on the final values of the parameters; it depends also on the route used to reach these values. No theoretical proof of this fact is available but this is a well accepted concept in dynamical systems theory and overwhelming experimental evidences are available; see for example the work of T. B. Benjamin and his collaborators in the case of the flow between rotating cylinders.

12.3 Large Scale Scientific Computing

Large scale scientific computing requires/allows:

- the utilization of a large number of spatial variables
- the integration of evolution equations on large intervals of time.

For example in fluid mechanics, it is not anymore exceptional to compute two-dimensional flows using 1024^2 ($\simeq 10^6$) modes or to compute three dimensional flows using 256^3 ($\simeq 17 \times 10^6$) modes.

An indication of the number of parameters that is needed to describe a turbulent flow is given by the Kolmogorov and Kraichnan theories of turbulence for space dimension 3 and 2 ([27], [28]). These figures are now confirmed by the estimates on the dimension of the attractor [9, 10], [8] (see also the estimates on the smallest scale [22], [3]).

We infer from these results that, with spectral methods, in order to completely resolve a turbulent flow, we need:

- Re modes in space dimension 2,
- $Re^{\frac{9}{4}}$ modes in space dimension 3,

where the Reynolds number Re varies between 10^4 and 10^8 for interesting applications. With finite differences, we need a mesh:

- $\Delta x \simeq L_\eta \simeq Re^{\frac{1}{2}}$ in dimension 2,
- $\Delta x \simeq L_d \simeq Re^{\frac{3}{4}}$ in dimension 3.

Here L_d is the Kolmogorov dissipation length and L_η is the Kraichnan dissipation length.

Of course such resolutions are still beyond the computing and memory capacities of the present computers and will remain so in a foreseeable future. However very fine resolutions are already used nowadays (10^6 modes is not an exceptional resolution as mentioned above) and such resolutions necessitate already some precautions.

For instance with spectral methods, it is common that half of the modes (the small ones) carry less than 10^{-4} to 10^{-6} of the total energy. *Hence the computer tends to be saturated by small wavelengths carrying little energy.* Similarly with finite differences the unknown varies little between two successive grid points when $\Delta x \simeq 10^{-3}$!

Figure 12.2 displays the result of a randomly chosen but typical computation. Here a two-dimensional space-periodic flow was computed with N^2 modes, $N = 128$. We call **y** the component of the velocity corresponding to the modes $e^{ijx}, |j_\alpha| \leq \frac{1}{2}N_1$ and **z** the component of the velocity corresponding to the other modes. Hence **y** involves N_1^2 modes and **z** involves $N^2 - N_1^2$ modes.

Figure 12.2 shows that the ratio of the L^2 norms of **z** and **y** is of order 10^{-6} for $N_1 = 80$ and of order 10^{-8} for $N_1 = 96$. *In the first case 51% of the modes carry about 10^{-6} of the total velocity and in the second case 44% of the modes carry about 10^{-8} of the total velocity.* If kinetic energies were to be considered then we must consider the ratio of the squares of the L^2-norms: for $N_1 = 80$, 51% of the modes carry 10^{-12} of the total kinetic energy and for $N_1 = 96$, 44% of the modes carry 10^{-16} of the total kinetic energy. Taking into account the round-off error of the machine the kinetic energy of **z** is essentially 0 in this case. Note also that the ratio

$$\frac{\| \mathbf{z} \|_{L^2}}{\| \mathbf{y} \|_{L^2}} \tag{12.3.1}$$

is much smaller than the Reynolds number equal, in this case, to 10^{-2}.

This remark is in no way a criticism of the computations that have been done in the past. The appearance of supercomputers is recent. At the beginning the emphasis was on software since it was necessary to become acquainted with the utilization of

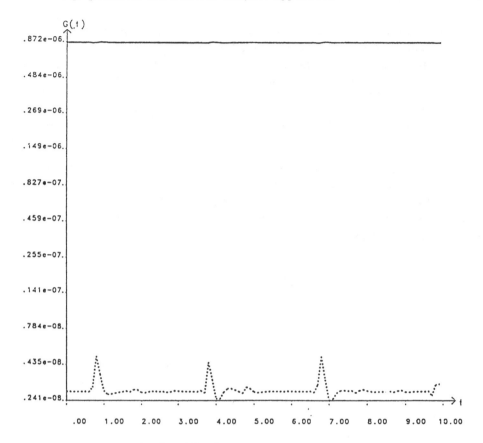

Figure 12.2: The ratio $\| z \|_{L^2} / \| y \|_{L^2}$ as a function of time for $N_1 = 80$ (solid line) and $N_1 = 96$ (dotted line).

the supercomputers and with vectorization or parallelization of the codes. These were necessary steps, but it may be time now to reconsider the numerical algorithms.

In conclusion, the ratio (12.3.1) being so small *there is, in large scale computing, a stiffness built into the discrete system. It is our belief that this stiffness problem should be addressed directly, and that the components of the unknown function* **u** *should be decomposed into arrays of different orders of magnitude that are treated differently.*

12.4 Small and Large Eddies

The interactions of small and large eddies in a turbulent flow are an important part of turbulence and, trying to understand these interactions, is important for the understanding of turbulence. There is some analogy here with the above mentioned difficulty in supercomputing of the handling of arrays of unknowns of different orders of magnitude.

The decomposition of the vector field **u** into small and large wavelengths is obvious when a spectral representation of **u** is available, i.e.,

$$\mathbf{u}(x,t) = \sum_{j=1}^{\infty} g_j(t)\varphi_j(x). \qquad (12.4.1)$$

In (12.4.1) the φ_j can be eigenfunctions of any type, such as sines and cosines for the space periodic case, Tchebychev or Legendre polynomials [17], [5], the empirical eigenfunctions [2], [35] or the eigenfunctions of a linear nonself adjoint operator such as the Rossby and gravity waves in meteorology [21]. In all such cases a decomposition of **u** into small and large eddies

$$\mathbf{u} = \mathbf{y} + \mathbf{z} \qquad (12.4.2)$$

can be naturally obtained by choosing a cut-off number m and setting

$$\mathbf{y} = \sum_{j=1}^{m} g_j\varphi_j, \quad \mathbf{z} = \sum_{j=m+1}^{\infty} g_j\varphi_j. \qquad (12.4.3)$$

When finite elements or finite differences are used decompositions of type (12.4.2) can be obtained although they become then less obvious. With finite elements they are based on the concept of hierarchical bases that is recalled below while in finite differences decompositions such as (12.4.2) are based on the concept of Incremental Unknowns (introduced in [37] see Section 12.6 hereafter). However there is a difference between the latter decompositions of **u** and (12.4.3): for (12.4.3) **y** and **z** correspond to small and large wavelengths i.e., the decomposition occurs in the spectral space. In finite elements and finite differences the decomposition (12.4.3) into large and small eddies occurs in the physical space. We expect the wavelets to provide a decomposition into small and large eddies both in the spectral and physical spaces. We will not address the utilization of wavelets here; for a preliminary work see [18].

12.4.1 Functional Form of the Navier-Stokes Equations

The incompressible Navier-Stokes equations are written as usual

$$\frac{\partial \mathbf{u}}{\partial t} - \nu\Delta\mathbf{u} + (\mathbf{u}\cdot\nabla)\mathbf{u} + \nabla p = \mathbf{f} \qquad (12.4.4)$$

$$\nabla\cdot\mathbf{u} = 0 \qquad (12.4.5)$$

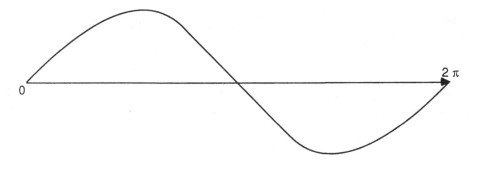

Figure 12.3: Small and large wavelengths for Fourier Series.

where $\nu > 0$ is the kinematic viscosity, p is the kinematic pressure and $\mathbf{f}$ represents volume forces; $\mathbf{u} = \mathbf{u}(x, t)$ is the velocity vector, $\mathbf{u} = (u_1, u_2)$ in space dimension $n = 2$ and $\mathbf{u} = (u_1, u_2, u_3)$ in space dimension $n = 3$. We either assume that the fluid fills a bounded domain Ω with a solid boundary Γ at rest so that

$$\mathbf{u} = 0 \quad \text{on } \Gamma ; \tag{12.4.6}$$

or that the flow takes place in $\mathcal{R}^n, n = 2$ or 3 and is space periodic with period 2Π in each direction x_1, x_2 and x_3; in this case we call Ω the period, $\Omega = (0, 2\Pi)^n$.

We know that these equations amount to an evolution equation for $\mathbf{u}$ in a suitable Hilbert space H,

$$\mathbf{u} : t\epsilon\mathcal{R} \longrightarrow \mathbf{u}(t)\epsilon H$$

$$\frac{d\mathbf{u}}{dt} + \nu A\mathbf{u} + B(\mathbf{u}, \mathbf{u}) = \mathbf{f} \tag{12.4.7}$$

where A is the Stokes operator associated to the boundary condition and B accounts for the inertial (nonlinear) term.

We restrict ourselves temporarily to the case where the φ_j in (12.4.1)–(12.4.3) are the eigenvectors of A, $\varphi_j = \mathbf{w}_j$,

$$A\mathbf{w}_j = \lambda_j \mathbf{w}_j, \quad j \geq 1, \quad 0 < \lambda_1 \leq \lambda_2 \leq \cdots, \quad \lambda_j \to \infty \quad \text{as} \quad j \to \infty. \quad (12.4.8)$$

Then if $P = P_m$ is the projection in H onto the space spanned by $\{\mathbf{w}_1, \cdots, \mathbf{w}_m\}$ and $Q = Q_m = I - P_m$, we can rewrite (12.4.7) as an equivalent coupled system for $\mathbf{y}$ and $\mathbf{z}$

$$\frac{d\mathbf{y}}{dt} + \nu A\mathbf{y} + PB(\mathbf{y} + \mathbf{z}, \mathbf{y} + \mathbf{z}) = Pf \quad (12.4.9)$$

$$\frac{d\mathbf{z}}{dt} + \nu A\mathbf{z} + QB(\mathbf{y} + \mathbf{z}, \mathbf{y} + \mathbf{z}) = Qf. \quad (12.4.10)$$

12.4.2   Approximate Inertial Manifolds

An inertial manifold is a manifold of equation

$$\mathbf{z} = \Phi(\mathbf{y}). \quad (12.4.11)$$

When it exists it provides an interaction law between the small and large eddies; physicists say that the small eddies $\mathbf{z}$ are slaved by the large ones. The existence of inertial manifolds has been proved for certain dissipative evolution equations but in the case of the Navier-Stokes equations their existence has not been proved nor disproved, even in space dimension two.

Instead the existence of Approximate Inertial Manifolds (AIMs) has been proven. A methodology was developed in [12] and [36] for the construction of AIMs. The simplest AIM appears in [12]; its equation is of the form (12.4.11), more precisely (compare to (12.4.10))

$$\nu A\mathbf{z} + QB(\mathbf{y}, \mathbf{y}) = Qf \quad (12.4.12)$$

or

$$\mathbf{y} = \Phi_1(\mathbf{y}) = (\nu A)^{-1}(Qf - QB(\mathbf{y}, \mathbf{y})). \quad (12.4.13)$$

The manifold $\mathcal{M}_1$ of equation (12.4.12), (12.4.13) is an approximation of the attractor describing the dynamics of (12.4.7) (or (12.4.9), (12.4.10)). It captures rapidly all orbits in a thin neighborhood and of course the attractor itself is included in this neighborhood. If η is the thickness of this neighborhood then the approximate interaction law that is enforced during the permanent regime is

$$\| \mathbf{z}(t) - \Phi_1(\mathbf{y}(t)) \|_H \leq \eta, \quad \forall t.$$

Another approximate inertial manifold (AIM) appears in [40] while [36] contains an infinite sequence of AIMs which give better and better approximations to the attractor.

The construction of the AIMs in [12] and [36] is based on the fact, proved in [12], that $\mathbf{z}$ is much smaller than $\mathbf{y}$ after a transient time, whatever the initial amplitude of $\mathbf{z}$:

$$\mathbf{z} \ll \mathbf{y} . \tag{12.4.14}$$

Of course (12.4.14) holds for m sufficiently large. For that reason

$$QB(\mathbf{y} + \mathbf{z}, \mathbf{y} + \mathbf{z}) \text{ is comparable to } QB(\mathbf{y}, \mathbf{y}) .$$

Also for other reasons (in particular statistical ones), $\frac{d\mathbf{z}}{dt}$ can be neglected in (12.4.10). In this way (12.4.10) yields (12.4.12).

Another approximate inertial manifold $\mathcal{M}_2$ close to the previous one is a manifold containing the stationary solutions (see [13], [40]) and its equation reads

$$\nu A\mathbf{z} + QB(\mathbf{y} + \mathbf{z}, \mathbf{y} + \mathbf{z}) = Q\mathbf{f} . \tag{12.4.15}$$

The numerical algorithms that are obtained by projecting the Navier-Stokes equations (or other evolution equations) onto an Inertial Manifold are called Inertial Projections. In the spectral case the method is called Nonlinear Galerkin Method. In the finite differences case the Incremental Unknowns are used and the method is called the IU Method.

In Section 12.5 we describe the Nonlinear Galerkin Method obtained by projecting the Navier-Stokes equations onto $\mathcal{M}_1$. Then in Section 12.6 we describe the inertial projections in the context of finite differences. Approximate inertial manifolds and inertial projections in the context of finite elements can be constructed by using hierarchical bases; we refer the reader to [32] for general equations and to [29] for the Navier-Stokes equations.

12.5 The Nonlinear Galerkin Method

We consider here the space periodic case. Fourier series are used and in particular we write the Fourier series expansion of $\mathbf{u}$

$$\mathbf{u}(x, t) = \sum_{j \varepsilon Z^n} \mathbf{u}_j(t)e^{ijx} . \tag{12.5.1}$$

For a pseudo-spectral computation we truncate the series at some fixed integer N and consider the approximate function

$$\mathbf{u} \simeq \mathbf{u}_N = \sum_{\substack{|j_\ell| \le N \\ \text{for each } \ell}} u_j(t)e^{ij \cdot x} . \tag{12.5.2}$$

The usual Galerkin Method (spectral method) based on the modes $| \ell | \leq N$ consists in solving

$$\frac{\partial \mathbf{u}_N}{\partial t} - \nu \Delta \mathbf{u}_N + P_N((\mathbf{u}_N \cdot \nabla)\mathbf{u}_N) + \nabla p_N = P_N f, \qquad (12.5.3)$$

where P_N is the truncation for the modes $| \ell | \leq N$. The pseudo-spectral method implies also a collocation treatment of the nonlinear term that will not be discussed here.

Let now $N_1 < N$ be another integer and consider the decomposition of $\mathbf{u}_N$ of the form

$$\mathbf{u}_N = \mathbf{y}_N + \mathbf{z}_N, \qquad (12.5.4)$$

$$\mathbf{y}_N = \sum_{\substack{| j_\ell | \leq N_1 \\ \text{for each } \ell}} \mathbf{u}_j(t) e^{ij \cdot x},$$

$$\mathbf{z}_N = \sum_{\substack{| j_\ell | > N_1 \text{ for some } \ell \\ | j_\ell | \leq N \text{ for each } \ell}} \mathbf{u}_j(t) e^{ij \cdot x}. \qquad (12.5.5)$$

Rewriting (12.4.2) in terms of $\mathbf{y}_N$ and $\mathbf{z}_N$ we obtain the coupled system

$$\frac{\partial \mathbf{y}_N}{\partial t} - \nu \Delta \mathbf{y}_N + P_{N_1}\{ ((\mathbf{y}_N + \mathbf{z}_N) \cdot \nabla)(\mathbf{y}_N + \mathbf{z}_N) + \nabla p_N\} = P_{N_1}\mathbf{f}, \qquad (12.5.6)$$

$$\frac{\partial \mathbf{z}_N}{\partial t} - \nu \Delta \mathbf{z}_N + (P_N - P_{N_1})\{ ((\mathbf{y}_N + \mathbf{z}_N) \cdot \nabla)(\mathbf{y}_N + \mathbf{z}_N) + \nabla p_N\} = (P_N - P_{N_1})\mathbf{f}. \qquad (12.5.7)$$

In the Nonlinear Galerkin Method based on $\mathcal{M}_1$ (see [31]), equations (12.5.6), (12.5.7) are replaced by

$$\frac{\partial \mathbf{y}_N}{\partial t} - \nu \Delta \mathbf{y}_N + P_{N_1}\{ (\mathbf{y}_N \cdot \nabla)\mathbf{y}_N + (\mathbf{y}_N \cdot \nabla)\mathbf{z}_N + (\mathbf{z}_N \cdot \nabla)\mathbf{y}_N + \nabla p_N\} = P_{N_1}\mathbf{f}, \qquad (12.5.8)$$

$$- \nu \Delta \mathbf{z}_N + (P_N - P_{N_1})((\mathbf{y}_N \cdot \nabla)\mathbf{y}_N + \nabla p_N) = (P_N - P_{N_1})\mathbf{f}. \qquad (12.5.9)$$

The solution $\mathbf{y}_N + \mathbf{z}_N$ of (12.5.8), (12.5.9) belongs to $\mathcal{M}_1$ or more precisely to the projection of $\mathcal{M}_1$ on $P_N H$.

If we use instead the manifold $\mathcal{M}_2$ in (12.4.15), then (12.5.6) is unchanged and (12.5.7) is replaced by

$$- \nu \Delta \mathbf{z}_N + (P_N - P_{N_1})\{((\mathbf{y}_N + \mathbf{z}_N) \cdot \nabla)(\mathbf{y}_N + \mathbf{z}_N)\} = (P_N - P_{N_1})\mathbf{f}. \qquad (12.5.10)$$

In practical computations, several technical aspects are involved:

1. The treatment of $\mathbf{y}$ and $\mathbf{z}$ is systematically different and this was one of the initial motivations.

2. Since the evolution of **z** is slower than that of **y**, the **z** equation is integrated less often than the **y** equation.

3. The value of N_1 varies along the time following a procedure described below.

4. The time discretization of (12.5.8) and (12.5.9) is explicit in $\mathbf{y}_N$ and implicit in $\mathbf{z}_N$ for the nonlinear terms. It follows (see [24]) that the stability condition is that of the large wavelengths and this allows a larger time step.

For the choice of N_1 and N the procedure is the following (see [11]): the total size of **y** + **z**, i.e., N, is chosen according to available computing resources and to physical consideration (e.g. to be able to cover the full inertial range). The size of **y** and **z**, i.e., N_1 is not fixed during the computation (as in [23, 24]) since the appropriate value of N_1 is not known beforehand and furthermore it can vary as time evolves. Hence an adaptive numerical procedure has been implemented (see [11]) where N_1 is chosen by numerical tests that are done repeatedly:

- Compare $\mathbf{z}_N$ to the discretization error of the scheme.

  Indeed we do not want $\mathbf{z}_N$ to be smaller than the discretization error of the scheme.

- Compute the ratio of the norm of $\mathbf{z}_N$ to that of $\mathbf{y}_N$ and compare it to some desirable value or to some expected value.

For the time discretization, a Runge Kutta method of the third order accuracy was used for the nonlinear terms. An indication on the stability analysis is given by [24] (see also [38]) where a two-level time discretization is considered, implicit in the linear terms, explicit in the nonlinear terms. It appears that the stability condition is that based on the N_1 modes and not on the N modes; hence a larger time step is allowed.

Another experimental test of stability of the nonlinear Galerkin method was used. For the 2D space periodic Navier-Stokes equations with 256^2 modes ($N = 128$), a time step $\Delta t = 0.0015$ was chosen close to the limit of stability in that example. Figure 12.4 shows that the solution explodes around $t = 1.5$ i.e., after about $1,000$ time steps for the usual Galerkin (pseudo-spectral) method. When the nonlinear Galerkin method is used, explosion occurs around $t = 10.21$ i.e., after about $7,000$ time steps.

Another advantage of the Nonlinear Galerkin Method over the usual Galerkin Method is the gain in accuracy and computing time. Examples reported in [23, 24] and [11] show that we can recover the same accuracy with a gain in CPU computing units of 40 to 60%; this is very significant since large scale computing with the Navier-Stokes may necessitate tens or hundreds of CPU computing units on a Cray II or on similar computers.

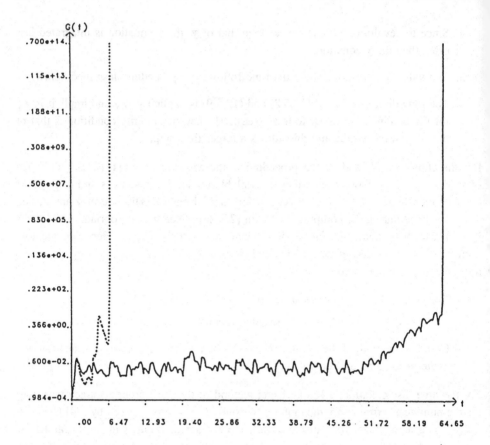

Figure 12.4: Enhanced stability: evolution of the error for the usual (dotted line) and the nonlinear Galerkin methods (solid line).

12.6 The IU Method

The concept of Incremental Unknowns (IU) was introduced in [37] in order to develop similar algorithms in the context of finite differences.

Incremental Unknowns can be defined when two or more nested meshes are used like in multigrid methods. For the sake of simplicity we will emphasize the case of two nested grids called the coarse grid and the fine grid but we do have in mind the multilevel case. A fuzzy definition of the incremental unknowns is the following one: they consist of the nodal values of the unknown function at the coarse grid points; and at the fine grid points not belonging to the coarse grid the incremental unknown is an increment to the averaged value of the function of the closest coarse grid points.

Let us start with the very simple but instructive case of space dimension one. We consider the Dirichlet two point boundary value problem on $(0,1)$:

$$-\frac{d^2u}{dx^2} = f, \qquad 0 < x < 1, \tag{12.6.1}$$

$$u(0) = u(1) = 0. \tag{12.6.2}$$

If N is an integer, we set $h = 1/2N$ and consider the discretization of (12.6.1), (12.6.2) with mesh h: the approximate values of u, $u_j \simeq u(jh)$, $j = 0, \ldots, 2N$ satisfy

$$2u_j - u_{j-1} - u_{j+1} = h^2 f_j, \qquad j = 1, \ldots, 2N-1, \tag{12.6.3}$$

$$u_0 = u_{2N} = 0, \tag{12.6.4}$$

where $f_j = f(jh)$.

The matrix of the linear system (12.6.3) is the well known tridiagonal matrix

$$A = \begin{pmatrix} 2 & -1 & \cdots & 0 \\ -1 & \ddots & \ddots & \\ & \ddots & \ddots & -1 \\ 0 & \cdots & -1 & 2 \end{pmatrix}$$

and the resolution of (12.6.3) relies on an LU factorization of A.

The incremental unknowns for this problem consist of the numbers $y_{2i} = u_{2i}, i = 0, \cdots, N$ and at points $2i+1$, the numbers

$$z_{2i+1} = u_{2i+1} - \frac{1}{2}(u_{2i} + u_{2i+2}), \qquad i = 0, \ldots, N-1. \tag{12.6.5}$$

Thus, as indicated above, z_{2i+1} is the increment of u to the average of the values at the neighboring points $2i$ and $2i + 2$. If proper interpolation functions are used (step or piecewise linear functions) then we obtain in this way a decomposition of the approximate function u_h in the form

$$u_h = y_h + z_h. \tag{12.6.6}$$

Here, as is shown hereafter, $y_h \simeq u_{2h}$ is a $2h$ approximation and z_h is small.

From the linear algebra viewpoint it is surprising that the utilization of the incremental unknowns yields a decoupling of system (12.6.3), (12.6.4). Indeed, at $j = 2i+1$, (12.6.3) reduces to

$$z_{2i+1} = \frac{1}{2}h^2 f_{2i+1}, \qquad i = 0, \ldots, N-1, \tag{12.6.7}$$

so that these incremental values are now explicit. At $j = 2i$, (12.6.3) using (12.6.5) and (12.6.7) becomes

$$
\begin{aligned}
2u_{2i} - u_{2i-1} - u_{2i+1} &= h^2 f_{2i}, \\
2y_{2i} - y_{2i-2} - y_{2i+2} &= 2h^2 f_{2i} + 2(z_{2i+1} + z_{2i-1}), \\
2y_{2i} - y_{2i-2} - y_{2i+2} &= h^2(f_{2i} + f_{2i+1} + f_{2i-1}).
\end{aligned}
\tag{12.6.8}
$$

The system consisting of (12.6.8) and (12.6.4) is similar to the system consisting of (12.6.3), (12.6.4), but involves twice less unknowns. We can of course repeat the procedure. If we start with $h = 2^{-\ell}$, $\ell \in \mathcal{N}$, then after ℓ steps all the incremental unknowns are computed and therefore u itself.

Although less simple, the same procedure can be applied in higher dimensions [37]; for linear elliptic problems the details of the implementation appear in [6, 7]. The discretization of a linear elliptic problem with mesh $h = h_\ell = 1/2^\ell N, \ell, N \in \mathcal{N}$, leads to a linear system

$$
A_{h_\ell} U_{h_\ell} = b_{h_\ell}.
\tag{12.6.9}
$$

We define in a similar appropriate fashion the incremental unknowns

$$
\overline{U}_{h_\ell} = \begin{pmatrix} Y^\ell \\ Z^\ell \end{pmatrix}, \qquad Z^\ell = \begin{pmatrix} Z_1 \\ \vdots \\ Z_\ell \end{pmatrix},
$$

and we denote by S_{h_ℓ} the transfer matrix

$$
U_{h_\ell} = S_{h_\ell} \overline{U}_{h_\ell}.
$$

Hence

$$
A_{h_\ell} S_{h_\ell} \overline{U}_{h_\ell} = b_{h_\ell}
$$

and we obtain the new system

$$
\overline{A}_{h_\ell} \overline{U}_{h_\ell} = \overline{b}_{h_\ell},
\tag{12.6.10}
$$

where

$$
\begin{aligned}
\overline{A}_{h_\ell} &= {}^t S_{h_\ell} A_{h_\ell} S_{h_\ell} \\
\overline{b}_{h_\ell} &= {}^t S_{h_\ell} b_{h_\ell}.
\end{aligned}
$$

The drawback is that $\overline{A}_{h_\ell}$ is a full matrix while A_{h_ℓ} was sparse. However this difficulty can be overcome and we can solve (12.6.10) in a very efficient way by using the conjugate gradient method (see [6, 7]). The implementation of inertial projections based on (12.6.6) has been described in [37]; effective applications to fluid mechanics will appear elsewhere.

References

[1] F. Abergel and R. Temam, "On some control problems in fluid mechanics," *Theoret. Comput. Fluid Dynamics*, 1, 1990, pp. 303–325.

[2] N. Aubry, P. Holmes, J. L. Lumley, and E. Stone, "The dynamics of coherent structures in the wall region of a turbulent boundary layer," *J. Fluid Mech.*, 192, 1988, pp. 115–173.

[3] M. Bartuccelli, C. Doering, and J. D. Gibbon, "Ladder theorems for the 2D and 3D Navier-Stokes equations on a finite periodic domain," preprint.

[4] C. H. Bruneau and C. Jouron, "An efficient scheme for solving steady incompressible Navier-Stokes equations," *J. Compp. Phys.*, to appear.

[5] C. Canuto, M. Y. Hussaini, A. Quarteroni and T. A. Zang, *Spectral Methods in Fluid Dynamics*, Springer-Verlag, New York 1987.

[6] M. Chen and R. Temam, "Incremental unknowns for solving partial differential equations," *Num. Math*, 59, 1991, pp. 255–271.

[7] M. Chen and R. Temam, "Incremental unknowns in finite differences: condition number of the matrix," *Siam J. of Matrix Anaysis* (SIMA-X), 1991, to appear.

[8] P. Constantin, C. Foias, O. Manley, and R. Temam, "Determining modes and fractal dimension of turbulent flows," *J. Fluid Mech.*, 150, 1985, pp. 427–440.

[9] P. Constantin, C. Foias, and R. Temam, "Attractors representing turbulent flows," *Memoirs of AMS*, 53, no 314, 1985, 67 + vii pages.

[10] P. Constantin, C. Foias, and R. Temam, "On the dimension of the attractors in two-dimensional turbulence," *Physica D*, 30, 1988, pp. 284–296.

[11] T. Dubois, F. Jauberteau, and R. Temam, "Solutions of the incompressible Navier-Stokes equations by the Nonlinear Galerkin Method," submitted to *J. Compp. Phys.*.

[12] C. Foias, O. Manley, and R. Temam, "Sur l'interaction des petits et grands tourbillons dans des écoulements turbulents," *C. R. Acad. Sc. Paris, Série I*, 305, 1987, pp. 497–500. "Modelling of the interaction of small and large eddies in two-dimensional turbulent flows," *Math. Mod. and Num. Anal. (M²AN)*, 22, 1988, pp. 93–114.

[13] C. Foias and R. Temam, "Remarque sur les équations de Navier-Stokes stationnaires et les phénomènes successifs de bifurcation," *Annali Scuola Norm. Supp. Pisa, Série IV*, Vol. V, 1978, pp. 29–63 (and volume dedicated to J. Leray, Pise, 1978).

[14] A. Fortin, M. Fortin, and J. J. Gervas, "A numerical simulation of the transition to turbulence in a two-dimensional flow," *J. Compp. Phys.* 70, 1987, p. 295.

[15] U. Ghia, K. N. Ghia, and C. T. Shin, "High resolution of Navier-Stokes equations in primitive variables," *J. Compp. Phys.* 65, 1982, pp. 387–411.

[16] J. W. Goodrich, K. Gustafson, and K. Halasi, "Hopf bifurcation in the driven cavity," *J. Compp. Phys.*, to appear.

[17] D. Gottlieb and S. Orszag, *Numerical Analysis of Spectral Methods; Theory and Applications*, SIAM, Philadelphia, 1977.

[18] O. Goubet, "Construction of approximate inertial manifolds using wavelets," to appear.

[19] M. Gunzburger, L. Hou, and T. Svobodny, "Analysis and finite elements approximation of optimal control problems for the stationary Navier-Stokes equations with Dirichlet controls," *Math. Model. and Num. Anal.* (M²AN), 1991, to appear.

[20] K. Gustafson and K. Halasi, "Cavity flows dynamics at higher Reynolds number and higher aspect ratio," *J. Compp. Phys.*, 70, 1987, pp. 271–283.

[21] G. Haltiner and R. Williams, *Numerical Weather Prediction and Dynamical Meterology*, 2nd edition, Wiley, New-York, 1980.

[22] W. D. Henshaw, H. O. Kreiss, and L. G. Reyna, "Smallest scale estimates for the incompressible Navier-Stokes equations," preprint.

[23] F. Jauberteau, C. Rosier, and R. Temam, "The nonlinear Galerkin method in computational fluid dynamics," *Applied Numerical Mathematics*, 6, 1989/90, pp. 361–370.

[24] F. Jauberteau, C. Rosier, and R. Temam, "A nonlinear Galerkin method for the Navier-Stokes equations," in Proceedings on "Spectral and High Order methods for partial differential equations," ICOSAHOM'89, Como, Italie, 89, *Computer Methods in Applied Mechanics and Engineering*, 80, 1990, pp. 245–260.

[25] M. S. Jolly, I. G. Kevrekidis, and E. S. Titi, "Approximate inertial manifolds for the Kuramoto-Sivashinsky equation: analysis and computation," *Physica* D, 44, 1990, pp. 38–60.

[26] M. S. Jolly, I. G. Kevrekidis, and E. S. Titi, "Preserving dissipation in approximate inertial forms," *J. of Dynamics and Differential Equations*, 3, 1991, pp. 179–197.

[27] A. N. Kolmogorov, "The local structure of turbulence in an incompressible fluid with very large Reynolds numbers," *Dokl. Akad. Nauk. SSSR* 30, 1941, pp. 301–305. "Dissipation of energy under locally isotropic turbulence," *Dokl. Akad. Nauk SSSR*, 32, 1941, pp. 16–18.

[28] R. H. Kraichnan, "Inertial ranges in two-dimensional turbulence," *Phys. Fluids*, 10, 1967, pp. 1417–1423.

[29] J. Laminie, F. Pascal, and R. Temam, "Nonlinear Galerkin Methods with finite elements approximations," *Proceedings of the 11th ICNMFD*, K. W. Morton ed., Springer-Verlag, Lecture Notes in Physics, 1990.

[30] J. Liou and T. E. Tezduyar, "Numerical simulation of a periodic array of cylinders between two parallel walls," *Theor. and Compp. Fluid Dynamics*, to appear.

[31] M. Marion and R. Temam, "Nonlinear Galerkin Methods," *SIAM J. of Num. Anal.*, Vol. 26, No. 5, 1989, pp. 1139–1157.

[32] M. Marion and R. Temam, "Nonlinear Galerkin Methods: The finite elements case," *Numerische Mathematik*, 57, 1990, pp. 205–226.

[33] P. Moin and J. Kim, "On the numerical solution of time-dependent viscous incompressible fluid flows involving solid boundaries," *J. Compp. Phys.*, 35, 1980, pp. 381–392.

[34] J. Shen, "Hopf bifurcation of the unsteady regularized driven cavity flows," *J. Compp. Phys.*, 95, 1991, pp. 228–245.

[35] L. Sirovich, "Empirical eigenfunctions and low dimensional dynamical systems," Center for Fluid Mechanics, Preprint No. 90-202, 1990.

[36] R. Temam, "Attractors for the Navier-Stokes equations, Localization and Approximation," *J. Fac. Sci. Tokyo*, Sec. IA, 36, 1989, pp. 629–647.

[37] R. Temam, "Inertial manifolds and multigrid methods," *SIAM J. Math. Anal.*, Vol. 21, No. 1, 1990, pp. 154–178.

[38] R. Temam, "Stability Analysis of the Nonlinear Galerkin Method," to appear in *Math. of Comp.*

[39] R. Temam, "Infinite-Dimensional dynamical systems in mechanics and physics," Springer-Verlag, New York, Applied Mathematical Sciences Series, Vol. 68, 1988.

[40] E. S. Titi, "Une varieté approximate de l'attracteur universel des équations de Navier-Stokes non linéaires de dimension finie," *C. R. Acad. Sci. Paris*, 307, Serie I, 1988, pp. 383–385.

13 The Finite Element Method for Three Dimensional Incompressible Flow

R. W. Thatcher

13.1 Introduction

The finite element method has been an established method for approximating incompressible flow for a number of years. The primitive variable, velocity/pressure formulation, is the most popular way to implement the method although it is not the only possibility. Much theoretical work has been done to establish convergence and error estimates and there is a large amount of literature on the topic, see for example Temam [1984], Thomasset [1981], Girault and Raviart [1986], Gunzburger [1989]. Effort has been concentrated on two-dimensional flow and although mathematically three-dimensional flow is no more difficult, in practice the current state of computer hardware makes the implementation of three-dimensional elements much more problematic. It is only the availability of modern supercomputers that has allowed the approximation of such flows to be attempted. However, an element and method of solution that works well on a large vector processor may be quite inefficient on a fine-grained parallel computer thus the concept of the "best element" or even a "good element" may be highly dependent on the computer on which it is to be implemented.

Most methods of solution involve at least one iteration of one form or another, the innermost loop being the solution of a set of linear equations. For practical three-dimensional flow problems a direct method of solution is unlikely to be a feasible proposition for almost all situations and this inner system of equations will have to be solved iteratively. "How accurately do we need to solve this system?" and "What are the interactions between the various iterations taking place?" are two questions that have to be faced by anyone implementing the finite element method for three-dimensional flow.

In the following sections the basic theory underpinning the finite element method for incompressible flow is briefly outlined, some elements that are being used are discussed and some ideas on iterative methods of solution introduced. Attention is focussed on low order finite elements and simple to implement solution processes in the belief that such an approach, if it can be made to work, is likely to produce the most efficient algorithm.

427

13.2 Theoretical Considerations

To approximate the solution of the stationary Navier-Stokes equation in some bounded spatial domain Ω using finite elements, the most common approach is to use the so-called primitive variable, velocity-pressure formulation. A velocity interpolation space $V_h \subset (L^2(\Omega))^3$ and a pressure interpolation space $P_h \subset L^2(\Omega)$ are chosen where h is the usual finite element mesh parameter giving an indication of the fineness of the mesh. Although it is common to use the same mesh for both velocity and pressure this is not essential; often there is a "pressure" mesh with the velocities defined on a sub-mesh. The interpolation spaces V_h and P_h have to be restricted in order that numerical solutions can be obtained and error estimates derived. "Conforming elements" are those for which the velocity space satisfies $V_h \subset (H^1(\Omega))^3$ but non-conforming elements are also occasionally used.

Both V_h and P_h will usually have further conditions enforced on them such as boundary conditions. For example, for an enclosed flow in which all the normal velocities are prescribed around the boundary $\partial\Omega$ of Ω these conditions, or an approximation to them, are imposed on V_h. Furthermore, in this case the pressure is only determined up to an arbitrary constant thus some pressure condition is imposed, for example a reference pressure level is imposed at some node.

The two spaces V_h and P_h cannot be chosen arbitrarily since they have to satisfy some compatibility condition. It is well-known, for example, that the same approximation for pressure and the components of velocity on the same grid cannot be used. The condition for compatibility of the velocity and pressure spaces in the context of the Navier-Stokes equations is identical to that for the far simpler, linear Stokes equations and consequently much analysis of these equations appears in the literature.

13.2.1 Stokes and Navier-Stokes Equations

The time dependent Navier-Stokes equations for incompressible flow can be written as

$$\nabla \cdot v = 0 \qquad \text{in } \Omega \tag{13.2.1}$$

$$\frac{\partial v}{\partial t} + (v \cdot \nabla)v - \nu\Delta v + \nabla p = f \qquad \text{in } \Omega \tag{13.2.2}$$

where v and p are respectively the velocity and pressure and satisfy the differential equation in Ω over some time interval $[0,T]$. The data required for solution are the external forces f, the viscosity ν (or the reciprocal of the Reynolds number) with initial velocity at time $t = 0$ and velocity boundary conditions on $\partial\Omega$ for all time. The steady state Navier-Stokes equations are given by the incompressibility condition (13.2.1) with

(13.2.2) replaced by

$$(v \cdot \nabla)v - \nu\Delta v + \nabla p = f \qquad \text{in } \Omega \tag{13.2.3}$$

where now the solution is sought in some spatial region Ω. The data f and ν are again required and boundary conditions are given on $\partial\Omega$. Finally, the stationary Stokes equations are given by (13.2.1) together with

$$\alpha v - \nu\Delta v + \nabla p = f \qquad \text{in } \Omega. \tag{13.2.4}$$

The Stokes equations arise in at least two ways; physically they are important for the case of a slow moving, highly viscous fluid (in which case $\alpha = 0$) and computationally when the Navier-Stokes equations are approximated by solving a series of Stokes problems. For the numerical analyst, the Stokes equations are of interest since a full mathematical analysis of the primitive variable, finite element method is available in the literature, see for example Girault and Raviart [1986]; this is an interesting example of the so-called "mixed finite element method." The analysis of the Stokes equations has important consequences for the Navier-Stokes equations and some outline of the analysis is given below. The solution (v, p) of the Stokes problem with fixed homogeneous boundary conditions satisfies the weak equations

$$a(v, u) + b(u, p) = (f, u)_\Omega \qquad \forall \ u \epsilon V \tag{13.2.5}$$

$$b(v, q) = 0 \qquad \forall \ q \epsilon P \tag{13.2.6}$$

where

   i. $(s, t)_\Omega = \int_\Omega s.td\Omega$

  ii. $a(u, v) = \alpha(u, v)_\Omega + \nu(\nabla u, \nabla v)_\Omega$

 iii. $b(u, p) = (\nabla.u, p)_\Omega$

 iv. $V = (H_o^1(\Omega))^3$.

  v. $P = L_o^2(\Omega)$.

Applying the primitive variable finite element method to these weak equations leads to the numerical problem, find $(v_h, p_h)\epsilon V_h \times P_h$ where $V_h \subset V$ and $P_h \subset P$ are finite element velocity and pressure spaces depending on a mesh parameter h such that

$$a(v_h, u) + b(u, p_h) = (f, u) \qquad \forall \ u \epsilon V_h \subset V \tag{13.2.7}$$

$$b(v_h, q) = 0 \qquad \forall \ q \epsilon P_h \subset P \tag{13.2.8}$$

A crucial part of the analysis is the compatability requirement between the spaces V_h and P_h , namely

$$\inf_{q \epsilon P_h} \left\{ \sup_{u \epsilon V_h} \left\{ \frac{b(u,q)}{\| u \|_V \| q \|_P} \right\} \right\} \geq \beta > 0 \qquad (13.2.9)$$

with β independent of h. This condition implies that the numerical approximation to the Stokes equation satisfies

$$\| v - v_h \|_V + \| p - p_h \|_P \leq C \left\{ \inf_{u \epsilon V_h} \| u - v \|_V + \inf_{q \epsilon P_h} \| q - p \|_P \right\} \qquad (13.2.10)$$

with C independent of h. Condition (13.2.10) naturally leads to the conclusion that a good pair of finite element spaces will use local interpolation of order l for the velocity and order $l - 1$ for the pressure. This leads to an $O(h^l)$ estimate of the numerical error in the L^2 norm for pressure and the H^1 norm for velocity, which is optimal in the sense that it is the same order as the interpolation errors for pressure and velocity in these spaces. The same compatibility condition between V_h and P_h seems to be a necessary condition for constructing a primitive variable approximation of the Navier-Stokes equations.

13.2.2   A Patch Test for Compatibility

It was not until the patch test of Boland and Nicolaides [1983] had been formulated that relatively simple proofs of the compatibility (or stability) of velocity and pressure approximations were established, i.e. that the spaces V_h and P_h satisfied condition (13.2.9). A similar patch test was devised by Stenberg [1984] and refined, Stenberg [1990a], to an easily used statement that is outlined below.

The notation C_h is used to denote a grid of elements which satisfies the usual regularity condition for tetrahedral or hexahedral elements. A patch of elements of the grid C_h is denoted by M. Topologically, the region M is the interior of the region formed by taking a union of a number of closed subregions (that is, a number of elements) and M must be simply connected. Two patches M_1 and M_2 are said to be equivalent if one can be mapped continuously onto the other. We denote by ξ the class of equivalent patches each element of which satisfies the required regularity condition, namely the same regularity condition for the elements in C_h. The space P_M denotes the unrestricted space of pressures over the patch M, the space $V_{0,M}$ denotes the space of velocities over the patch that are zero on the boundary of the patch and the space N_M is defined by

$$N_M = \{ q \epsilon P_M \mid (\nabla . u, q)_M = 0 \ \forall \ u \epsilon V_{0,M} \} . \qquad (13.2.11)$$

Finally the notation M_h denotes a splitting up of the elements of C_h into patches which may overlap. Then if

1. there is a fixed number q of equivalence classes ξ_i for $i = 1, 2, .., q$,

2. each $M \epsilon M_h$ belongs to one of the classes ξ_i for $i = 1, 2, .., q$,

3. each element of C_h belongs to one and not more than L patches of M_h where L is some fixed integer,

4. each face common to two elements of C_h is contained in one and not more than L patches of M_h,

then the compatibility condition (13.2.9) is satisfied with a constant C independent of h.

13.3 Three-Dimensional Elements

13.3.1 Taylor-Hood Elements

Historically one of the most important group of elements for approximating the Navier-Stokes equations is the so-called "Taylor-Hood" group of elements and they remain widely used. One of the first successful finite element approximations was calculated by Hood and Taylor [1974] using continuous, piecewise quadratic velocities and continuous, piecewise linear pressures on a grid of triangular elements. Currently the name "Taylor-Hood" element is used to describe any finite element approximation using the same grid for pressure and velocity, which provide continuous approximations for velocity and pressure based on Lagrange type interpolation, where the order of approximation is one less for pressure than for the components of velocity. Thus, for example, for three-dimensional flow this would include:

1. quartic velocities and cubic pressures on a grid of tetrahedral elements.

2. tricubic velocities and triquadratic pressures on a grid of hexahedral elements.

In practice, the only elements that are widely used are based on quadratic velocities and linear pressures on tetrahedra and triquadratic velocities and trilinear pressures on hexahedra and it is the latter that are most often used for three dimensional flow. An analysis of these elements is given by Stenberg [1990a].

A variation on the Taylor-Hood group of elements is to use serendipity rather than full Lagrange type elements. An element with serendipity, triquadratic velocities and trilinear pressure is used by Smith [1984].

Another variation on the quadratic velocity, linear pressure tetrahedral is derived using an idea suggested by Bercovier and Pironneau [1979] and is based on approximating the quadratic velocity by a piecewise linear one. Thus, a linear velocity approximation on a tetrahedral sub-grid is obtained by splitting each tetrahedron of the pressure grid into 8 smaller tetrahedra by connecting the mid-points of sides. A similar but rather simpler to visualise idea can be used with hexahedral elements. The appeal of this approach is that only linear functions are used which are much easier to handle and although the number of unknown parameters remains the same as for the equivalent Taylor-Hood element, the matrices involved are much more sparse.

13.3.2   Bubble Functions

Several elements have been proposed that use the idea of "bubble functions" to supplement the velocity space. Bubble functions are often used to produce an element that satisfies condition (13.2.9) from one that did not originally satisfy it.

A velocity bubble function is any function that has non-zero components inside an element and zero components around the boundary of that element. Thus for a triangle there is a cubic bubble function, for a tetrahedron a quartic bubble function and for a hexahedral element a triquadratic bubble function. It is sometimes convenient to define bubble functions in a piecewise manner, thus for any element it is possible to construct a piecewise linear bubble function. For example, consider the quadrilateral element ABCD of Figure 13.1 where G is, say, the centroid. A piecewise linear bubble function can be defined in the element ABCD as the function linear in each triangular region GBA, GCB, GDC and GAD equal to 1 at G and zero on that part of the boundary of ABCD that is common with the boundary of the triangle. This idea can of course be used to construct bubble functions in tetrahedral and hexahedral regions.

13.3.3   The Mini Element

A second continuous pressure element that has received a lot of attention is the Mini element, see Arnold, Brezzi, Fortin [1984]. The three-dimensional version is based on a tetrahedral element and is motivated by the use of low order interpolation. The basic interpolation space is linear for both velocity and pressure providing continuous approximations for both. The velocity space is supplemented by a quartic bubble function. This high order bubble function does detract from the natural simplicity of linear functions inherent in this approach, but since it can be eliminated by static condensation, this is not a major obstacle. Instead of the high order bubble, a piecewise linear bubble could be introduced in the manner described above which is equivalent to using a tetrahedral grid for the pressure and a tetrahedral sub-grid for the velocities and means that only linear functions would have to be handled.

13.3.4   Discontinuous Pressures

For two-dimensional flows there are several reports of poor approximations obtained by the use of both Taylor-Hood and the Mini elements, see for example Gresho *et al* [1980], Debbaut and Crochet [1986], Tidd *et al* [1988], Nafa and Thatcher [1989].

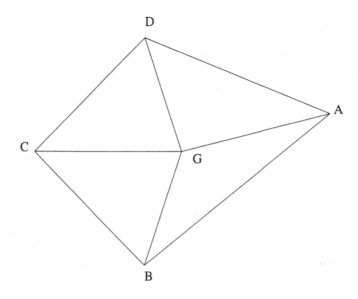

Figure 13.1: A piecewise linear bubble function

Thus some care should be taken when using elements that have continuous pressure approximations. Such reports have led researchers to advocate the use of discontinuous pressure approximations, although no incontrovertible proof has been provided that the poor results arose solely from the use of a continuous pressure. One of the reasons suggested for the poor performance was that continuity for the approximate velocity was only satisfied over the region as a whole and that fluid could be created and/or destroyed within each element.

A way of overcoming this problem for Taylor-Hood elements is to supplement the pressure space (i.e. a continuous Lagrange approximation space) with the space of piecewise constant functions, constant in each element. This sort of idea was first described by Gresho *et al* [1980] as a means of overcoming the poor results obtained with continuous pressure approximations and was analysed by Thatcher [1990a]. By supplementing the pressure space in this way the approximate velocity solution obtained satisfies continuity over each element.

Some methods of solution of the resulting discrete equations require the use of the inverse of the pressure mass matrix. For such a method then clearly a fully discontinuous pressure space in which the pressure in one element is completely unrelated to pressure in any other element have an advantage since they lead to a block diagonal matrix with very small diagonal blocks.

13.3.5   Piecewise Constant Pressure Approximation

Much attention in the literature has been focused on the use of linear velocities and constant pressure for triangular and tetrahedral elements (and bilinear or trilinear velocities and constant pressure for quadrilateral and hexahedral elements respectively). They are optimal in the approximation sense mentioned in Section 13.2 and their simplicity would make them computationally very attractive. Unfortunately they do not pass the compatibility test (13.2.9) and on regular grids exhibit spurious pressure modes. Attempts to filter these modes have been made (see for example Sani *et al* [1981]) but what seems a more satisfactory approach is to put constraints on the pressure jumps between neighbouring elements, see Hughes and Franca [1987], Silvester and Kechkar [1990]. This remains a very active area of development and interesting new results are being produced.

The use of a piecewise constant pressure with quadratic or higher order velocities, whilst satisfying compatibility (13.2.9), numerically produces very disappointing results. This is not surprising since the potential high order approximation for velocity is lost because of the error bound (13.2.10) and thus it is not a recommended approach.

13.3.6 $Q_l - P_{l-1}$ Hexahedral Elements

This is a group of hexahedral elements which, for a given $l > 1$, provides an lth order continuous approximation for the components of velocity and an $(l-1)^{th}$ order discontinuous approximation for pressure. They can easily be shown to satisfy the compatibility condition (13.2.9), see for example Girault and Raviart [1986]. The case $l = 2$ is by far the most widely used. This element provides a high order approximation, is relatively easy to use but can be rather expensive. It leads to a less sparse system of equations than linear elements and, in the context of a non-linear problem, the time processing elements becomes significant but it remains one of the best methods available.

By using velocity bubble functions, a similar idea can be used for tetrahedral elements. The only occasion that this is likely to be of interest is when the three-dimensional region is much more easily split into tetrahedra than hexahedra.

13.3.7   Linear Pressure with Linear Velocities on a Sub-Grid

The basic idea behind this approach is to use linear elements to "approximate" the $Q_2 - P_1$ element described above, see Nafa and Thatcher [1989]. Thus a hexahedral region is split up into 48 tetrahedra as illustrated in Figure 13.2 with each tetrahedron having a vertex at the centroid C of the hexahedron and its base on the boundary of the hexahedron. The tetrahedral subgrid is used to provide a continuous, piecewise linear approximation to the components of velocity and the original hexahedral grid provides a discontinuous,

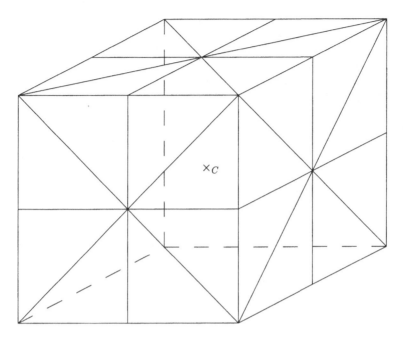

Figure 13.2: A 48 tetrahedral macro-element, where C is the centroid

piecewise linear approximation to the pressure.

The advantages of using this approach over a grid of $Q_2 - P_1$ elements is that the low order makes the individual elements much easier to process and it leads to a much more sparse system of linear equations halving the bandwidth. The disadvantages are that the high order approximation and high order accuracy for many problems is lost and there are a great many more elements to assemble. Vectorising over the elements is very efficient because of the simplicity of the linear elements, and is a way of reducing this latter difficulty. This approach can be used as a method in its own right or can be used as some preconditioner for the higher order scheme.

The proof that this method satisfies the compatibility condition (13.2.9), see Nafa and Thatcher [1988], only requires the mid-face nodes and not the mid-side nodes. Thus a discontinuous, piecewise linear pressure approximation on a hexahedral grid could be used together with a continuous, piecewise linear approximation on the tetrahedral subgrid obtained by splitting each hexahedron into 24 tetrahedra as illustrated in Figure 13.3, again with each tetrahedron having a vertex at C with its base on the boundary of the hexahedron. This approach leads to a much smaller number of unknown velocities but although the error will be different for the 24 and 48 tetrahedral subgrid methods the

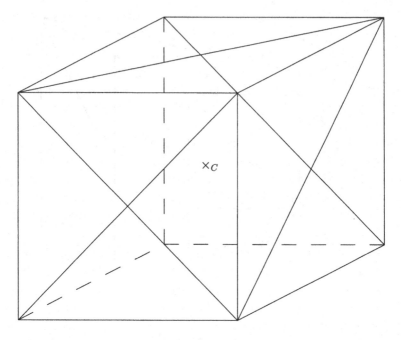

Figure 13.3: A 24 tetrahedral macro-element, where C is the centroid

order of the error remains the same. Numerical experiments suggest that the actual error of the two approaches are not dissimilar provided the velocity gradients are not too great.

In fact, the use of hexahedra themselves is not essential. If the three-dimensional region is split up into a number of flat faced, polyhedral shapes a discontinuous, piecewise linear pressure approximation over this polyhedral grid may be used. The approximation for velocity is obtained by placing a mid-face node on each face of each polyhedron and a mid-element node in each polyhedron and constructing a tetrahedral grid in the spirit of that illustrated in Figure 13.3. If no such interior node can be found that allows a tetrahedral grid satisfying the regularity condition to be obtained then this polyhedral grid can not be used, i.e. it is not a regular polyhedral grid. A continuous, piecewise linear approximation for each component of velocity is used on the tetrahedral grid. The pressure space is a set of discontinuous, piecewise linear functions determined by their value and the values of their first derivatives at the central node of each polyhedron in the grid. This approach can be shown to satisfy the compatibility condition (13.2.9) in the following way. To pass the patch test as described in Section 13.2.2 it is first necessary to consider a patch of two polyhedral elements. Thus we will consider a

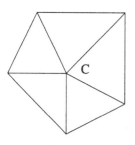

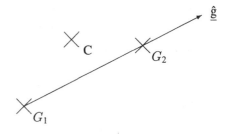

a) common face of two
polyhedral elements
K_1 and K_2

b) the plane of the central nodes
G_1 and G_2 of K_1 and K_2
and the node C

Figure 13.4: Example polyhedra illustrating patch test

patch M obtained from two polyhedral elements K_1 and K_2 that have the common face illustrated in Figure 13.4a, the precise shape of which is not important.

By taking as a test velocity u in $V_{0,M}$ that velocity which is zero except for the component in the x direction at the central node G_1 of the polyhedron K_1 and considering the condition (13.2.11), we find that the x-derivative of pressure at G_1 must be zero. By considering all three components individually at both G_1 and G_2 then the only terms that can be in N_M are discontinuous, piecewise constant over the patch M. To show that N_M is one-dimensional (that is containing only functions that are constant on M) we consider the plane determined by C, the central node on the interface and the central nodes G_1 and G_2 of the polyhedra K_1 and K_2 respectively as illustrated in Figure 13.4b. The direction from G_1 to G_2 is denoted by $\hat{\underline{g}}$. By taking a test velocity u with its only non-zero component as u_g (that is, the velocity in the $\hat{\underline{g}}$ direction) at C then

$$\nabla . u = \frac{u_g}{g_1} \quad \text{in the elements of } K_1 \text{ with a node at } C$$
$$= -\frac{u_g}{g_2} \quad \text{in the elements of } K_2 \text{ with a node at } C$$
$$= 0 \quad \text{in other tetrahedral elements of } M$$

where

  i. g_1 is the perpendicular distance from G_1 to the common face of K_1 and K_2.

  ii. g_2 is the perpendicular distance from G_2 to the common face of K_1 and K_2.

Denoting by $|\ F\ |$ the area of the common face, that is the area of the pentagon of Figure 13.4a then

$$(\triangledown.u, q) = \frac{u_g}{g_1} \ (\frac{1}{4} \ g_1 \ |\ F\ |)P_1 - \frac{u_g}{g_2} \ (\frac{1}{4} \ g_2 \ |\ F\ |)P_2$$

where P_i the constant pressure in K_i. Thus for $(\triangledown.u, q)$ to be zero then $P_1 = P_2$. This establishes that N_M is of the correct form for a patch of two polyhedra. The proof is completed by considering the set of larger patches one for each vertex of each polyhedron of the pressure grid that is in the interior of the domain of the problem; each of these patches having that node in its interior. Thus the polyhedral grid satisfies the compatibility condition with a constant independent of h provided a fixed number of polyhedral shapes are used, the number of equivalence classes of patches does not increase on grid refinement, and that the polyhedral grid is regular, in the sense described above. The use of general polyhedral regions provides great flexibility in this approach with an ability to cover quite complex regions with a relatively simple grid.

13.3.8   Other Elements

A range of other elements have been proposed including non-conforming elements, see for example Hecht [1983], and the use of terms, like bubble functions, associated with the normal component of velocity at a mid-face node, see Bernardi and Raugel [1985]. This latter idea seems a particularly useful and attractive method for producing simple, low order elements. One of the most important of which uses an enhanced trilinear velocity approximation with a piecewise constant approximation for pressure. It is often referred to as the $Q_1^+ - P_0$ element.

In addition, approximating not only the velocities and pressures but also the stresses by finite element spaces has been analysed by Stenberg [1990b].

13.4   Solving the Discrete System

Although a suitable choice of element is a major concern for three-dimensional flows, of equal and perhaps greater concern is the method for solving the very large and very sparse system of equations that results from a finite element discretization. Many of the standard methods of solution for two-dimensional flow involve the solution of a large system of equations on an innermost loop of some iterative process. Modern computing power has allowed most users to solve this inner system directly by elimination for two-dimensional problems, but for almost all practical problems and for almost all users this approach is not available for three-dimensional flows. At best an iterative method of solution has to be employed on the equations in the innermost loop and this inevitably introduces arbitrary parameters into the solution process. Solving the innermost loop to a reduced accuracy has a major interaction with the overall solution process.

Many recent papers have addressed this problem including Cahouet and Chabard [1988], Engelman *et al* [1988], Brown and Saad [1990] and this will continue to be an actively debated topic for some years. It can be argued that relatively simple methods with few or no arbitrary parameters in them are the most attractive. Inevitably iterative methods contain some parameters associated with the stopping criteria but methods for solving the discrete Stokes and Navier-Stokes equations often have other parameters present. If a parameter is present then a solution process that is fairly insensitive to its choice over quite a wide range near its optimal value are advantageous. In this section some ideas and results on these issues are presented with an emphasis on the use of a very small fixed number of iterations of the solver on the innermost loop.

These ideas have been tested on the two-dimensional lid-driven cavity problem using the two-dimensional version of the macro-element idea described in Section 13.3.7. Performing a comprehensive set of test runs with three-dimensional elements is expensive and hard to justify; by choosing a relatively small number of sample cases their behaviour is seen to be very similar to the two-dimensional elements.

13.4.1 Solution of the Stokes Equations (Splitting Pressure from Velocity)

Much attention has been given to the solution of the Stokes equations since the Navier-Stokes equations can be solved as a sequence of Stokes problems, see Gunzburger [1989]. The discrete Stokes equations are given by

$$
\begin{aligned}
Au + B^T p &= f \\
Bu &= 0
\end{aligned}
\tag{13.4.1}
$$

where u and p are the unknown parameters of the velocity and pressure spaces respectively. Moreover the matrix A is a symmetric, positive definite matrix and for three-dimensional problems may be written as a 3×3 block diagonal matrix with the diagonal blocks being identical. One of the earliest papers to address the problem of solving an innermost loop to reduced accuracy is by Verfurth [1984] in the context of the following algorithm:

Method 1

i. choose $p^{(0)}$

    a. solve $Au^{(0)} = f - B^T p^{(0)}$ for $u^{(0)}$

    b. $g^{(0)} = -Bu^{(0)}$

ii. iterate

    a. solve $Az^{(i)} = B^T g^{(i)}$ for $z^{(i)}$

    b. $\rho_i = -(u^{(i)})^T B^T B u^{(i)} / (z^{(i)})^T B^T B u^{(i)}$

    c. $p^{(i+1)} = p^{(i)} - \rho_i g^{(i)}$

    d. $u^{(i+1)} = u^{(i)} + \rho_i z^{(i)}$

    e. $\sigma_{i+1} = (u^{(i+1)})^T B^T B u^{(i+1)} / (u^{(i)})^T B^T B u^{(i)}$

    f. $g^{(i+1)} = -B u^{(i+1)} + \sigma_{i+1} g^{(i)}$

This algorithm is theoretically equivalent to solving the equation (13.4.1) by eliminating the velocity and applying the conjugate gradient method to the set of equations

$$BA^{-1}B^T p = BA^{-1} f = g \qquad (13.4.2)$$

Verfurth [1984] used a multigrid algorithm for the solution of the inner system rather than a direct method. In equation (13.4.2) the vectors p and g formally represent functions **p** and **g** $\in P_h$ and there exists an operator L such that

  i. $L : P_h \rightarrow P_h$

  ii. equation (13.4.2) can be written as L **p** $=$ **g**

Moreover, L is a symmetric operator and Verfurth showed that

$$\beta^2 \|\varphi\|^2_{L^2(\Omega)} \leq (L\varphi, \varphi) \leq \|\varphi\|^2_{L^2(\Omega)} \qquad (13.4.3)$$

where β is the constant from (13.2.9). That is, the operator is well-scaled with its eigenvalues bounded above and below independent of h. This translates into the matrix $(M_p^{-\frac{1}{2}} B) A^{-1} (M_p^{-\frac{1}{2}} B)^T$ (where M_p is the pressure mass matrix) being well-scaled with its spectrum of eigenvalues bounded independent of h. The above algorithm when recast to take account of the pressure mass matrix, does converge in a number of iterations which is essentially grid independent for fine enough grids. This remains a good algorithm when iterative solvers are used for the innermost loop although quite accurate solutions are required. This does introduce a parameter, the stopping criterion of the iterative solver, into the solution process a poor choice of which can either produce non-convergence or a much less efficient process.

    A second, rather simpler algorithm related to Uzawa's algorithm, the iterated penalty method and the augmented Lagrangian method is given by:

Method 2

  i. choose $q^{(0)}, r$ and ρ

ii. iterate

    a. solve $(A + r\tilde{B}^T \tilde{B})u^{(i)} = f - \tilde{B}^T q^{(i)}$ for $u^{(i)}$

    b. $q^{(i+1)} = q^{(i)} + \rho \tilde{B} u^{(i)}$

where either

$$\tilde{B} = B \text{ and } p^{(i)} = q^{(i)}$$

or

$$\tilde{B} = M_p^{-\frac{1}{2}} B \text{ and } p^{(i)} = M_p^{-\frac{1}{2}} q^{(i)}$$

The optimal value of ρ lies somewhere between r and $2r$ but seems to be close to r for fine grids and indeed $\rho = r$ is a common choice. With a direct inner solver this method works best for very large r (and ρ) but the system is known to become progressively poorly conditioned as r is increased. A mathematically equivalent method, for fixed ρ, but which can also be shown to give to a minimum residual or steepest descent algorithm for the solution of the set of equations (13.4.1) when ρ is allowed to vary from one iteration to the next is given by:

Method 3

    i. choose $q^{(0)}$ and r then solve $(A + r\tilde{B}^T \tilde{B})u^{(0)} = f - \tilde{B}^T q^{(0)}$ for $u^{(0)}$

    ii. iterate

        a. solve $(A + r\tilde{B}^T \tilde{B})z^{(i)} = -\tilde{B}^T \tilde{B} u^{(i)}$ for $z^{(i)}$

        b. choose ρ_i

        c. $q^{(i+1)} = q^{(i)} + \rho_i \tilde{B} u^{(i)}$

        d. $u^{(i+1)} = u^{(i)} + \rho_i z^{(i)}$

where $\tilde{B}$ and q are as defined for Method 2.

For further details of this algorithm see Fortin and Glowinski [1983], Girault and Raviart [1986]. By considering a number of inner-iterative schemes, for example Jacobi, Gauss-Seidel and conjugate gradient methods, it seems that using Method 2 with $r = 0$, a suitable value of ρ and one or at most two inner-iterations gives a good solution process. Moreover, there seemed little to choose between the solvers used. Conversely, Method 3 seemed to work very poorly no matter what scheme for fixing ρ was chosen and Method 1 was only satisfactory when an accurate solution of the linear equations was obtained. These are the conclusions drawn in Thatcher [1990b] and Chandler [1990] based on the use of the two-dimensional version of the macro-element idea discussed in Section 13.3.7. The advantages of using $r = 0$ are that it removes one of the parameters of the method, and that the matrix A is much more sparse than the matrix $A + r\tilde{B}^T\tilde{B}$. Moreover, as discussed by Cahouet and Chabard [1988], only one diagonal block of A need be stored, namely that derived by considering the x-component of velocity, so there is an important saving in storage requirements. The use of the mass matrix M_p to scale B was important for these results because the presence of pressure and its derivatives as coordinates of the pressure space meant that B was badly scaled. The scaling by M_p is also important when Lagrange interpolation for pressure on strongly graded grids is used, this can be a disadvantage for continuous pressure approximation.

These ideas have only been developed in the context of a particular element. Despite this, it does seem that the advantages of using the parameter r in the augmented Lagrangian and iterated penalty methods may be lost when iterative solvers are used.

13.4.2   A Solution of the Navier-Stokes Equations

A second example that illustrates how the approximate solution of the innermost loop can affect the behaviour of the solution process is given in this section. Both the standard Newton and Picard iterations, to solve the discrete system for the Navier-Stokes equations, can be written in the form

$$K(u^{(n-1)})u^{(n)} + B^T p^{(n)} = f(u^{(n-1)})$$
$$Bu^{(n)} = 0 \qquad (13.4.4)$$

where the matrix B is the same as in equation (13.4.1). Newton's method has second order convergence and works well when (13.4.4) is solved exactly or at least to high accuracy. Since the matrix $K(u^{(n-1)})$ is not symmetric, some of the better methods for solving the Stokes system are not strictly appropriate. For Picard's method $K(u^{(n-1)})$ is independent of $u^{(n-1)}$ and identical to A of (13.4.1), that is all the non-linearities are removed into $f(u^{(n-1)})$. The resulting method is first order and rather slow but does allow the use of special methods available for the Stokes system with the matrix A being much more sparse than the matrix $K(u^{(n-1)})$ of the Newton's method.

Some results given here which illustrate these points were obtained by using a Schwarz iteration, which is a domain decomposition method with overlapping domains. One iteration of the Schwarz method is similar to one iteration of a block Gauss-Seidel method. The Schwarz process is well suited to parallel computation and is particular attractive here since a really large system of equations is not required to be assembled and solved. The results quoted are for a two-dimensional lid-driven cavity problem using the two-dimensional version of the macro-element described in Section 13.3.7. Results

	4×4		8×8	
	A	B	A	B
Solution of Stokes problem	1	23	1	223
Newton's method				
Solution of Navier-Stokes problem	4	95	4	640
Solution of N-S problem (1)	21	21	228	228
Solution of N-S problem (2)	11	22	114	228
Solution of N-S problem (5)	5	25	46	230
Solution of N-S problem (10)	4	40	23	230
Picard's method				
Solution of Navier-Stokes problem	23	454	*	*
Solution of N-S problem (1)	23	23	233	233
Solution of N-S problem (2)	25	50	118	236
Solution of N-S problem (5)	29	145	273	1865

The number in brackets indicates the fixed number of Schwarz iterations
The * means that this case was not computed
Column A - the number of Newton/Picard iterations
Column B - the total number of Schwarz sweeps

Table 13.1: Solution of Schwarz method applied to a two-diminsional lid-driven cavity problem.

were obtained using a 4×4 and an 8×8 square grid for the pressures with each pressure element split into a "union jack" of eight triangular elements to give the velocity grid. The Schwarz iteration is based on 2×2 blocks of pressure elements (and corresponding velocity elements). The Schwarz iteration was stopped when either two consecutive approximations agreed, in the infinity norm, to 10^{-6} or when some maximum number of iterations was hit. The whole process was stopped when the infinity norm of the difference between the approximate solution and the exact solution on a given grid was less than 10^{-4} and the initial approximation was taken to be zero. The results obtained for Reynolds number 100 (i.e. $\nu = 0.01$) are given in Table 13.1.

It is quite remarkable how the total number of Schwarz iterations remained fairly insensitive to a few iterations of Schwarz process before updating the right-hand-side or indeed the matrix $K(u^{(n-1)})$ and it did not seem to matter whether the basic process was a Picard or Newton method. As expected, increasing the Reynolds number changed this picture. The Picard process did not converge at Re = 200 and Newton's method, with one Schwarz iteration per Newton solve, did not converge at Re = 500 whereas Newton's method with a direct solver on the 8 × 8 grid gives a solution for Reynolds number bigger than 2000.

Thus for a solution of the three-diminsional Navier-Stokes equations at low Reynolds number it looks as if a Picard iteration with the Stokes system solved to very reduced accuracy can provide an efficient and simple to implement algorithm. The major cost in such a scheme will be the repetitive recalculation of $f(u^{(n-1)})$ and the above comment assumes that such a term can be quickly and efficiently calculated. For a higher Reynolds number, one seems to be faced with searching for high accuracy in the solution of systems of equations of the form (13.4.4) in order to get a convergent algorithm. It may be possible to treat the time dependent equations and march out in time to a steady state using a sequence of Stokes problems, see the discussion in Gunzburger [1989], but this has not yet been investigated.

References

[1] Arnold, D., Brezzi, F. & Fortin, M. (1984). A stable finite element for Stokes equations. *Calcolo* **21**, 337–344.

[2] Bercovier, M. & Pironneau, O. (1979). Error estimates for finite element method solution of the Stokes problem in primitive variables. *Numer. Math.* **33**, 221–224.

[3] Bernardi, C. & Raugel, G. (1985). Analysis of some finite elements for Stokes problems. *Math. Comp.* **44**, 71–80.

[4] Boland, J. & Nicolaides, R.A. (1983). Stability of finite elements under divergence constraints. *SIAM J. Numer. Anal.* **20**, 722–731.

[5] Brown, P.N. & Saad, Y. (1990). Hybrid Krylov methods for non-linear systems of equations *SIAM J. Sci. Stat. Comp.* **11**, 450–481.

[6] Cahouet, J. & Chabard, J. P. (1988). Some fast 3D finite element solvers for the generalised Stokes problem. *Int. J. Num. Meth. Fluids* **8**, 869–895.

[7] Chandler, A. (1990). *A study of Langrangian methods for the Stokes problem.* M.Sc Dissertation, University of Manchester.

[8] Debbaut, B. & Crochet, M. (1986). Further results on the flow through an abrupt contraction. *J. Non-Newt. Fluid Mech.* **20**, 173–185.

[9] Engelman, M. S., Haroutunian, V. & Hasbani, I. (1988). A proposed finite element algorithm for incompressible flows, in *Finite Element Analysis in Fluids* (ed. T. J. Chung and G. R. Kerr) UAH Press, 790–804.

[10] Fortin, M. & Glowinski, R. (1983). *Augmented Lagrangian methods: Applications to the numerical soluton of boundary-value problems.* North Holland, Amsterdam.

[11] Girault, V. & Raviart, P. A. (1986). *Finite element methods for Navier-Stokes equations.* Springer Series in Computational Mathematics, Vol. 5, Springer-Verlag, New York.

[12] Gresho, P. M., Lee, R. L., Chan, S. T. & Leone, J. M. (1980). A new finite element for Bouissinesq fluids, in *Proc. Third Int. Conf. on Finite Elements in Flow Problems.* Wiley, 204–215.

[13] Gunzburger, M. D. (1989). *Finite element methods for viscous incompressible flows.* Computer Science and Scientific Computing, Academic Press, Boston.

[14] Hecht, F. (1983). A non conforming P^1 basis with free divergence in R^3. *RAIRO serie analyse numerique*.

[15] Hood, P. & Taylor, C. (1974). Navier-Stokes equations using mixed interpolation, in *Finite element methods in flow problems*. UAH Press, 121–132.

[16] Hughes, T. J. R. & Franca, L. P. (1987). A New finite element formulation for CFD: VII The Stokes problem with various well-posed boundary conditions. *Comp. Meth. Appl. Mech. Eng.* **65**, 85–96.

[17] Nafa, K. & Thatcher, R. W. (1988). *Analysis of some low order macroelements for two and three dimensional flow.* Joint University of Manchester/UMIST NA Report 164.

[18] Nafa, K. & Thatcher, R. W. (1989). Equal order mixed finite element approximation, in *Finite Element Analysis in Fluids* (ed. T. J. Chung and G. R. Kerr), UAH Press, 1080–1085.

[19] Sani, R. L., Gresho, P. M., Lee, R. L. & Griffiths, D. F. (1981). The cause and cure (?) of spurious pressures generated by certain finite element method solutions of the incompressible Navier-Stokes equations. *Int. J. Num. Meth. Fluids* **1**, 17–43 (Part 1) and 171–204 (Part 2).

[20] Silvester, D. J. & Kechkar, N. (1990). Stabilised bilinear-constant velocity-pressure finite elements for the conjugate gradient solution of the Stokes problem. *Comput. Meth. Appl. Mech. Eng.* **79**, 71–86.

[21] Smith, R. M. (1984). A practical method for two equation turbulence modelling using finite elements. *Int. J. Num. Meth. Fluids* **4**, 321–336.

[22] Soulaimani, A., Fortin, M., Ovellett, Y., Dhatt, G. & Bertrand, F. (1987). Simple continuous pressure elements for two and three dimensional flow. *Comp. Math. in Appl. Mech. Eng.* **62** 47–69.

[23] Stenberg, R. (1984). Analysis of mixed finite elements for the Stokes problem. *Math. Comp.* **42**, 9–23.

[24] Stenberg, R. (1990a). Error analysis of some finite element methods for the Stokes problem. *Math. Comp.* **54**, 495–508.

[25] Stenberg, R. (1990b). Some new families of finite elements for Stokes equations. *Numer. Math.* **56**, 827–838.

[26] Temam, R. (1984). *Navier-Stokes equations.* North Holland, Amsterdam.

[27] Thatcher, R. W. (1990a). Locally mass conserving Taylor-Hood elements for two nd three dimensional flow. *Int. J. Num. Meth. Fluids* **11**, 341–353.

[28] Thatcher, R. W. (1990b). *The use of iterative solvers in an augmented Lagrangian solution of Stokes equations*, Joint University of Manchester/UMIST NA Report 193.

[29] Thomasset, F. (1981). *Implementation of finite element methods for Navier-Stokes equations.* Springer Series in Computational Physics. Springe-Verlag, New York.

[30] Tidd, D. M., Thatcher, R. W. & Kaye, A. (1988). The free surface flow of Newtonian and non-Newtonian fluids trapped by surface tension. *Int. J. Num. Meth. Fluids* **8**, 1011–1027.

[31] Verfurth, R. (1984). A combined conjugate gradient-multigrid algorithm for the numerical solution of Stokes problem. *IMA J. Num. Anal.* **4**, 441–455.

14 A Posteriori Error Estimators and Adaptive Mesh-Refinement Techniques for the Navier-Stokes Equations

R. Verfürth

14.1 The Adaptive Mesh-Refinement Philosophy

In computational fluid dynamics, as well as in other problems of physics or engineering, one often encounters the difficulty that the overall accuracy of the numerical solution is deteriorated by local singularities such as, e.g., singularities near re-entrant corners, interior or boundary layers, or shocks. An obvious remedy is to refine the discretization near the critical regions, i.e., to place more grid-points where the solution is less regular. The question then is how to identify these regions automatically and how to guarantee a good balance of the number of grid-points in the refined and un-refined regions such that the overall accuracy is optimal.

Another, closely related problem is to obtain reliable estimates of the accuracy of the computed numerical solution. A priori error estimates, as provided, e.g., by the standard error analysis for finite element or finite difference methods, are in general not sufficient, since they only yield asymptotic estimates and since the constants appearing in the estimates are usually not known explicitly. Morover, they often require regularity assumptions about the solution which, for practical problems, are hardly satisfied.

Therefore, a computational fluid dynamics code should be able to give reliable estimates of the local and global error of the computed numerical solution and to monitor an automatic, self-adaptive mesh-refinement based on these error estimates. Such an algorithm will have the following general form:

1. *Construct a coarse initial discretization T_0 of the physical problem representing sufficiently well the geometry of the problem. Put $k := 0$.*

2. *Solve the discrete problem on T_k.*

3. *For each element T in T_k compute an estimate η_T of the error of the computed numerical solution.*

4. *If the estimated error is sufficiently small then* **stop**. *Otherwise decide which elements of T_k have to be refined. Construct the next discretization, set $k := k+1$, and return to step* **2**.

447

The above algorithm is best suited for stationary problems. For transient calculations, some changes have to be made:

1. The accuracy of the computed numerical solution has to be estimated every few time-steps.

2. The refinement process in space should be coupled with a time-step control.

3. A partial coarsening of the mesh might be necessary.

4. Occasionally, a complete re-meshing could be desirable.

In both stationary and transient problems, the refinement / un-refinement process might also be coupled with or replaced by a moving-point technique, which keeps the number of grid-points constant but changes their relative location, see, e.g., Arney and Flaherty [1990] and Ribbens [1989]. Besides an efficient algorithm for the numerical solution of the discrete problems, one needs in any case two major ingredients:

1. an *a posteriori error estimator* and

2. a *mesh-refinement facility* based on the information of the error estimator.

Of course, both should be less expensive than the process of solving the discrete problems.

In what follows, we will first give a survey of a posteriori error estimators for scalar, second order, elliptic equations. Then we present two a posteriori error estimators for mixed finite element discretizations of the stationary, incompressible Navier-Stokes equations. They are based on a suitable local evaluation of the residual of the finite element solution with respect to the strong form of the differential equation and on the solution of local Stokes problems. We will show that both yield global upper and local lower bounds for the true error of the computed solution. In a second part, we will then comment on different mesh-refinement strategies and on the data structures needed for their implementation. Finally, we present some numerical examples.

We will work in a finite element framework for stationary, viscous, incompressible flow problems. For time dependent problems see, e.g., Adjerid and Flaherty [1988], Arney and Flaherty [1989 and 1990], Bieterman and Babuška [1982a and 1982b], Johnson *et al.* [1990], and Strouboulis and Oden [1990]. An overview on the adaptive mesh-refinement philosophy and on related questions is given, e.g., in Babuška [1986], Babuška and Gui [1986], and Rheinboldt [1980]. Different error estimators for the Navier-Stokes equations are compared in Bank and Welfert [1990a and 1990b] and Welfert [1990].

14.2   Error Estimators for Scalar, Second Order, Elliptic Equations

In order to get an idea about the different techniques of constructing a posteriori error estimators, we consider the following simple model problem:

$$-\Delta u = f \text{ in } \Omega$$
$$u = g \text{ on } \Gamma$$

(14.2.1)

in a bounded, connected, polygonal domain $\Omega \subset \Re^2$ with boundary Γ. Let u_h be the solution of a Galerkin finite element discretization of Problem (14.2.1).

The simplest and, perhaps, most straightforward candidate for an a posteriori error estimator is ∇u_h. This means that the mesh is refined where the gradient ∇u_h is large. This strategy is investigated in Benner *et al.* [1987] for one dimensional problems. When used in computational fluid dynamics, it corresponds to refining the mesh where the gradient of some important physical quantity such as the velocity, pressure, density, entropy, or Mach number is large. Which quantity is taken, depends on the problem. This strategy is quite simple, but requires some knowledge and intuition about the physical problem. Results are reported, e.g., in Bristeau *et al.* [1988].

Another very simple approach is to take the residual of the finite element solution with respect to the strong form of the differential equation. Intuitively, this means for Problem (14.2.1) that the element T is refined when $\int_T |\Delta(u - u_h)|^2$ is large. The example of a linear finite element approximation of an harmonic function, i.e., $\Delta u = 0$, however shows that this quantity alone is not sufficient. One also has to incorporate into the error estimator jumps of the normal derivative of u_h across ∂T. In Section 14.4, we will introduce an a posteriori error estimator for the Navier-Stokes equations which is based on this principle. This approach was originally considered in Babuška and Rheinboldt [1978] for Problem (14.2.1) in one dimension.

Babuška and Rheinboldt [1976] propose an error estimator for Problem (14.2.1) which is based on the solution of local Dirichlet problems of the same type using higher order elements. More precisely, consider a node x of the finite element partition and denote by ω_x the union of all elements having x as a node (see Figure 14.1).

Let $\tilde{u}_h$ be the weak solution of the local Dirchlet problem

$$-\Delta \tilde{u}_h = f \text{ in } \omega_x$$
$$\tilde{u}_h = u_h \text{ on } \partial \omega_x$$

where $\tilde{u}_h$ has a higher polynomial degree than u_h. Babuška and Rheinboldt prove that $\int_{\omega_x} |\nabla(u_h - \tilde{u}_h)|^2$ yields local lower and global upper bounds for the error of the finite element approximation in the H^1-norm. However, their technique is quite expensive when applied to the Navier-Stokes equations. Assume for example that the Navier-Stokes equations are solved using the Taylor-Hood element consisting of continuous, piecewise quadratic velocities and continuous, piecewise linear pressures and that ω_x consists of 6 triangles. For the local problem on ω_x, one then has to use at least continuous, piecewise

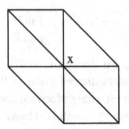

Figure 14.1: Node x and domain ω_x.

cubic velocities and continuous, piecewise quadratic pressures yielding a system of 80 equations with 80 unknowns per grid-point.

Bank and Weiser [1985] introduce an error estimator for Problem (14.2.1) which is based on the solution of local problems of the same type with Neumann boundary conditions. Let T be an element of the finite element partition and denote by $\hat{u}_h$ the weak solution of the local Neumann problem

$$-\Delta \hat{u}_h = f + \Delta u_h \text{ in T}$$
$$\hat{u}_h = [\frac{\partial u_h}{\partial \vec{n}}]_{\partial T} \text{ on } \partial T$$

where $[\frac{\partial u_h}{\partial \vec{n}}]_{\partial T}$ denotes the jump of the normal derivative of u_h across ∂T and where $\hat{u}_h$ has a higher polynomial degree than u_h. Bank and Weiser prove that $\int_T |\nabla \hat{u}_h|^2$ yields local lower and global upper bounds for the error of the finite element approximation in the H^1-norm. When applied to the Navier-Stokes equations their approach leads to local problems of a much smaller size than the one of Babuška and Rheinboldt. For the Taylor-Hood element for example, one now has to solve a system of 21 equations with 21 unknowns per triangle. We will investigate an error estimator based on these ideas in Section 14.5.

Finally, Eriksson [1985] and Eriksson and Johnson [1988] analyze an error estimator for Problem (14.2.1) which combines a priori L^∞-norm error estimates of the form

$$||u - u_h||_{L^\infty} \leq ch^\alpha ||u||_{W^{2,\infty}}$$

with higher order finite difference approximations of $||u_h||_{W^{2,\infty}}$. To our knowledge, this approach has not yet been applied to the Navier-Stokes equations.

14.3 Mixed Finite Element Discretization of the Navier-Stokes Equations

We consider the stationary, incompressible Navier-Stokes equations

$$
\begin{aligned}
-\Delta \vec{u} + Re(\vec{u} \cdot \nabla)\vec{u} + \nabla p &= \vec{f} \text{ in } \Omega \\
\nabla \cdot \vec{u} &= 0 \text{ in } \Omega \\
\vec{u} &= 0 \text{ on } \Gamma
\end{aligned}
\tag{14.3.1}
$$

in a connected, bounded, polygonal domain $\Omega \subset \Re^d$, $d = 2, 3$, with boundary Γ. Here, $\vec{u}$, p, and $Re \geq 0$ denote the velocity, pressure, and Reynolds number, respectively. Note, that $Re = 0$ corresponds to the Stokes equations. The restriction to polygonal domains and to homogeneous Dirichlet boundary conditions is made in order to simplify the exposition. Domains with curved boundaries and different boundary conditions can be treated similarly.

For any open subset ω of Ω with Lipschitz boundary γ, we denote by $L^2(\omega)$, $H^k(\omega)$, and $L^2(\gamma)$, $k \geq 1$, the standard Lebesgue- and Sobolev-spaces, resp., equipped with the norms $\|\cdot\|_{0,\omega} := \|\cdot\|_{L^2(\omega)}$, $\|\cdot\|_{k,\omega} := \|\cdot\|_{H^k(\omega)}$, and $\|\cdot\|_{0,\gamma} := \|\cdot\|_{L^2(\gamma)}$. The inner products of $L^2(\omega)$ and $L^2(\gamma)$ are denoted by $(\cdot, \cdot)_\omega$ and $(\cdot, \cdot)_\gamma$, respectively. Since no confusion can arise, we use the same notation for the corresponding norms and inner products on $L^2(\omega)^d$, $H^k(\omega)^d$, and $L^2(\gamma)^d$.

Put

$$
\begin{aligned}
X &:= H_0^1(\Omega)^d &:= \{\vec{u} \in H^1(\Omega)^d : \vec{u} = 0 \text{ on } \Gamma\} \\
Y &:= L_0^2(\Omega) &:= \{p \in L^2(\Omega) : (p, 1)_\Omega = 0\} .
\end{aligned}
$$

The standard weak formulation of Problem (14.3.1) then is to find $\vec{u} \in X$, $p \in Y$ such that for all $\vec{v} \in X$ and $q \in Y$:

$$
\begin{aligned}
(\nabla \vec{u}, \nabla \vec{v})_\Omega + Re((\vec{u} \cdot \nabla)\vec{u}, \vec{v})_\Omega - (p, \nabla \cdot \vec{v})_\Omega &= (\vec{f}, \vec{v})_\Omega \\
(\nabla \cdot \vec{u}, q)_\Omega &= 0 .
\end{aligned}
\tag{14.3.2}
$$

Problem (14.3.2) has at least one solution, which is unique if $Re\|\vec{f}\|_{0,\Omega}$ is sufficiently small. Moreover, it is well-known that the following inf-sup condition holds (cf. Girault and Raviart [1986]):

$$
\inf_{[\vec{u},p]\in X\times Y\setminus\{0\}} \sup_{[\vec{v},q]\in X\times Y\setminus\{0\}} \frac{(\nabla \vec{u}, \nabla \vec{v})_\Omega - (p, \nabla \cdot \vec{v})_\Omega - (\nabla \cdot \vec{u}, q)_\Omega}{\{\|\vec{u}\|_{1,\Omega} + \|p\|_{0,\Omega}\}\{\|\vec{v}\|_{1,\Omega} + \|q\|_{0,\Omega}\}} \geq \beta_c > 0 .
\tag{14.3.3}
$$

Denote by $\mathcal{T}_h$ a partition of Ω into polyhedral domains, which satisfies the usual compatibility conditions for finite elements. The partition $\mathcal{T}_h$ must be shape regular, i.e., the ratio of the circumscribed ball for $T \in \mathcal{T}_h$ to that of the largest inscribed ball is bounded independently of T and h. Note, that this is satisfied for all local refinement strategies used in practice (compare also Section 14.7). In two dimensions, the shape regularity is equivalent to a minimal angle condition.

Denote by $\mathcal{E}_h$ the set of all inter-element boundaries in $\mathcal{T}_h$. The shape regularity implies that the ratio h_E/h_T is bounded independently of h, T and $E \subset \partial T$, where h_T and h_E denote the diameter of T and E, respectively. For $E \in \mathcal{E}_h$ with $E = T_1 \cap T_2$, $T_1, T_2 \in \mathcal{T}_h$, and $\varphi \in L^2(\Omega)$ with $\varphi_{|T_i} \in C(\bar{T}_i)$, $i = 1, 2$, we denote by $[\varphi]_E$ the jump of φ across E.

Let
$$X_h \subset X \quad \text{and} \quad Y_h \subset Y$$
be finite element spaces corresponding to $\mathcal{T}_h$. We assume that X_h and Y_h contain functions which are piecewise polynomials of degree at least 1 and 0, resp., and of degree at most ℓ and m, respectively. Put

$$k := \begin{cases} \max\{m-1, \ell-2, \ell(\ell-1)\} & \text{if } Re > 0 \\ \max\{m-1, \ell-2, 0\} & \text{if } Re = 0 \end{cases} \qquad (14.3.4)$$

and denote by $\vec{P}_k$ the orthogonal projection of $L^2(\Omega)^d$ onto $M_k := \{ \vec{\varphi} \in L^2(\Omega)^d : \vec{\varphi}_{|T}$ is a polynomial of degree at most $k \}$.

From Clément [1975] we know that there is an interpolation operator $\vec{I}_h : X \longrightarrow X_h$ such that the following approximation estimates hold for all $T \in \mathcal{T}_h$ and $E \in \mathcal{E}_h$ (E = $T_1 \cap T_2$ with $T_1, T_2 \in \mathcal{T}_h$)

$$\begin{aligned} \|\vec{u} - \vec{I}_h \vec{u}\|_{0,T} &\leq c_1 h_T \|\vec{u}\|_{1,T} & \forall \vec{u} \in H^1(T) \\ \|\vec{u} - \vec{I}_h \vec{u}\|_{0,E} &\leq c_2 h_E^{1/2} \|\vec{u}\|_{1,T_1 \cup T_2} & \forall \vec{u} \in H^1(T_1 \cup T_2) . \end{aligned} \qquad (14.3.5)$$

Here and in what follows $c_1, c_2, \ldots$ are constants which are independent of h.

The mixed finite element approximation of Problem (14.3.2) then is to find $\vec{u}_h \in X_h$ and $p_h \in Y_h$ such that for all $\vec{v}_h \in X_h$ and $q_h \in Y_h$:

$$\begin{aligned} (\nabla \vec{u}_h, \nabla \vec{v}_h)_\Omega + Re((\vec{u}_h \cdot \nabla)\vec{u}_h, \vec{v}_h)_\Omega - (p_h, \nabla \cdot \vec{v}_h)_\Omega &= (\vec{f}, \vec{v}_h)_\Omega \\ (\nabla \cdot \vec{u}_h, q_h)_\Omega &= 0 . \end{aligned} \qquad (14.3.6)$$

In order to guarantee that Problem (14.3.6) has at least one solution, which is unique if $Re\|\vec{f}\|_{0,\Omega}$ is sufficiently small, the spaces X_h and Y_h have to satisfy the well-known *Babuška-Brezzi condition*, i.e., there is a constant β, *which does not depend on h*, such that

$$\inf_{p_h \in Y_h \setminus \{0\}} \sup_{\vec{u}_h \in X_h \setminus \{0\}} \frac{(p_h, \nabla \cdot \vec{u}_h)_\Omega}{\|p_h\|_{0,\Omega} \|\vec{u}_h\|_{1,\Omega}} \geq \beta . \qquad (14.3.7)$$

The above assumptions are satisfied, e.g., for

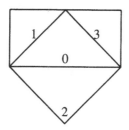

Figure 14.2: Triangle T and domain ω_T.

1. the *Taylor-Hood elements* consisting of continuous velocities and pressures, which are piecewise polynomials of degree k and $k - 1$, resp., with $k \geq 2$ (see Brezzi and Falk [1988], Scott and Vogelius [1985], and Verfürth [1984]),

2. the *modified Taylor-Hood element* consisting of continuous, piecewise linear pressures and velocities, the latter on a refined grid (see Verfürth [1984]), or

3. the *mini-element* consisting of continuous, piecewise linear pressures and velocities, the latter augmented by *bubble*-functions vanishing on the element boundaries (see Arnold, Brezzi, and Fortin [1984]).

More examples of stable mixed finite element spaces for the Navier-Stokes equations can be found in Girault and Raviart [1986].

Note, that the restriction to the simplest, central difference-type discretization of the convection term $(\vec{u} \cdot \nabla)\vec{u}$ in Problem (14.3.6) is made in order to simplify the exposition. More sophisticated approximations, such as, e.g., upwind, streamline-diffusion or transport-diffusion discretizations, can be handled quite similarly.

14.4 An Error Estimator Based on the Evaluation of Local Residuals

Our first error estimator is based on the evaluation of local residuals with respect to the strong form of the differential equation: For any $T \in \mathcal{T}_h$ we define

$$\eta_{R,T} \;\; := \;\; \Big\{ h_T^2 \|\vec{P}_k \vec{f} + \Delta \vec{u}_h - Re(\vec{u}_h \cdot \nabla)\vec{u}_h - \nabla p_h\|_{0,T}^2$$
$$+ \sum_{E \subset \partial T \cap \Omega} h_E \|[\frac{\partial \vec{u}_h}{\partial \vec{n}} - p_h \vec{n}]_E\|_{0,E}^2 + \|\nabla \cdot \vec{u}_h\|_{0,T}^2 \Big\}^{1/2}, \qquad (14.4.1)$$

where $\vec{n}$ is a unit vector orthogonal to ∂T.

Denote by ω_T the union of all elements in $\mathcal{T}_h$, which have a common face with T (see Figure 14.2).

Proposition 14.4.1 *Assume that* $Re||\vec{f}||_{0,\Omega}$ *is sufficiently small such that Problems (14.3.2) and (14.3.6) have a unique solution* $[\vec{u}, p] \in X \times Y$ *and* $[\vec{u}_h, p_h] \in X_h \times Y_h$, *respectively. Then there are two constants* c_3 *and* c_4, *which do not depend on h, such that the estimates*

$$||\vec{u} - \vec{u}_h||_{1,\Omega} + ||p - p_h||_{0,\Omega} \leq c_3 \Big\{ \sum_{T \in \mathcal{T}_h} [\eta_{R,T}^2 + h_T^2 ||\vec{f} - \vec{P}_K\vec{f}||_{0,T}^2] \Big\}^{1/2} \qquad (14.4.2)$$

and

$$\eta_{R,T} \leq c_4 \Big\{ ||\vec{u} - \vec{u}_h||_{1,\omega_T} + ||p - p_h||_{0,\omega_T} + h_T ||\vec{f} - \vec{P}_K\vec{f}||_{0,\omega_T} \Big\} \qquad (14.4.3)$$

hold for all $T \in \mathcal{T}_h$.

Proof: Let $[\vec{v}, q] \in X \times Y$ with $||\vec{v}||_{1,\Omega} + ||q||_{0,\Omega} = 1$. From Equations (14.3.2), (14.3.5), and (14.3.6) we conclude that

$$
\begin{aligned}
(\nabla(\vec{u} &- \vec{u}_h), \nabla\vec{v})_\Omega - (p - p_h, \nabla \cdot \vec{v})_\Omega - (\nabla \cdot (\vec{u} - \vec{u}_h), q)_\Omega \\
&= (\nabla(\vec{u} - \vec{u}_h), \nabla(\vec{v} - \vec{I}_h\vec{v}))_\Omega - (p - p_h, \nabla \cdot (\vec{v} - \vec{I}_h\vec{v}))_\Omega - (\nabla \cdot (\vec{u} - \vec{u}_h), q)_\Omega \\
&\quad + Re((\vec{u} \cdot \nabla)\vec{u} - (\vec{u}_h \cdot \nabla)\vec{u}_h, \vec{v} - \vec{I}_h\vec{v})_\Omega - Re((\vec{u} \cdot \nabla)\vec{u} - (\vec{u}_h \cdot \nabla)\vec{u}_h, \vec{v})_\Omega \\
&= \sum_{T \in \mathcal{T}_h} \Big\{ (-\Delta(\vec{u} - \vec{u}_h) + \nabla(p - p_h) + Re(\vec{u} \cdot \nabla)\vec{u} - Re(\vec{u}_h \cdot \nabla)\vec{u}_h, \vec{v} - \vec{I}_h\vec{v})_T \\
&\quad + (\frac{\partial(\vec{u} - \vec{u}_h)}{\partial\vec{n}} - (p - p_h)\vec{n}, \vec{v} - \vec{I}_h\vec{v})_{\partial T} + (\nabla \cdot \vec{u}_h, q)_T \Big\} \\
&\quad - Re(((\vec{u} - \vec{u}_h) \cdot \nabla)\vec{u} + (\vec{u}_h \cdot \nabla)(\vec{u} - \vec{u}_h), \vec{v})_\Omega \\
&\leq c_5 \sum_{T \in \mathcal{T}_h} h_T ||\vec{v}||_{1,T} ||\vec{f} + \Delta\vec{u}_h - Re(\vec{u}_h \cdot \nabla)\vec{u}_h - \nabla p_h||_{0,T} \\
&\quad + c_6 \sum_{E \in \mathcal{E}_h, E = T_1 \cap T_2} h_E^{1/2} ||\vec{v}||_{1,T_1 \cup T_2} ||[\frac{\partial\vec{u}_h}{\partial\vec{n}} - p_h\vec{n}]_E||_{0,E} \\
&\quad + \sum_{T \in \mathcal{T}_h} ||\nabla \cdot \vec{u}_h||_{0,T} ||q||_{0,T} + c_7 Re||\vec{u} - \vec{u}_h||_{1,\Omega} ||\vec{f}||_{0,\Omega} ||\vec{v}||_{1,\Omega} \\
&\leq c_8 \Big\{ \sum_{T \in \mathcal{T}_h} [\eta_{R,T}^2 + h_T^2 ||\vec{f} - \vec{P}_K\vec{f}||_{0,T}^2] \Big\}^{1/2} + c_7 Re||\vec{u} - \vec{u}_h||_{1,\Omega} ||\vec{f}||_{0,\Omega} .
\end{aligned}
$$

Together with inequality (14.3.3) and the assumption about $Re||\vec{f}||_{0,\Omega}$, this implies estimate (14.4.2).

Let $T \in \mathcal{T}_h$. Enumerate T and its neighbouring elements such that $\omega_T = \bigcup_{i=0}^{n} T_i$ with $T_0 := T$ (see Figure 14.2). Put $E_i := T_0 \cap T_i$, $1 \leq i \leq n$. Then there exist functions ψ_{T_i},

$0 \le i \le n$, and ψ_{E_j}, $1 \le j \le n$, which are suitable products of barycentric co-ordinates and which have the following properties:

$$\psi_{T_i} = 0 \text{ on } T_j, j \ne i, \quad , \quad \psi_{T_i} > 0 \text{ in } T_i \quad , \quad \int_{T_i} \psi_{T_i} = 1$$

and

$$\psi_{E_i} = 0 \text{ on } E_j, j \ne i, \quad , \quad \psi_{E_i} > 0 \text{ on } E_i \quad , \quad \int_{E_i} \psi_{E_i} = 1 \ .$$

Put

$$
\begin{aligned}
\vec{w}_{\text{T}} \ := \ & -\sum_{i=1}^{n} h_{E_i} [\frac{\partial \vec{u}_h}{\partial \vec{n}} - p_h \vec{n}]_{E_i} \psi_{E_i} \\
& + \sum_{j=1}^{n} \left\{ \sum_{i=1}^{n} h_{E_i} [\frac{\partial \vec{u}_h}{\partial \vec{n}} - p_h \vec{n}]_{E_i} \int_{T_j} \psi_{E_i} \right\} \psi_{T_j} \\
& + \left\{ \sum_{i=1}^{n} h_{E_i} [\frac{\partial \vec{u}_h}{\partial \vec{n}} - p_h \vec{n}]_{E_i} \int_{T_0} \psi_{E_i} \right. \\
& \left. + h_{T_0}^2 (\vec{P}_K \vec{f} + \Delta \vec{u}_h - Re(\vec{u}_h \cdot \nabla)\vec{u}_h - \nabla p_h) \right\} \psi_{T_0} \\
q_{\text{T}} \ := \ & \nabla \cdot \vec{u}_h \ .
\end{aligned}
$$

We then obtain

$$
\begin{aligned}
\eta_{R,T}^2 \ \le \ & \sum_{j=0}^{n} (\vec{w}_{\text{T}}, \vec{P}_K \vec{f} + \Delta \vec{u}_h - Re(\vec{u}_h \cdot \nabla)\vec{u}_h - \nabla p_h)_{T_j} \\
& - \sum_{i=1}^{n} (\vec{w}_{\text{T}}, [\frac{\partial \vec{u}_h}{\partial \vec{n}} - p_h \vec{n}]_{E_i})_{E_i} \ + \ (q_{\text{T}}, \nabla \cdot \vec{u}_h)_{\text{T}} \\
= \ & \sum_{j=0}^{n} (\vec{w}_{\text{T}}, -\Delta(\vec{u} - \vec{u}_h) + Re(\vec{u} \cdot \nabla)\vec{u} - Re(\vec{u}_h \cdot \nabla)\vec{u}_h + \nabla(p - p_h))_{T_j} \\
& - \sum_{j=0}^{n} (\vec{w}_{\text{T}}, \vec{f} - \vec{P}_K \vec{f})_{T_j} \\
& + \sum_{i=1}^{n} (\vec{w}_{\text{T}}, [\frac{\partial(\vec{u} - \vec{u}_h)}{\partial \vec{n}} - (p - p_h)\vec{n}]_{E_i})_{E_i} \ - \ (q_{\text{T}}, \nabla \cdot (\vec{u} - \vec{u}_h))_{\text{T}} \\
= \ & (\nabla(\vec{u} - \vec{u}_h), \nabla \vec{w}_{\text{T}})_{\omega_{\text{T}}} - (p - p_h, \nabla \cdot \vec{w}_{\text{T}})_{\omega_{\text{T}}} - (\vec{f} - \vec{P}_K \vec{f}, \vec{w}_{\text{T}})_{\omega_{\text{T}}} \\
& + Re((\vec{u} \cdot \nabla)\vec{u} - (\vec{u}_h \cdot \nabla)\vec{u}_h, \vec{w}_{\text{T}})_{\omega_{\text{T}}} - (\nabla \cdot (\vec{u} - \vec{u}_h), q_{\text{T}})_{\text{T}}
\end{aligned}
$$

$$\leq \left\{ ||\vec{u} - \vec{u}_h||_{1,\omega_{\mathrm{T}}} + ||p - p_h||_{0,\omega_{\mathrm{T}}} + c_9 Re||\vec{u} - \vec{u}_h||_{1,\omega_{\mathrm{T}}}||\vec{f}||_{0,\Omega} \right.$$

$$\left. + c_{10} h_{\mathrm{T}} ||\vec{f} - \vec{P}_{\vec{k}} \vec{f}||_{0,\omega_{\mathrm{T}}} \right\} \left\{ ||\vec{w}_{\mathrm{T}}||_{1,\omega_{\mathrm{T}}} + ||q_{\mathrm{T}}||_{0,\mathrm{T}} \right\} .$$

On the other hand, a simple homogeniety argument implies

$$||\vec{w}_{\mathrm{T}}||_{1,\omega_{\mathrm{T}}} + ||q_{\mathrm{T}}||_{0,\mathrm{T}} \leq c_{11} \eta_{\mathrm{R,T}} .$$

This proves Estimate (14.4.3). □

The proof of Proposition 14.4.1 shows that one cannot expect to obtain from $\eta_{\mathrm{R,T}}$ *local* upper bounds for the true error, since this would require control of $\frac{\partial \vec{u}}{\partial n} - p\vec{n}$ on the element boundaries.

The restriction to small Reynolds numbers simplifies the proof of Proposition 14.4.1 considerably. Using the techniques of Brezzi *et al.* [1980] for the approximation of non-linear problems, it can be extended to higher Reynolds numbers provided $\vec{u}_h$ is approximating a regular branch of solutions of Problem (14.3.2).

An error estimator similar to $\eta_{\mathrm{R,T}}$ was introduced in Abdalass [1987] for the mini-element discretization of the Stokes equations without proving the upper and lower bounds (14.4.2) and (14.4.3). The estimator (14.4.1) was analyzed in Verfürth [1989a and 1989b] for the mini-element and modified Taylor-Hood element discretizations of the Navier-Stokes equations.

14.5   An Error Estimator Based on the Solution of Local Stokes Problems

We use the same notations as in Section 14.4. Let $\mathrm{T} \in \mathcal{T}_h$. Denote by n the maximal number of vertices of all $\mathrm{E}_i \subset \partial \mathrm{T}$. Then the functions ψ_{E_i} can be chosen to be elements of M_n. Put

$$X_{\mathrm{T}} := \mathrm{span}\{\psi_{\mathrm{T}} \vec{w} , \psi_{\mathrm{E}_i} \vec{v} : \vec{w} \in \mathrm{M}_{\vec{k}+n} , \vec{v} \in \mathrm{M}_{\vec{k}+1} , 1 \leq i \leq n\}$$

$$Y_{\mathrm{T}} := \mathrm{span}\{\psi_{\mathrm{E}_i} q : q \in \mathrm{M}_{\max\{0,\ell-1\}} , 1 \leq i \leq n\} .$$

Lemma 14.5.1 *There is a constant $\beta_{\mathrm{T}} > 0$, which only depends on the ratio of the largest inscribed ball to the radius of the circumscribed ball of* T, *such that*

$$\inf_{[\vec{u},p] \in X_{\mathrm{T}} \times Y_{\mathrm{T}} \setminus \{0\}} \sup_{[\vec{v},q] \in X_{\mathrm{T}} \times Y_{\mathrm{T}} \setminus \{0\}} \frac{(\nabla \vec{u}, \nabla \vec{v})_{\mathrm{T}} - (p, \nabla \cdot \vec{v})_{\mathrm{T}} - (\nabla \cdot \vec{u}, q)_{\mathrm{T}}}{\{||\vec{u}||_{1,\mathrm{T}} + ||p||_{0,\mathrm{T}}\}\{||\vec{v}||_{1,\mathrm{T}} + ||q||_{0,\mathrm{T}}\}} \geq \beta_{\mathrm{T}} . \quad (14.5.1)$$

Proof: Since every $\vec{u} \in X_{\mathrm{T}}$ vanishes at the vertices of T, there is a constant α, which only depends on the ratio of the largest inscribed ball to the radius of the circumscribed ball of T, such that

$$||\nabla \vec{u}||_{0,\mathrm{T}} \geq \alpha ||\vec{u}||_{1,\mathrm{T}} \quad \forall \vec{u} \in X_{\mathrm{T}} . \quad (14.5.2)$$

Let $p \in Y_T \setminus \{0\}$. Thanks to the definition of X_T, we have $\vec{u} := \psi_T \nabla p \in X_T$. Since p vanishes at the vertices of T, we conclude that

$$\sup_{\vec{v} \in X_T \setminus \{0\}} \frac{(p, \nabla \cdot \vec{v})_T}{||\vec{v}||_{1,T}} \geq \frac{(p, \nabla \cdot \vec{u})_T}{||\vec{u}||_{1,T}} \geq \tilde{\gamma} h_T ||\nabla p||_{0,T} \geq \gamma ||p||_{0,T} , \qquad (14.5.3)$$

where the constant γ only depends on the ratio of the largest inscribed ball to the radius of the circumscribed ball of T. The inf-sup condition (14.5.1) immediately follows from Inequalities (14.5.2) and (14.5.3) by a standard argument. □

Because of Lemma 14.5.1 the following local Stokes problem has a unique solution. Find $\vec{u}_T \in X_T$ and $p_T \in Y_T$ such that for all $\vec{v} \in X_T$ and $q \in Y_T$:

$$\begin{aligned}
(\nabla \vec{u}_T, \nabla \vec{v})_T - (p_T, \nabla \cdot \vec{v})_T &= (\vec{P}_K \vec{f} + \Delta \vec{u}_h - Re(\vec{u}_h \cdot \nabla)\vec{u}_h - \nabla p_h, \vec{v})_T \\
&\quad + \sum_{i=1}^{n}([\frac{\partial \vec{u}_h}{\partial \vec{n}} - p_h \vec{n}]_{E_i}, \vec{v})_{E_i} \\
(q, \nabla \cdot \vec{u}_T)_T &= -(q, \nabla \cdot \vec{u}_h)_T .
\end{aligned} \qquad (14.5.4)$$

Now, we define an error estimator by

$$\eta_{S,T} := ||\vec{u}_T||_{1,T} + ||p_T||_{0,T} . \qquad (14.5.5)$$

Proposition 14.5.2 *There are two constants c_{12} and c_{13}, which only depend on the ratio of the largest inscribed ball to the radius of the circumscribed ball of T, such that the inequalities*

$$c_{12} \eta_{R,T} \leq \eta_{S,T} \leq c_{13} \eta_{R,T} \qquad (14.5.6)$$

hold for all $T \in \mathcal{T}_h$.

Proof: Let $T \in \mathcal{T}_h$. Since the functions in X_T and Y_T vanish at the vertices of T, we obtain from Equations (14.4.1), (14.5.4), and (14.5.5) and from Lemma 14.5.1 that

$$\begin{aligned}
\eta_{S,T} &\leq \sup_{[\vec{v},q] \in X_T \times Y_T, ||\vec{v}||_{1,T} + ||q||_{0,T}=1} \{(\nabla \vec{u}_T, \nabla \vec{v})_T - (p_T, \nabla \cdot \vec{v})_T - (\nabla \cdot \vec{u}_T, q)_T\} \\
&= \sup_{[\vec{v},q] \in X_T \times Y_T, ||\vec{v}||_{1,T} + ||q||_{0,T}=1} \{(\vec{P}_K \vec{f} + \Delta \vec{u}_h - Re(\vec{u}_h \cdot \nabla)\vec{u}_h - \nabla p_h, \vec{v})_T \\
&\qquad + \sum_{i=1}^{n}([\frac{\partial \vec{u}_h}{\partial \vec{n}} - p_h \vec{n}]_{E_i}, \vec{v})_{E_i} - (q, \nabla \cdot \vec{u}_h)_T\} \\
&\leq c_{13} \eta_{R,T} .
\end{aligned}$$

Define $\vec{v}_T \in X_T$ and $q_T \in Y_T$ by

$$\vec{v}_T \;:=\; -\sum_{i=1}^{n} h_{E_i} [\frac{\partial \vec{u}_h}{\partial \vec{n}} - p_h \vec{n}]_{E_i} \psi_{E_i}$$

$$+ \Big\{ \sum_{i=1}^{n} h_{E_i} [\frac{\partial \vec{u}_h}{\partial \vec{n}} - p_h \vec{n}]_{E_i} \int_{T_0} \psi_{E_i}$$

$$+ h_{T_0}^2 (\vec{P}_K \vec{f} + \Delta \vec{u}_h - Re(\vec{u}_h \cdot \nabla) \vec{u}_h - \nabla p_h) \Big\} \psi_{T_0}$$

$$q_T \;:=\; \sum_{i=1}^{n} \psi_{E_i} \nabla \cdot \vec{u}_h \ .$$

With the same arguments as in the proof of Proposition 14.4.1, we then obtain

$$\eta_{R,T}^2 \;\leq\; (\vec{v}_T, \vec{P}_K \vec{f} + \Delta \vec{u}_h - Re(\vec{u}_h \cdot \nabla) \vec{u}_h - \nabla p_h)_T$$

$$- \sum_{i=1}^{n} (\vec{v}_T, [\frac{\partial \vec{u}_h}{\partial \vec{n}} - p_h \vec{n}]_{E_i})_{E_i} + (q_T, \nabla \cdot \vec{u}_h)_T$$

$$= \; (\nabla \vec{u}_T, \nabla \vec{v}_T)_T - (p_T, \nabla \cdot \vec{v}_T)_T - (q_T, \nabla \cdot \vec{u}_T)_T$$

$$\leq \; c_{14} \eta_{S,T} \{ \|\vec{v}_T\|_{1,T} + \|q_T\|_{0,T} \} \ .$$

On the other hand, a simple homogeniety argument implies

$$\|\vec{w}_T\|_{1,T} + \|q_T\|_{0,T} \leq c_{15} \eta_{R,T} \ .$$

This completes the proof.   □

For the same reasons as for $\eta_{R,T}$, one cannot expect to get from $\eta_{S,T}$ *local* upper bounds for the true error.

When applying the error estimator $\eta_{S,T}$ to the Taylor-Hood element or to the mini-element on a triangular mesh, Problem (14.5.4) results in a system of linear equations with 63 or 74 unknowns, respectively. When adapting the arguments of this section to these special elements, the size of the linear problems can be reduced to 21 or 12 unknowns, resp. (see Verfürth [1989a]). Other error estimators based on the solution of local Stokes problems are investigated in Bank and Welfert [1990a and 1990b] and Welfert [1990].

14.6 Non-Conforming and Petrov-Galerkin Discretizations

Up to now, the finite element spaces X_h and Y_h used for the discretization of Problem (14.3.1) had to satisfy the following two essential conditions:

1. *conformity*, i.e., $X_h \subset H^1(\Omega)^d$ and

2. *stability*, i.e., the Babuška-Brezzi condition (14.3.7).

Although there is known a broad range of finite element spaces of arbitrary order satisfying these conditions (see, e.g., Girault and Raviart [1986]), there are important examples of methods used in practice which do not fit into this framework.

As a first example, we consider non-conforming methods such as the *Crouzeix-Raviart element* consisting of piecewise constant pressures and piecewise linear velocities, whose normal components are continuous in the barycentres of the element faces (see, e.g., Crouzeix and Raviart [1973], Dobrowolski and Thomas [1984], and Griffiths [1979]). For these elements, the discrete velocities are only contained in H(div;Ω), but not in $H^1(\Omega)^d$. Their main advantage is the possibility to construct solenoidal basis functions of the velocity space having a local support. This allows one to compute the discrete velocity and pressure seperately, thus reducing the size of the discrete problems considerably.

In Verfürth [1989c], the error estimator $\eta_{R,T}$ is applied to the Crouzeix-Raviart element. It still yields local lower and global upper bounds for the error of the finite element solution up to higher order terms, which stem from the consistency error. The techniques of Verfürth [1989c] can also be applied to other non-conforming finite element discretizations of the Navier-Stokes equations.

Another example are the so-called *Petrov-Galerkin* methods. They approximate the velocity and pressure by finite element spaces, which do not satisfy the Babuška-Brezzi condition (14.3.7) such as, e.g., the popular linear / constant and bilinear / constant elements, and enforce the stability of the discretization by adding consistent penalty terms to the discrete continuity equation (see, e.g., Brezzi and Douglas [1988], Hughes *et al.* [1986], and Verfürth [1990]). The resulting discrete problem is to find $\vec{u}_h \in X_h$ and $p_h \in Y_h$ such that for all $\vec{v}_h \in X_h$ and $q_h \in Y_h$:

$$(\nabla \vec{u}_h, \nabla \vec{v}_h)_\Omega - Re((\vec{u}_h \cdot \nabla)\vec{u}_h, \vec{v}_h)_\Omega - (p_h, \nabla \cdot \vec{v}_h)_\Omega = (\vec{f}, \vec{v}_h)_\Omega$$

$$(q_h, \nabla \cdot \vec{u}_h)_\Omega + \delta \sum_{T \in \mathcal{T}_h} h_T^2 (\nabla p_h, \nabla q_h)_T$$

$$+ \delta \sum_{T \in \mathcal{T}_h} h_T^2 (-\Delta \vec{u}_h + Re(\vec{u}_h \cdot \nabla)\vec{u}_h, \nabla q_h)_T$$

$$+ \delta \sum_{E \in \mathcal{E}_h} h_E ([p_h]_E, [q_h]_E)_E = \delta \sum_{T \in \mathcal{T}_h} h_T^2 (\vec{f}, \nabla q_h)_T , \quad (14.6.1)$$

where δ is a small parameter, *which is independent of h.*

A close inspection of the proofs of Propositions 14.4.1 and 14.5.2 shows that the error estimators $\eta_{R,T}$ and $\eta_{S,T}$ carry over to the Petrov-Galerkin formulation without

modification, since the momentum equation of Problem (14.6.1) is identical to the one
of Problem (14.3.6).

14.7 Refinement Strategies

Babuška and Rheinboldt [1976] and Rheinboldt [1980] give an heuristic argument for a
linear finite element discretization of Problem (14.2.1) that among all partitions with a
given number of elements that one is optimal which equilibrates the error, i.e., the errors
in all elements are the same. Many practical examples demonstrate the efficiency of their
strategy.

Based on the strategy of Babuška and Rheinboldt, there are two main possibilities to
decide, based on an error estimator, which elements have to be refined:

1. Let T be an element of $\mathcal{T}_h$ and let τ be obtained by subdividing T. It is quite
   reasonable to assume that the errors E_T and E_τ in T and τ behave like ch_T^λ and
   ch_τ^λ with unknown constants c and λ. Having computed error estimators η_T and
   η_τ for E_T and E_τ, one can then roughly determine c and λ and thus approximately
   predict the error in an element τ' obtained by a further subdivision of T.

2. Suppose that one has computed for each element T of the partition $\mathcal{T}_h$ an estimator
   η_T of the error in T. Put $\eta := \max_{T \in \mathcal{T}_h} \eta_T$. Then an element T is subdivided if
   $\eta_T \geq \gamma\eta$, where $0 < \gamma < 1$ is a prescribed threshold; typically $\gamma = 0.5$.

The first possibility clearly is more sophisticated. In practice, however, the second one,
which is much cheaper, gives satisfactory results.

Jarausch [1986] proposes a different refinement strategy. He introduces a cost functional and minimizes the error, approximated by an error estimator, subject to a prescribed
bound for the cost functional. This strategy allows the construction of optimal meshes
with fewer intermediate refinement steps.

A close inspection of the proofs of Propositions 14.4.1 and 14.5.2 shows that it is
mandatory to preserve the shape regularity of the partition during the refinement process.
When using a quadrilateral mesh in two or three dimensions, this condition is satisfied
when subdividing the elements by joining their midpoints of edges. However, this process introduces hanging nodes, i.e., midpoints of edges where not all of the neighbouring
elements have been divided (see Figure 14.3). At those points the compatibility conditions for finite element partitions are violated. There are two possibilities to overcome
this difficulty:

1. Treat the unknowns corresponding to hanging nodes as spurious degrees of freedom, i.e., their values are fixed to be a suitable interpolation of the unknowns
   corresponding to neighbouring regular nodes.

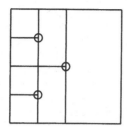

Figure 14.3: Hanging nodes (marked by ∘).

2. Introduce additional triangular or tetrahedral elements such that the resulting partition satisfies the regularity assumptions for finite element partitions.

In practice, the second possibility, however, seems to be used only occasionally.

For tetrahedral meshes it is much more complicated to preserve the shape regularity during a refinement process, since a tetrahedron cannot be divided into similar subtetrahedra. A refinement algorithm for tetrahedral meshes preserving the shape regularity is described in Bänsch [1989]. Different refinement strategies for Problem (14.2.1) are compared by Mitchell [1989].

In order to simplify the exposition, we will only discuss in what follows the most commonly used refinement strategies for triangular meshes:

1. Bisect triangles by joining the midpoint of the longest edge with the vertex opposite to this edge (*longest edge bisection*).

2. Bisect triangles by joining the midpoint of a marked edge with the vertex opposite to this edge (*marked edge bisection*).

3. Divide triangles into four by joining the midpoints of edges (*regular refinement*).

The longest edge bisection strategy is described in detail in Rivara [1984a and b]. In order to obtain an admissible triangulation, one has to add the following two rules:

1a. A triangle having two hanging nodes on edges, which are not the longest one, is divided by joining the midpoints of edges (*red refinement*, see Figure 14.4).

1b. A triangle having exactly one hanging node on an edge, which is not the longest one, is divided by joining the hanging node to the midpoint of the longest edge and by joining the latter one to the vertex opposite to the longest edge (*blue refinement*, see Figure 14.4).

Figure 14.4: Red, green, and blue refinement of a triangle.

Figure 14.5: Bisection of marked edges (indicated by ∘).

Rosenberg and Stenger [1975] prove that, when using the longest edge bisection, the smallest angle in all refined triangulations is at least half of the smallest angle in the initial triangulation.

The marked edge bisection is described in Sewell [1972]. It obeys the following rules:

2a. A triangle has exactly one marked edge.

2b. When bisecting a triangle its un-marked edges become the marked edges of its two sons (see Figure 14.5).

2c. An edge may only be bisected, if it is the marked edge of both its adjacent triangles.

Rule 2b ensures that there is only a finite number of different types of triangles in the triangulation. Thanks to rules 2a and 2c triangles are refined only by bisection.

The regular refinement of triangles is described, e.g., in Bank [1983], Bank *et al.* [1983], and Verfürth [1989a and d]. It has the advantage that the sons of a regularly refined triangle are similar to their father. On the other hand it introduces hanging nodes. In order to ensure the compatibility of the triangulation, the refinement process has to obey the following rules:

3a. A triangle having at least two hanging nodes is refined regularly.

3b. A triangle having exactly one hanging node is bisected by joining the hanging node to the vertex opposite to it (*green refinement*).

3c. A triangle having a hanging node on an edge, which has been previously bisected, must be refined regularly.

Figure 14.6: Rule 3c′, hanging node marked by ∘.

When refining a triangle with two hanging nodes according to rule 3a, a new hanging node is introduced. One can prove, however, that the refinement process obeying the above rules is finite. In order to avoid the introduction of new hanging nodes and to keep the refinement zone as narrow as possible, one can modify rule 3a as follows:

   3a′ A triangle having three hanging nodes is refined regularly; one having two hanging nodes is refined according to rule 1b (*blue refinement*).

Rule 3c tries to keep the number of bisected triangles as small as possible and ensures that the minimal angle in all refined triangulations is at least half of the minimal angle in the original triangulation. In order to keep the number of bisected triangles as small as possible and the refinement zone as narrow as possible, rule 3c may be modified in two ways:

   3c′. A triangle having a hanging node on an edge, which has been previously bisected, must be refined according to rule 1b (*blue refinement*).

   3c″. Let τ be obtained by bisecting T and let x be a hanging node of τ on the previously bisected edge. Then τ is removed from the triangulation, T is refined regularly, and the son of T, which has x as a hanging node, is bisected.

Rule 3c″ was introduced by Bank [1983]. It is illustrated in Figure 14.6. It has the disadvantage of introducing new hanging nodes and of leading to triangulations which are not nested.

   For some applications such as, e.g., semi-conductor device simulation, one is interested in having a discrete maximum principle, which requires weakly acute triangulations. In order to preserve the weak acuteness of a triangulation during the refinement process, rule 3c can be sharpened in the sense that a triangle may only be bisected along its perpendicular bisector.

14.8   Data Structures

Data structures for adaptive mesh-refinement are described, e.g., in Bank [1983], Bank *et al.* [1983], Rheinboldt and Mesztenyi [1980], Rivara [1984a], and Verfürth [1989d].

Their implementation depends among others on the refinement strategy and on the programming language used and on whether the underlying data structure of the finite element code is triangle- or node-oriented. As an example, we shortly describe the data structure used in the FORTRAN-code FEMFLOW for the mixed finite element discretization of the stationary, incompressible Navier-Stokes equations in general two-dimensional domains (see Verfürth [1989d]).

FEMFLOW is strictly triangle-oriented. The connection between local and global enumeration of the nodes is given by an array ITNODE where

$$\text{ITNODE}(i,j) = \text{global number of node } i \text{ of triangle } j.$$

The neighbourhood-relations of the triangles are described by an array ITEDGE where

$$\text{ITEDGE}(i,j) = k \text{ if and only if edge } j \text{ of triangle } i \text{ is adjacent to triangle } k.$$

Here, edge i is the edge opposite to the vertex with the local number i. The information stored in ITEDGE is not only used for the refinement algorithm. It is also needed for the multi-grid algorithm and it is useful for algorithms such as, e.g., the transport diffusion algorithm, which require the efficient search for triangles having a given node as vertex.

The information particularly needed for the refinement process is stored in an array ITMARK where

$$\text{ITMARK}(1,j) \quad = \quad \text{number of the father of triangle } j,$$

$$\text{ITMARK}(2,j) \quad = \quad \begin{cases} 0 & \text{if triangle } j \text{ is not refined,} \\ i & 1 \le i \le 3, \text{ if edge } i \text{ of triangle } j \text{ is bisected,} \\ 4 & \text{if triangle } j \text{ is regularly refined,} \\ 4k+i & \text{if triangle } j \text{ is refined blue by bisecting edges } i \\ & \text{and } (i+k) \text{ modulo } 3, \end{cases}$$

$$\text{ITMARK}(3,j) \quad = \quad \text{number of a son of triangle } j.$$

The remaining sons are numbered according to Figure 14.7, where ITMARK(3,j) = JS and ITMARK(2,j) = i for the green refinement. In that case, the numbers i+1 and i+2 have to be taken modulo 3. Figure 14.7 also shows the local enumeration of the vertices in the refined triangles. A blue refinement will be treated as if it is the result of two successive green refinements.

14.9 Numerical Examples

We consider three numerical examples:

1. The Stokes equations in a circular segment of radius 1 and angle 270°.

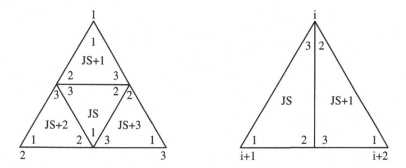

Figure 14.7: Enumeration of nodes and triangles for red and green refinement.

2. The Stokes equations in a circular segment of radius 1 and angle 360°.

3. The Navier-Stokes equations at Reynolds number 50 in a backward facing step with channel width/length of 1/3 units upstream and 2/19 units downstream and unit parabolic in- and outflow.

In examples 1 and 2, the exact solutions are given in polar co-ordinates by

$$\vec{u} = \begin{pmatrix} r^\alpha[(1+\alpha)\sin(\varphi)\psi(\varphi) + \cos(\varphi)\partial_\varphi\psi(\varphi)] \\ r^\alpha[\sin(\varphi)\partial_\varphi\psi(\varphi) - (1+\alpha)\cos(\varphi)\psi(\varphi)] \end{pmatrix}$$

$$p = -r^{\alpha-1}[(1+\alpha)^2\partial_\varphi\psi(\varphi) + \partial_\varphi^3\psi(\varphi)]/(1-\alpha)$$

with

$$\psi(\varphi) = \sin((1+\alpha)\varphi)\cos(\alpha\omega)/(1+\alpha) - \cos((1+\alpha)\varphi)$$
$$- \sin((1-\alpha)\varphi)\cos(\alpha\omega)/(1-\alpha) + \cos((1-\alpha)\varphi) ,$$

$$\alpha = 856399/1572864 \quad , \quad \omega = 3\pi/2$$

for example 1 and

$$\psi(\varphi) = 3\sin(0.5\varphi) - \sin(1.5\varphi) ,$$

$$\alpha = 0.5 \quad , \quad \omega = 2\pi$$

for example 2. The velocity fields are shown in Figures 14.8 and 14.9.

All examples are discretized using the modified Taylor-Hood element described in Section 14.3. The error estimator is $\eta_{R,T}$. A triangle is refined if its estimated error is at least half of the estimated maximal error. We use the regular refinement strategy described in Section 14.7 according to the rules 3a', 3b, and 3c'. Figures 14.10–14.22

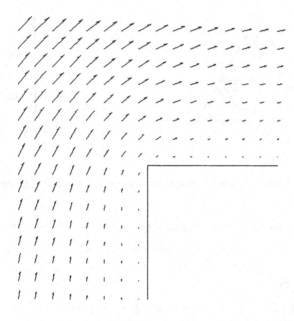

Figure 14.8: Example 1, velocity field.

show for all examples the initial triangulations together with two refined triangulations and zooms of the refinement near the critical regions. When using the estimator $\eta_{S,T}$, the results are very similar.

Table 14.1 gives a comparison of uniform and adaptive refinement strategies for examples 1 and 2. Here L, NT, NN, $\epsilon_{\vec{u}}$, ϵ_p, q, Tt, and Ts are the number of refinement steps, the number of triangles, the number of unknowns, the relative error of the velocity in the H^1-norm, the relative error of the pressure in the L^2-norm, the ratio of the predicted value of $\{\|\vec{u} - \vec{u}_h\|_{1,\Omega} + \|p - p_h\|_{0,\Omega}\}^{1/2}$ to the true one, the time in seconds needed for mesh-generation, and the time in seconds required for the solution of the discrete problems. All computations were done on a MacIntosh IIx. Note, that the solution algorithm could be accelerated, since it presently does superfluous work when only few unknowns are added during the refinement process.

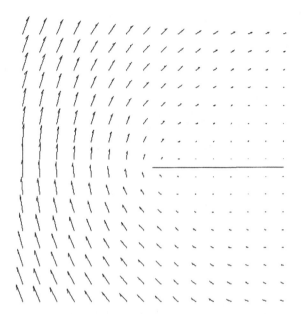

Figure 14.9: Example 2, velocity field.

Ex.	Ref.	L	NT	NN	$\epsilon_{\vec{u}}$	ϵ_p	q	Tt	Ts
	uniform	5	3072	3338	0.0375	0.074		10	322
1	$\eta_{R,T}$	7	2298	2534	0.0211	0.0437	0.6	14	316
	$\eta_{S,T}$	7	3108	3489	0.0313	0.0663	0.8	51	572
	uniform	5	4096	4466	0.0828	0.2580		13	431
2	$\eta_{R,T}$	7	2832	3138	0.0428	0.0811	0.5	17	310
	$\eta_{S,T}$	7	2678	2986	0.0434	0.0829	0.7	42	297

Table 14.1: Comparison of uniform and adaptive refinement.

Table 14.1 shows that $\eta_{S,T}$ gives a slightly better information about the numerical value of the error than $\eta_{R,T}$. Figures 14.10 – 14.22, however, show, that the latter efficiently detects the regions, where the mesh must be refined. When compared with the results of Babuška and Dehao Yu [1987], Babuška and Rheinboldt [1976], and Bank and Weiser [1985] for Problem (14.2.1), one should note, that, for indefinite problems

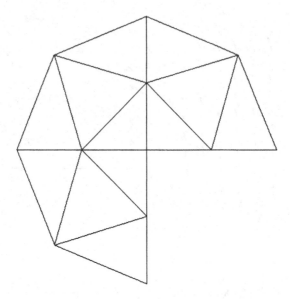

Figure 14.10: Example 1, initial triangulation.

such as the Stokes equations, one cannot expect the ratio of the predicted error to the true one to approach 1, since this quotient will always contain the Brezzi constant β_c of Equation (14.3.3).

References

[1] Abdalass, E. M. (1987). *Resolution performante du problème de Stokes par mini-éléments, maillages auto-adaptifs et méthodes multigrilles - applications*. Thèse de 3me cycle, Ecole Centrale de Lyon.

[2] Adjerid, S. and Flaherty, J. E. (1988). A local refinement finite element method for two-dimensional parabolic systems. *SIAM J. Sci. Stat. Comput.* **9**, 792–811.

[3] Arney, D. C. and Flaherty, J. E. (1989). An adaptive local mesh-refinement method for time dependent partial differential equations. *Appl. Numer. Math.* **5**, 257–274.

[4] Arney, D. C. and Flaherty, J. E. (1990). An adaptive mesh-moving and local refinement method for time dependent partial differential equations. *ACM Trans. Math. Software* **16**, 48–71.

[5] Arnold, D. N., Brezzi, F., and Fortin, M. (1984). A stable finite element for the Stokes equations. *Calcolo* **21**, 337–344.

[6] Babuška, I. (1986). Feedback, adaptivity, and a posteriori estimates in finite elements: aims, theory, and experience. *Accuracy estimates and adaptive refinements in finite element computations* (Babuška, I. *et al.*; eds.) Wiley, New York, pp. 3–23.

[7] Babuška, I. and Dehao Yu (1987). Asymptotically exact a-posteriori error estimator for biquadratic elements. *Finite Elem. Anal. Des.* **3**, 341–354.

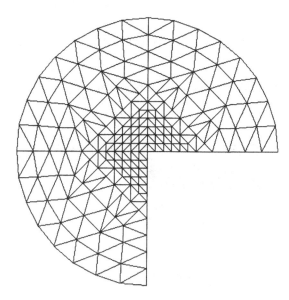

Figure 14.11: Example 1, 4 refinement steps, 318 triangles, 183 nodes.

[8] Babuška, I. and Gui, W. (1986). Basic principles of feedback and adaptive approaches in the finite element method. *Comp. Meth. Appl. Mech. Engrg.* **55**, 27–42.

[9] Babuška, I. and Rheinboldt, W. C. (1976). Error estimates for adaptive finite element computations. *SIAM J. Numer. Anal.* **15**, 736–754.

[10] Babuška, I. and Rheinboldt, W. C. (1978). A posteriori error estimates for the finite element method. *Int. J. Numer. Methods in Engrg.* **12**, 1597–1615.

[11] Bänsch, E. (1989). *Local mesh refinement in 2 and 3 dimensions.* Report No. 6, SFB 256, Universität Bonn.

[12] Bank, R. E. (1983). The efficient implementation of local mesh refinement algorithms. *Adaptive Computational Methods* (Babuška, I. *et al.*; eds.) SIAM, Philadelphia, pp. 74–81.

[13] Bank, R. E., Sherman, A. H., and Weiser, A. (1983). Refinement algorithms and data structures for regular local mesh refinement. *Scientific Computing* (Stepleman, R. *et al.*; eds.) Noth Holland, Amsterdam, New York, Oxford, pp. 3–17.

[14] Bank, R. E. and Weiser, A. (1985). Some a posteriori error estimators for elliptic partial differential equations. *Math. Comput.* **44**, 283–301.

[15] Bank, R. E. and Welfert, B. D. (1990a). A posteriori error estimates for the Stokes equations: a comparison. *Comp. Meth. Appl. Mech. Engrg.* **82**, 323–340.

[16] Bank, R. E. and Welfert, B. D. (1990b). *A posteriori error estimates for the Stokes problem.* Preprint, Univ. of California San Diego.

[17] Benner, R. E., Davis, H. T., and Scriven, L. E. (1987). An adaptive finite element method for steady and transient problems. *SIAM J. Sci. Stat. Comput.* **8**, 529–549.

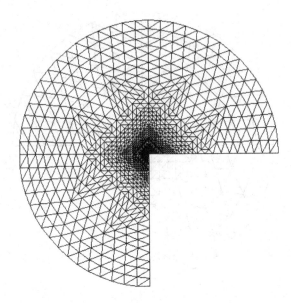

Figure 14.12: Example 1, 7 refinement steps, 2298 triangles, 1209 nodes.

[18] Bieterman, M. and Babuška, I. (1982a). The finite element method for parabolic equations. I A posteriori error estimation. *Numer. Math.* **40**, 339–371.

[19] Bieterman, M. and Babuška, I. (1982b). The finite element method for parabolic equations. II A posteriori error estimation and adaptive approach. *Numer. Math.* **40**, 373–406.

[20] Brezzi, F. and Douglas, J. (1988). Stabilized mixed methods for the Stokes problem. *Numer. Math.* **53**, 225–235.

[21] Brezzi, F. and Falk, R. S. (1988). *Stability of a higher order Hood-Taylor element.* Preprint, Rutgers University.

[22] Brezzi, F., Rappaz, J., and Raviart, P. A. (1980). Finite dimensional approximation of non-linear problems I. Branches of non-singular solutions. *Numer. Math.* **36**, 1–25.

[23] Bristeau, M. O., Glowinski, R., Mantel, B., Periaux, J., and Rogé, G. (1988). Adaptive finite element methods for three dimensional compressible viscous flow simulation in aerospace engineering. *Proc. 11th Int. Conf. on Numerical Methods in Fluid Mechanics*, Williamsburg, Springer, Berlin, Heidelberg, New York.

[24] Clément, P. (1975). Approximation by finite element functions using local regularization. *RAIRO Anal. Numèr.* **9**, 77–84.

[25] Crouzeix, M. and Raviart, P. A. (1973). Conforming and non-conforming finite element methods for solving the stationary Stokes equations. *RAIRO Anal. Numér* **7**, 33–76.

[26] Dobrowolski, M. and Thomas, K. (1984). On the use of discrete solenoidal finite elements for approximating the Navier-Stokes equations. *Application of Mathematics in Technology*, Rome, pp. 246–262.

[27] Eriksson, K. (1985). Improved accuracy by adapted mesh-refinement in the finite element method. *Math. Comput.* **44**, 321–343.

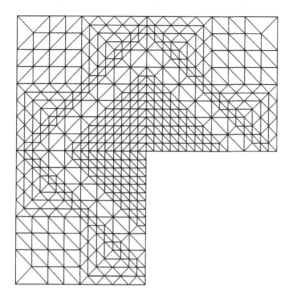

Figure 14.13: Zoom of $[-\frac{1}{8}, \frac{1}{8}] \times [-\frac{1}{8}, \frac{1}{8}]$ in Figure 14.12.

[28] Eriksson, K. and Johnson, C. (1988). An adaptive finite element method for linear elliptic problems. *Math. Comput.* **50**, 361–383.

[29] Girault, V. and Raviart, P. A. (1986). *Finite Element Approximation of the Navier-Stokes Equations.* Series in Computational Mathematics, Springer, Berlin, Heidelberg, New York.

[30] Griffiths, D. F. (1979). Finite elements for incompressible flow. *Math. Meth. in the Appl. Sci.* **1**, 16–31.

[31] Hughes, T. J. R., Franca, L. P., and Balestra, M. (1986). A new finite element formulation for computational fluid dynamics: V. Circumventing the Babuška-Brezzi condition: A stable Petrov-Galerkin formulation of the Stokes problem accommodating equal order interpolation. *Comput. Meths. Appl. Mech. Engrg.* **59**, 85–99.

[32] Jarausch, H. (1986). On an adaptive grid refining technique for finite element approximations. *SIAM J. Sci. Stat. Comput.* **7**, 1105–1120.

[33] Johnson, C., Yi-Yong Nic, and Thomée, V. (1990). An a posteriori error estimate and adaptive timestep control for a backward Euler discretization of a parabolic problem. *SIAM J. Numer. Anal.* **27**, 277–291.

[34] Kornhuber, R. and Roitzsch, R. (1990). On adaptive grid refinement in the presence of initial or boundary layers. *Impact* **2**, 40–72.

[35] Mitchell, W. F. (1989). A comparison of adaptive refinement techniques for elliptic problems. *ACM Trans. Math. Software* **15**, 326–347.

[36] Rheinboldt, W. C. (1980). On a theory of mesh-refinement processes. *SIAM J. Numer. Anal.* **17**, 766–778.

[37] Rheinboldt, W. C. and Mesztenyi, C. K. (1980). On a data structure for adaptive finite element refinement. *ACM Trans. Math. Software* **6**, 166–187.

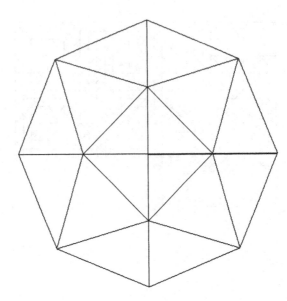

Figure 14.14: Example 2, initial triangulation.

[38] Ribbens, C. J. (1989). A fast adaptive grid scheme for elliptic partial differential equations. *ACM Trans. Math. Software* **15**, 179–197.

[39] Rivara, M. C. (1984a). Design and data structure of fully adaptive, multigrid, finite element software. *ACM Trans. Math. Software* **10**, 242–264.

[40] Rivara, M. C. (1984b). Algorithms for refining triangular grids suitable for adaptive and multigrid techniques. *Int. J. Numer. Meth. Engrg.* **20**, 745–756.

[41] Rosenberg, I. G. and Stenger, F. (1975). A lower bound on the angles of triangles constructed by bisecting the longest side. *Math. Comput.* **29**, 390–395.

[42] Scott, L. R. and Vogelius, M. (1985). Norm estimates for a maximal right inverse of the divergence operator in spaces of piecewise polynomials. *RAIRO M^{2}AN* **19**, 111–143.

[43] Sewell, E. G. (1972). *Automatic generation of triangulations for piecewise polynomial approximations.* Ph. D. Thesis, Purdue University, West Lafayette.

[44] Strouboulis, T. and Oden, J. T. (1990). A posteriori error estimation of finite element approximations in fluid mechanics. *Comp. Meth. Appl. Mech. Engrg.* **78**, 201–242.

[45] Verfürth, R. (1984). Error estimates for a mixed finite element approximation of the Stokes equations. *RAIRO Anal. Numér.* **18**, 175–182.

[46] Verfürth, R. (1989a). A posteriori error estimators for the Stokes equations. *Numer. Math.* **55**, 309–325.

[47] Verfürth, R. (1989b). A posteriori error estimators and adaptive mesh-refinement for a mixed finite element discretization of the Navier-Stokes equations. *Numerical Treatment of the Navier-Stokes Equations* (Hackbusch, W. and Rannacher, R.; eds.) Notes on Numerical Fluid Mechanics, Vol. 30, Vieweg, Braunschweig, pp. 145–152.

Figure 14.15: Example 2, 4 refinement steps, 424 triangles, 240 nodes.

[48] Verfürth, R. (1989c). *A posteriori error estimators for the Stokes Equations. II Non-conforming methods.* Report, Universität Zürich (Numer. Math., to appear 1991).

[49] Verfürth, R. (1989d). *FEMFLOW - user guide. Version 1.* Report, Universität Zürich.

[50] Verfürth, R. (1990). *On the stability of Petrov-Galerkin formulations of the Stokes equations.* Report, Universität Zürich.

[51] Welfert, B. D. (1990). *A posteriori error estimates for the Stokes problem.* PhD Thesis, Univ. of California San Diego.

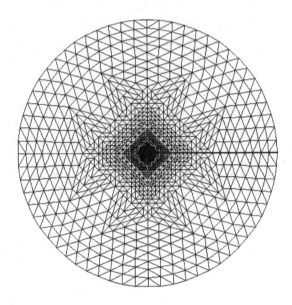

Figure 14.16: Example 2, 7 refinement steps, 2832 triangles, 1482 nodes.

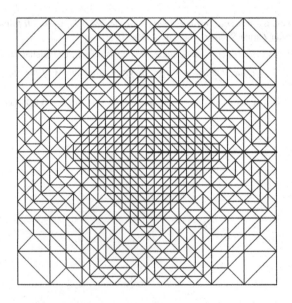

Figure 14.17: Zoom of $[-\frac{1}{8}, \frac{1}{8}] \times [-\frac{1}{8}, \frac{1}{8}]$ in Figure 14.16.

Figure 14.18: Example 3, initial triangulation.

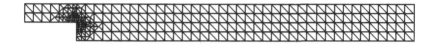

Figure 14.19: Example 3, 3 refinement steps, 503 triangles, 285 nodes.

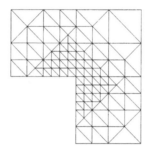

Figure 14.20: Zoom of $[-1, 1] \times [-1, 1]$ in Figure 14.19.

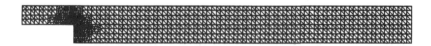

Figure 14.21: Example 3, 6 refinement steps, 2399 triangles, 1319 nodes.

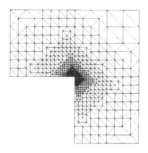

Figure 14.22: Zoom of $[-1, 1] \times [-1, 1]$ in Figure 14.21.

Index